FROM BIOTECHNOLOGY TO GENOMES
The Meaning Of The Double Helix

FROM BIOTECHNOLOGY TO GENOMES

The Meaning Of The Double Helix

PHILIPPE GOUJON

Université Catholique de Lille
France

World Scientific
Singapore • New Jersey • London • Hong Kong

Published by
World Scientific Publishing Co. Pte. Ltd.
P O Box 128, Farrer Road, Singapore 912805
USA office: Suite 1B, 1060 Main Street, River Edge, NJ 07661
UK office: 57 Shelton Street, Covent Garden, London WC2H 9HE

British Library Cataloguing-in-Publication Data
A catalogue record for this book is available from the British Library.

FROM BIOTECHNOLOGY TO GENOMES
The Meaning of the Double Helix

Copyright © 2001 by World Scientific Publishing Co. Pte. Ltd.

All rights reserved. This book, or parts thereof, may not be reproduced in any form or by any means, electronic or mechanical, including photocopying, recording or any information storage and retrieval system now known or to be invented, without written permission from the Publisher.

For photocopying of material in this volume, please pay a copying fee through the Copyright Clearance Center, Inc., 222 Rosewood Drive, Danvers, MA 01923, USA. In this case permission to photocopy is not required from the publisher.

ISBN 981-02-4328-6

Printed in Singapore by Mainland Press

Foreword

At the time of writing this foreword (June 1998), the genome of 15 unicellular species have been completely sequenced and made available to the scientific community: ten *Eubacteria*, four *Archaebacteria* and yeast, the only eucaryotic cell so far.

During the last three years, the amino acid sequences from over 35,000 proteins uncoded by these genomes have been deciphered. More than 10,000 proteins thus unveiled have an unknown function and are members of new protein families yet to be identified. The sequence of each new bacterial genome (with the possible exception of *E. coli*) provides the microbiologist with a huge amount of fresh metabolic information, far in excess of that already in existence. A surprisingly large biochemical diversity in the bacterial world has been revealed. If over a million bacterial species exist, of which only approximately 3,000 have so far been cultured, then the enormous task of identifying the bacterial world through the sequencing of its genomes is still to be accomplished. Dozens of sequencing centers (rapidly expanding to become hundreds in number) all over the world are undertaking this task, which will result in the discovery of thousands of new proteins. These figures do, however, call into question the exploitation of this data.

Each one of these bacterial genomes forms a tier of biotechnological information, the exploitation of which will create new catalysts, hopefully more efficient and more specific, but less detrimental to the environment than most of the chemical processes currently in use. When the plant and animal kingdoms have been fully exploited by the human race, the domestication of a still largely-unexplored and immense world will remain — that of the *Eubacterial* and *Archaebacterial* kingdoms.

In the next few years, a considerable portion of the sequencing of bacterial genomes could be performed in developing countries, where labor costs are low. It is, however, to be expected that the industrial exploitation of this newly available knowledge will continue to be carried out primarily by large multinational corporations. Whatever the outcome, the race for the identification and exploitation of new microbial enzymes is on; we can only hope that, *in fine*, it will be beneficial to mankind as well as to the preservation of other living species. Equally important as the production of new industrial enzymes, newly acquired panoramic knowledge of pathogenic microbial genomes will considerably speed up the development of new vaccines and antibiotics.

In addition to harnessing of the bacterial world, it is easy to imagine the quantitative and qualitative increases in plant productivity which will result from the complete knowledge of the genomes of some one hundred plant species, which currently constitute the core of our food source. Moreover, it is possible to predict the elimination of the potent parasites currently responsible for the untimely deaths of millions of people each year by thoroughly understanding of their metabolism. This knowledge will not be complete until their genomes have been sequenced.

So, what will the repercussions on the medical world would be, that will be brought about by knowing the sequence of all the proteins constituting over 200 cell types which make up the human body? All the large pharmaceutical companies are investing heavily in medical genomics, from which they are hoping to create new drugs, new diagnostic tools and new genetic treatments. The extent of these investments is somewhat surprising, as mentioned by Philippe Goujon in his book, since the financial return appears risky and, at best, will only be obtainable in the long term. I am, nevertheless, convinced that the reading of the human genome is a necessary (though not, of course, the only) step in the molecular biology of all essential human biological functions, which sooner or later will be used to the benefit of medical knowledge and ultimately mankind.

This optimism is justified by my belief in the intrinsic value of scientific knowledge and by the extraordinary progress in molecular genetics in the past 25 years. It has flourished from the first genetic transformation of bacteria to the first complete sequence of microbial genomes, including yeast, and will lead to the complete sequence of the human genome, hopefully, by the year 2005.

In the following pages, the science historian Philippe Goujon describes, in detail, the action taken by the European Commission in the domain of biotechnology. He has compiled a comprehensive amount of documentation and has dug deep into the archives and the memories of pioneers such as Fernand Van Hoeck, Dreux de Nettancourt, Etienne Magnien, Ati Vassarotti and Mark F. Cantley, who were (or are still) key players in the birth and development of biotechnology programs at the European Commission. Philippe Goujon has placed this European effort in a scientific, sociological and industrial context going back to the beginning of the 20th Century. Although the European effort has become much more prolific than initially predicted, Philippe Goujon highlights the still dominant contributions of the United States of America and multinational corporations to the massive industrial exploitation of molecular biotechnology which, after a first half-century of progressive implementation, will undoubtedly triumph during the 21st Century. Philippe Goujon, an informed observer, meticulously describes the succession of scientific discoveries which have resulted in the recent discovery of the complete sequence of genomes. Finally, as a scientific philosopher, he makes his position clear on the future of genomics and its possible implications. He also gives a concluding analysis of the role of science in our society.

This comprehensive analysis, based on remarkable documentation, is a massive and unique undertaking. I am pleased to have been able to foster the writing of a book which, I believe, will become a classic in the contemporary History of Science.

Professor André Goffeau
Université catholique de Louvain
Unité de Biochimie Pysiologique
2, Place Croix du Sud
1348 Louvain-la-Neuve
Belgium

Contents

Foreword   v

Introduction: The Essence of Life and the Labyrinth of
the Genome   xiii

1   The Invention of Biotechnology   1
    1.1 The Origins of Biotechnology   5
    1.2 The Emergence of a New Concept of Life   11
    1.3 From Zymotechnology to Biotechnology   13
    1.4 The Engineering of Nature – Towards the Best of all
         Possible Worlds   21
    1.5 Technique, Biology and the Development of Biotechnics   23
    1.6 The Recognition of Biotechnology by the Institutions   43

2   Political Interpretations of Biotechnology and the Birth of
    the First Research Programs   62
    2.1 The Example of the United States   64
    2.2 Biotechnology in Japan: Economic Success and
         Ecological Failure   73
    2.3 Germany and the Political Aspect of Biotechnology   78
    2.4 The British Development of Biotechnology: Delayed
         Political Reaction   84
    2.5 The French Reaction   90
    2.6 The European Community and Biotechnology — The
         Emergence of the First European Biotechnology Programs   92

3	The Foundations of the Heralded Revolution	116
	3.1 From the Frontiers of Genetics to the Birth of Molecular Biology	116
	3.2 The Secret of Life: DNA	126
	3.3 The First Sequencing of a Protein: Insulin	135
	3.4 Techniques of DNA Sequencing	139
	3.5 Gene Money, or the Miracles Expected of Biotechnology	161
	3.6 The Japanese Threat and the Human Frontier Science Program	174
4	Attack on the Genomes: The First Genetic and Physical Maps	182
	4.1 The Problem of Gene Localization	182
	4.2 Polymorphic Markers, Gene Mapping and the Great Gene Hunt	185
	4.3 Towards a Complete Linkage Map	194
	4.4 Physical Genome Mapping: The Reconstruction of a Complicated Puzzle	202
	4.5 The First Physical Maps of Large Genomes	205
	4.6 Towards a Physical Map of the Human Genome	215
5	The Human Genome Project and the International Sequencing Programs	217
	5.1 The Ultimate Challenge: The Human Genome Project	218
	5.2 The Department of Energy Initiative	230
	5.3 The NIH Genome Project	245
	5.4 HUGO, or the Difficulties of International Coordination	252
	5.5 The Importance of Model Organisms	267
	5.6 The International Dimensions of Genome Research: The First Stirrings in other Countries	273
6	European Biotechnology Strategy and Sequencing the Yeast Genome	327
	6.1 Towards a New European Research Policy for Biotechnology	328

6.2	The 1980s: An Implementation of the 1983 Strategy?	341
6.3	BAP's First Year	353
6.4	The Revision of the BAP Program	359
6.5	The Origins and Nature of the Yeast Genome Sequencing Project	369
6.6	Critical Discussions and the Adoption of the Yeast Genome Sequencing Project	382

7 The Decryption of Life — 421
   7.1 The Structure and Organization of the European Yeast Genome Sequencing Network — 421
   7.2 A World First — The Sequence for a Whole Eucaryote Chromosome: Chromosome III of the Yeast *Saccharomyces cerevisiae* — 434
   7.3 The Complete Sequence of the Genome and the Intensification of European Efforts — 451
   7.4 After the Sequence — The Challenge of Functional Analysis — 487
   7.5 Sequences, Sequences and More Sequences — 511
   7.6 From Science to Economics — 606

8 Conclusion: The Dreams of Reason or the New Biology's Dangerous Liaisons — 643
   8.1 Fascination but Anxiety Concerning Progress in the Life Sciences — 643
   8.2 Reductionism *vis-a-vis* the Complexity of Life — 648
   8.3 From Science to Ideology — The Dangers of "The Genetic All" — 658
   8.4 The Health Excuse — A New Utopia? — 666
   8.5 Behind Gene Therapy — The Dangerous Liaisons of the New Biology — 676
   8.6 Convenient Reductionism — 683
   8.7 The Reasons Behind an Ideology — 684

EPILOGUE	690
Dreams or Nightmares? Man Reasoned out by His Genes	690
Bibliography	693
Acronyms	751
Author Index	755
Subject Index	773

Introduction:
The Essence of Life and the Labyrinth of the Genome

During this century, all scientific domains have been marked by important discoveries, often made by extraordinary researchers and by a dynamism which can truly be considered a permanent revolution. Since 1900, scientists and their ideas have brought changes so sweeping in their scope that they have modified our perceptions of the nature of the universe. The first of these changes was in physics, with biology hot on its heels. The revolution in physics started in the beginning of the 20th Century with the theories of quantum mechanics and relativity. It was concerned with the inside of the atom and the structure of space-time, continuing well into the 1930s and throughout the development of quantum mechanics. Most of what has happened in physics, at least until recently, has been the result of those three decades of work. With the Manhattan Project, physics was restructured at the deepest level. A new form of research called "Big Science" had been invented.

In biology, the modern revolution began in the mid-1930s. Its initial phase, molecular biology, reached a plateau of maturity in the 1970s. A coherent, if preliminary, sketch of the nature of life was set up during these decades, in which a mastery of the mechanisms of life was increasingly sought after, particularly for industrial use. There would be a second phase to the biological revolution, that of genetic engineering and genome research. The consequence of this contemporary revolution was the advent of a new form of research in biology. It is the harbinger of the eminent role that biology will have in the evolution of society, and the effect this will have on our way of life and thinking.

Several decades after physics, biology has also become "Big Science" with the genome projects and is undergoing a revolution of a structural, methodological, technological and scientific nature. Less than half a century after the 1953 discovery of the structure of deoxyribonucleic acid (DNA) by James Watson and Francis Crick, the great dream of decoding the entire sequence of the genome of an organism became a reality. After viruses and other "simple" organisms, scientists were starting to work on the genomes of "complex" organisms.

In 1989, the United States initiated the Human Genome Project and several nations, including Japan, became involved in genome research, setting up large biotechnology and sequencing programs. At the same time, the biotechnology programs funded by the European Economic Community (the European Union) provided support for a consortium of European laboratories to sequence the genome of the yeast *Saccaromyces cerevisiae*.

Consequently in 1992, for the first time, an entire eucaryote chromosome, in the form of chromosome III of the aforementioned baker's yeast, was sequenced. In addition, the European consortium and its international collaborators managed in 1996 to break another record and obtain the entire sequence for the yeast genome, some 14 million base pairs. During this gene race, Europe managed to maintain its position as a leader in world genome research and provided the international scientific community with a tool whose usefulness, especially for our understanding of the human genome, becomes increasingly clear.

Scientific and industrial competition continues. The number of organisms for which the entire genome sequence has been read is increasing. Along with the yeast genome and smaller genomes that are nevertheless of biological or economic importance, we expect a forthcoming completion of the sequencing of the 100 million base pair genome of the nematode *Caenorhabditis elegans*. The first pilot programs for systematic sequencing of the human genome's three billion base pairs have been completed, making way for larger sequencing projects of whole human chromosomes.

A technological, scientific and economic revolution is under way, a revolution whose effects are already being felt and which will shake the fields of human and animal health, the pharmaceutical business, the agro-food industry, our environment and society as a whole. As announced by the scientists and publicized by the mass media, the prowess of the new biotechnology, the cloning of genes and even of entire animals (sheep and monkeys) as well as the prophecies of several great scientist pundits bring out fascination and concern within us. Science also produces perplexity.

Despite the importance of this new biology, of genome research and of the aura that surrounds them, it seems as if the general public is somewhat in the dark as to the real issues, progress, possibilities and risks. The history of these achievements is often misunderstood, even by some of the people involved in them. In view of the ongoing revolution and its consequences, whether scientific, political, economic, social or ethical, and the public's general misunderstanding of them, there was need for a book presenting the origins and the history of genome research, its current status and its future. This book hopes to fill that need.

At the source of this book there is a wonder, a passion, but also the chance meetings that led me to write it. The wonder is that of the philosophy of science and the window it provides on our world and its secrets, Man and his place in the universe. The passion is that of understanding science as it evolves, the challenges that human intelligence seeks and the risks linked to the discoveries and revelations of this increasingly powerful field of biotechnology. I first thought of writing this book during my post-doctoral period between 1993 and 1996 at the Université catholique de Louvain. It was during this time that I was fortunate enough to meet Professor André Goffeau who was then coordinating the European yeast genome sequencing project. During many conversations he opened up his archives and his laboratory to me, as well showing me the fascinating and multidisciplinary field of genome research, which lies at the frontier between fundamental and applied science, and where technology and biology, research and industry, economy, politics, law and science all meet.

Fascinated by the work under way and carried by my enthusiasm for the research despite the obstacles and difficulties involved in this endeavor, naturally I was taken by the idea of a book on the European and world effort to sequence the genome of *Saccharomyces cerevisiae*, but also on genome research as a whole, its origins and developments. I was immediately supported by Professor Goffeau who, prompted by some of his colleagues, was looking for someone to write the "little tale" of the European and world effort to sequence the yeast genome. As I browsed through the archives, I became more and more convinced that the origins, the size, and the consequences of the changes in biology and society in general, wrought by the initiators and managers of the genome project, had to be communicated to the world at large.

Academic work always uses a strict method of approach. This means an ordered process of disciplined progress. Like any study this work needed a perspective, an approach. But because it is a story, a rational approach would have been limiting and unsatisfactory, which accounts for the anecdotes included in this work. Ideas are often born of happy coincidences, during a conversation, while reading an article or meeting someone, or during a conference, and most often if not always in a propitious scientific, economic and sociological context. They generally begin as an attractive hypothesis. If the idea is stimulating enough, work begins in a new field or domain. It's often not a linear process. Various eddies influence the flow of scientific construction as it is taken away from the official paths and then often seems, to use a term from physics, chaotic and subject to sudden changes of direction. But that means surprise, and surprise is what researchers live for and from.

Besides contributing to an understanding of modern biology and biotechnology, this work also hopes to serve to explain scientific practice, its means, its rules, and its presence amongst us, in our individual and collective lives. In particular, it seeks to underline, through the history of biotechnology and the genesis and development of genome research, the deep relationship that links these domains, while stressing that they face not only the material world but society as a whole. Biotechnology and genome research goes beyond the purely operational perspective to embed itself resolutely in human history, in communication, negotiation and communal creation.

This work shows that the bases of biotechnology and the genome projects lie within historic projects in their own particular political, socio-cultural and economic context. It also demonstrates through its investigation of the history of this ongoing biological and particularly biotechnological and genome research revolution, the human aspect of science and the creativity which is inherent to it. It attempts to, through its philosophical and ethical reflections, clarify the manner in which they are products of society as well as an influence upon it.

To trace the story of biotechnology and the genome programs, in particular that of the yeast genome, one has to take into account the various groups affected, watch the development and change of their various scientifico-socio-economic aspects in reaction to conceptual and technical modifications and the rapprochement of previously quite separate disciplines. One single method would therefore not allow us to understand the highly multidisciplinary history and evolution of biotechnology and the genome programs. The multidisciplinary components involve:

– The scientific and technical bases
– The multisectoral applications
– The industrial and economic dimensions
– The national, European and international political levels
– The institutional and international organizations
– The legal dimensions
– The ethical dimensions
– The "public" dimensions

Each of these aspects must be taken into account.

Since it will be the record of some of the most modern developments in biology, in places this book will be unavoidably somewhat didactic. As for approach and methodology, it should be pointed out that for the purposes of this work, scientific creativity is not just seen in an abstract fashion, either conceptually or philosophically but mainly seen through the people at the root of it, through their social and intellectual biographies, the influences they have had, their motivations and the spirit in which they undertook their studies. Great projects cannot be separated from the destinies of the people involved, nor can the intellectual process be separated from the many meetings and opportunities that happen in the course of a life. Scientific creation is, more so today than ever, incompatible with a closed society. Any rigid approach might lead to the pitfall of revisionism, and as Martin Heidegger wrote, "the inability to produce any point of view other than that of the author". It is necessary to point out to what extent apparently objective, upstanding arguments are personal and often empty, their rigorous formality masking the lightness of the ideas. For this reason, and in order to respect the story of modern biology and in particular its multidimensional nature, the historian-philosopher must go to meet those actually doing the science, dive into their culture, follow them in their paths and pilgrimages, understand the web of their community, their intellectual, cultural and social evolution and accept the humility needed to understand pure fact.

This book, on the development of biotechnology and the genome programs, owes a great deal of its substance to the author's interaction with the work and personality of Professor André Goffeau. It also depends on the author's observations as a listening, watching, external observer right inside the coordinating laboratory and community of the yeast genome project, much as an anthropologist would live with a far-flung tribe. Aside from these interviews, it also gained much from the archives of the yeast genome project and DG XII (Directorate General for Science, Research and Development of the European Commission in Brussels). Many documents, both biographical and scientific, were sent to the author by various laboratories and international institutions (DOE, NIH, HUGO, and the embassies) which helped to reconstitute the scientific, technical and economic environment to which the first genome projects were born.

This work is directed not only to the scientific community, but also to the lay reader. In its progression from the origins of biotechnology and its political and economic interpretation by the great nations, this book retraces the birth of the first national and international biotechnology projects. On reaching the end of the 1960s, it reveals the foundations of the modern biological revolution, in particular the techniques of genome sequencing and analysis and highlights the importance of Japan in this context on the international research stage. This book also describes the completion of the first genetic and physical maps and the political and scientific genesis of the American Human Genome Project. It then follows with a detailed analysis of the establishment of European biotechnology strategy, the birth of the first European biotechnology programs and the yeast genome project.

After describing the main stages of technical, administrative, political and scientific successes leading to the yeast genome project and its major aspects in Europe and world-wide, this work mentions the other genome projects in order to provide the reader with a wide view of the importance and consequences of this science ongoing at the dawn of the 21st Century, as well as greater knowledge of current activities at the national and international level. In the last chapters, it considers the challenges to be overcome as well as the perspectives opened to us by systematic sequencing projects.

The conclusion, which is not meant to be a once-and-for-all answer, brings the reader clearer understanding of genome research at the biological, biotechnological, and current socio-economic level. It underlines their meanings, justifications and their perspectives. Through an epistemological analysis of the "dreams of the rational", it voices our confusion at the new definitions of life. It tries, with a critical eye, to find limits and lacunae, taking into account the ethical, social and political problems linked to the advent of this bio-society heralded since 1970 and which raises such hopes and fears.

This work is also a testimony to a new way of carrying out biological research. For some veterans who have survived from past generations of researchers, this change in the way science is being done, particularly in biology, is surprising and even shocking. The underlying determination to progress is still — but for how long? — that of curious scientists probing the mechanisms of nature. Of course, it is also the hope that some of the new knowledge will serve to better the human condition. But the main element, the main permanent and irresistible drive is most often the simple desire to discover. In fact, to work on hard biological problems, is for the biologist the greatest pleasure in the world, their reason for existence. However, when you hear the description of the strategies, successes and risks of genome research, I fear that part of this pleasure is being taken away, in this current socio-economic context that restricts the opportunity to express and accomplish all that is possible to each of us.

It is true that some aspects of research have not changed, for example the long and feverish hours of waiting for an experiment's result, the surprises, the disappointments, the joys of attaining what was hoped for, of finding something, of contributing a snip of new understanding, of participating in the launch of an idea, of being part of that brotherhood sharing the same obsessions and speaking the same private language, the tedious task of writing articles and giving conferences... But other aspects have become very different than they were in the past. Competition has come knocking at the gates.

Of course, there has always been competition in research, especially when the expectations of the new information, are as high as they are today in the domain of genome research. But competition today sometimes looks more like a war at the national level and although cooperation continues and even develops science it does not bode well. Whereas before, science happened at the laboratory level, the new projects are at national level, involve colossal budgets and have fundamental economic consequences. In these projects individuals fade into the mass, strategies overlap with each other. The process looks like that previously seen in industry; from small firms to groups, chains and international holding companies.

At the local level, there is a real problem with jobs. There are more young researchers than before, in the early stages of their careers, who face a hierarchical pyramid with a limited number of university posts and stable positions in industry, a hierarchy often conservative and typified by a certain inflexibility. The breathtaking rate at which they work and publish is the consequence of this as well as the result of the new methods being used. These methods have allowed discovery in the field of life sciences to accelerate to an unprecedented pace, but they also have deep consequences in the organization and work patterns of laboratories, leading to ever larger teams of young researchers, mostly at the thesis or post-doctoral level, each with a task linked to a minuscule element of the global picture, and each considering the success of that task as the hook from which their future career depends. The fight to survive is therefore more ferocious, positions increasingly rare and financial support ever more scarce. Worse, a large number of laboratory directors and professors neglect their students, being preoccupied with their own ambitions. In this context, as an independent and privileged witness, the author pays due homage to Professor Goffeau for his devotion to his laboratory, his researchers and to his students.

Another source of worry, money, has become a far more crucial problem. The story this book tells, shows that for some scientists, faith still can move mountains, that the pleasure of discovery can still drive scientific activity with the joy or even hope of a result, that setbacks, dead ends, unsuccessful experiments, can still be forgotten. But this pleasure has a cost and the cost can often be high. There are sacrifices made, long periods away from the family, career uncertainty (for many years), the dominance of science in a researcher's life often to the detriment of family life — has not more than one scientist's wife said "we are a *ménage à trois*, my husband, science and myself"? But most of all, there is the risk that this scientific activity be blunted by the increasingly tough competition currently reigning in scientific circles and which pervades far beyond the academic sphere. In the pages of this book it becomes apparent that there are new considerations on the scene, considerations of an economic nature, the large biotechnology companies running in the gene race are continually throwing more and more money at targeted research themes. The pressure exerted by the Member States on the European Union, during the negotiations for the Vth Framework Program, for a far more applied orientation to funded research, is symptomatic of this global movement in which research must be productive economically, and furthermore productive in the short term.

In less than 20 years since the end of the 1980s and with unprecedented acceleration, life science has migrated from being a pure science to being a hard science with endless applications and with fundamental industrial, economic and social expectations that will change our lives, with of course both the advantages and the dynamism it will bring but also the deep modifications in the way science is built and carried out, modifications that will also have counterproductive effects. Not least of these will be the risk that communication of information within the scientific community, even through chatting, will be compromised.

If anything takes away the pleasure and dynamics from research, it will be the excess of confidentiality. The scientific networks described in this book, constituted of national and especially international cooperation, only work because the researchers and policy makers share what they know, at congresses and during private conversations... Telling other scientists what you have found is a fundamental part of the fun in being a scientist. Now that private firms are investing massively in genome research, it is easy to worry that events might end this privilege, a worry justified by the fact that life science information is more and more applicable to Man, with all the risks that that may mean for the future.

This book hopes to provide the public, often kept in the dark, the opportunity to keep current on the new progress in biology and the risks linked to the new direction science is taking. The only way to control them is to transcend both the approach of the geneticist and the industrials, and to remember that living beings are more than just vectors for the transfer of genetic information from one generation to the next, that human life, and life in general, is much more than the running of a computer program written in DNA. But maybe this is all an illusion? Might there still be a place for science with a conscience, when economic interest rules and profit is king? In any case, science in order to save itself must build stronger links with society and no longer remain in an elitist ivory tower. That is the only way it will be able to link up once more with a political and ethical conscience. What is a knowledge you cannot share, that remains esoteric, that can only be damaged by popularization or be used by industry in destructive processes, that influences the future of societies without control of itself and which condemns citizens to be the subjects of a rationality and technique that they no longer understand even as they are ignorant of the problems of their destiny? As Edgar Morin points out in "*Science avec Conscience*", "empirical science deprived of forethought, like purely speculative philosophy, is insufficient, conscience without science and science without conscience are radically mutilated and mutilating." In genome research, a science without conscience risks to ruin Man, to paraphrase Rabelais. A new alliance has to be born. The author modestly hopes with this book to bring the reader the chance to understand what is going today in the field of genome research and therefore contribute to communication at the boundaries of three cultures: the scientific, the humanist and that of the citizen through the media. He hopes to have forged a link between scientific problems and problems of the citizen, who more than ever needs a vision of the world but also to debunk science from its fetish-religion position.

****

This work could only have been brought to term with the help of those persons or institutions who have supported the effort and trusted me: the Université catholique de Louvain (Belgium) where this work began at the Institut Supérieur de Philosophie and its Philosophy of Science Center where I was welcomed by Professor Bernard Feltz, and of course in André Goffeau's physiological biochemistry laboratory. Mark Cantley, head of the Biotechnology Unit at the Science, Technology and Industry Directorate of the OECD, Fernand van Hoek, then the Director of the Life Sciences Directorate of DG XII, who agreed to an interview, entrusted me with documents and shed light on the story of the decisions that led to the adoption of the first biotechnology programs as well as the arcane world of community science policy. Dreux de Nettancourt, the man "behind the scene" who, with great modesty and altruism, has initiated and implemented the biotechnology programs of the European Union. The European Commission, in particular the people at DG XII and its management. Madame Anne-Marie Prieels, secretary of the Yeast Industry Platform. Professor Bernard Dujon, head of the Molecular Genetics of Yeast unit at the Institut Pasteur in Paris, who was kind enough to lend me a large number of documents on the work carried out by his team during the "yeast program". The Plant Genome Data and Information Center of the US Department of Agriculture, the National Institute of Health (NIH) and the Department of Energy (DOE) in the United States, the British Medical Research Council, the now-defunct French GREG venture, and the HUGO organization's European and Pacific offices, all of which gracefully provided me with the documents I needed for this work. The embassies of Germany, Denmark, the United Kingdom, Norway and Sweden who answered my requests for information. I also thank my translator, Marianne, for agreeing to undertake the difficult task of translating this work. Alban de Kerckhove d'Exaerde, a real friend, devoted long hours to showing me the techniques of modern biology. Heartily thanks to the administrative and laboratory personnel of the Physiological Biochemistry Unit of the Université catholique de Louvain, Belgium, who were always very welcoming to this outsider... I would like to express my profound gratitude to Professor Jean-Claude Beaune for his help. Thanks are also due to my thesis director, Professor Jean Gayon, and to Professor Claude Debru, both of whom provided constant support and encouragement. The Université Catholique de Lille, under the leadership of the rector, Gaston Vandecandelaere, welcomed me to its center for contemporary ethics. I would like to acknowledge Monsieur Jean-Marc Assié, Monsieur Michel Falise, Monsieur Bruno Cadoré, Monsieur Bertrand Hériard, as well as the teaching and administrative staff of the center for ethics.

But more than anyone else, this work is owed to André Goffeau, who opened his laboratory and archives to me, encouraged and guided me, and who constantly supported the writing and translation of this work.

My warmest thanks to the Fondation pour le Progrès de l'Homme, in particular to Monsieur Yves de Bretagne whose financial support contributed to the present publication.

All research work necessarily eats into the private life of the researcher and those who surround him. This work was no exception, and I would like to thank my wife Clara and my parents who have withstood the invasion of work, even though I am quite sure that they know it's worth the trouble...

to Virgilio and Nahui-Ollin
with love

1
The Invention of Biotechnology

Although it is often seen as a recently developed label, the term biotechnology itself dates from 1917. Today, its best-known definition is the one used by the Organization for Economic Cooperation and Development (OECD). It is a vague definition[1], but the many attempts to formulate a more precise one, and the synthetic definitions they have produced, have proved inadequate. Over time, the various ideas and interests related to biotechnology have not replaced each other, but have become layered, forming a strata of connotation during the course of the 20th Century. Some of these connotations have been reworked, forgotten or even consciously rejected, but in such a worldwide and cosmopolitan science they have not been lost forever.

A historical study would untangle and clarify the knots of semantic associations and meanings rarely distinguished from each other, leaving us with a picture of how apparently quite disparate interpretations and concepts ended up with the same label of biotechnology. Such a study would also review the development of the concept of biotechnology, which despite its roots in the more historical idea of biotechnology, now focuses far more on genetic engineering.

The alliance of genetic engineering techniques with industrial microbiology has been such a new beginning, that there is now unfortunately a common misconception of biotechnology as the science behind genetic

---

[1] "Biotechnology is the application of scientific and engineering principles to the processing of materials by biological agents to provide goods and services".

manipulations, merely the child of genetic engineering. Biotechnology is not distinguished in the public mind from the technical revolution of the 1970s and 1980s, when scientists learnt to delicately alter the genetic constitution of living organisms. This technique was rapidly considered capable of "turning DNA into gold"[2].

This concept of biotechnology was highly praised. It was hoped that a better understanding of DNA[3], that magical acronym, would be the key to a better world. At the same time, the layman public worried about such esoteric techniques. These underlying worries were caused by the Nazi and racist eugenics theories, their terrible consequences, the cynicism of the Vietnam war period, the fear of the military industrial complex and of course the terrifying precedent of atomic fission. As biologists pointed out the apparently unlimited potentials opened up by better understanding of DNA and ever-widening techniques to manipulate it, they themselves recalled that physicists built the first atomic bomb without getting a chance to ask about its long-term consequences. The biologists hoped for a better conclusion to their endeavors.

The technical parallels that can be drawn between nuclear technologies and biotechnology always lurk in the background of the biotechnology debate. There were growing hopes for scientific commercial applications of the new biotechnology of genetic engineering, but also deep worries about its potential risks. With all this the science took on such notoriety that even the most extravagant proclamations of its power could be believed. There was an even stronger parallel when the scientists themselves, in 1974[4], alerted public and policymakers in calling for regulation of experimentation. A technological evaluation process was initiated to ask what regulation should be made of this new and powerful technology. The biggest, most sensational promises of the new alliance between biotechnology and genetic engineering

---

[2] Title of the first chapter of Sharon McAuliffe and Kathleen McAuliffe's *Life for Sale*, published in 1981 by Coward, McLann and Georgehan, New York.

[3] Deoxyribonucleic acid: a molecule in a double helix structure that is the chemical base of heredity. It resides in the chromosomes but can also be found in mitochondria and chloroplasts.

[4] Paul Berg *et al.*, "Potential biohazards of recombinant DNA molecules", *Science,* vol. 185, 1974, p. 303, and Paul Berg *et al.*, "Potential biohazards of recombinant DNA molecules", *Nature,* 1974.

were made, publicized and supported with the help of considerable advertising.

Scientists, industry, and of course the governments, were met with deep public concern as, during the 1970s, they linked the power of the new genetic engineering techniques, especially recombinant DNA, to uses, where precautions were to be taken. Until then the borderline between biology and engineering had been commercially underexploited. But given the alleged spectacular opportunities, the public's estimation of the distance between a realistic short-term benefit and the more speculative benefits gradually shrank in the wash of announcements of a new industrial revolution that would provide stunning advances in fields as far apart as medicine, food, agriculture, energy and ecology.

According to these prophecies, through the revelation of the mystery of life, the constant progression of our understanding of the material nature of genes (accelerated by the national and international sequencing programs) and the growing clarification of mechanisms of inheriting genes, biotechnology would enable us to cure incurable diseases. It would provide more abundant and healthy meat, milk products, fruit and vegetables, and allow us to improve pesticides, herbicides and irrigation procedures. Environmentalists saw biotechnology as a way to process oil wastes and other ecological damage more efficiently and more safely. Industrials saw biotechnology as a route to economic revitalization and new markets. This is why a large number of firms of various sizes have over the last twenty years launched biotechnology ventures, and why ministries, government institutions, international institutions, universities, research centers and private investors have also contributed to research into the skyrocketing field of bio-industry.

Other fields in science and technology, such as computers, communications, spatial technologies and robotics, have also risen and soared during the second half of the 20th Century, pushed by an acceleration in innovation. But the bio-industry is peculiar among them because of its multidisciplinary aspect. The technologies themselves, that use living cells to degrade, synthesize or modify substances have to be multidisciplinary as their products are used in many sectors of human society, such as:

- Microbiological engineering, i.e. the search for, collection, selection and conservation of microbial strains, and the study of the way they operate
- Biochemical and industrial engineering, i.e. the fine-tuning of bioreactors in which biochemical conversion is taking place, controls of production procedures, the optimization of techniques to extract and purify the products in question
- Enzyme engineering, i.e. using enzymes as optimally as possible in solution or immobilized on a solid base, or producing enzymes that remain stable in unusual physical conditions
- And of course genetic engineering, a generic term for a group of techniques discovered and perfected since the early 1970s and also called genetic manipulation. These techniques involve the isolation and transfer of genes into microbial, animal or plant cells, allowing them to produce large quantities of different substances.

Depending on the application, one or other of these four technologies might be dominant. The most marked progress in bio-industry has been accomplished through the conjoined efforts of the four technologies, but it is first and foremost genetic engineering that is associated with the miracles, prophecies and promises in interviews, articles, and documentation. This manipulation means that genetic engineering is associated with mastery of heredity, mastery of what makes us what we are. In addition to this we should note the part played by the large-scale genome sequencing programs launched in various countries and by the EEC[5] during the 1980s. These large-scale programs, especially the Human Genome Project, popularized the idea that DNA was the key to a new radiant future, to the new "bio-society"[6], and to the conquest of a new frontier that would allow the full realization of modern biotechnology's potential.

The aims of this chapter on biotechnology are twofold: firstly, to better situate the birth of the new biotechnology after the marriage of biotechnology

---

[5] European Economic Community

[6] From the FAST subprogram C: Bio-society, *European Commission FAST/ACPM/79/14-3E*, 1979.

and genetic engineering in the 1970s, and secondly, to return to biotechnology those hopes and motivations founded in it but more specifically in genetic engineering. In our journey from the origins of biotechnology to the inception of the large-scale sequencing programs, this chapter will clarify the ideological, economic and political contexts during its development, and how they made modern biotechnology what it is today.

1.1 The Origins of Biotechnology

The most important element in the history of biotechnology is the process of making alcohol, or fermentation. In the Middle Ages distilling alcohol was still a combination of metaphysical theory and practical skill, but 500 years later the alcohol-related industries were integrated into science and presided over the birth of modern biotechnology. It all began with the establishment, in the 19th Century, of a new and vital subject called "zymotechnology."

From the Greek *"zyme"*, meaning leaven, zymotechnology could involve all sorts of industrial fermentation, not just its main connotation, brewing beer. Because of its newfound applicability to a wide field of uses, from tanning to making citric acid, zymotechnology was thought in the early 1900s to be the new economic panacea. A Danish pioneer, Emil Christian Hansen, proclaimed that with this new discipline "In this entire field, a new era has now commenced[7]".

The meaning of zymotechnology was ingested by biotechnology, and fed its growth, providing a practical continuity as important as the intellectual continuity. The Berlin *Institut für Garungsgewerbe*, for example, set up in 1909 in a magnificent building funded by government and industry, had a fundamental role in the 1960s in the establishment of biotechnology in Germany.

---

[7] Emil Christian Hansen, *Practical Studies in Fermentation,* translated by Alese K. Miller, Spon, London, 1896, p. 272.

Zymotechnology is therefore a decisive stage in the progression from the ancient heritage of biotechnology to its more modern associations and connotations. Most of the work to set up zymotechnology as a discipline was also to prove important later. Small zymotechnology expertise bureaus were set up to carry out specific research projects for the various industries that used fermentation.

The central importance of alcohol production in the history of biotechnology is fully recognized by biotechnologists, who point to the demonstration of the microbial origin of fermentation by Louis Pasteur. For Pasteur's biographers, and perhaps also for his fellow biotechnologists, Pasteur made the science of the microbe - microbiology - the only way to study fermentation and its industrial applications. It is perhaps a bit simplistic to say that one man was the founder an entire discipline. History is always more complex than that. Most of the measures of hygiene that we associate today with Pasteur are in fact the fruit of earlier work, as we are told by Bruno Latour[8].

In Germany, it was chemistry, in particular, that had a significant hold over the processes of fermentation. Pasteur was a chemist by training, as it was one of the more important sciences of the 19th century, and his lessons and skills were useful in various ways to prove his theories on the special properties of microorganisms. An all-inclusive discipline design of "microbiology" competed with other disciplinary formulations such as bacteriology, immunology, technical mycology and biochemistry. The uncertain relationship between chemistry and biology was not specific to this particular group of studies either. It was uncertain throughout medicine and physiology, and still to this day remains unresolved. Furthermore, whether the focus is on chemistry or microbiology, the successes of basic science as a measure of the progress of industries such as brewing is often overestimated. Industrial requirements play a large part too.

Based as it was on the practical application of any relevant science, zymotechnology was a crucible for skills and knowledge at the service of

---

[8] Bruno Latour, *Microbes: Guerre et Paix, Suivi de irréduction*, A.M. Métilié, Paris, 1984.

supply and demand. As a group of several disciplines, zymotechnology is a science typical of the last century, but it is also a clear descendent of German chemistry in the Century of Enlightenment.

Despite its ancient appearance, even the root word *zymotecnia* is a recent invention. It was invented by the father of German chemistry, Georg Ernst Stahl, in his work *Zymotechnia Fundamentalis* of 1697. Stahl maintained that the applied study of fermentation - *zymotechnia* - would be the basis of the fundamental German industry of *Garungskunst* - the art of brewing. Fermentation, for us today in itself an expression of life, could be scientifically analyzed, because it was "like movement" and because like putrefaction it happened to materials as they became disorganized and when they were removed from "the empire of the vital force". This interpretation of course has roots of its own which can be found in Thomas Willis' writings of the beginning of the 17th century.

Zymotechnica Fundamentalis can be considered not only as an explanation, but as one of the origins of biotechnology. Such an appraisal is always a matter of opinion, but Stahl's work marks the foundation of the subject of biotechnology in a time when its specific characteristics — the process of fermentation and the potential of the science-technology relationship — were themselves still developing. He was the first to express the now long-standing hope that an understanding of the scientific basis of fermentation could lead to improvements in its commercial applications.

Zymotechnology was popularized by Stahl's protégé Caspar Neumann, and the translation of *Zymotechnica Fundamentalis* into German, which rescued it from Stahl's obscurantism. It even became an internationally recognized concept. A sign of this recognition is indicated by the 1762 acceptance of the term *zymotechnie* for the French dictionary of the *Académie Française*.

The Stahlians constantly had to draw the distinction between their science and the work of charlatan alchemists. It is therefore interesting to note that chemistry has effectively, in the long run, fulfilled the alchemist's promise of prosperity and health. Its apparently unlimited power and potential were already commonly understood by the 19th century. In her novel *Frankenstein*, published in 1817[9], Mary Shelley, a member of the British intellectual elite, managed to convey and amplify the conventional concepts of her time, and she often expressed her admiration for the successes of chemistry. Beyond the realm of science fiction, and despite the fact that Frankenstein pays for his ambitious disruption of the categorical division between the living and the dead with his own life, medical doctors, particularly in France, were increasingly exploring the physiology and chemistry of the sick. Historians have identified a clear change of focus in medicine at the beginning of the 19th century, a chemistry-driven change of interest that brought medicine from simply nursing the sick to treating their illnesses. The idea that living matter was alive due to a divine vital force was eroded and chemistry progressively became the main way to technologically exploit vital processes by interpreting what was going on[10].

Freiderich Wohler's synthesis of urea in 1828 is a clear and possibly final landmark in the erosion of the distinction between natural and chemical products. Wohler's demonstration that a natural product — urea — could be synthesized had implications which were studied by his friend Justus Liebig. With a Stahlian faith in practical applications, and his feeling that the potential of the "vital force" could not be explained, Liebig progressively outstripped his predecessor of the previous century in his concepts of the applicability of science. Initially he also supported the existence of a "vital force", but during the 1850s the "vital force" was secularized[11] and became just another natural force like the others of the inorganic world[12]. Liebig left pure

---

[9] Mary Shelley, *Frankenstein*. Available in French from ed. Garnier Flammarion, 1979.

[10] For a history of biochemistry, see Claude Debru's *L'esprit des protéines (Histoire et philosophie biochimique)*, éditions Hermann, Paris, 1983.

[11] J. Liebig, *Die Organische chemie in ihrer anwendung auf physiologie und pathologie*, Braunschweig, 1842.

[12] For Liebig, the vital force was not a force that passed all understanding. It is a natural cause that must be studied like the other forces in the inorganic world. "If doctors want to deepen their understanding of the nature of the vital force", he writers in his *Letters on Chemistry*, "if they want to understand its effects, they must follow exactly the path that has been shown to them by the great successes of physics and chemistry". Quoted by Claude Debru in *L'esprit des protéines (Histoire et philosophie biochimique)*, éditions Hermann, Paris, 1983. From this point of view, the vital force is no different from electricity, whose relationship with magnetism and light had been conquered with great difficulty.

chemistry at the end of the 1830s, and came to be more and more associated with plant chemistry and physiology. His contribution to, and his mark on, the science was a radical reduction of physiological processes, of fertilization and digestion and the transformation of substances. So although his fundamental interest was in chemical inputs and outputs, Liebig agreed with Stahl that fermentation was a propagation of a molecular collapse into decomposition by matter becoming disorganized and moving away from "the empire of the vital force", from the organized[13]. Liebig thought that chemists could control the process of fermentation by deploying explanatory theory along with the key skills of temperature measurement, hygrometry and analysis.

As a great European discipline, however, chemistry did not follow the route that Liebig took in his own studies. It rather became more and more dominated by his first interest, organic chemistry.

Organic chemists sought to replace the laborious and costly extraction of natural products by their synthesis in laboratories. In his attempts to synthesize quinine, a real challenge in the mid-1800s, Hoffman's pupil William Perkin managed to discover the first synthetic organic dye, mauveine. Simultaneously with Caro, Craebe and Liebermann in Germany, in May 1889, Perkin found commercially viable synthetic varieties of an important natural dye called alizarine. Adolf von Baeyer, Liebig's successor in Munich, set up a research institute dedicated to the study of natural products. Chemistry had rejected Stahl, but it sanctified Liebig and became one of the biggest success stories of the 1800s. The rapid growth of the big German chemical firms testified to the vision of a science of the artificial.

---

[13] This transmission of the internal movement of decomposition inside matter is at the heart of Liebig's theory. It denies any essential role to the vitality of the microscopic vegetable world, since it is because it dies and is restructures that it transmits a movement that no other organic matter can transmit while decomposing. Pasteur's break with this thought was that he took fermentation studies away from the field of decomposition, degradation's and other morbid processes and brought it into that of living activities. With regard to Liebig, E. Duclaux writes that "Liebig only had to pick up the ideas of Willis and Stahl on the internal movement of fermenting masses, and attribute the movement to the fermentation". E. Duclaux, *Traité de Microbiologie, vol. 1, Microbiologie générale,* Paris 1989, p. 43. Claude Bernard, in *Leçons sur les phénomènes de la vie,* vol. 1, p. 159, says the same thing: Liebig is the heir of the iatrochemists. Quoted by Claude Debru in *L'esprit des protéines (Histoire et philosophie biochimi*que), Hermann, ed. des Sciences et des Arts, Paris, p. 39.

The power of organic chemistry was acclaimed but it was also feared. Sometimes it looked as if the results were rather disappointing compared to all of Nature's complexity. Towards the end of the 19th century a greater comprehension of natural products helped develop the public's respect for Mother Nature herself.

Emil Fischer had an intense interest in, and admiration for, the subtlety of the chemical processes in living creatures. He was the first to synthesize a polypeptide and explored carbohydrates and proteins. The chemistry of living organisms was also being studied more and more, and the limits of human chemistry being elucidated, as its potential was also being realized.

Chemistry, justifying its hold over the living world, found itself up against the new dynamic disciplines of biology and physiology, and zymotechnology was no longer simply a facet of applied chemistry. It became a vague catch-all of a term that included the theories and techniques linked to fermentation. It was a key interface between science and industry. As chemistry took a tighter hold over the world of living processes, physiology, backed by powerful medical interests, became more and more reductionist. Zymotechnology was typically handled by the Institutes of Physiology such as that of Bois-Reymond, founded in 1877 in Berlin, which had a chemistry section as well as groups that studied the higher levels of biological organization. The tendency in physiology to reduce life to chemistry and physics were complemented by other perspectives that underlined the importance of ecology. This insistence on the living world was embodied in biology.

Gottfried Rheinhold Treviranus[14], another of Stahl's admirers, and with Lamark and Oken one of the first people to use the term "biology", began his 1801 work entitled *Biologie oder philosophie der lebenden natur fur naturforscher und aertzte* with the sentence "it is exploitation, and not study, that brings a treasure its worth". He went on to point out that biology's value is in its combination with pharmacy and economy.

---

[14] Gottfried Rheinhold Treviranus, *Biologie oder philosophie der lebenden nature für naturfoscher und aertzte,* ed. Rower, Göttingen, 1802.

At the beginning of the 20th century, while biology was breaking free from other disciplines such as zoology and botany, the problems of its new definition were not the most practical of all the worries it was facing. Colloquia and symposia were dominated instead by the evolutionary question and the mechanism-vitalism debate.

1.2 The Emergence of a New Concept of Life

Biology was often claimed to be of practical application. Biological sciences such as botany already had applications. An ecological concept of plants was providing a specifically botanical perspective, helping with the reduction of plant physiology to chemical components. The famous German professor Julius Wiesner, author of *Die Rohstoffe des Pflanzenreiches*, thought that further studies in the botanical sciences would lead to the discovery of exotic materials in the tropics and improve agricultural production at home. In a manner typical to his time, he maintained that a technological approach to the agricultural production of raw materials could and should be adopted by the great technical schools.

As chemical technology was the interface between chemistry and its industries, Wiesner's *Rohstofflehre* would become the mediator between technology and natural history. In France, the barrier between the divine essence of life and the secular province of technology was demolished by microbiology in a way very similar to that of the *Rohstofflehre*. Microbiology also had strong pretensions to being of practical use, particularly in fermentation.

Since Stahl, chemistry in general had been offering more and more technical possibilities, generating increasingly powerful theories and producing a series of extremely high-performance techniques for the control of specific fermentation processes. But compared to the detailed development of organic chemistry, fermentation chemistry's path was that of a pauper, borderline and empirical. It was Pasteur who created a discipline with fermentation at the center of its interests, and who, with microscopy, explored the processes of fermentation, as well as its inputs and outputs. With his

1857 demonstration that lactic acid fermentation was the result of live bacteria, Pasteur created microbiology[15]. He himself defined subjects such as brewing, *vinification* and hygiene as the main fields of application for this new discipline. In 1887 an Institute was set up in Paris bearing the name of the national hero, and since then other Pasteur Institutes have been set up all over the world. Pasteur's own contribution to the French wine and silk industries is of course legend.

In France Pasteur's influence dominated these Institutes. But elsewhere, microbiology had to rub alongside other disciplines in institutes set up to fulfill local needs. In Germany, for example, the lessons of microbiology were taught in practical contexts such as bacteriology. In medicine, physiology was a sort of federating concept, and in industry zymotechnology had a similar role allowing chemistry to cohabit with more recent disciplines such as microbiology, bacteriology, mycology and botany.

The interconnection of disciplines is often the result of circumstances at an institution. In zymotechnology's case the connection was because of an industrial and research context focused on the rapid development of agriculture, and in particular the improvement of productivity and of brewing techniques. On the agricultural side, although research stations and institutes were being set up everywhere, agricultural productivity was improving and surplus was being used more judiciously in industry, this success was being attributed to the improvement of agricultural education. But the relationship between chemists with their laboratory experiments and the growers and breeders was not as simple as that. Although it would be wrong to say that the research institutes and chemists were of no help at all, the farmers were often disappointed with their results. Artificial fertilizers were often not as effective as natural ones, and research stations had to take constant feedback from those using their products in the field.

---

[15] More than the German approach, Pasteur's studies of fermentation opened the field of microbiology more than that of chemistry, because of his resistance to the hypothesis of diastasic action and contact actions. In the German school of thought, doctrines inspired by those of Liebig, or Berzélius' notions of catalysis, proved themselves more easily translated into a language of chemical molecular actions, a language that had no counterpart in Pasteur's science.

As for brewing beer, the chemists were offering precise technical and scientific methods to control the manufacturing processes. As the role of the agricultural college went from educational to being influential on research, the evolution also occurred in brewing, a process in which zymotechnology had a very pragmatic role. It was an eclectic mix of techniques and skills from chemistry, microbiology and engineering, and its roots were sufficiently scientific to extend beyond the realm of brewing to all arts and processes based on fermentation. Zymotechnology, with its almost strategic vagueness that implicated it in a specific industrial area but allowed it to cross the traditional boundaries of the structured market, at that time held the role biotechnology would take at the end of the century.

In the 19th century, zymotechnology was just a subdivision of chemistry, but at the beginning of this century it came to mean a technological competence with its roots in a variety of sciences that nevertheless had a practical side way beyond the simple appliance of science. Retrospectively, although the new science of microbiology might seem to have brought fundamental advances to fermentation technologies, the process of applying the science was a complex one and was only a part of the more general development of zymotechnology.

However, as a brilliant series of microbiologists and bacteriologists displayed their abilities, it became more common to see biology considered in technological terms, if only vaguely. Its first applications in hygiene and alcohol were joined, during the First World War, by the production of chemical substances such as lactic, citric and butyric acids, and yeast cultures. Gradually, the microbiology industry came to be considered an alternative to conventional chemistry, instead of just a peripheral variant of brewing technologies. This new concept of biology also had a new name: biotechnology.

1.3 From Zymotechnology to Biotechnology

There are as many opinions on the applications of microbiology as there are in the more familiar example of chemistry. You could have found chemists who thought that industrial processes should be studied in themselves. This approach was called chemical technology. Others thought that these processes were the application of a pure science and should be recognized that applied chemistry had key importance as a science in itself. The technical aspect was not their concern, but that of the engineers. A third school of thought focused on the birth of a new, distinct and increasingly sophisticated chemical technique.

Although a specific form of engineering would develop later, in the first half of the 20th century the study of fermentation technologies encouraged developments similar to those happening in chemical technology and applied chemistry. Fermentation was increasingly considered part of the applied science of economic microbiology. Although academic supporters of zymotechnology took a more rigidly technological stance on micro-organisms (Max Delbrück began a conference in 1884 with the sentence "yeast is a machine"[16]), "technologists" had to take into account the growing importance of biological perspectives, and zymotechnology evolved into biotechnology.

This evolution was reflected in another growing development, that of the concept of microbiological centers as sources of learning the technological and scientific bases of fermentation industries, centers that would provide advice on micro-organisms, keep culture collections, and carry out research. These microbiological centers varied from country to country, but there was a general progression from brewing and technique to a more general insistence on science, and in particular microbiology.

In Germany the situation did not seem very satisfactory, and there were pressing calls for a theory of bacteriology, which until then had only been an applied science and was not seen as the application of a pure science.

Despite this, the overall situation was changing. Technologies were increasingly being considered the applications of fundamental science. This gradual widening of interest from brewing to science in general, be it microbiology, bacteriology or biochemistry, and the diversity of possible applications, was an underlying trigger of zymotechnology's evolution into biotechnology.

---

[16] Max Delbrück, "Ueber hefer und gärung in Der bierbrauerei", *Bayerischer Bierbrauer*, vol. 19, 1884, p. 304. "Die hefe ist eine arbeitsmachine, wenn ich mich so aus-drücken darf".

This widening of meaning was most notable in two towns outside Germany in which zymotechnology had been actively promoted: Chicago and Copenhagen. The transitions were probably quite independent, but pressures and opportunities in the two cities were very similar. It can easily be seen in Copenhagen's case how a specific interest in zymotechnology became a general interest in "biotechnique".

The close links between fermentation studies and agricultural research were explored in Denmark in the early 20th century, when it had the highest performance agriculture in the world. Characteristically, the Rector of the Professoral College of Copenhagen, who had turned his institution into a polytechnic school, wanted to encourage the development of new agricultural industries. In 1907, the Polytechnic School granted the Associate Professor of agricultural chemistry tenure of a Chair, but in a new subject: the physiology of fermentation and agricultural chemistry. The University prospectus, in its justification of this new applied science, explained that fermentation physiology had developed so far that it should be split from chemistry, although it would continue to be taught to all future chemical engineers. It looks as if the Polytechnic was simply ratifying Jorgensen's laboratory's concept of zymotechnology as a separate science.

Orla Jensen, who had been a student of Jorgensen's, was an expert on the micro-organisms involved in cheesemaking. He was also considered one of the great microbiologists of his time. He had worked at the *Institut Pasteur*, and had then spent several years in Switzerland where he became the Director of the Central Institute of Cheese. This long experience gave him a perspective and philosophy that ranged far beyond the borders of fermentation physiology.

In 1913, taking advantage of some other changes happening at the Polytechnic, Orla Jensen changed his title to Professor of Biotechnical Chemistry. This widening of the subject from pure zymotechnology was a deliberate act, as can be seen from the introduction to his conference notes of 1916. Orla Jensen defines biotechnical chemistry as being linked to the food and fermentation industries, and as a necessary basis to the physiology of nutrition and that of fermentation. The vital processes he was trying to define, such as protein metabolism, underlie these studies. From zymotechnology, Orla Jensen had developed the concept of biotechnical chemistry.

In Chicago, American linguistic creativity engendered the word biotechnology from the same roots. In 1872, John Ewald Siebel founded an analytical laboratory in Chicago, which grew into an experimental station, and by 1884 a brewing school called the Zymotechnical College. Just like Jorgensen, who had set his own institute up a year earlier, Siebel used the term zymotechnics to designate a field wider than just brewing, perfectly aware that it would have to be used in very diverse practical industrial uses.

Siebel's work was carried on by his four sons. Three of them continued with their father's institute, but in 1917[17], the year Prohibition was voted in by Congress, Emil set up on his own. He concentrated on the provision of services, advice and apparatus to the producers of new non-alcoholic drinks. At the end of Prohibition in 1932, he found himself back in the same business as his father, teaching brewers and bakers and giving them advice and expertise. He set up an expert's bureau under his own name. It seems that it was not long before his school took on a very different character than that of his father. Instead of a Zymotechnical Institute, he called his own office the Bio-Technology Office. He probably wanted to attenuate the link with brewing, which was still heavily implied by the word zymotechnology, because of Prohibition. Siebel also took advantage of his good relationship with federal inspectors. So it is clear that the name Emil Siebel chose for his office was more for commercial reasons than education-related ones. There was no perceptible impact on academia. It may however have had quite some effect on British commerce.

When the Murphy chemical analysis office decided to open a branch offering microbiological expertise in 1920s London, they called it the Bio-Technology bureau, like the one in Chicago. The British office had more academic ambition than its opposite number in Chicago, and published a bulletin of the results of its inquiries. This bulletin was sent out to university libraries and the Natural History Museum. The Brewer's Journal referred to

---

[17] E.A. Siebel C°, *Western Brewer and Journal of Barley, Malt and Hop Trades,* vol. 25, January 1918.

its transatlantic title with disdain[18]. The editor of this bulletin, Murphy's expert Frederik Mason, published articles supporting the use of the microscope for industrial discoveries[19]. Although its origin is clearly in zymotechnology, the bureau also seems to have worked on the microbial aspects of tanning leather, and was mentioned in an Italian tanning journal, introducing the word biotechnology into the Italian language[20].

But the main source of the word biotechnology was neither in the United States, nor Britain, but Hungary. The word biotechnology was really invented by a Hungarian agricultural engineer called Karl Ereky, who was trying to turn his country into another Denmark. Denmark exported agricultural produce to its industrial neighbors, Germany and Britain, and Hungary, whose capital Budapest had grown very quickly indeed, had become the agricultural center of the Austro-Hungarian empire. Hungarian practices of intensive cattle breeding had triggered such international interest that before the Great War several hundred experts, including 18 American veterinarians, had taken time off from a colloquium in London to visit Hungarian cattle breeding associations.

Ereky invented the term biotechnology as part of his campaign to modernize agricultural production. Between 1917 and 1919 he wrote three declarations of his faith. The last one was titled *Biotechnologie der fleisch, fett und milcherzeugung im landwirtschaftlichen grossbetreibe*[21] Ereky was no ivory tower intellectual; after the war he was appointed the Minister for Food Questions in Horthy's counter-revolutionary government. Later he pioneered efforts to promote the conversion of leaves to protein and tried to attract British investment. In 1914 he persuaded two banks to support an industrial-scale agricultural enterprise consisting of an installation that could

---

[18] The criticism by the *Brewers' Journal* of 15/12/1920 was reprinted in "Some Press Comments" of the *Bulletin of the bureau of biotechnology*, vol. 1, 1921, p. 83.

[19] F.A. Mason, "Microscopy and biology in industry", *Bulletin of the Bureua of Biotechnology*, vol. 1, 1920, pp. 3–15.

[20] E. Andreis, "Il bureau per le ricerche biologiche e l'industria delle pelli", *La concesia*, vol. 29, 1921, p. 164.

[21] Karl Ereky, *Biotechnologie der fleisch, fett und milcherzeugung im landwirtschaftlichen grossbetreibe*, Paul Arey, Berlin, 1919.

handle a thousand pigs per day and an intensive pig-fattening farm for 50,000 pigs at a time and up to 100,000 pigs per year. This enterprise was enormous and became one of the largest and most profitable operations of this sort in post-war Europe.

Ereky's motivations, divulged in his second manifesto of 1917 [22 and 23] are easily understood. He wanted to replace the "archaic peasant economy" with a "capitalist agricultural industry based on science". What Ereky wanted was an industrial revolution in agriculture and the abolition of the peasant class. The difference, for Ereky, between an industrial approach and peasant approach in pig breeding, was not the use of new technologies such as electrical pumps and automated feeding. It was in the underlying scientific basis, and Ereky called this "biotechnology". For Ereky, pigs should be considered as machines that could convert a calculated amount of input (food) to a certain quantity of output (meat). He called pigs a *"Biotechnologische Arbeitsmachine"*.

Ereky's third paper [24] developed a theme that was to be widely repeated. During the war the main problem had been the food shortage. Ereky thought that the big chemical industries could help the peasants through biotechnology. For him, biotechnology was the process by which raw materials were biologically validated. He expressed his great hopes for this biotechnology, which would open up a new biochemical era.

Of course, the parallels between Orla-Jensen and Ereky's concepts were quite clear. Unlike the Anglo-Saxon pioneers of biotechnology, Ereky was an influential leader at the center of intellectual life. His ideas were noticed and widely publicized, and as they were disseminated, biotechnology came more and more to be associated with the use of micro-organisms, especially in the accounts of his work *Biotechnology*. Paul Lindner, editor of the

---

[22] Karl Ereky, *Nahrungsmittelproduktion und landwirtschaft*, Freiderich Kilians Nachfolger, Budapest, 1917.

[23] Karl Ereky, "Die Großbetwebsmäbige entwicklung der Schweïnemast in Ungarn", *Mitteilungen der deutschen landwirtschafts-gesellechaft*, vol. 34, 25 August 1917, pp. 541–550.

[24] Karl Ereky, *Biotechnologie der fleisch, fett und milcherzeugung im landwirtschaftlichen grossbetreibe*. Paul Arey, Berlin, 1919.

Zeitschrift fur Technische Biologie, gave a favorable account of this book[25], followed by an editorial by Orla-Jensen on lactic bacteria, describing them as the quintessence of what biotechnology should be. This juxtaposition inexorably associated "biotechnology" with the use of micro-organisms.

In German, there is a fundamental difference between the words *"technik"* and *"technologie"*. The latter is a branch of academic knowledge, while the former is more descriptive of activity. The difference was explored in a 1920 article in the paper *Zeitschrift fur technische biologie*[26]. Hase, the article's author, worked at the *Kaiser Wilhelm Institut für Biologie* in Berlin. Reflecting on the wide field of application for biology during the course of the recent war, including the development of pesticides as well as fermentation technologies, he also felt it needed a new name. He considered the use of the term biotechnology, but rejected it, because he felt that much of what was being done was still more of a skilled craft. He therefore proposed the name *"biotechnik"*.

Linder's redefinition and Hase's enthusiasm reflected the growth in importance of micro-organisms in industry. Three production processes developed during the First World War revealed this trend, and suggested a radiant future in the fermentation industry. In Germany, two processes were developed. W. Connstein and K. Ludecke's processes for the production of glycerol, and more efficient yeast culture processes developed by Hayduck and Wohl contributed, along with the use of yeast as animal fodder, to the prevention of famine during the First World War. In Britain, the production of acetone and butanol by the Weizmann procedure implied the use of a new bacteria that could turn a starchy raw material directly into acetone and butanol. This became a vital part of the chemical industry during the war. These three processes were the result of pre-war studies and had considerable economic impact, especially the production of butanol for the synthetic

---

[25] Paul Lindner, "Allgemeines aus dem bereich der biotechnologie", *Zeitschrift für technische biologie*, vol. 8, 1920, pp. 54–56.

[26] Albrecht Hase, "Ueber technische biologie: Ihre aufgaben und ziele, ihre prinzipielle und wirtschaftliche bedeutung", *Zeitschrift für technische biologie,* vol. 8, 1920, pp. 23–45.

rubber industry, and acetone for the producers of solvents and explosives, that were very important during the war and the immediate post-war period.

Although there was slow commercial development, there was, at the time of the First World War, debate on questions that remain familiar at the end of the 20th century: in particular, the idea of biotechnology as a separate science. Ereky called it the technology of the future, while others, such as William Pope[27], saw the production of chemical substances with micro-organisms as an alternative to the chemical industry that focused on the high-energy transformation of coal and oil. The idea, that one way or another, renewable resources and more particularly agricultural products could provide a more elegant approach to a new biochemical industry, was becoming more and more attractive. Economically it meant competition between agricultural products and crude oil. The American agricultural crisis also showed that industry could use agricultural surplus, thus saving agriculture and giving hope back to the farmers driven to desperation by the fall in price of agricultural produce that made them destroy their crops at a time when many were starving.

As Europe and the United States were searching for industrial uses for farming produce, yet another term came into use in the United States[28]. William J. Hale, reflecting on the tragic problem of agricultural surplus, and wondering whether it could be used to produce chemical substances, invented the word "chemurgy" meaning chemistry at work and taken from *chem* and the Greek word *ergon*. Hale did not have a very clear view of where the limits of chemistry were, and he used his new term much as others were using the term zymotechnology. The main idea was the production of alcohol[29] from fermented starches, to replace oil-based petrols, with the hope that this new industry could put an end to the farming crisis.

---

[27] Sir William Pope, "Address by the President", *Journal of the Society of Chemical Industry,* vol. 40, pp. 179T to 182T.

[28] As for a history of the invention of the term chemurgy, see Wheeler McMillen, *New Riches from the Soil: The Process of Chemurgy,* van Nostrand, New York, 1946.

[29] Hale was not thinking of using alcohol as a power source, as in Europe, but as a stock source for organic chemical substances, such as acetic acid, essential for plastics.

Despite all the prophecies it has to be admitted that this chemurgy, this zymotechnology, was still on the fringe as far as the developing chemistry industry was concerned. The great corporations of the 1920s, I.G. Farben, du Pont de Nemours, and the British Imperial Chemical Industry, had but small confidence in biological processes. Why should they, when it was easy to extract aromatic chemical components from coal tar, and more and more aliphatic chemicals, such as alcohols and glycols, could be produced from crude oil, especially in the United States?

1.4 The Engineering of Nature — Towards the Best of all Possible Worlds

Between the Wars, biotechnology acquired another connotation, rather different from that it had as "chemurgy".

With the idea of integrating biology with engineering, a combination seen as fundamental to an entirely new phase of human civilization, the composite idea of "biotechnology" came to mean the application of biology to humanity, a humanity where the individual is forgotten when the industrial liberal utopia becomes an engineering/medical utopia where the human and political bodies are one and the same automaton, a machine that works very well on its own, if watched and occasionally purged.

Health was not just seen as a matter of occasional medical care, but as the result of an environment harmonized with the needs of society. Alongside non-polluting industrial food technologies based on renewable natural resources, this philosophy also harbors eugenic thought.

Despite the apparent divergence from the traditions of zymotechnology, these two traditions converged in 1930 in a concept of a new, benevolent technology[30]. For idealists, the idea of a new technology for a new age was very attractive, and took hold very easily. In its apparent confusion of categories that until then had been held apart, it resembled other aspects of

---

[30] This apparently abstract perspective was often to be found in eugenicist contexts.

European culture that had been radically reformed following the chaos of the First World War: art, music, physics, philosophy. As in other fields, these ideas remained relevant long after the knowledge and techniques designated by the term biotechnology, their essential concepts, had changed so as to be unrecognizable. At the heart of these concepts is the confusion of the categories made possible by biotechnology and amplified by the idea of the robot as created by the Czech writer Karel Capek in his 1920 play R.U.R. Today a robot is mechanical, but Capek's were not. His robots had enzymes, hormones, and were the motor for commercial profit. In a certain way, this was a vision of machines as strangers, linked to Man only through commercial profit.

The concept of biotechnology, in contrast, integrated the contemporary dream of environment and humanity-conscious production, with an unshakable belief in a privileged vision of life. The metaphor of the machine was being used a lot by the organic philosophers of the time, but they felt the machine was the symbol of a system greater than the sum of its parts, and which in itself was irreducible. So at the same time there were strictly mechanistic models of biological organisms that compared life to a chemical machine, and the neovitalist models in which biological systems had organizing principles or specific developments that could not be reproduced in artificial systems, and these both led to biotechnical concepts. Although these concepts were apparently far further away from our modern ideas of biotechnology than zymotechnics, and might appear strange, they were to become the cornerstones of later ideas.

For researchers interested in practical application, the biological domain was attractive because the raw material is everywhere, with the challenge of exploiting it practically. For others, the idea of this type of technology was still fraught with danger. The terrifying image of industries and pollution of nature was but the most visible facet of a break-up in traditional ways of life, a break-up that for many was the result of the social impact of "techniques". For more than a century, technology shook the Western world, unleashing a plethora of emotional, literary, political and cultural responses. During this slow process of domesticating technology that is part of Man's history, and part of Man's natural history too, biotechnology itself was built.

1.5 Technique, Biology and the Development of Biotechnics

As the host city for the world congress on zoology, Berlin was the appropriate site for the construction of biotechnology. A German participant, Gustave Tornier, pointed out how many articles were drawing analogies between biological and mechanical systems. In his own contribution he tried to define the category of technology formally, saying that it should be applied to living organisms in general, which would be called *Bioten*, and that the process of modifying them or using them technologically should be called *biontotechnik*[31].

Tornier's article was well received, and his new term was included in Roux's 1910 dictionary. The word looks innovative, but the biology behind it remained conservative. Decades before, the analogies between living being and human technology had been drawn in studies that showed how joints, bones and organs could be considered sophisticated machines. In 1877, Ernest Kapp brought these analogies together by seeing technology as a result of the concept that tools were an extension of the hand. His work transformed specialized physiological literature into a philosophy of technology, a shift of ideas at a speed that by the beginning of this century was difficult to emulate[32]. It provided a reference for those trying to bring meaning to diverse concepts and preoccupations, such as the scientific community's efforts to combine evolutionary thinking with mechanistic models of the body.

Analogies between skeletons and machines were deepened by the growing field of embryology. Biologist Wilhelm Roux found that the development of organisms could be altered by removing a cell from the fertilized egg. Roux coined the term *Entwicklungsmechanik* for the mechanical quality of an embryo's development.

---

[31] Gustave Tornier, "Ueberzahlige bildungen und die bedeutung der pathologie für die biontotechnik (mit demonstrationen)", in *Verhandlung des V Internationalen Zoologen-Congresses zu Berlin, 12–16 August 1901*, ed. Paul Matschie, Gustave Fischer, Jena, 1902, pp. 467–500.

[32] Friederich Rapp, "Philosophy of technology: a review", *Interdisciplinary Science Reviews*, vol. 10, 1985, pp. 126–139.

Jacques Loeb[33], a young émigré to the United States, was greatly influenced by his mentor Roux in his search for a technique of living material. He himself made a decisive step forward when he managed to chemically induce the division of an unfertilized sea urchin egg. The press was certain that this technique of parthenogenesis would soon lead to the laboratory production of babies. Loeb was more prudent than Roux, although they shared a common arbitrary view of the threshold between the animate and the inanimate. His technical approach to biotechnology deeply influenced a generation of disciples such as Hugo Muller and Gregory Pincus who were themselves highly influential in the 1960s.

The distinction between the French and German schools of physiology was made by historian of medicine Owsei Temkin in the brilliant epigram "In France, for the vitalist materialists, man is but an animal! For the mechanistic materialists in Germany, he is but a brief constellation of lifeless particles of matter."[34] A similar distinction underlay the debate on the relationship between culture and technology. Man should, by all rights, be reducible to a machine, but the creation of machines should be considered as a characteristic trait of the human animal.

From the French pioneers in this field, we should pick out Jean-Jacques Virey[35], a popularizer of science and a contemporary of Lamarck's. He developed a theme which was to become very important for evolutionists, that is, that Man developed techniques to compensate for the loss of his animal instincts. The term used by Virey in 1828 to describe this inborn human trait was "biotechnie". Virey's contribution was not just the coining of a term, but also of a concept behind it, which was to crop up again. While he himself was soon forgotten, his idea that intelligence replaced instinct in Man was furthered by Henri Bergson. Bergson's work, and particularly his

---

[33] For a detailed study of Loeb's philosophy, see Philip Pauly's *Controlling Life: Jacques Loeb and the Engineering Ideal in Biology*, Oxford University Press, 1987.

[34] Owsei Temkin, "Materialism in French and German physiology of the early nineteenth century", *Bulletin of the History of Medicine*, vol. 20, 1946, pp. 322–327.

[35] On Virey, see Alex Berman's "Romantic Hygeia: J.J. Virey (1775–1846) Pharmacist and philosopher of nature", *Bulletin of the History of Medicine,* vol. 39, 1965, pp. 134–142.

most well-known book *L'évolution créatrice,* published in 1907, were very successful. Bergson developed three themes; life, the evolution of consciousness and that of technology. He is most famed for having pointed out the importance of *élan vital.*

Bergson's philosophical work, shared by his contemporaries, is less well remembered. He felt that Man might continually recreate himself through work, and thus the use of tools is a part of the spiritual expression of humanity. Bergson suggested that life was specific because its infinite potential is denied to the inanimate, which is constrained by the circumstances of its manufacture. For him, technology is a symbol of Man's skill at creating new worlds, and for this reason he identified the human species as *Homo faber.* Following the ideas of 19th Century archaeologists who classified prehistoric Man by their tools, he classified more recent history by its characteristic technology. For Bergson, the technology of a time defined a historical era. Our technological status defined our level of consciousness, and perhaps it is due to the concept of this self-reinventing *Homo faber* that we owe the idea of a third industrial revolution that has been diversely associated with information, nuclear power, and of course biotechnology. This concept of eras defined by technologies led even in Bergson's time to a view of "biotechnique" as a technology characteristic of its time. So long before the advent of molecular biology and genetic engineering, Bergson based this view on the close entwinement of two other themes of the biological debate with his model of history; the question of evolution and whether biological products should be seen in the same way as manmade ones.

According to Darwin's formulation of evolution by natural selection, natural selection explains the gradual process of evolution without finalism and without any tendency towards a Lamarkian perfect organized being. In addition, natural selection allows evolution to move on from the determinist model to a theory of probability into which, at the beginning of our own century[36], Mendel's rediscovered laws and mutationism could be spliced. Darwin's French predecessor, Lamark, thought that animals could evolve

---

[36] For a deeper study of Darwinism, read J. Gayon's *Darwin et l'après-Darwin (une histoire de l'hypothèse de séléction naturelle),* ed. Kimé, Paris, 1992.

during their own lifetimes in response to the environment. He got carried away by his theory of general transformation, formulating two laws[37] which are, in fact, merely hypotheses of the fundamental mechanism of the evolutionary process.

On the basis of Lamark's theory, neo-lamarkians would suggest that herein lay a possibility for the improvement of living creatures. Biological systems could therefore be considered machines with the specific ability to improve themselves, making them particularly worthy of respect and emulation. It was in this way that the theory of orthogenesis popular at the turn of the 20th century suggested that evolution was not a random force but on the contrary was progressing in a specific direction. In human evolution in particular, the implication was that technological and human progress could go hand in hand.

For those who supported organicism, furthermore, the body could not be reduced to the sum of its parts since it was necessary to take into account the additional organizing principle. The most charismatic organicist was August Pauly, author of *Lamarkismus und Darwinismus*,[38] an influence on an entire generation of embryologists and an idol of Freud's. Pauly thought that evolution could not be explained unless you took into account an additional psychic principle, characteristic of each cell. His friend the botanist Raoul Francé, editor of the *Journal de Physiologie*, agreed that living organisms could be considered as mechanisms only when this principle was taken into account. The term Biotechnik was created by Francé and again independently by Rudolf Goldscheid, a protégé of Wilhelm Ostwald. Ostwald was the founder of physico-chemistry and had outlined a new philosophy in which energy was the fundamental and unified component of the universe. Francé and Goldscheid can themselves be considered the precursors of distinct, although linked, biotechnological developments in British thought between 1920 and 1930.

---

[37] Lamark, *Philosophie zoologique*, Bibliothèque 10/18, 1968, p. 204.

[38] A. Pauly, *Lamarkismus und Darwinismus*, Rheinhardt, Munich, 1905.

Rudolf Goldscheid[39], respected novelist and sociologist, developed the first biotechnical philosophy. His 1911 work *Hoherentwicklung und Menschenokonomie* bases the need for improvement of the present generation for the benefit of future generations on Lamarkian principles. The Darwinian image of the survival of the fittest was seen as wasteful, and offensive in its application to Man. Goldscheid really felt that improvements in living conditions would improve genetic heritage. Even an "average" person should have some economic potential. Like many socialists of his time, he thought that engineering would help to raise the level of the "average". This was expressed in the various terms such as *Hoherentwicklung, Menschenokonomie, socialbiologie*, and as mentioned before, the concept of *biotechnik*. Goldscheid strongly believed in the impact of biotechnology. His vision of biotechnik was as the implementation of the program of *socialbiologie* which had origins going back to the 18th Century.

One of the founding problems had been how to improve human reproduction by looking at the transmission, adaptation and selection of traits. There was an urgent need to understand why the reproduction of characteristics is unstable and the factors underlying variations in fertility. The application of such knowledge, it was then thought, would free humanity from the erratic process of quantitative production, with the concomitant evils of alcoholism, prostitution and little chance for the majority to attain a level of qualitative production. Goldscheid often used an analogy between industrial production using techniques with a sound scientific base, and the social production of individuals. The reference to technique and to machines was not pejorative, but showed Goldscheid's respect for the organic wholeness and functionality of machines, being more than the individual parts of the mechanisms. Goldscheid suggested that biotechnik should be based on the respect of living organisms, of the human being, and he hoped that a complementary technology could be developed that would be organic, spiritual and ethical.

---

[39] For an analysis of Goldscheid's terminology, read Paul Weindling, *Health, politics and German politics between national unification and Nazism, 1870–1945*, Cambridge University Press, 1988.

The details of *biotechnik* were never clearly laid out. It is clear that Goldscheid was very impressed with the developments in embryology, biology and bacteriology, and that he wanted to bring a technological aspect to the concept of social hygiene which became a fundamental question between the wars.

Numerous debates on eugenics, feminism, birth control, and also the implications for the nutrition and food politics of the new concept of vitamins, led to support for social hygiene. This would be horribly underlined by the Nazi perversions of the idea, but Goldscheid's *Menschenokonomie* also impressed Julian Moses, the Socialist orator for the hygiene campaign in Weimar. What Goldscheid was saying was, in one form or another, popular throughout Europe.

In France, where the watchwords were "quality with quantity", there were discussions of the importance of human zootechnics, the last stage of hygiene. If Man is an animal machine, an industrial material, an economic element, then the hygienist is an engineer of the human machine.

Goldscheid's ideas had great influence, and from France to their later interpretation in the Soviet Union they crystallized into a variation of new technologies suggested by conventional concepts of social hygiene, during a time when there were strong eugenic concerns. At the end of it all, Goldscheid was inspired by the fundamental importance of biology to the conventional eugenic obsessions of the engineers[40].

Francé shared Goldscheid's conviction of the fundamental value of natural phenomena, and granted fundamental importance to biotechnics. For him, biology and its understanding of natural phenomena was a model for technology in its own immense territory. His axiom was that life could be seen as a series of technical problems for which living organisms presented optimum solutions. The relationship could be described technically. His

---

[40] For a detailed history of eugenics, see D.J. Kevles, *In the name of eugenics, Genetics and the uses of human heredity,* 1985. Galton defined eugenics as "the science of all the influences improving the innate qualities of a race and those which develop them to their maximum advantage", Galton, *Inquiries into human faculty and its development,* ed. Macmillan, London, 1983.

thesis was that we could not only conquer by destroying disturbing influences but also by compensating in a world-harmonious fashion. It is to this end, he felt, that the cogs of the world turn [41].

By following this line of thought, the meaning of biological example is clear. For Francé, it augurs the removal of many obstacles, a redemption, a hope for the resolution of many problems in harmony with the forces of the world. Francé had thus coined his term by combining the concept of a new sort of technology with harmony with nature. He saw himself as the prophet of the new world and predicted a new era in cultural development. His ideas spread quickly in Germany. Alfred Giessler, the director of the *Biotechnik* group in Halle, wrote a paper entitled *Biotechnik* [42], which appropriated Francé's arguments as German science (albeit granting him the odd reference).

Stylistically, Francé and Goldscheid's philosophies can seem very far away from the pragmatic or even technocratic concepts of Orla-Jensen and Ereky. They do however meet at several points. They express a whole new constellation of ideas about biology that transcend particular techniques. The new science seemed the basis for technologies still in gestation, technologies better described by their collective potential and meaning than by their individual worths. Furthermore, these ideas shared the metaphor of a new revolution, a new industrial age. At a time when the First World War and the Russian revolution were marking the end of the ideas of the 19th Century, these philosophies were a technological expression of the new century and had great influence, especially because they were intended for a lay audience.

Francé and Goldscheid's concept of *biotechnik* successfully precipitated a number of German ideas. The English translation also brought meaning to similar concepts, and the term into everyday usage. The archaic and declining industrial empire that had constituted the background of Francé and Goldscheid's Austro-Hungarian culture was counterbalanced on the other side of Europe by the British Empire. There too, people were thinking of

---

[41] R.H. Francé, *Plants as inventors,* Simpkin and Marshall, London, 1926.

[42] Alfred Giessler, *Biotechnik,* Zuelle and Meyer, Leipzig, 1939.

new sources of vitalism. Well before the First World War, philosophies that enthroned the natural and living, and condemned death and pollution, were already widespread. Of course Britain, as the only long-established industrial country also a world empire, had its own problems, in particular that of pollution and industrial destruction of the environment. This problem generated anti-industrial feeling and increasingly lively concern for the environment. More concisely, the public's conscience and eyes were being joined by their tongues in the rebellion against black and desolate countryside. At this time, the agricultural base of the empire, when considered as a whole, still provided alternatives.

The synthetic rubber solution underlines clearly that in Britain, as in Austria, natural means of production were still considered adequate answers to the chemical challenge presented by Germany. In practice, the economic development of agriculture and the reconstruction of damaged British environment were often linked. Biology, which became a university course at the end of the 19th Century, was now a new profession exceptionally well-placed to handle such problems. Unfortunately there were still very few academic posts, and they were particularly nonexistent in industry. However, popularizing works and an appeal to public interest, which provided their authors with alternative revenue from their knowledge, contributed to a growing public respect for the newborn profession. The explanatory model of evolution *de facto* was held in high esteem. After all, Britain boasted Darwin[43], Thomas Huxley, and Herbert Spencer[44] who marked the narrow line between sociology and biology. Human evolution had become a specific concern since the work of Francis Galton[45], the father of militant eugenics, and a remarkable succession of geneticists. Karl Pearson, W.F.R. Weldon and Fischer built a research center at the University College of London.

---

[43] Charles Darwin, *The Origin of Species*.

[44] Herbert Spencer, *Les bases de la morale évolutionniste*, ed. G. Bullière, 1981, and Herbert Spencer, *L'individu contre l'état*, ed. Félix Alcan, 1901.

[45] F. Galton, *Hereditary Genius, an inquiry into its laws and consequences*, Macmillan, London, 1869. For more on Galton, see Derek W. Forrest's *Francis Galton: the life and work of a Victorian genius*, Taplinder, 1974.

Bastion families of the scientific aristocracy such as the Huxleys and Haldanes[46] continued to distinguish themselves. During the 1920s their youth Julian Huxley[47] and J.B.S. Haldane[48] disdained academic convention by energetically communicating their ideas directly to the public. And if the British public was not well-known for the attention it paid to intellectuals, it was still intellectually close to the German public. Although scientists generally read German articles and works, German philosophy was not very well thought of, which is understandable given the nationalism of the time. The German heroes Friedrich Nietzsche and Arthur Schopenhauer were suspect and most of the other authors unknown.

There was however a lot more intellectual exchange than is currently recognized. During these exchanges, ideas underwent complex transformation. We thus have to decipher the relationship between the English usage of the words "biotechnics" and "biotechnology" and their German precedents, a relationship that was not particularly investigated by British writers. However, none of these British writers claimed to have coined the term. Anglo-German relations in fields as borderline as biotechnics were complex and difficult, but they were the key to intellectual survival (or even physical survival as far as the many refugees were concerned). Francé's speculative thought crossed the Channel in an English translation of his popular works. Psychobiology was adopted by the neo-Lamarkian E.S. Russel and gradually disseminated. Francé's ideas on the analogy between engineering and biological form were found in Britain in the work *On Growth and Form*[49] by d'Arcy Thompson. A few months after the publication of Francé's *Bios: der Gesetze der Welt*[50] in 1921, another Scottish biologist, Patrick Geddes,

---

[46] For a biography of J.B.S. Haldane, see the work edited by K.R. Dronamraju, *Haldane's Daedalus Revisited*, Oxford University Press, New York, 1995, and in particular the article written by the editor, "Chronology of J.B.S. Haldane's Life".

[47] J. Huxley, *L'homme cet être unique*, traduction de Jules Castiers, La Presse Française et étrangère, Oreste Zeluck, éditeur, Paris, 1947.

[48] J.B.S. Haldane, *Heredity and politics*, Allen and Unwin, London, 1938. As for Haldane's ideas on genetics and evolution, see his work *The causes of evolution*, London, 1932.

[49] D'Arcy Thompson, *On Growth and Form*, Cambridge University Press, Cambridge, 1942.

[50] R.H. Francé, *Bios: der Gesetze der Welt*, Hofstaengli, 1921.

used the term biotechnics, although he did not acknowledge a source, and even a lot later, when he was using the term very often, he never gave a reference.

Geddes left a deep imprint on British thinking through his successors, particularly Lewis Mumford. He was also a key figure in the transformation of the term biotechnics from a revolutionary biological idea to a characteristic type of technique. At the beginning of our century, he was an *avant-garde* thinker on sexuality and evolution, subjects that were being linked ever more closely to the problems of the evolution of society and towns. A botanist by training, he had studied under T. Huxley and been to conferences given by Herbert Spencer, a thinker considered by some historians to be the father of social Darwinism. In the heart of the city of Edinburgh, Geddes drew up his point of view by which all of human knowledge could be used to handle the problem of towns. It would require a grand synthesis of even deeper understandings, founded on the triptych of the French geographer Le Play, "crowd, space and work". The two interests - biology and sociology - were too distant to grasp simultaneously in the growing professionalisation of the beginning of the 20th Century, and so Geddes took on two co-workers, biologist J. Arthur Thompson, Professor of Biology at Aberdeen University, and the sociologist Victor Brandford.

From 1895 onwards, Geddes had been thinking about the distinctions drawn by the archaeologist Evans between Paleolithic and Neolithic, and he described how the industrial age could be equally divided. He outlined a paleotechnical civilization and a neotechnical one divided by what in 1915 he called the second industrial revolution[51]. These two civilizations were characterized by different technologies: the first by steam engines concentrated around coal mines, the second by the technology of electricity. Geddes did not think that such technologies led civilization, but rather that they were one of its expressions.

Before the War, he became interested in more idealistic concerns. He went as far as describing a new age in the future development of technology,

---

[51] See Patrick Geddes, *Cities in Evolution: An introduction to the town planning movement and the study of civics*, Williams and Norgate, London, 1915.

a geotechnical age during which technology would be in harmony with the needs of the earth, with the environment. Many years later, Benton Mackaye[52] recalled a conversation of 1923:

"'Geography' said Geddes, 'is descriptive science (*geo* earth *graphy* describe); it tells what is. Geotechnics is applied science (*geo* earth *technics* use) it shows what ought to be. This would lead up to the eutechnics, which would characterize "eutopia", that is, a practical utopia'"

Initial public characterizations did not mention biotechnology, but this does not mean that Geddes was not interested in it. It crops up in *The Coming Polity*[53] which he published with Brandon in 1917 and modified in 1919. The latter publication had an extra chapter called "the Post-Germanic University", which had lines of thought somewhat similar to those of Francé.

The word biotechnics was available from 1921 when Francé's *Bios* was published. Geddes used it for his sketch of a historical progression with a key role for biotechnics. This concept of biotechnics was incorporated in a great historical outline that Geddes called the Transition IX to 9. This foresaw the translation of a society as described by Combe and interpreted by Geddes in a 9-box grid describing the kingdom of war with the key words militantism, state, individual and industry (mechanotechnics) and its transformation to a future kingdom of peace also defined in a 9-box grid described by the key words biotechnics, synergy in geotechnics and crowd, work, politics. In 1931, Geddes triumphed the onset of the new age when he saw Krupp investing in tractors rather than arms, and using his profits for the improvement of towns.

Despite its importance for the concept of a transition between two types of society, Geddes first published the term biotechnics in brackets and without explanation in a chapter introducing the short preliminary text that he wrote with Thompson in 1925 entitled *Biology*[54]. Again, although manuscripts show that the concept was meant to run through the entire volume on *Life*,

---

[52] Benton Mackaye, *From Geography to Geotechnics*, University of Illinois Press, 1968, p. 22.

[53] V.V. Branford and P. Geddes, *The Coming Polity,* Play House, London, 2nd edition, 1919.

[54] P. Geddes and J.A. Thompson, *Biology*, Home University Library, London, 1925.

it only appears sporadically at the end, without any introduction or explanation. So *Life* stands as a barely visible spark in the history of biology, rather than an explosion of affirmation.

An article on biology by Thompson for the supplement to the *Encyclopedia Britannica* of 1926 [55] is probably a more important attempt to popularize the subject. Thompson attributes the invention of the term biotechnics to Geddes and explains that it means the use of biological organisms for Man's benefit. If this appears a very simplistic application of Geddes' ideas, and is very far off those of Francé, it is nonetheless without a doubt a demonstration of how their ideologically laden terms were rendered more acceptable to the Anglo-Saxon public.

Although Geddes seemed to be rather anachronistic to the British professionals of the 1930's and his work *Life* was not held in much esteem academically upon its publication, this does not mean that his work did not influence the intellectual environment of his time. In order to truly understand the acceptance of his concept of biotechnics, we should look at the thoughts of the younger biologists of his time. In the same way as a confused Geddes borrowed ideas from his predecessors, younger biologists such as J.B.S. Haldane, J. Huxley, and L. Hogben absorbed the biotechnical concepts of Geddes, Francé and Goldscheid.

In 1902 when H.G. Wells published *Anticipations* [56], flight and the radiotelegraph were the height of scientific ideas. By 1920 his ideas were on the market. For Haldane, the scientific risks and problems of his time were not ones of technique but really ones of biology [57]. In answer to a conference given in 1914, John Burdon Sanderson Haldane presented in 1924 a scientific formulation of utopic eugenics in a little work entitled "*Daedalus or science and the future*." He observed that the people who had invented things within the realm of technics were like Prometheus. "The

---

[55] J.A. Thompson, "Biology", *Encyclopaedia Britannica, Addendum to the 11th edition,* vol. 1, 1926, pp. 383–385.

[56] H.G. Wells, *Anticipations of the Reaction of Mechanical and Scientific Progress upon Human Life and Thought,* Harper, London, 1902.

[57] J.B.S. Haldane, *Daedalus or science and the future*, re-edited in K.R. Dronamraju's *Haldane's Daedalus Revisited*, Oxford University Press, New York, 1995, pp. 28.

chemical or physical inventor is always a Prometheus. There is no great invention from fire to flying, which has not been hailed as an insult to some god". This was not the case with the first inventor in the biological field, Daedalus, the first genetic engineering inventor as we would have called him, who presided over the creation of the Minotaur by allowing the coupling of Pasiphae and the Cretan bull. Haldane points out that he was not punished for this manipulation, the most monstrous and artificial of all human manipulations, either in this or the next world. "But if he escaped the vengeance of the gods he has been exposed to the universal and agelong reprobation of a humanity to whom biological inventions are abhorrent". "But if every physical and chemical invention is a blasphemy, every biological invention is a perversion". In other words, for most observers, they appeared "inconvenient and unnatural, not just offensive to some gods but to Man himself" [58].

Furthermore Haldane expected the public to recoil from his enthusiastic proposal of a new Daedalian accomplishment, the creation of children by ectogenesis, a technique which coupled with in vitro fertilization and the development of the embryo outside the uterus would separate sexual love and reproduction. In his *Daedalus*, Haldane prophesied that if reproduction were completely separated from sexual love, humanity would be freed in a totally new fashion. Man would have nothing to fear from the gods, but only from himself. "The scientific worker of the future will more and more resemble the lonely figure of Daedalus as he becomes conscious of the ghastly mission and proud of it".

> Black is his robe from top to toe,
> His flesh is white and warm below
> All through his silent veins flows free
> Hunger and thirst and venery
> But in his eyes a still small flame
> Like the first cell from which he came

---

[58] J.B.S. Haldane, *Daedalus or science and the future*, re-edited in K.R. Dronamraju's *Haldane's Daedalus Revisited*, Oxford University Press, New York, 1995, pp. 23–54.

Burns round and luminous, as he rides
Singing my song of deicides"[59].

In the year following its publication, *Daedalus* sold about 15,000 copies and attracted much attention from left-wing intellectuals concerned by reports on science and society. Ectogenesis was not presented in a very flattering light in the main book that *Daedalus* inspired, Aldous Huxley's *Brave New World*, which can be seen as a critique of *Daedalus*. And though Aldous Huxley strongly opposed the Daedalian vision of humanity, his brother Julian, himself an eminent biologist, showed great enthusiasm for the engineering of humanity.

If Haldane, like Hogben, had refused to become involved with the British eugenics society, Julian Huxley was one of its pillars. There is no doubt that Julian Huxley, the first director of UNESCO from 1946 to 1948, was a great biologist, humanist and socialist. In 1935, J. Huxley and A.C. Hadden (latterly an anthropologist at Cambridge University) published a book called *We Europeans: A Survey of Racial Problems*[60] in which they criticized Nazi theory on race, and in particular works such as that of Madison Grant, *The Passing of the Great Race*. Despite all this J. Huxley was still capable of thinking that different human groups had innate genetic differences when it came to intelligence[61].

It is true that J. Huxley had insisted on the importance of the environment and the need to associate eugenics with social reform. He had highlighted this in a famous conference given in 1936 to the British Eugenics Society, at which he declared unambiguously that a system based on private capitalism and nationalist politics would be *de facto* dysgenic. It would be incapable of using the existent reserves of high-value genes and would lead directly to the maximum dysgenism: war. It would be impossible, said J. Huxley, to really put eugenics into practice while we have not more or less levelled out

---

[59] J.B.S. Haldane, *Daedalus or science and the future.*

[60] J.S. Huxley and A.C. Haddon, *We Europeans, a survey of racial problems*, Jonathan Cape, 1935.

[61] J.S. Huxley, *L'homme cet être unique*, traduction de Jules Castiers, La Presse Française et étrangère, Oreste Zeluck, éditeur, Paris, 1947.

living conditions for everybody, a leveling upwards[62]. It is not certain that J. Huxley was explicitly influenced by Goldscheid, but whatever the case, his thoughts were taking on a clearly parallel direction.

The link with Goldscheid's philosophy is even clearer in a 1934 article by J. Huxley entitled *"The Applied Science of the Next Hundred Years: Biological and Social Engineering"*[63], in which he called for a plan for leisure (social engineering) and control of the quantity and quality of the population (biological engineering). For him biological engineering was the more necessary of the two.

The use of the idea of biological engineering seems to have been quite common at the time. At a meeting of the Workers' Educational Association in Cambridge in 1931, the social biologist Joseph Needham in his comments on Aldous Huxley's *Brave New World* proposed to his audience that the concerns of the day were mainly biological engineering and the possibility of a world dictatorship. Needham's concept of biological engineering was more Loebian than after Kammener, who had fallen from grace.

Ectogenesis was an increasingly common idea[64]. Polyembryonics, in which a fertilized egg is subdivided to produce genetically identical descendants, was being investigated in rabbits at the time. Needham insinuated to his audience that the social implication would be on the one hand a rational perpetuation of characteristic traits such as submissiveness or a capacity for routine work, and on the other hand the possible development from one same egg of factory staff strictly identical to each other. In other terms, standardization right down to the biological base and mass-production nature, with the underlying idea that Man would not only be product

---

[62] As commented by D.J. Kevles in *In the name of eugenics, Genetics and the uses of human heredity*, 1985.

[63] J.S. Huxley, The applied science of the next hundred years: biological and social engineering, *Life and letters*, vol. 11, 1934, pp. 38–46.

[64] In the decade following the publication of *Daedalus*, Hermann Muller and the British eugenicist Herbert Brewer independently pointed out that it was already possible to take first modest steps in the direction indicated by Haldane. "Brewer furthermore dubbed test-tube fertilization "penectogenesis" because he thought it was a great step in the direction of Haldane's ectogenesis" D.J. Kevles, *In the name of eugenics, Genetics and the uses of human heredity*, 1985.

consumers but mass products themselves. Although it already existed in principle, practice on humans however remained in the realms of science fiction.

At the same time, biological understanding was producing ideas of more immediately applicable social interest. The fall in population and increasing evidence of malnutrition in Britain in the years before the 1930s depression led to the development of a British social biology. Although you can find the term social biology in 19th century English, it is considered foreign, perhaps because it had been popularized in Austria by Goldscheid. Nevertheless, as a center for social sciences, the London School of Economics was very much *au fait* with the German ideas of social hygiene, and had a copy of Goldscheid's *Hoherentwicklung* from the beginning of the 1920s onwards.

The school's from 1919 to 1936, William Beveridge, worked hard to create a bridge of social biology over the gap between the social sciences and the biological sciences through biology-based studies of human behavior, including genetics. Even before the First World War, he had been very interested in the German solutions to the problem of social well-being. Later he was to become concerned about social and demographical problems.

In 1920 he gave a lecture to the British Association on the dangers of overpopulation. It greatly impressed Beardsley Ruml, the director of the Laura Spelman Rockefeller Foundation. The two men met and Beveridge presented his vision of a social science based on natural sciences, which would include a whole set of understandings in anthropology, eugenics, nutrition and psychology. This vision, which resembles Goldscheid's biotechnics, impressed the Rockefeller Foundation, particularly with its reference to a social biology. During the years to follow, Beveridge persuaded his colleagues and the Foundation how important the new subject was. He needed to work with a biologist interested in economics and politics, and finally recruited J. Huxley's friend Lancelot Hogben, of the University of Cape Town.

Hogben was a hard-baked opponent of eugenics and all confused philosophical thought on biology. He was desperate for recognition as a serious eminent scientist, and plotted with Huxley to be elected a Fellow of the Royal Society. At other times he felt that it was even more important to promote the public's understanding of science. These two commitments sometimes threatened to tear him apart. He wanted people to understand the importance of zoology, as well as his own.

The Social Biology Chair seemed to have given Hogben an opportunity to promote the significance of his science. He developed his interest in the social implications of biology. His inaugural lecture underlined the well-established fear of depopulation and linked it to Haldane's idea of biological invention. "The declining birth rate has brought us face to face with the fact that we are entering upon the era of biological invention"[65]. Hogben developed a birth rate measurement and took on the assistance of Rene Kuczynski, a refugee from Germany and a demographer worried about the falling population. In his *Political Arithmetic*[66] of 1937, which summarizes the department's work, Hogben wished with all his heart for applications of biology that would be as practical as applications of chemistry[67].

It was thus that at the beginning of the 1930s, as Geddes and Goldscheid died, social biology had become a university subject. While Geddes looks anachronistic, at least in his personal beliefs, the concept of biotechnics which he was associated with had become a part of the cultural substructure of biology. On the first anniversary of Geddes' death, *Nature* magazine published an editorial entitled *"Biotechnology"*[68]. Except for its commercially motivated use by Siebel and Masson, this was the first British use of the term. The author of the editorial was the chemist Rainald Brightman but the title and the ideas expressed in the editorial are clearly those of *Nature*

---

[65] Lancelot Hogben, "The foundations of social biology," *Economica,* no. 31, February 1931, pp. 4–24.

[66] Lancelot Hogben, "Prolegomenon to political arithmetic", in *Political Arithmetic, A Symposium of population studies,* ed. Lancelot Hogben, Allen and Unwin, London, 1938, pp. 13–46.

[67] Hogben, following the realization in the early 1930s of the progress made in mathematical methods and that through this progress the future for the development of human genetics as an exact science was very bright, called for the establishment of a vast research program in human genetics: studies of twins to evaluate the respective roles of heredity and the environment, the measurement of variability in hybrid populations to see if there were specific racial characteristics, genealogical research, especially in medical archives, and studies on consanguinity. Hogben himself, with ill-equipped resources at the London School of Economics, could not put his program into practice on his own. The collection and analysis of facts required organized work on a massive scale.

[68] Rainald Brightman, "Biotechnology", *Nature,* vol. 131, 29 April 1933, pp. 597–599.

editor R.A. Gregory, an old friend of Thompson's, an admirer of Geddes', a technocrat and an enthusiast of eugenics. The editorial claims that scientific technology would contribute to the improvement of humanity and that biology, through contraception and eugenics, could both remedy the fall in the birth rate and improve the quality of the British population.

In an echo of Goldscheid's concepts twenty years previously, the article expressed feelings close to those of Julian Huxley when he spoke of biological engineering. An eclectic interpretation was made three years later, in a 1936 conference, "The retreat from reason"[69], given by Hogben in London's Conway Hall. J. Huxley, as chairman, introduced his friend's lecture by predicting that in the future biotechnology would be as important as technologies based on physics and chemistry had been in the past. With the benefit of hindsight this is a remarkable prophecy.

In his speech, Hogben did more than simply reiterate the ideas on biological engineering laid out by Huxley, or repeat Goldscheid's arguments or Geddes' twenty-year-old prophesies for biotechnics. Instead he combined them, amalgamated them and transformed them. For him, the ideal modern society would be found not in the factory but in the countryside. Biotechnology was a socialist and aesthetic answer to the mechanical and polluting technologies that his generation had inherited. In his lecture he announced that social science could no longer accept that evolution's work was done. Influenced by his wife, Enid Charles, Hogben did some thinking on the need to reverse the population decline. It was also clear that agricultural production could not continue at the rate it was going. As a biologist, he was in favor of planned production based on biotechnology because he could not conceive how one could plan consumption without planning production. A final component of biotechnology was to be chemical transformations carried out by bacteria.

Hogben's use of the concept of biotechnology synthesized the ideas of Ereky, Pope, Goldscheid and more clearly those of Geddes. For Hogben, biotechnology was a facet of a far greater bio-aesthetic utopia, in which the

---

[69] Lancelot Hogben, "The retreat from Reason: Conway Memorial lecture delivered at Conway Hall," Watts, London, 20 May 1936.

life of small communities would depend on hydroelectric power, light metals, fertilizers and of course on the application of biochemistry and genetics to control qualitative and quantitative evolution in the population. J. Huxley's strongly eugenicist interpretation of this was only one aspect. The dream also involved production through fermentation, a link back to the traditions of zymotechnology.

Hogben repeated his vision of biotechnology as an aspect of productive agriculture in his work *Science for the Citizen*, which was a bestseller. In the meantime, he had obtained the Regius Professorship of Natural History at Aberdeen through the help of the nutritionist John Boyd-Orr, director of the nearby Rowett agricultural research station. Hogben, a man who never paid compliments, later admitted that Boyd-Orr had had a deep influence on his ideas and values. Boyd-Orr[70] was a fundamental figure in the development of British food policy, and later founder of the United Nations Food and Agriculture Organization. From 1930 onwards he tried to solve the problems of low agricultural prices and urban malnutrition simultaneously by promoting free school milk. For Hogben, better and more efficient agriculture was also linked to health and a growth in the birth rate. Boyd-Orr's work was just an element in a world current of opinion in the 1930s that focused on the primordial importance of good nutrition. The importance of vitamins and a balanced diet were much vaunted in industrialized countries and through the agency of the League of Nations. They pointed out how biology could be used in the improvement of agriculture and population health.

New concepts in genetics were also beginning to cast doubts on the simplistic attitudes of the first eugenicists and racial hygienists. Population genetics was beginning to show how complex the links between the characteristics of successive generations could be. A new understanding of environmental problems, the increasing complexity of genetics and the use of genetics by the Nazis led to the 1939 declaration by Anglo-American geneticists including J.B.S. Haldane, J. Huxley and L. Hogben, that the improvement of the environment was the major factor in the improvement of populations.

---

[70] Boyd-Orr, *As I recall*, McGibbon and Kee, London, 1966.

These promoters of biotechnology gradually gravitated to the center of British science. In 1933, Haldane took a professorship at the University College of London. J. Huxley was elected a Fellow of the Royal Society in 1935, and became a secretary of the Zoological Society the same year. Even Hogben was finally elected a Fellow of the Royal Society, and obtained his aforementioned Regius Professorship. On a more popular note, J. Huxley and J.B.S. Haldane were radio stars and ownership of a copy of Hogben's *Science for the Citizen* was a sure sign of the man of culture.

Their inspiration and reflections influenced a large number of students in the field, especially in view of the great events of the time, industrial change and malnutrition. N.W. Pirie, a friend of Haldane's, moved at the beginning of the Second World War from his studies on Tobacco Mosaic Virus to the production of proteins from leaves, taking up work where Ereky left off and making a change of scientific direction that lasted for the remainder of his life. Another protégé of Haldane's, Ernst Chain, had an important role in the development of penicillin.

As biologists, J. Huxley and L. Hogben elicited enthusiasm for natural production from chemists. They combined it with ideas on improving the environment and its effect on human nature, thus amalgamating biological engineering and biotechnology. At least one chemist was inclined to return the favor and make respectful noises about the potential of biotechnology, the British industrialist, Harold Hartley.

As he reflected on Hogben's praises of biotechnology, Hartley concluded "with the modern techniques of genetics and the closer association of the farmer and the manufacturer there is a fascinating prospect of new strains that will yield the ideal products for industry almost to standard specification. Then indeed we should have Bacon's ideal of commanding nature in action"[71].

Concepts of biotechnology as a descendent of zymotechnology and biotechnique crystallized, especially in the United States and in Sweden, although there they had a specific translation and development.

---

[71] Harold Hartley, "Agriculture as a source of raw materials for industry", *Journal of the textile institute*, vol. 28, 1937, p. 172.

1.6 The Recognition of Biotechnology by the Institutions

Hartley's euphoria was stimulated by developments that had occurred in the United States. There engineers were beginning the crucial move towards the alliance with biology in the institutions and bureaucracies. Postwar Europe was in chaos and had lost all political hegemony. Haldane's visions, Francé and Geddes' biotechnics, and Ereky's biotechnology were all aspects of cultural creativity being born of a disaster. For the United States of America, the end of the First World War was not the cause a feeling of cultural crisis but rather one of supreme triumph, and a confirmation that the 20th Century would indeed be that of the USA.

If in 1918 the United States were clearly a leading nation, by 1939 however a major cultural and economic crisis had hit the nation very hard. As it damaged traditional optimism and values, the Depression also damaged trust in the very technologies that were increasingly becoming a cause of unemployment. In the United States this loss of confidence was fought with the presentation of biotechnics as a promise of new technique and technology intimately linked to human and biological needs. This engendered a concept of biotechnology which would develop during the two decades after the Second World War. For the first time, biotechnology took on a precise meaning, rooting itself in the American will to promote the status and extent of engineering as a scientific discipline that could ensure a certain prestige, institutional and social positions as well as the bestowing of grants[72].

In the early 1930s two very influential publications contributed to the translation of European concepts to the American context: *Technics and civilization*[73] by Lewis Mumford, and William Wickenden's final report on the teaching of engineering[74].

---

[72] See Lawrence J. Fogel's *Biotechnology, Concepts and Applications,* Prentice Hall, Englewood Cliffs, 1963. Also see Craig G. Taylor and LMK Boelter, "Biotechnology: a New Fundamental in the Training of Engineers", *Science*, vol. 105, 28 February 1947, pp. 217–219.

[73] Lewis Mumford, *Technics and Civilization,* Harper Brace and World, New York, 1934.

[74] W.E. Wickenden, Final report of the director of investigations, June 1933, in *Report of the Investigations of Engineering Education, 1923–1929,* vol. 1 and 2, Society for the promotion of engineering education, Pittsburgh, 1934.

In *Technics and civilization*, published in 1934, Mumford adopted the Geddesian historical categories of the paleotechnical, neotechnical and biotechnical eras. He also felt that the biotechnical era in which things would be made with respect for the biological needs of workers for clean air and light and with respect for the basic biological needs of consumers would be a considerable step forward. Mumford's passion was for urban architecture. Here, above all, the distinction between paleotechnical industrial cities and the vast residential towns of the biotechnical age underlined the nature of progress. Mumford approached technique like a talented journalist. His passion for town planning continued all his life. In a similar way Geddes had been a biologist approaching the problem of industrial production in an abstract fashion. Despite their praises of technology, they were both considering it from the outside. Nevertheless, biotechnique and the prophecies that had been made about it also impressed some American engineers. As the science of engineering evolved through the 1920s and became a professional category represented by an increasingly serious university degree, the image and concept of an appropriate curriculum changed.

The sciences of the engineer had originally developed as a separate group of specialties and professions. But distinguished engineer and future President of the United States, H. Hoover, maintained after the First World War that this image was obsolete and that coherent and unified courses were necessary. William Wickenden, a researcher at ATT who was later to go on to become the director of the Case Institute of Technology, was recruited by the Society for the Promotion of Engineering Education to undertake a fundamental re-evaluation of the courses for engineering science. His voluminous reports published between 1925 and 1934 were key elements in the directed movement towards the recognition of engineering science, not only as a university subject, but also as a career that could guarantee intellectual and social prestige equal to that of the scientific or medical professions.

Wickenden was also concerned about technology's place in human history. He gave many conferences under the title *"Technology and culture"*[75]. In

---

[75] W.E. Wickenden, "Technology and culture, Commencement address", *Case School of Applied Science*, 29th May 1929 and 1933, pp. 4–9, and Technology and Culture, *Ohio College Association Bulletin*.

the early 1930s, he was concerned and worried by the significance of the Depression that led to the destruction of foodstuffs at a time when the world was hungry. The problem of hunger could not be left out of the engineer's concerns. He progressively turned to biology to support his vision of the scientific and expansionist engineer. The psychology of work also became part of the management-oriented responsibilities of the engineer, and psychotechnics came along with it. It was thus that when in June 1933 Wickenden submitted his final report[76], underlining the importance of considering the implications of biotechnics and psychotechnics, he provided engineers with a new scientific foundation, a more human direction and a way to exploit general interest for biological discovery.

The job of implementing the conclusions of the Wickenden report was given to the Engineering Council for Professional Development (ECPD), set up in 1932, and which remains one of the most durable results of the Wickenden report. The chair of this Council was entrusted to Karl Compton, the President of the famous Massachusetts Institute of Technology (MIT) and the most celebrated leader of the chemurgic movement. He was particularly interested in the biological aspects of engineering, and in his own University he hoped to revitalize the old programs for food technology and bacteriology which had lost speed during the advance of biochemistry. The result of this reform movement was that well-esteemed universities including the MIT in 1939 and the University of California (UCLA, Los Angeles) in 1947 set up units on Biological Engineering and Biotechnology. At the MIT, the name was chosen following careful consideration of alternatives, and had the support of Vice-President Vannevar Bush for its resonance with the science of engineering which was defined as "the art of organizing and directing Man, and controlling the forces and materials of nature for the benefit of the human race". The field, known as Biological Engineering, was defined as "the art of applying knowledge obtained from research in biological problems with the help of physics, chemistry and

---

[76] W.E. Wickenden, Final report of the director of investigations, June 1933, in *Report of the investigations of engineering education, 1923–1929*, vol. 1 and 2, Society for the promotion of engineering education, Pittsburgh, 1934.

other allied sciences to problems of human health". Five sectors were to be covered in the laboratories: bioelectrical engineering, electrophysiology, biophysics, microbiology and nutritional biochemistry. What was foreseen turned out to be the description of a new academic sector, whose field of application according to Bush would extend from agriculture to fermentation.

Despite the enthusiasm of Compton and Bush and the vibrant call for subject reform, MIT remained a university with firmly entrenched traditional demarcations between the disciplines. Biological engineering rapidly became "just biology". In California, the reform had better success. L.M.K. Boelter was asked to direct the new school of engineering science for the UCLA., on the campus of the City of the Future. Boelter was a thermodynamicist mainly interested in heat transfer, but also by the interface between people and machines. During the war he had worked on problems of human performance at high altitude in airplanes. Boelter, a friend of Wickenden's, was determined to run a school that would treat engineering science as a unified and integrated whole. Within this whole specializations could of course appear, but these would be considered technologies that when taken together, reinforced engineering. Boelter's own war experience had highlighted the interaction between Man and machine. This was what he and his associate Craig Taylor called biotechnology. For them the engineer needed to be trained from a more homocentric point of view.

The UCLA program was generally respected. Although Boelter's interests were mainly of a philosophical nature, they were linked to the man/machine interface problem and biotechnology came to be associated mainly with the study of what was in any case attached to human factor research already. It was a growth area, since the military and space engineers needed to optimize the conditions for rapid pilot reaction. In 1962, some five hundred engineers and psychologists in the United States were working with the new subject description [77].

At the same time, the need to enlarge the intellectual base for engineering by including the life sciences was being felt by many educational institutions.

---

[77] J.A.R. Koraft, "The 1961 picture of human factors research in business and industry in the United States of America", *Ergonomics,* vol. 51, 1962, pp. 293–299.

Some fifty American colleges or universities included the science of interfaces in the engineering curriculum. Although they did not all use the name biotechnology, and the courses at the MIT equivalent to the UCLA's "Machines and Systems Biotechnology" were often entitled "Sensory Communication and Man-Machine Systems", biotechnology remained a popular term.

In 1966, the term bioengineering was officially chosen by the engineers of the Joint Council Committee in Engineering Interaction with Biology and Medicine to designate "the application of the knowledge gained by cross-fertilization of engineering and the biological sciences so that both will be more fully utilized for the benefit of man". This large category included a whole set of projects a little different to those foreseen by Bush, and closer to the pioneers of biotechnology and biotechnics, medical engineering, the improvement of health through the improvement of the environment, agricultural engineering, bionics[78], fermentation engineering and the engineering of human factors.

"Bionics," for example, is the English translation of Francé's interest in the study of the formation and principles of operation of living things to apply the knowledge gained to the development of physical systems. This interest was also translated as "cybernetics" by visionaries such as Norbert

---

[78] The name bionics was invented by Major Jack E. Steele of the aerospace medical division of the US Air Force on an August evening in 1958. Its first official use took place at the end of 1960, spectacularly. Seventy biologists, engineers, mathematicians, physicists and psychologists were invited by the Air Force to a congress between 13 and 15 September 1960 in Dayton, Ohio: thirty speakers presented bionics. A large 500-page volume reported the ceremony: Bionics Symposium - Living Prototypes - the key to new technology, 13, 14 and 15 September 1960, Wadd Technical Report 60-600, 30 reports, 499 pages. To be totally accurate, bionics had been mentioned several months beforehand at the 12th Annual National Aeronautical Electronics Conference in early May 1960. One of the sessions of the meeting was dedicated to bionics under the chairmanship of Dr John E. Keto of the US Air Force. Four studies on bionics were read, including that of Major Steele. The first time the word bionics was used was in Waveguide, Daytona Section IRE publication, 39 North Torrence Street, August/September 1960. There are the four studies, including Steele's stating that bionics is the science of systems that have a function copied from natural systems, or which present the specific characteristics of natural systems or that are analogous to natural systems. More precisely, one can define bionics as the art of applying the understanding of living systems to the solution of technical problems.

Wiener[79] whose own parallels between machines and organisms harked back to Francé's time. Agricultural engineering recalled Ereky's concerns, fermentation engineering those of Orla-Jensen, while the engineering of human factors was only another name for Boelter's research interests.

From a research perspective these subjects might have seemed extremely diverse. It is also true that their commercial and technological significance were quite different, as were their dynamics. However they shared a same educational base and were brought together by the belief that the life sciences should play a crucial role in the education of engineers. Faith was kept in the importance of biotechnology for the future of all of engineering. Furthermore, whatever the diversity of interpretation, there was consensus that whatever was taught should underline the unity of functions in living organisms and, whenever appropriate, treat materials quantitatively.

So, in principle, biological engineering as it was understood in the 1960s included an infinitely wide field of potential specializations. Debate as to the name and character of biological engineering outlined two main aspects. Bioengineering emerged as one single category of engineering because it was an ideal vehicle for bringing the life sciences into teaching. It was more a teaching category than a research category. It was also a domain that obtained its integrity and identity from its functional role as an interface between the historically quite separate disciplines of biology and engineering.

In the 1960s, while more and more interfaces were being opened up between biology and engineering sciences, different terms were being applied to these new disciplines in various ways. In Britain the approach was different to that of the United States. Human factors engineering and human engineering, which Boelter had called biotechnology, were called ergonomics instead, and a Society for Ergonomics was created in 1949. While biological engineering was a term that harbored wide connotations in the United States, in Britain it was first used only to designate that specialization which the Americans called medical engineering.

---

[79] Norbert Wiener, *Cybernetics or Control and Communication in the Animal and the Machine*, MIT Press, Cambridge, Massachusetts, 1948.

Physiology of course had a well-established engineering slant in any case, and this was particularly well-represented at the University College of London biophysics research unit, founded by physiologist A.V. Hill [80], whose work on anti-aircraft battery targeting is considered a classic precursor of cybernetics. The experience gathered during the war in fields from prostheses to radar technology had created a whole community of engineers working at the interface with the human body, and more and more researchers concerned with these different interests came to work together. Nevertheless, action at the national level was left to the American researcher Wladimir Zworykin, well-known for his contribution to electronic microscopy and television technology and who had directed a medical electronics laboratory at the Rockefeller Institute. Like Bush, he found that the scientific community was working in an isolated fashion and often on parallel subjects in ignorance of the other research. In June 1958, Zworykin organized the first conference on medical electronics in Paris, leading to the birth of the International Federation of Medical Electronics in 1959. This stimulated the British to create organizations that could be affiliated, the first one being the medical electronics section of the British Institution of Radio Engineers.

In his speech launching the group, A.V. Hill [81] complained that the term "medical" was too restrictive. In its place, the wider term of "biological" should be used to describe applications for the sensitive detectors and amplifiers that engineers were developing. The problem evidently went deeper than mere electronics. This was recognized in June 1960 when a group of specialists was set up at a meeting mainly attended by doctors, physiologists, electronic engineers, mechanical engineers and physicists. They named it the Biological Engineering Society [82]. The objectives of the association were to bring together members of the different disciplines to be found in hospitals, research institutes and industry, and to favor the application of engineering science to biological and medical problems. The first scientific meeting took

---

[80] A.V. Hill, "Biology and Electronics", *Journal of the British Institution of Radio Engineers*, vol. 19, 1959.

[81] A.V. Hill, *as above*, p. 80.

[82] Biological Engineering Society, *The Lancet*, vol. 2, 23 July 1960, p. 218.

place in October of the same year at the NIMR, and it called for a fusion of biology and engineering as disciplines[83]. It was thus, quite independently of American usage of the term, that biological engineering was chosen once more as the most appropriate federating concept.

The British usage of the word was recognized by the International Federation, and later in 1963 when the Federation launched a review, the title they chose for it was *Medical Electronics and Biological Engineering*. However, the work of biological engineers was not described as a revolution at the heart of teaching as it had been in the United States. The promoters were consciously specialists, bringing together several skills to bring life to a new era of research. Furthermore, there was a significant aspect in which the sense of biological engineering was different: the aspect of fermentation technology. This might be incorporated in the American model of integrated biology and engineering, in principle at least, but the British insistence on the human applications of the fusion excluded the field. These two alternative visions of the limits of the field of biotechnology, control of which is simultaneously sought by the physiological engineers and the microbiologists, were sufficiently distinct to be the subject of debate in Sweden, resulting in the redefinition of the category of biotechnology as being clearly centered on bacteria and not on Man.

Sweden, open to other cultures, provided an intellectual crucible for the melting of different biotechnical traditions. Curiously, the American movements of biological engineering and biotechnological engineering arrived in Sweden separately and were interpreted independently of each other. Their significance was formally specified. Biological engineering (bioteknik) emerged as quite a vast term, and biotechnology as a specific term involving human factors engineering. In 1942 the word bioteknik was institutionalized for the first time. The Royal Swedish Academy of Engineering Sciences (IVA) had turned to biological methods in its research in energy, food and medical supplies during the war and, in December 1942, a new section was created at the Academy under the name of Bioteknik. Two separate influences were clearly at work here. On the one hand there was the interest of

---

[83] Biological Engineering Society, *The Lancet*, vol. 2, 12 November 1960, p. 1097.

experimentalists involved in the various food industries such as brewing and bread making and who were not really represented in the organization. On the other, there was that of the "engineer-philosophers" who were directing the IVA. On the death of Almgren, who had suggested that the new section be set up, it was taken over by the recently-retired former director, Axel Enström, who had founded the Academy.

Enström was the prototype cosmopolitan engineer. He was progressively absorbed by the problems of technological evolution, unemployment and the commercial cycle. He led debates that occupied many intellectuals in the capital of this newly industrialized country. Sweden had suffered badly during the Depression, and Enström had represented his homeland at several international meetings to discuss the various technologies, including agriculture. At a world conference on energy in Berlin, he had reflected with his American colleagues on the importance of engineering as a solution to modern problems. In 1940, Enström was replaced as director of the IVA by Edy Velander, an equally cosmopolitan engineer with a Harvard degree and another from MIT who managed to keep in touch with his American colleagues, even during the war.

During 1943 Velander paid a formal visit to the States. Like the Americans, he was interested by the discovery of new fields for engineering and as a result of this and his interests and research during the war, he started a journal of technology and food in the 1950s. The developments in Stockholm reflected the creative tension that existed between these cosmopolitan engineers and industrialists. At the Section's first meeting on 19th June 1943, Velander presented a preliminary document which he had written 20 months earlier on biotechnological research. This text started off with general reflections on the problems of contacts between biology and engineering, reflected on the importance of nutrition and hygiene and suggested some very general spheres of interest. Strangely it also contains exactly the same mention of the importance of questions of clothing, nutrition and urbanism that his old MIT professor Vannevar Bush had made some time earlier at Rutgers in his call for a true recognition of biological engineering, thereby proving that he was clearly using an American model. In contrast, the Section's discussion centered on the diversity of agro-alimentary techniques and particularly on aspects such as hormones which were brought in under the new department.

Literally the term "bioteknik" was a translation of Bush's biological engineering. We should note that the word already existed as part of the name of Orla-Jensen's department, Bioteknisk kemi. Whichever term had the most influence, it is certain that the two researchers supported the importance of bioteknik for engineering as a whole and in its practical applications. For Velander, bioteknik brought together the applications that appear when you learn to influence biological processes scientifically and exploit them technologically as a whole and in practical application. As in the USA, this development was seen to provide new opportunities for the engineer. A wide field was opening up for industry and technical agricultural work in which the engineer would have a major role to play, being as always the interpreter of scientific theory and discovery to the users and economists, thus greasing the wheels of assimilation into practice.

Initially, the new Bioteknik Section of the Academy was dominated by the interests of the brewers, but its activities were gradually influenced by more medically oriented engineers whose understanding of the term bioteknik reflected the technology of physiology, known in Britain as biological engineering.

At the end of the 1950s, the Section included bacteriologist Carl-Goran Heden, then Assistant Professor of bacteriology at the Karolinska Institute in Stockholm, was interested in fermentation technology. Heden was not, and still is not, just interested in bacteriology. For more than 40 years he, more than many other biologists, has passed on visions reminiscent of those of Hogben and Geddes to later generations. Heden, although not an engineer himself, was influenced by his own mentor, the cytologist Caspersson, who like Svedberg and Tiselius had underlined the importance of the use of automatic instrumentation. Heden had become fascinated during his studies by aspects of microbiology linked to technique. He even constructed a pilot installation for fermentation which was one of the most important academic installations in the world. So although Heden was not an engineer himself, he was, in practice, capable in the field. It is certain he understood bioteknik to be the study of the control of micro-organisms. In 1956 he persuaded the

Director of the Commission for Technical Nomenclature of his vision. Two years later he enthusiastically encouraged the IVA's Nomenclature Commission to organism two colloquia based on the meaning of the term bioteknik. Those invited to the first colloquium were strictly of the original category of industrially interested parties and were divided between those interested in proteins (such as microbiologists, plant physiologists and brewers, as well as cellulose industrials) and those more interested in fats (such as milk product industrials).

Although Heden called his own specific field of interest "bacteriological biotechnics" to distinguish it from medical biotechnics, he far preferred the term biotechnology. In Sweden, Heden was frustrated because Boelter's concept of biotechnology had been imported as an alternative to the new English term of ergonomics. This usage was introduced by the influential professor, Sven Forssman of the Institute of Work Studies (ASTI). Furthermore this was a subject in which the Swedish were pre-eminent - in fact the first world congress on ergonomics was held in Sweden in 1962. In a contradiction of the pretensions of these "engineers of Man", Heden wrote to the Secretariat of the IVA pointing out that biotechnology had for some time been associated with his interest for the industrial application of biological principles (an interest which almost always had microbiological examples where he was concerned) and the biological aspects of technology. He deeply resented the manner in which it had been taken by the practicians of human engineering, and pressed the Commission for Technical Nomenclature to accept the British term of ergonomics or a more Swedish-sounding variant such as adaptation technology. Despite his efforts, Forssman's influence was too strong, and biotechnology kept its Boelterian connotation.

Heden can be considered to be the first to have separated the problems of applied microbiology from the physiological aspects of bioengineering with which they had been entwined for twenty years. He also tried to reverse the dominant US interpretation of his subject as being purely one of engineering, by insisting that attention should be centered henceforth on the biological side of things. An international perspective shows that his influence was not limited to his own country, and it was thus an American and not a

Swede he managed to convert to his concept of biotechnology. After a conversation with his friend Elmer Gaden, the editor of *the Journal of Microbiological and Biochemical Engineering and Technology*, founded in 1958, that the review's name was changed to *Biotechnology and Bioengineering*.

Whether in the States or in Sweden, the vision of a new dynamic technology sustained the federating concept of biological engineering. Enthusiasts all tried to co-opt the term for their favorite specialization. Boelter chose the term to designate what others called ergonomics and what Heden called biochemical engineering. All this debate on denomination might appear a little confused, but it is symptomatic of the effervescence of contemporary thought on the relationship between biology and engineering. Men such as Mumford, Wickenden, Compton, Bush, Boelter and their institutions in the States, and Velander and Enström and the IVA in Sweden, were providing links between previous concepts of the transcendent significance of biotechnics, biological engineering and biotechnology. Although distanced by time and intention, it seems that when you consider practical biological philosophy that in some way previous abstract thoughts provided language and a structure of thought for the post-war engineers.

At the end of the Second World War, there was international interest in integrating biology and engineering, an interest variously represented by the terms bioengineering, biotechnics and biotechnological. This interest had a very large field of application and was promoted simultaneously in philosophical and educational spheres. While visionary and idealistic enthusiasts hoped to promote an integrated vision, the truth of the matter is that engineering sciences had an entirely different and far more fragmented image. In the years following the Second World War, individual disciplines were proudly autonomous, and although ergonomics or biomedical engineering had been important, it was at the interface between chemical engineering and microbiology that examples of commercial implications of this alliance of biology and engineering were to be seen.

While in Sweden this alliance had not yet taken place, in the United States the term was adopted with success and benefited from its inheritance of the fantastic dynamism of chemical engineering, the remarkable success of the antibiotics industry (in particular penicillin) and the general and philosophical ambitions of the attempts to integrate engineering with biology. Even before the Second World War, Compton had already foreseen that the deployment of biological engineering could contribute to medical technology. We should recall that society between the wars was much more conscious of the problems of health. Many countries' vision of the Western world's priorities was indicated by the establishment of free or low-cost medical check-ups by the government. The wartime discovery of microbial antibiotics and especially penicillin only strengthened the close link between medical care and chemical engineering. It was thought chemical engineering could help with the handling and treatment of large quantities of biological material. It also appeared likely to help with research into the nature of biological systems themselves.

Two major domains of activity were regrouped under the term of biotechnology, following its choice as the new 1961 title for the review *Biotechnology and Bioengineering*.

— Area 1 comprises extraction, separation, purification and processing of biological materials.
— Area 2 embraces use of complex biological systems (e.g. cells and tissues) or their components to effect directed and controlled chemical or physical changes [84].

This insistence on bioprocess technology and the dominance of microbiology in its partnership with chemical engineering were sustained by the continued fundamental importance of two wartime efforts, antibiotics production and response to the threat of bacteriological warfare.

The importance of antibiotics production in the stimulation of a new industry is well-known, but the significance of the second effort is perhaps under-estimated. It is however certain that military research on continuous fermentation, in particular at the Microbiological Research Establishment at Porton Down, was to have implications for the whole industry even if

---

[84] This second field of biotechnology is often known as bioprocess technology.

retrospectively the breweries' great excitement at continuous fermentation was largely premature. The savings made were not that attractive, and the control and technical competence standards much higher than for normal installations. Furthermore, although the advantages of continuous production were very clear when a large quantity of one single product was being handled and the process could be allowed to work for a long period of time, most breweries make a great variety of beers and thus needed to be able to switch from one product to another as demand changed. The beer was not always very good, either.

The insistence on bioprocess technologies was not the only characteristic of the thoughts of mid-20th century leaders of biotechnology such as Heden or Elmer Gaden. Biotechnology was for them not only an advanced technology for developed societies. In 1970 the most appropriate use for continuous fermentation seemed to be the culture of hundreds of millions of tons of single cell proteins (SCPs) to feed populations with malnutrition. Biotechnology, unlike chemical engineering, seemed particularly benign and even futuristic seeing as it appeared perfectly adapted to the majority of humanity not living in developed societies and too poor to import oil to produce vital consumer goods, but whose agriculture produced great quantities of biomass for conversion.

More generally, after fifty years of unsuccessful declarations from fringe intellectuals that biotechnology would be an innovative technology to fill new sorts of needs, the right wave of support appeared in the 1960s.

In the period after the Second World War, biotechnology meant a lot more than another set of techniques. It had come to represent an ideal alternative to the list of new but destructive technologies associated with the military industrial complex. Goods producers and consumers could refuse these new products for themselves out of conservatism, but they could not refuse to react against world poverty, overpopulation and famine. Such evils were appropriate targets for a revolutionary technology that transformed agricultural products.

During the 1960s and 1970s biotechnology was promoted as the use of rich countries' scientific resources to solve poor countries' problems. It brought to mind a new industrial revolution and represented a symbol of hope and an answer to famine, sickness and the dwindling of resources. It was not only felt that it would provide solutions for the third world, but increasingly be the fundamental technology for the development of even industrialized nations.

The dramatic end of the Second World War did indeed engender fears of future disasters, but in reaction to this it also nurtured optimism guided by the urgent sense of hope for a new world order. This was confirmed in the creation of the United Nations and gradual decolonization. There were prophesies of another industrial revolution but little agreement as to what it would revolve around. For some, nuclear power and the promise of clean unlimited energy illuminated the future. Others dreamt of Man's escape from his prison planet to reach new worlds. Some thought that modern microbiology was a way for humanity to finally banish evils such as famine and sickness. The Cold War chilled these optimistic concepts of science with its threat of using space technology, nuclear physics and progress in microbiology to develop terrifying weapons. The nuances of natural idealism translated to difficult decisions. Many scientists involved in high-level science in the 1950s found themselves confronted with a Manichean dilemma: either to use the knowledge, skill and abilities for weapons production or to use them for the good of humanity as a whole.

Elmer Gaden and Carl-Goran Heden were the precursors of the movement for the use of biotechnology for Man's good. The most celebrated visionary was Leo Szilard, a symbol of the rejection of nuclear warfare and its association with their science by the military physicists after the war. Once the Nazi menace was banished from the ruins of Berlin, Szilard became very worried about the prospect of an arms race with the Soviet Union. He was one of the founders of the Pugwash conferences that brought together American and Soviet scientists. He was also one of those remarkable physicists who turned to biology and created the new discipline of molecular biology. His classic article of 1950 [85] on continuous fermentation came from his interest in mass-producing cells for phage studies.

---

[85] A. Novick and Leo Szilard, "Experiments with the chemostat on spontaneous mutations of bacteria", *Proceedings of the National Academy of Sciences,* vol. 36, 1950, pp. 708–7019.

A decade later, Szilard's efforts were more on social concerns. In his science-fiction work of 1961, *The Voice of the Dolphins*[86], he imagined a better technology invented by the dolphins, very much like what was then being called biotechnology. Imagining a time of better international relations, he related how the Americans and Russians consolidated their new cooperation by setting up a new International Institute of Molecular Biology. The scientists of the Institute show that dolphins' brains are better than human ones, and teach them science. The dolphins start to win Nobel prizes. Their greatest discovery is a seaweed that fixes nitrogen straight from the air, produces antibiotics and can be turned into a food protein source. The result of this invention is that the poor of India no longer need to have large families, thus preventing the population explosion. Even the product's name, Amruss, symbolizes its international roots.

A year after his publication of *The Voice of the Dolphins*, Szilard set up a civil rights organization, The Council for a Livable World. The Council identified crucial problems in human development. Despite Szilard's death in 1963, his work was continued into the 1970s by his assistant and biographer J.R. Platt.

In 1972, Platt and Cellarius wrote an article in *Science* calling for the creation of Task Groups to identify the fundamental problems for the future and look for solutions to them. They thought that problems of overpopulation, environment and health could be solved through biotechnology. This hopeful vision was the antithesis of nuclear technology. Several post-war idealists predicted new possibilities in fermentation and enzyme technology industries and those using micro-organisms, all part of what was coming to be called biotechnology. It seemed particularly appropriate for developing countries rich in biological raw materials and needing products such as fermented foods, power alcohol and biogas for energy needs and nitrogen-fixing bacteria. These would be provided by small locally-adapted firms, not distant multinationals. This concept of biotechnology that then emerged had a strangely close resemblance to the bioaesthetic concepts of Mumford and Hogben before the Wars.

---

[86] Leo Szilard, *The Voice of the Dolphins,* Simon and Schuster, 1961.

The biotechnical dream may well have mainly been the dream and concern of Western scientific intellectuals, but it is not for all that commercially inconsequential. It showed to big firms the link between an emerging technology and potential solutions for a world problem that was more and more worrying, at a human, economic and political level. This problem had a name that had a dramatic military ring. It was the demographic explosion, and its corollary of starving populations leading to a secondary implication of political instability. There was no doubt about the scale of the problem.

To fight this threat radical measures were called for. Biotechnology looked as if it might bring some remedies. Unfortunately the solutions foreseen proved to be failures. Banishing hunger by engineering nitrogen fixation into cereals[87] remained a pipe dream for a long time. The hope of beating this threat of world famine through new foods created by biotechnological advances was no more successful. Alfred Champagnat and Jacques Senez had spectacular experimental results in growing micro-organisms, in particular yeast, on oil derivates. Multinationals like B.P., Hoescht, and international organizations such as the United Nations, as well as nations such as the Soviet Union and Japan took great interest in the new food that Scrimshaw at the MIT dubbed Single Cell Protein (SCP) in 1966, but it did not have the expected success[88]. Ironically, market studies eventually showed that the best consumers of SCP were rather the animals of Europe than the poor peoples of the third world, who were very conservative about what they ate and in any case had even less access to oil derivatives than their richer neighbors. Here also, hopes were killed off by the sudden oil crisis and the fall in price of competing products, in particular Soya.

---

[87] Sir Harold Hartley has called the mystery of how the bacteria *Rhizobia* that lives in nodules on the roots of leguminous plants converts atmospheric nitrogen into ammonia one of the last great unsolved mysteries, vital for the world's economy. It was thought then that if only Man could copy this process in cereals, there would no longer be any need for artificial fertilizers. Despite great progress in research carried out by Shell and Dupont de Nemours, we still have not developed cereal strains that fix their own nitrogen.

[88] In developed countries, these new foods were met with general skepticism from the general public, especially in Japan and Italy where there were uncertainties about its safety. Clearly just producing the food was not enough. You also had to consider equity, rural needs, national specificities and dependencies on complex technologies as important factors.

The success of the green revolution through the sale of new varieties of conventionally-growing but high-yield wheat, maize and rice also brought a certain cynicism. Whereas in terms of food production the consequences had been remarkable and world famine had been fended off, the new strains not only needed excellent irrigation, but large quantities of pesticides and fertilizers. In a poor and traditionally-based agricultural system this meant significant investment and borrowing. So the new strains ironically favored rich farmers rather than the small-scale agriculturist who could not afford either the ever-increasing quantities of fertilizer nor the servicing of their debts, thereby perversely increasing the number of farmers who had lost their lands.

At the beginning of the 1960s biotechnology, from an environmental point of view, was benign, well-meaning, and aimed for the good of humanity. Twenty years later it was considered an additional menace. This reversal of opinion is partially due to anxiety over the new science of recombinant DNA technology, but it has deeper roots. Biotechnology was being gradually more and more considered to be unnatural, and its distinction from chemical technology was gradually wiped out.

At the end of the 1970s, the energy crisis replaced that of food. Biotechnology once more seemed to have the answer to the problem, for the poor countries as well. But here too, this hope ended in failure because of production costs. In practice, the technological implications could never really be tested. Despite this failure, dreams of biotechnology coming to the rescue of starving people in the third world were the key strength in the development and preservation of international technology-oriented networks. Men like Heden obtained commitment as well as funding from agencies such as the United Nations, UNESCO, The Food and Agriculture Organization, the International Association of Microbiological Societies and the World Academy of Arts and Sciences, for the promotion of biotechnology. Their contributions were also decisive in the support of this science in which advances announced by scientists are always delayed. The change of target from developing countries to developed ones in the 1970s tragically marks the failure of the new techniques to integrate ethical and rational considerations.

Recombinant DNA techniques that appeared in the middle of the 1970s however offered a miraculous reversal. These techniques caused the rebirth of the humanitarian arguments on biotechnology, arguments that came to be used for the support of the great biotechnology programs, and in particular the great sequencing programs. These arguments took on connotations very close to those used by the founders of biotechnology, such as Ereky, Goldscheid, France, Geddes, Haldane, Hartley, Mumford, Hogben, and Boelter, who saw biotechnology as the technology of the future, the miraculous technology which would come to the rescue of humanity, society and economy, and cure them of their greatest ills.

2
Political Interpretations of Biotechnology and the Birth of the First Research Programs

In the 1960s, the champions of the biological revolution were more prophets than managers or politicians. The political and industrial worlds could not but be interested in biologists' recurrent affirmations of the advent of a biological revolution and undoubted technical progress. During the 1970s, biotechnology became the target of national policymakers [1]. In the United States, Japan, Germany, the United Kingdom, the European Commission,

---

[1] Let us recall that in the first half of the 20th Century, a series of new concepts put forward by biologists like Francis Galton, Jean Rostand (*L'homme, introduction à l'étude de la biologie humaine*, ed. Gallimard, Paris, 1926), Alexis Carrel (*L'homme, cet inconnu*, ed. Plon, Paris, 1941), Hermann J. Müller (*Hors de la nuit, vues d'un biologiste sur l'avenir*, trad. J. Rostand, ed. Gallimard, Paris, 1938), Julian Huxley, J.B.S Haldane, Lancelot Hogben, Rudolf Goldscheid, to name but a few, reached the whole of society, which reflected them in legislation. Nazi Germany was not the only country to ratify eugenics laws: democratic countries ratified them before Germany, and gave it their examples. The United States were the first to legislate on the sterilization of criminals and people with various diseases. Initially, the states of Indiana in 1907, Washington, Connecticut and California in 1909 ratified such laws, and by 1950, thirty-three of the states had followed suit. In 1928, Switzerland and Canada joined the bandwagon. In 1929, Denmark. In 1934, Norway and Germany. In 1935, Finland and Sweden, and in 1937 Estonia also passed similar laws. They were joined by several countries in Central America in 1941. Japan followed suit in 1948. It should be noted that these laws were not just passed spontaneously. They were requested by associations led by medical doctors and biologists, which, through propaganda and intensive lobbying, brought pressure to bear on the legislators. Scientific support and prestige from great names of biology, particularly genetics, were also a determining factor. There is also an undoubted influence of the biology of the beginning of the century on Nazi ideology. In Germany, the first eugenics law was voted through on 14 July 1933 and became effective on 1 January 1934.

and countries all around the world, apparently governments grabbed at the promise of biotechnological solutions to the paradoxical needs of revitalizing industry and economy, and simultaneously taking better care of the environment.

A century of prophesies were finally being listened to. The existing technological system had to be replaced. It would be easy to blame the two oil crises of 1974 and 1980, which brought a tenfold multiplication of the West's energy costs, but even before the crises, the richest nations were already worried that their most productive industries had reached their maximum growth. Added to this fear of a decline in the industries at the root of post-war growth (shipbuilding, chemical, metallurgical and automobile industries), public concern was increasingly alerting the policymakers to how polluting these industries were. Policymakers and executives armed themselves with the promise of biotechnology against these public concerns, raising the subject to a higher level and including it in their programs of industrial and economic development.

Small investors and the general market were to be the ones who developed the economic potential of biotechnology as well as undertaking the financing of research and industrial activity linked to research.

Biotechnology was included in national programs, typically represented as an infant industry bound by commercial, administrative and political limitations. It was promoted in widely broadcast announcements of the imminent advent of a new industrial revolution, announcements made alongside more familiar ones for the new era of information technology. Remarkable scientific progress seemed to justify these prophesies and reassure the public that the biologists' promises would be kept, embodied in what was called "life science" in Japan, "biotechnology" in Europe, and "bioresources" in the United States.

Most histories of biotechnology have a tendency to see in this progress the cause of the change that occurred in political, social and economic perceptions of biotechnology. This must be taken with a pinch of salt. The development of a scientific technique is a reflection of pre-existing social, economic and political aspirations, and when it comes into action, perceptions of it are a reflection of previous aspirations. Bobbing in the wake of the genetic engineering revolution, widely shared hopes in biotechnology sustained the image that had grown during the 1970s from economic and political concerns.

2.1 The Example of the United States

During the 1970s, these economic and political concerns offered a challenge set by the United States, then at the top of the world's economies. Other nations took up the challenge. The United States provided leading trends to other industrialized nations: the environmental movement, an appreciation of the importance of enzymes (which were emerging as integral components of the development and exploitation of bioresources), the scale of life science research in general and the Silicon Valley model.

Problems in Agriculture

Historically, problems in agriculture have always triggered reflection on, and hopes in, biotechnology. This pattern occurred again in the 1970s. At that time, concerns and interests were being loudly expressed outside the establishment from the environmental movement, the birth of which generally dates from Rachel Carson's attack against the use of the pesticide DDT in her book *Silent Spring*[2]. This work gained worldwide notoriety and is considered an informal marker of the beginning of ecological awareness in the West today.

The consequences of intensive and unlimited use of non-renewable resources provoked an extensive set of analyses during the creation of the *Whole Earth Catalog*. Undoubtedly, the most well-known analysis was the

---

[2] Rachel Carson, *Silent Spring*, Houghton Mifling, Boston, 1962. The emotions raised in the general public by this book, which was a complete condemnation of the use of chemical fertilizer in agriculture, led President John F. Kennedy to set up a commission of public inquiry, which several months later confirmed the author's conclusions. Carson's exposé was followed, in 1968, by the revelation of another time bomb by the American biologist Paul Ehrlich, the "P." Bomb, for "Population". (Paul Ehrlich, *The Population Bomb*, Ballantine, New York, 1968).

first report of a series of studies on the theme of growth called *The limits to growth*[3], which made the Club of Rome[4] world famous.

Limits to Growth and the Club of Rome's Analysis

This report intended to demonstrate mathematically (with the help of a model developed by Jay W. Forrester[5] of the Massachusetts Institute of Technology) that exponential population and economic production growth were impossible in the long term, because they supposed an unlimited quantity of exploitable non-renewable resources, arable lands and environmental capacity to absorb pollution. The various simulations all ended sooner or later, but generally sooner, in a brutal drop in the quality of life and/or life expectancy. This was either due to a fall in agricultural productivity, rising costs in increasingly rare raw materials or increased pollution. The simulations also showed that a purely sectoral policy could not prevent a major crisis.

---

[3] D.H. Meadows, D.L. Meadows, Jorgen & Behrens Randers, W. William, *The Limits to Growth*, University Books, New York, 1972, trad. fr. *Halte à la croissance*, ed. Fayard, Paris, 1971.

[4] The Club of Rome was created in April 1968, during an interdisciplinary meeting organised at the behest of Aurelio Peccei (an Italian consultant and businessman) and Alexander King (the director general of scientific affairs at the OECD). The Club of Rome originally gathered together thirty industrials, economists, scientists and high level civil servants from about ten countries. Amongst them, other than A. Peccei and A. King, the Nobel Prize-winning physicist Dennis Gabor, and the Director of the Battelle Institute (Geneva) Hugo Thiemann were prime motivators. The objectives of the club were to develop an understanding of interactions between the economic, political, natural and social components of the global system that make up our environment, to enlighten policymakers and world public opinion with regard to these interactions and to promote new political initiatives.

[5] Initially, in September 1969, the club decided to entrust the Turkish-born Californian cyberneticist and economics planning expert Hasan Ozbekhan with the task of setting up a formal model of world dynamics. The project he submitted was met with many objections and criticisms and encountered funding difficulties, and it was eventually abandoned. During a meeting in Bern in June 1970, the club then turned to Professor Jay W. Forrester who, in a mere month, and with the aid of studies in systems dynamics gained in other contexts, constructed a rough model of the world that had five key variables: population, invested capital, use of non-renewable resources, food production and pollution. At the end of July, the members of the club met at MIT in twenty-day work periods during which Forrester presented the results of the computer test of his model World 1, as well as a modified model called World 2. The club then decided to ask one of Forrester's young assistants, Dennis Meadows, to set up a more complex model for which funding was obtained from the Volkswagen Foundation. This was World 3, whose publication had a resounding impact and made the name of the Club of Rome.

The authors argued that what was needed was a stabilizing demographic and economic system based on the maintenance of population and productive capital at its current "zero growth" level. The extraordinary impact of this report (an impact notably measurable by the virulent criticism it elicited, mainly from university-based economists) can only be explained by the cult following that economic growth enjoyed amongst governments and intellectuals.

Consequences of the Oil Crisis

Almost immediately, the readers of the report were to experience the implications of a shortage of a key natural resource with the first oil crisis. They were able to see for themselves much earlier than foreseen the effects of the shortages predicted by the Club of Rome.

In response to the crisis, the American midwest offered vast quantities of agricultural products for conversion into biofuels. In 1971, even before the oil crisis, tests linked to biofuels[6] by the Agricultural Products Industrial Utilization Committee of Nebraska showed that the old pre-war 10% mix of ethanol in car petrol was still of some use. A year later, the term "gasohol" was invented by William Scheller, Professor at the University of Nebraska. Seizing on his work for its own advantage, the state of Nebraska took the term for an official seal. Gasohol fuel grew in reputation, gaining presidential and congressional approval. In 1979, as the Soviet Union sent troops into Afghanistan, the Carter administration's reprisal was to cease exporting agricultural products to the USSR to prevent their potential use as of bioresource products. As in 1930, the only solution for the survival of farmers was the use of excess agricultural products for industrial ends.

In the second oil crisis, gasohol seemed to offer an ideal solution to the oil shortage and agriculture. On 11 January 1980, a program on biofuels was set up aiming for a large increase in ethanol production within two years. In June, Congress approved a project which set up a US$1.27 million reserve

---

[6] Hal Bernton, William Kovarik and Scott Sklar, *The Forbidden Fuel: Power Alcohol in the Twentieth Century*, Boyd Griffin, New York, 1982.

of federal aid for the development of alcohol and other biomass-derived fuels[7]. Before anything of real significance took place, however, the political and economic climate changed. The Reagan administration came to power on 20 January 1981, and with the falling oil prices of the 1980s, encouragement and support of this fledgling industry ended before it had really taken off.

Federal support for the gasohol program is an anomaly in the history of American government policy on industry. The vitality of industrial research and development (R&D) policy and their fondness for the market economy had often inhibited national leaders from identifying with or supporting promising technologies. Instead of promoting revolutionary techniques for the future, the federal government had always favored the support of fundamental research. Such research would not disturb the market nor look like undue preferential treatment of individual companies. This self-inhibition, however, did not affect the sectors of health, space or the military-industrial complex. During the 1970s, therefore, under the aegis of what we could call corporate trends, action was taken for the promotion of alternative technologies.

In 1971, the National Science Foundation (NSF) set up a small program called RANN (Research Applied to National Need), and from the start, technologies based on the use of enzymes had part of the budget. The sums allocated were minimal, however, rising from half a million dollars at the beginning of the program to two million dollars in 1976.

The Importance of Fermentation and Techniques of Enzyme Engineering

The potential role of technologies based on fermentation and the use of enzymes in the manufacture of cheap gasohol and the energy efficiency

---

[7] Generally, alcohol fuels were at the centre of bioenergy concerns in the USA. A new national commission, the National Alcohol Fuels Commission (NAFC), chaired by Senator Birch Bayh of Indiana, has been studying their new potential. In the late 1970s, some experts thought that enough ethanol could be produced to replace 10% of car petrol consumed in the USA with 40% of the American maize crop.

of enzyme processes, brought the "enzyme program" into the "energy program" from 1977 on. At that point, its specific image as a discrete entity was lost, as well as the political support from which it had benefited. Although the USA had been supporting key fundamental research and US industries were central protagonists in the use of enzymes in the 1970s, the government was not a fervent sponsor of enzyme-based biotechnology.

The existence of the RANN program and industrial investments at the time, nevertheless, managed to promote many applications for enzymes in the agriculture, environment, industry and energy sectors. Colloquia and documents from RANN contributed to the development and distribution of models of the central role of industrial enzyme processes, models which a decade later would look very familiar.

Biomedical Research

Agriculture and its problems, as we will see, continued to provide a powerful motor for the development of biotechnology, but the life sciences became redirected as their fundamental medical significance was realized.

In the United States, sizable growth in research was supported by public interest in matters of health, an interest manifested by a society-wide involvement in medicine. In 1930, half the civilian chemists employed by the government worked in the agricultural departments. By 1978, this proportion had dropped to 13%, while twice as many worked for the Health, Education and Welfare Agency, an agency which had not existed in 1930.

The National Institute of Health (NIH), the administrative body founded in 1930 and entrusted with funding the medical domain, saw a budget increase from $3 million in 1946 to $76 million in 1953 and $1.1 billion in 1969. Expenditure in academic circles on fundamental research in the life sciences doubled between 1964 and 1972 while the budget for physical sciences only increased by 50% over the same period. Cardio-vascular disease, mental disorders and cancer became major challenges for American science, which had previously triumphed with technical solutions in more physical fields such as the moon landings and the atomic bomb.

In 1971, two years after the last of these successes — the Apollo program — President Nixon declared war on cancer. The new Cancer Act authorized him to spend $1.59 billion on the fight against cancer over three years. By 1975, more than 600,000 people were engaged in this scientific battle. During this time, molecular biology, especially its implication in the development of new drugs, as well as related regulatory problems, were vigorously explored. The scale of American research in the life sciences, both governmental and commercial, impressed and influenced the rest of the world.

The Silicon Valley Model

If the development of biomedical research highlighted the vitality of one of the major activities supported by the United States government, the electronics boom of Silicon Valley (the valley of Santa Clara in California) foretold change in the structure of industrial organization. Small companies specializing in avant-garde areas such as computers, robotics, office equipment and optical fibers, with minimal start-up capital and sustained with venture capital, looked as if they were the appropriate structure for future high-tech industries. In 1968, for example, only eight companies were producing semiconductors; but by 1970, there were about thirty-five. The turnover of this sector tripled in the 1960s and did so again between 1970 and 1973.

Venture capitalism proved to be particularly appropriate for the start-up of new industries. Its prestige came from its association with the development of Silicon Valley. An official study showed that seventy-two semi-conductor industries started in the 1970s with $209 million of venture capital, which were then floated on the stock exchange, had a total turnover in 1979 of more than $6 billion. Since most of the economies of the developed world had now reached maturity and were looking for new bases for future economic growth, the burning question was whether it was possible to simulate this Silicon Valley miracle, through which science turned to gold, in other activities.

At first, the answer was "Yes". Between 1971 and 1984, there was a period of growth in venture capital and avant-garde technology firms. In fact, support of high technology implies a specific financing structure, which allows for a delay in the return on funds (a delay due to the time needed for the perfecting of a new procedure or product and its entrance into the market) and includes the risk of failure. This requires maximum diversification of investors' portfolios and control of the firms in which they hold shares [8].

From 1971 to 1982, a large number of firms were set up with venture capital [9] (Table 1).

Table 1 Most representative beneficiaries of venture capital

Firm	Field of research	Date of creation	Status
Cetus	Genetic engineering, diversification	1971	On stock market
Bioresponse	Cell culture, hybridomas	1972	On stock market
Native Plants	Plants and genetic procedures	1973	
Agrigenetics	Genetic engineering and plants	1975	
Bethesda Research Labs	Restriction enzymes, genetic engineering, hybridomas	1975	Development
Genentech	Genetic engineering, human health, vaccines	1976	On stock market
Genex	Genetic engineering, diversification	1977	On stock market
Hybritech	Cell biology, antibodies, hybridomas	1978	On stock market
Biogen	Genetic engineering, diversification	1978	On stock market
Hanabiologics	Cell culture, diagnostics	1978	
Molecular Genetics	Genetic engineering, animal health, agriculture	1979	On stock market
Centocor	Antibodies, hybridomas, diagnostics	1979	On stock market
Monoclonal Antibodies	Hybridomas, antibodies	1979	On stock market
Applied Molecular Genetics	Genetic engineering, diversification, hybridomas	1980	
Phytogen	Genetic engineering, plant cell culture	1980	
Codon	Genetic engineering, enzymes	1980	Start-up
Plant Genetics	Improvement of plants with genetic processes	1981	Development
Integrated Genetics	Genetic engineering, human health, agriculture	1981	Development
Applied Biosystems	Gene and protein synthesizers and sequencers	1981	Start-up
Genetics Institute	Genetic engineering, human health	1981	Start-up
Repligen	Genetic engineering, hybridomas, human health, enzymes	1981	
Cytogen	Monoclonal antibodies, human health, diagnostics	1981	
Seragen	Services and products for cell biology	1982	

[8] The principle is that if the new company succeeds, it will grow and so will its capital. Then the capital can be liquidated in one of three ways; purchase by a large firm, the sale of all or part of the shares to more conservative investors or flotation — the aim at the end of the game remaining the reinvestment of the profits.

[9] Tables 1 and 2 give the names of typical firms that benefited from venture capital and a list of five major ventures carried out by the main independent venture capital firms in 1982 (Tables from P.J. Raugel's article "Nothing ventured, nothing gained", *Biofutur,* June 1983, p. 11).

Table 2 Largest input provided by venture capital firms

Firm	Capital input (millions of US$)
Kleiner, Perkins, Caulfield and Byers	150.0
John Hancock Venture Capital Fund	88.5
Concord Partners	84.5
M.L. Ventures	60.6
Morgan Holland Management Compnay	58.5
Hamoro International Venture Fund	50.5
Narrangansett	44.2
Matrix Partners	44.1
Acler and Company	43.5
Institutional Venture Partners	40.0
Interwest Partners	38.9
Charles River Partnership	38.3
Technology Venture Investors	38.4
Ascott Norton	34.9
Robertson, Coleman and Stephens	34.5

Spectacular growth of biotech companies had occurred by the end of 1981. Wall Street, for example, valued Genentech at more than $300 million while its profits were only $51,000 and its assets $14 million. This aberrant valuation only lasted a month, but the important point is that the Americans had a stock market that handled the shares of small firms and guaranteed that investment could be obtained when needed. By 1984, however, share prices had altered to such an extent that for 24 of the 26 companies floated in 1983, shares were on their way down. For 19 of the 26, this fall was over 50%. To counter the downward trend and channel funds towards research, the federal administration allowed groups of co-shareholders an almost complete tax deduction on funds invested in R&D partnerships. Some $500 million in 1983 and $1.5 billion in 1984 were invested in this manner. This extended the viability of the firms and temporarily sustained the market for research venture capital.

The arrival on the market of biotechnology products was much delayed and less spectacular than foreseen. It gave several multinational firms time to set up suitable research centers and buy commercial firms in key distribution sectors. From 1989 on, Dupont de Nemours dedicated $220 million to biotechnology R&D, and in 1982, it spent $85 million on a research center. Ciba Geigy set up a research institute at the Research Triangle Park. In 1984, Standard Oil of California opened a center for biotechnology R&D near the Richmond refinery. In November of the same year, Monsanto opened a center costing $150 million.

Many small biotechnology companies were taken over, but multinationals also contracted small specialized biotechnology firms to carry out research for them on the side. This was, of course, risky for the small firms. If the research was a success, the large company that had awarded the contract would dispense with the small firm's services and continue the work in its own centers. Alternatively, they might buy the small firm. However, if the project failed, the small firm might fail too. In 1984, there were several such cancellations of contracts: for example, Allied Co shut down its projects with Calgene and Genex; Monsanto did the same with Biogen on the plasminogen activator research; Grand Met shut down its contract with the same company for rennet and food technologies; and Green Cross did the same with Collaborative Research on the alpha interferon project.

The value of the Silicon Valley model should not be misunderstood. Biotechnology may, indeed, be a technology of the future and generate economic rebirth. It should be remembered, however, that biotechnology is by nature very different from electronics and that their respective "revolutions" will follow different paths. Between 1970 and 1980, many biotechnology firms had to downscale due to a lack of capital. This has not, apparently, damaged US government confidence in biotechnology since the level of public aid granted to the bioindustries and research has never ceased increasing (Tables 3 and 4).

Table 3 Minimal public aid to American bioindustries

Year	Million FF
1983	30
1984	100
1986	300
1985	410

Table 4 Federal funding of biotechnology research

Year	Billion dollars
1983	1.5
1984	1.75
1985	2
1986	2.25
1987	2.5
1988	3
1989	3.10
1990	3.25
1991	3.5
1993	4

(Source: Office of Science and Technology Policy)

In Japan and Germany, biotechnology was promoted as a technology parallel to chemistry, which, with the latter, would answer the calls for an industrial renewal that respected the environment.

2.2 Biotechnology in Japan : Economic Success and Ecological Failure

In 1970, Japan had the world's third largest Gross National Product (GNP) after the United States and the Soviet Union. This was due to industrial growth of more than 14% per year and the tripling of energy consumption during the 1960s. Because of the small size of the country and the limited amount of inhabitable land, industry is concentrated geographically. In 1969, the amount of copper used per square kilometer was nearly ten times that of the United States. The consequence of this extraordinary growth was considerable pollution to the environment. By 1970, almost all freshwater was contaminated with industrial waste. The rivers and estuaries that run through cities such as Tokyo, Osaka and Nagoya were poisoned. The contamination level was far higher than government-set norms. Evidence of the health impact of pollution had been gathering since 1960. Better known examples include *Minamata's* disease, caused by mercury poisoning from eating contaminated fish and the illness known as *Itai-Itai,* that comes from eating rice grown in water that contained cadmium. Although in previous decades, the Japanese public accepted and respected the government's decisions, by the end of the 1960s, this was no longer the case. The Japanese had become concerned by environmental damage and lack of respect for the environment. This concern led to discussions in the Japanese Ministry of Trade and Industry (MITI), the Science and Technology Agency and large corporations.

In 1977, the White Paper on Science and Technology[10] highlighted public demands that science and technology clean up their acts and respect and conserve the environment, instead of only considering speed, low cost, quantity, efficiency and ease of production. The emerging consensus suggested that there was a real need for the development of non-polluting technologies. In the early 1970s, Japan set up ambitious industrial reform programs, mainly because, having no oil deposits of its own, it felt the oil crisis badly.

This need for a new "clean" industry led to specific attention to the new information technologies; the Japanese conquest of the markets of the developed world in this area is now famous. Less famous are the new manufacturing philosophies which also evolved, for example, the creation of the 'mechatronics' concept to represent the integration of mechanics and electronics in 1960.

The Importance of Microbiology and Enzyme Engineering

The life sciences also benefited from the Japanese re-evaluation. As is often the case in Japan, new industrial developments were based on current interpretations of past experience. The Japanese conception of the interface between biology and technique was loosely similar to that in the West. Although the term used (*Hakkô*) has a similar meaning to 'fermentation', it

---

[10] Japan Science and Technology Agency, *Outline of the White Paper on Science and Technology: Aimed at making technological innovations in social development,* February 1977. Trans. Foreign Press Center, pp. 176–178.

includes a far larger set of phenomena. The essential idea of Hakkô is the production of substances useful for the destruction of unwanted products by living micro-organisms.

In 1970, applied microbiology in Japan had already been around for a very long time; it had a much higher profile than in Europe or the United States. Applied microbiology was seen as an ideal solution for the chemical industry, which was desperately trying to combat its reputation as a disrespectful environmental polluter. In 1971, Mitsubishi's chemical division set up a life sciences institute. The Council of Science and Technology pinpointed the life sciences as a key sector for the 1970s in its 1971 report to the Prime Minister, "Fundamentals of Comprehensive Science and Technology Policy for the 1970s". It stated that Japan should aim for strategic results through the improvement of research infrastructure. This advice was not ignored. Two years later, the Committee for the Promotion of the Life Sciences was set up to coordinate the activities of the government and the Science and Technology Agency. The Institute of Physics and Chemistry, RIKEN, became the host of a special office for the promotion of life sciences entrusted with the implementation of programs and directives. As recommended in the 1971 *White Paper on Science and Technology*, the government's commitment was maintained.

The political debate of the first half of the decade transformed specific experience into a vision of generic revolutionary technology for the future. The 1970s saw two parallel events: during the commercial development of the enzyme-based industry, the Committee for the Promotion of Life Sciences developed both a philosophy of technologies linked to enzymes and a program to implement the bioreactor revolution. A 1975 report published in *Nature* by Professor Akioyshi Wada, the director of the Committee for the Promotion of Life Sciences, reflected the contemporary interest in robots. Called *One step from chemical automations*[11], it expressed a holistic philosophy of

---

[11] A. Wada, "One step from chemical automations", *Nature*, vol. 257, 1975, pp. 633–634.

bioreactors which was repeated in two Japanese reports the following year.[12] This augury of a revolution in the future of industry was observed with great attention by the Europeans. The reports prophesied:

The industrial application of enzymatic reactions, which have been known to play a central role, in highly organized and efficient biological activities, has been recognized, in recent years, as one of the important and urgent tasks for the benefit of human welfare. Some of the greatest benefits, which society can expect once such application is made practical, are:

- Reduced energy consumption
- Chemical industry based on aqueous solutions under normal temperature and pressure
- Streamlined processing of complicated chemical reactions
- Self controlled chemical reactions
- Minimum disturbance to ecology

From this model came a complex analysis that showed the great potential of bioreactor technology. Although Professor Wada's outline was impressive, it was not radically new. It brought ideas that were already well established together into a coherent whole.

Aside from enzyme technology, there was industrial microbiology, the second feather in Japan's bioindustry cap. In this field, Japan had gradually acquired a very high reputation as one of the world's main producers of antibiotics as well as a great variety of fermentation derivates. Research in

---

[12] These two reports are: "Present and Future of Enzyme Technology" and the "Report of Current Advances in Research of Enzyme Technology", the titles and introductions of which may be read in Appendix 1 *"Report on the Current State of Planning of Life Sciences Promotion in Japan"* in A. Rörsch's *Genetic Manipulation in Applied Biology: A study of the necessity, content and management principles of a possible community action*, EUR 6078, Office for Official Publications of the EC, Luxemburg, 1979, pp. 59-63. One might also refer to Dreux de Nettancourt, André Goffeau and Fernand van Hoeck's "Applied Molecular and Cellular Biology, Background note on a possible action of the European Communities for the optimal exploitation of the fundamentals of the new biology", Commission of the European Communities, DG XII, XII/207/77.E, 15 June 1977, principally in the annexes where Figs. 1, 2 and 3 (Table 2) concern enzyme technology and the systematic approach taken by Japan for the research and development of enzyme technology, implied and applied research and the anticipated social impact.

the field of antibiotics had become very "officialized", supported by a number of government and private institutions at which fundamental research and research aimed at improving technology were harmoniously combined, in particular, the Institute of Microbial Chemistry in Tokyo, which was set up in 1962, the very large Microbiology Institute of Tokyo University, the antibiotics department of the NIH and the Kitasato Institute. Research in these centers contributed to the activities of a dozen very large industrial firms including Kanefugachi, Kyowa, Meiji, Seika, Shionagi and Takeda. With such a strong support network, Japan raised itself to be the second world producer of antibiotics after the United States.

The other component of applied Japanese microbiology sprang from the traditional food and alcohol fermentation industries and the biological production of a large number of metabolites and enzymes. It is certain that Japanese bioindustry plans and development tended mainly towards the refinement of production technologies and highly developed know-how. Japan chose to support its private sector more heavily and openly than the United States. This choice indicates how important the Japanese government thought biotechnological activity was. In 1985, while the nation's R&D budget increased by 4.5%, the budget set aside for bioindustries increased by 35.5% compared with 1984. Eyeing Japanese development with apprehension, pundits in the west gathered Japanese philosophy, good planning, industrial indicators and statistics on patent applications[13] as proof of the Japanese industrial threat to the economies of other nations, especially those of Europe.

The Japanese concept of the life sciences as a government policy category in its own right contrasted with the American idea of biotechnology as a part of applied science. But in the two countries, the concept of biotechnology itself remained largely unused; at the time, it wasn't used at all in Japan. It was Germany which brought the scientific and political aspects of biotechnology together and gave political meaning to the term biotechnology for the first time.

---

[13] Between 1965 and 1977, 4,539 patent applications in enzyme engineering were made, 67% of which were Japanese, 18% American and 15% from the member states of the EEC.

2.3 Germany and the Political Aspect of Biotechnology

In the two decades after the Second World War, Germany prospered thanks to the vitality of its chemical, metallurgical, automobile and electronics industries. However, by the end of the 1960s, this economic renewal seemed to be ending. At the same time, the Americans were dominating the development of technology in the new computer and aerospace industries. In addition to these new technologies from the United States, a second factor became increasingly significant during the German promotion of biotechnology at the end of the 1960s: the problem of environmental protection.

The Impact of the "Green" Movement

In Germany, the environmental movement has a more vigorous and intense history than in other countries. After the Vietnam war, nuclear sites were targeted for protests, whether civilian or military. The 1968 student generation also strongly criticized the chemical industry, which had been polluting the Rhine for a century. As in Japan, concern for the environment was linked to questions of political organization.

In response to the age of industrial pollution, the Germans rediscovered the romanticism of the 1920s. Although the Green Movement was not yet established and the name had not yet even been coined — that happened in the spring of 1978 — groups concerned about the environment had already formed years before. By 1972, there were already over 7,000 such groups in existence. At the root of their policies were eight basic points: decentralization, participation, reduction of power, the economical use of renewable natural resources, ecological behavior, clean technologies, freedom and direct democracy.

The emerging green tendency favored technologies that depended on renewable resources, used in low-energy processes, producing large quantities of biodegradable products that took account of world problems such as health and hunger. The movement cannot be identified as the promoter of biotechnology and biotechnics. It was not even really a political force of any power before the middle of the 1980s. It was, however, a challenge to the various established political parties, particularly to the Social Democrats, whose ranks would otherwise have been swelled by the socialist members of these smaller "green" groups.

Biotechnology as a Political Response

When Willy Brandt's new government came to power in 1971, he set up a department to support industry, the *Bundesministerium fur Forschung und Technik* (BMFT). This agency had a wide range of public interest duties including health, nutrition and the quality of the environment. After the success of the American space program, the mobilization of intellectual, economic and technological resources on large projects had come to be considered the only way to make real technological progress. Impressed, the Germans analyzed the needs and problems for which they needed solutions from the new technologies in a report called *Ester ergebnisbericht des ad hoc ausschusses neue technologien*[14], published in 1970. This identified three types of need:

- Fundamental needs such as food or raw materials
- Infrastructure needs such as transport
- Environmental needs.

In 1972, a biology and technology program was set up in the new BMFT. Six fields were declared priorities:

- The creation of sufficient sources of human and animal foodstuffs
- The reduction of environmental pollution
- The improvement of pharmaceutical production
- The development of new sources of raw material
- The production of metal and chemical substances
- The development of biotechnological processes through fundamental research

---

[14] "Ester ergebnisbericht des *ad hoc* ausschusses neue technologien, des beartenden ansschusses für forschungspolitik", *Schriftenreihe forschungsplanung,* vol. 6, Bonn, BMBW, December 1971.

The close association of these with the conclusions of the new technologies report was clear. At a time when priority was being granted to environmental problems, biotechnology was entrusted with their solution. However, the infrastructure was inadequate. Fermentation and biology were not at the level that chemistry had attained, and biochemical engineering was still an undeveloped science. In view of this, particular attention was given to integrating engineering with the biological sciences.

In this dynamic context, the term "biotechnology" was used as a symbolic term for the importance of developing a new type of industry.

The First Biotechnology Enthusiasts

The first advocate of biotechnology seems to have been Professor Karl Bernhauer of Stuttgart. In the 1930s, when he became interested in biochemistry, he held a teaching post in Prague. After the German invasion, his commitment to the teaching of chemistry, fermentation and nutrition grew. During the war, his knowledge of submerged fermentation allowed him to work on the production of penicillin. Three years after the German defeat, he was to be found at Hoffman-la Roche in Stuttgart studying Vitamin B and the production of cobalamines. In 1960, he received an honorary professorship in biochemistry; four years later, he renamed his group "Biochemistry and Biotechnology".

Bernhauer had close links with the penicillin industry, and it is reasonable to think that he adopted the term biotechnology from American usage. However, his prewar activity was such that he could well have learned the term in its German connotations. Bernhauer was not a modern molecular biologist. He was clearly a descendant of the technological school of Delbrück and Lindner. The history of fermentation chemistry was, as far as he was concerned, that of yeast and its use in alcohol and later in food production.

In 1967, Hanswerner Delweg, a colleague of Bernhauer, became the director of Berlin's Institut für Garüngsgewerbe. He felt that the institute, whose origins lay in brewing, should widen its range to include modern industries such as the penicillin industry. He followed Bernhauer's example and renamed Delbrück's Institute the "Institut für Garüngsgewerbe und

Biotechnologie". Now that it was part of the name of such a famous institution, the word biotechnology was effectively adopted in Germany.

The first symposium on industrial microbiology took place in 1969. The following year, microbiologist H.J. Rehm called for a meeting of biochemists unaware of technique and engineers unaware of biochemistry claiming, "A future aim would be to fill this gap with an appropriate training, to climb beyond classical fermentation technology and build the modern science of biochemical-microbiological engineering". At about the same time, he published an article in *Nachrichten aus Chemie und Technik* on "Biotechnik und Bioengineering" [15]. Pointing out that German applied microbiology and bioengineering were falling behind that of the United States, Japan and Great Britain, Rehm called for a reorganization of education and research to cover the new field, a name for which had as yet not been agreed.

These initiatives completed the industrial attempt to answer the ecological nature of national demands.

The Industrial Movement and Dechema's Study

In 1972, the chemical company Bayer set up Biotechnikum, a center for research in biological engineering. The same year, DECHEMA, the German association for chemical equipment manufacturers, which had already taken the initiative of setting up a working group, was formally commissioned [16] by the BMFT to carry out an inquiry into initiatives that should be taken in biotechnology. The report, which pointed out the variety of meanings associated with the word biotechnology and decided to exclude biomedical aspects, defined biotechnology in a way that was to influence European thought. Biotechnology, in the report, was the use of biological processes in technological processes and industrial production. It included the application

---

[15] Wilhelm Schwartz, "Biotechnik und Bioengineering", *Nachrichten aus Chemie und Technik*, vol. 17, 1969, pp. 330–331.

[16] DECHEMA (Deutsches Gesellschaft für chemisches Apparatewesen e.v.) 1974, *Biotechnologie: Eine studie uber forschung und entwicklung-möglichkeiten, Aufgaben und schwerpunkten der förderung*, DECHEMA, Frankfurt, 1974. This report was the base of DECHEMA's study funded by the BMFT, which was published in 1976.

of microbiology and biochemistry to chemistry, chemical engineering and process engineering. This report systematically analyzed the opportunities in the field and pointed out that, in contrast to the United States, Great Britain and Czechoslovakia, Germany had underestimated the potential of microbiology. From then on, the diversity of possible production uses and the non-polluting nature of such processes were urgent priorities, closely fulfilling recent political demands.

The impact of the report was heightened by its release in 1974, a time of oil crisis. It became the basis of a series of plans drawn up by the Federal Research Ministry to promote the development of biotechnology. Between 1974 and 1979, government support for biotechnology programs increased considerably, from DM18.3 million to DM41.3 million. This governmental support encouraged the chemical industry to seriously launch itself into the field of biotechnology at a time when petrochemical technologies seemed archaic, and when even chemistry seemed to be losing momentum.

The enthusiasm for biotechnology led to the construction of a large complex in Braunschweig in 1965 by the Volkswagen Foundation, a complex which was known as the Gesellschaft für Molekularbiologische Forschung (GMBF). The institute became independent in 1968. In parallel, the federal government considered the future of the institute and decided that a pilot installation should be set up doing research that would be oriented more towards downstream applications. In 1975, the state acquired the institute and renamed it the Gesellschaft für Biotechnologische Forschung (GBF); its role now being to encourage the optimum use of natural resources and protection of the environment.

In addition to these centers, the Federal Republic of Germany also boasted universities and large institutes such as the Max Plank Institute financed by the Deutsche Forschung Gemeinschaft (DFG) as well as other important centers such as the Institute of Food Technology and Packaging in Munich and the Institute of Surfaces and Biological Engineering in Stuttgart. Furthermore, there were large industrial centers, many of which were multinational firms that dedicated a large part of their budgets to various aspects of R&D in biological engineering and pharmacobiology.

Despite this socio-political sensitivity to environmental problems, and despite the "green" nature of biotechnology, the Germans undertook a serious policy of R&D in biotechnology, taking into account the growing importance of industrial microbiology and the new possibilities offered by the technologies of immobilized enzymes. Rather than aiming at revolutionary change, as envisaged earlier in the century and then later in the 1970s, this approach aimed at a gradual cumulative process based on the improvement of man's ability to control useful microbial activity.

Dechema and the Foundation of the European Federation of Biotechnology

The German and Japanese concepts of biotechnology were formulated at the beginning of the 1970s before the full impact of recombinant DNA techniques. They reflected the political and economic needs of the time more than administrative progress demanded by the onward march of science.

The field was interdisciplinary. DECHEMA's staff described biotechnology as the interface between chemical engineering, microbiology and biochemistry, but also as a well-recognized technology in its own right. Their model was based on that of chemistry, but it evolved through the traditional approaches of industrial research. It was stable and influential and changed slowly in the 1970s with the fantastic new promising results of molecular biology. It changed more radically in the 1980s. From the American point of view, Japanese biotechnology policy was seen as a terrible menace and German biotechnology policy as a strange case of corporatism. The latter's origins as a response to environmentalist demands were forgotten.

In the 1970s, Rehm's philosophy and report were of fundamental importance for all of Europe. Although Robert Finn's attempt to create a biotechnological society failed, an attempt which arose from a visit to the prestigious ETH University in Zurich, the idea had been passed on. DECHEMA was particularly well placed to use Finn's individual initiative to its advantage in order to give biotechnology an institutional link.

With the support of the British and French chemical industries, DECHEMA's initiative bore fruit at Interlaken (Switzerland), in 1978, on the occasion of the foundation of the European Federation of Biotechnology (EFB)[17]. Despite its pan-European character, the secretariat of this organization remained with DECHEMA. The first meeting of the EFB was held at Innsbruck. It was hosted by DECHEMA and organized by Rehm and Fichter of the ETH. Its principal themes were immobilized enzymes, bioreactors and biochemical engineering, themes which clearly came from the German concept of biotechnology in the early 1970s:

> "Biotechnology is the integrated use of biochemistry, microbiology and engineering sciences in order to achieve technological application of the capacities of microorganisms cultured cells and parts thereof"[18].

DECHEMA's 1976 report, funded by the BMFT, had a similar definition: "Biotechnology concerns the use of biological processes in the context of technical processes and industrial production. It implies the application of microbiology and biochemistry in conjunction with technical chemistry and production technique engineering"[19].

The second international meeting was held in Eastbourne, England and reflected the importance of the United Kingdom in this European organization.

2.4 The British Development of Biotechnology: Delayed Political Reaction

The UK is an interesting parallel with Germany. British scientists had closely followed the development of biotechnology. Since 1972, the chemical industry

---

[17] I refer to Mark F. Cantley's *The regulations of modern biotechnology: A historical and European perspective (A case study in how society copes with new knowledge in the last quarter of the twentieth century)*, Chapter 8, vol. 12: Legal, economic and ethical dimensions, Treaty in several volumes entitled *Biotechnology*, ed. V.C.H., 1995.

[18] DECHEMA, *op. cit.*, p.7.

[19] Quoted by Mark Cantley, *op.cit.* We must note that these definitions do not make any connection with the scientific innovations of genetic engineering, in particular the techniques of recombinant DNA. It seems that for many years, there was a deliberate distancing of the traditional fermentation industry and the new genetics. This distance was even reflected at the beginning of the EFB (see M. Cantley's study, *op.cit.*, Section 6.1).

conglomerate had retitled its publication *Journal of Applied Chemistry* the *Journal of Applied Chemistry and Biotechnology*. British firms such as Beecham played a determining role in the development of antibiotics. Furthermore, British firms were among the world leaders in the attempt to produce new protein foods. Institutions such as the Institut für Garungsgewerbe and the Gesellschaft für Biotechnologische Forschung in Germany and the British Microbiological Research Establishment, respectively, held different pragmatic views of biotechnology. While the Germans considered it the basis of future development, for the British it was an economic inheritance.

However, more notable differences between them became clear. While in Germany the government reacted rapidly, in Britain nothing happened until 1979. The consequence of this delay was that the British government was able to take into account the progress in genetic techniques (in particular in techniques of DNA modification, which from 1975 onwards deeply influenced the concept of biotechnology) although most of the government's programs were developed from ideas that dated from the previous decade.

Britain's biotechnological developments were undoubtedly influenced by its heritage of empire and world power, that left it with sizable government research institutes and large oil and chemical companies seeking diversification. These developments were characterized by the national obsession with economic decline and counterbalanced with hopes of technological greatness.

Biotechnology, or the Hope for an Economic Rebirth

Compared to Germany and Japan, the British were less ecologically concerned, but the strong tradition of biology research, its history, and hopes for its industrial potential led biotechnology to be considered a means to industrial and economic rebirth. From the start, British enzyme technology was developed at the industrial level. At this time, BP, Shell, ICI and Rank Hovis MacDougall were expressing their interest in single cell proteins [20],

---

[20] Single Cell Proteins (SCPs) term invented at MIT by Scrimshaw: one might consult R.I. Mateles and S.R. Tannenbaum's *Single Cell Protein*, MIT Press, Cambridge, Mass. 1968.

indicating that Britain had a key role in what then looked like a central market. There was also fundamental expertise in the *avant-garde* field of bioreactor use in the continuous fermentation technique developed at Porton Down. Behind the scenes, the country's excellence in molecular biology fed the hopes of the new generation. However, although the molecular biology carried out in Britain since the 1950s was the best in the world, it remained stuck at the laboratory level (except perhaps in some areas of medicine: virology, cancerology, hematology, prenatal diagnosis of hereditary disease and immunotherapy). British industry, whose interests traditionally lay in chemistry, approached the techniques of biological engineering with caution. Despite the fact that it had won the war, the industries of Britain in the postwar world suffered from a certain malaise which caused some failures. For biotechnology, the failure was a lot less serious than for others.

An Old Tradition

Chemical engineering developed in Britain at a speed second only to that of the United States, allowing the rapid establishment of an antibiotics industry. Britain also had a strong tradition of applied microbiology. The postwar decades witnessed enthusiasm for the merger of these two fields even though the opportunity was often lost whenever it was put forward by the Microbiological Research Establishment. Many biochemical engineering departments were set up during the 1950s. In 1958, T.K. Walkers, a pupil of Weizmann's, renamed Weizmann's flagging department at Manchester Biochemical Engineering. This success was countered by the failure of the government chemical research team (another part of the Weizmann inheritance) which was restructured and had its budget reduced in the wake of government spending cuts.

As a result, by the mid-1960s, the British had a variety of institutions that dealt with biochemical engineering. Most of them had developed from departments that trained specialized personnel and had neither the size nor stature to take a major role like that of the Institut für Garungsgewerbe or the new GBF.

Demands for an Impossible National Institute

There was, of course, a candidate, the Microbiological Research Establishment at Porton Down, whose military activities had been greatly reduced. The successful test of the hydrogen bomb and a particular defense report suggested that, in future, it would be less necessary to ensure that Britain was the leader in bacteriological weapons research. In 1959, the disappearance of its founding department, the Resources Ministry, gave the Microbiological Research Establishment the opportunity to lessen the military element of the laboratories' activities and, finally, set up a National Institute of Applied Microbiology. This plan was not to succeed, mainly because of the Cuban missile crisis. Nevertheless, Porton Down provided not only an idea, but also researchers who dispersed the discipline of applied microbiology amongst British universities.

When the Medical Research Council (MRC) closed its penicillin production center in Cleveden in 1962, Porton Down inherited the job of providing specialized chemical products for research. That same year, the British decided not to set up a national institute, but five years later, Harold Hartley launched a new initiative. He was confident that what was still known as biochemical engineering was the way of the future and had been for fifteen years, but he conceded that he had, perhaps, overestimated the role of the chemical engineer and underestimated that of the biologist. Porton Down may have been mentioned again as providing the right mix of biology and engineering, but the idea was not taken up, and no professor was so influential that his laboratory could provide a plausible alternative.

Ernest Chain, who joined Hartley in his fight, managed to persuade the Science Research Council to set aside a special budget for applied microbiology. This also failed, upon the closure of the program after four years, due to the poor quality of the projects, a reflection of the dominance of biochemists and molecular biologists.

The Spinks Report

The weakness of biochemical engineering in Britain was highlighted in a report in 1976. At the time, there were only four world class departments in Britain: at the University College of London, and in Birmingham, Manchester and Swansea. In 1978, the Advisory Board for the Research Councils, (ABRC) along with the Royal Society and the Advisory Council for Applied Research and Development (ARCAD), sponsored a working party to define, by the end of the year, which lines of research in biotechnology were the most promising and to propose a series of concerted actions to ease the so-called "transfer stages" towards application.

Shirley Williams, the Secretary of State at the time, set up the inquiry to review the following subjects: photosynthesis, genetic recombination, biomass fermentation and cellular fusion. The applications particularly targeted pharmaceutical products, recycling waste, food technologies and the production of oil. The inquiry concluded in 1979 with the Spinks report[21], so named after the chairman of the working party. (Since 1975, there had been a series of long-term actions in agro-food research and development called the "Food from our own resources" program). The themes that the working party considered a priority were the following:

- The production of fertilizers through atmospheric nitrogen fixation;
- The production of proteins of animal and vegetable origin;
- The production of hydrogen from plants, using plants selected with the aid of genetics.

The Spinks report can be considered a successor to the DECHEMA study published seven years earlier; it shared the same general approach. Biotechnology was defined as "the application of biological organisms, systems or processes to industries of production or services". This definition was similar to Rehm's definition of biotechnology, a definition which had already been explored in Britain in a study by three microbiologists, Alan Bull who had worked for Glaxo, Derek Ellwood, a Porton Down scientist and Colin Ratledge of the University of Hull, an expert on developing

---

[21] *Advisory Council for Applied Research and Development, Advisory Board for the Research Councils and the Royal Society, Biotechnology: Report of a joint working party,* HMSO, London, 1980.

countries [22]. The 1979 Spinks report differed from the German approach in its allocation of greater importance to the new developments in genetic engineering, in particular, the new recombinant DNA techniques [23]. As in Germany, these new technical advances had not been considered separately. Although they were already very important, they were still considered just another technology among all the other biotechnologies, including enzyme engineering and bioreactors, which had all been of fundamental importance in the prevailing political and economic context loaded with energy problems and ecological concerns. The key fields identified concerned the application of genetic and enzyme engineering, monoclonal antibodies, immunoglobins, waste treatment and more generally, bioremediation, the artificial production of food biomass and the production of biofuels from biomass.

Barely out of a serious recession, but with its considerable scientific expertise intact, Britain realized the importance of biotechnology for the future. By the end of the 1970s, it also looked as if Britain's industrial and economic biological revolution might well succeed. Britain was to benefit from then on from simple systems that improved interaction between research and industry and which allowed frequent personnel exchanges. There were many forms of cooperation, since the big industrial companies (BP, ICI, Shell, IBM, etc). willingly offered partnership and financial support to the universities, and many university researchers acted as consultants for them. What one might call the biotechnology reflex was not fully developed in this country, which boasted a large amount of expertise, but the British model was, in its turn, to become very influential.

The Influence of the British Revolution

Early in the 1980s, the OECD, based in Paris, was trying to pinpoint which technologies were going to have a fundamental impact on the world. Their

---

[22] "The changing scene in microbial technology", in *Microbial technology, Current state, future, prospects* ed. A.T. Bull, D.C. Ellwood and C. Ratledge, Society of General Microbiology, Symposium no. 29, Cambridge University Press for the Society for General Microbiology, 1979.

[23] On the advent of these techniques, please refer to Chapter 3 "The foundations of the heralded revolution".

analyst, Salomon Wald, thought that biotechnology should be one of the many selected. Three names had been recommended to him for the production of a report: the British strategists, Alan Bull, of University of Kent, T.G. Holt at Central London Polytechnic, and Malcolm Lilly of the biochemical engineering unit of the University College of London. He put them together in a team. Their report[24] was published in 1982. It reflected a half-century of European and American thinking and became a well-quoted classic. Biotechnology was defined as "the application of scientific and engineering principles to the processing of materials by biological agents to provide goods and services".

Parallel discussions on the eminent role that science, and especially biotechnology, would have to play in society's evolution, took place in other European countries such as the Netherlands and France. These discussions focused in particular on its effect on medicine, pharmacy, chemistry, agriculture, food, energy production and the protection of the environment.

2.5 The French Reaction

There was general public interest in France in the life sciences and their impact. In accordance with this, the president, Valéry Giscard d'Estaing was concerned that the status of French science should be reinforced. He noted that just as the physical sciences were helping to form social and industrial organization, the life sciences would also be required to exert a determining influence. France, he saw, was amongst those countries that had the calling and the means to contribute to the increase of knowledge and master the modifications that were needed for the bioindustrial revolution. He asked three eminent French biologists, François Gros, François Jacob and Pierre Royer to study the consequences that the discoveries of modern biology might have on the organization and workings of society, choose the biotechnological applications which were most useful for human progress and well-being and propose means for the establishment of these applications.

---

[24] A. Bull, T.G. Holt and M. Lilly, *Biotechnology: International Trends and Perspectives*, OECD, Paris, 1982.

The Report on Life Sciences and Society Presented to the President of the French Republic

The report, published in 1979[25], was a clear successor to the German and British reports. Although it took into account genetic engineering and its perspectives (prenatal diagnosis through gene analysis, production of proteins or peptides of biological interest and the transformation of physiological properties of plants), the biological engineering themes identified in the summary and recommendations of the report were, nevertheless, along the lines of the biotechnological philosophy developed in the previous decades:

- Enzyme engineering and bioreactors;
- Applied microbiology (bioremediation, biometallurgy, chemical industry, nitrogen fixation);
- Bioenergy aspects (bioconversion and biomass, biofuels);
- The artificial production of food biomasses;
- Nitrogen fixation.

Through these themes the French showed their preoccupation with a revitalization of the chemical and agro-food industries (due to economic worries), their intention of preparing an alternative to oil consumption and their concern for improvement and respect of the French environment whilst allowing France to continue its pattern of consumerism.

As in many other countries, the idea of biotechnology came to include many strategies and technologies for resolving the main socio-economic difficulties of our time, in particular, the food and energy crises and economic development. This report was preceded by two other reports[26] (published in conjunction with it and as microfiche by *La documentation française*). It was followed by another report targeting the fundamental priorities for France and yet a further report on the bureaucratic implementation that would be

---

[25] François Gros, François Jacob and Pierre Royer, *Sciences de la vie et société, Rapport présenté à Monsieur le Président de la République V.G. d'Estaing*, La documentation française, Paris, 1979.

[26] Joël de Rosnay, *Biotechnologies et Bio-industries*, Jean-Paul Aubert, *Microbiologie générale et appliquée*.

required for its execution. It contributed to the creation of a national biotechnology program, the Essor biotechnology program, launched in 1982. Essor was endowed with a substantial budget and was intended to be a four-year mobilization strategy.

In addition to these national reflections and initiatives[27], during the 1970s, there was also the emergence of an important European dimension.

2.6 The European Community and Biotechnology — The Emergence of the First European Biotechnology Programs

In the 1970s, the European Commission[28] was also examining the requirements that profound technological change would have of society. This was a new train of thought for the organization, which until now, had mainly concerned itself with customs tariffs, cross-border protectionism and increasingly expensive agricultural policies. It was certain that through its subsidiary organizations, the European Community could play a very important role as a catalyst for the launch of special R&D programs. Prompted by some of its members, European scientists and civil servants of the Commission of the European Communities, who were responsible for biology, radioprotection and medical research, had undertaken a series of studies that led to the design and adoption of the multi-year Community Biomolecular Engineering Program (BEP).

Throughout the 1970s and into the first years of the 1980s, the political objectives of research and development tended to be rather *ad hoc* and unsystematic. The integration of Euratom's research activity, including that section which led to the various establishments of the Joint Research Center, and the political repercussions of the 1974 oil crisis, led to an initial predominance of research geared towards energy.

---

[27] We should note the allocation of 4.7 billion FF in Italy for biotechnology projects linking the private and public sectors, 3.3 billion FF over five years in West Germany, 87 million FF in the Netherlands for biotechnology projects on the environment... From the middle of the 1980's onwards, many European countries were investing in biotechnology.

[28] For a diagram of the various administrative divisions of the European Commission see Fig. 1.

Rachel Carson's *Silent Spring* (1962), the Stockholm conference of 1972 and the apocalyptic predictions of the successive publications from the Club of Rome (including *Limits to Growth* and *Mankind at the Turning Point*) brought environmental research back into fashion.

Throughout the 1970s and the early 1980s, the relative weakness of European industrial performance compared to that of the United States and Japan became more and more evident. At this point, the Commission did not have any biologists, except (quite by chance) those working in the Euratom program on the effects of radiation on living tissue and man. These included R. Appleyard, F. van Hoeck, A. Bertinchampts, D. de Nettancourt and A. Goffeau. In fact, the biologists had been employed to bring some scientific opinion in from external laboratories that could give impartial advice on certain independent actions.

As the program for the use of radiation in agriculture seemed to be failing, the big research center on atomic energy use in agriculture, in Wageningen (the Netherlands), set up to bring together the scientists united under Euratom, was gradually winding down its activities. Some of these scientists accepted an offer to move to the European Commission in Brussels to take on new administrative tasks, and they included D. de Nettancourt and A. Goffeau.

Dreux De Nettancourt had agreed to move from Wageningen to Brussels to work on contract management with van Hoeck during 1962–63. He began to worry that there was no future in radiobiology and no community program on the basic biology of species important to man. D. de Nettancourt persuaded van Hoeck of the need for a European initiative in this area.

The Biologists, or Recognizing a Need for New European Action

Dreux De Nettancourt's idea, at this point, was to study cultivated plants, in particular, the biology of their reproduction, with the hope of finding applications and creating an applied program on plants, a program outside the aegis of the Euratom treaty. At the time, the creation of such a program at the Commission was not an easy task. He had to engender innovative

94 From Biotechnology to Genomes: A Meaning for the Double Helix

opinion, have it accepted by all the national delegates and liberate some funding. Geneticist de Nettancourt convinced Goffeau, then delegated to the Université catholique de Louvain, where he was studying the effects of radiation on cellular membrane, and Etienne Magnien, a plant biologist working on cell culture, to join him in this new program and bring the basic scientific competence required.

In 1975, D. de Nettancourt, A. Goffeau and F. van Hoeck drew up the first draft document for a European biotechnology program. As suggested by A. Goffeau, it had two sections, one on enzyme engineering and another on genetic engineering, both of which were very *avant-garde*. To appreciate the originality and impact of this document, it should be noted that the first genetic transformation of bacteria had only taken place two years previously, in 1973. Once the report was drafted, the three men tried to convince the commission hierarchy, the ministers of Europe, scientists and industrialists that they should receive support to create this biotechnology program on enzyme and genetic engineering.

Proposal for a First Program for "the Optimum Exploitation of the Fundamentals of the New Biology"

This draft document was followed by a more detailed document, in 1977 [29], entitled *Applied molecular and cellular biology (Background note on a possible action of the European Communities for the optimal exploitation of the fundamentals of the new biology)*. The document took into account the general objectives [30] of community research and technology policy defined

---

[29] D. de Nettancourt, A. Goffeau and F. van Hoeck, *Applied molecular and cellular biology (Background note on a possible action of the European Communities for the optimal exploitation of the fundamentals of the new biology)*, Report DGXII, Commission of the European Communities, XII/207/77-E, 15 June 1977.

[30] Com. Doc. (77) 283 Final, *The common Policy in the field of science and technology*, 30 June 1977.

by the Commission[31] as being:

- The long term provision of resources (raw materials, energy, agriculture and water);
- The promotion of internationally competitive economic development;
- The improvement of living and working conditions;
- The protection of nature and the environment.

The document's objectives were to evaluate the potential of modern biology to promote this policy and sketch the general lines of a community program in biology, molecular biology and cellular biology, a program that, through a better liaison between fundamental and applied research, could later lead to new possibilities in agriculture, medicine and industry such as:

- The development of improved products;
- The determination of more efficient means of production;
- The reduction of energy consumption and commercial deficit;
- The reduction of the quantity of waste;
- The improvement and reduction of the cost of medical care.

Although the biological revolution so long heralded by the promoters of biotechnology had not as yet provided applications for everyday life, the document defined the following areas[32] as having important implications:

- The development of enzyme technology; i.e., the creation of new bio-industrial methods using biological entities in bioreactors;

---

[31] We should remember that at this point and until the 1989 implementation of the Single European Act, there was no specific legal basis in the founding treaties of the European Community for research and development programs other than a reference to the co-ordination of agricultural research in Article 41 of the Treaty of Rome founding the European Economic Community and the research objectives under the Euratom Treaty. Most of the R&D programs proposed by the commission during the 1970s and 1980s had to use the following very general provision buried in Article 235 as a legal basis: "If action by the community should prove necessary to attain, in the course of the operation of the common market, one of the objectives of the community and this treaty has not provided the necessary powers, the council shall, acting unanimously on a proposal from the commission and after consulting the European Parliament, take the appropriate measures." This inconveniently required unanimous approval from the member states.

[32] D. de Nettancourt, A. Goffeau and F. van Hoeck, *op.cit.*, p.5.

96 *From Biotechnology to Genomes: A Meaning for the Double Helix*

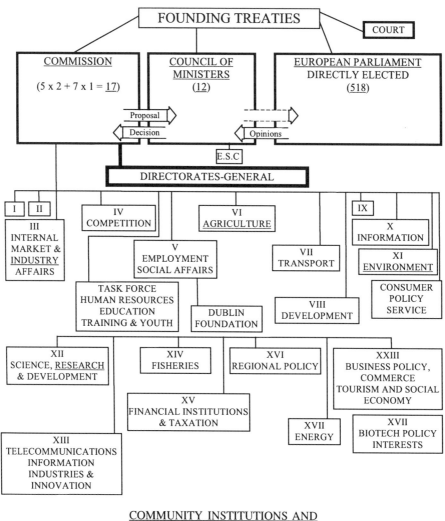

COMMUNITY INSTITUTIONS AND
THE D.G.s. OF THE COMMISSION

Fig. 1 Community institutions and the D.G.s of the commission (source: M. Cantley, 1990)

- The recently acquired human ability to break biological insulation and transfer biological information between different organisms (genetic engineering);
- The development of molecular approaches to the detection and therapy of pathogenic conditions in living cells (molecular pathology).

According to the authors, the importance of these three areas, mentioned throughout the document, was underlined by two facts. On the one hand, they felt there was a convergence, a real need and a possibility that through the use of applied molecular biology, some of the crucial problems of humanity could be solved. On the other hand, they thought that research into molecular and cellular biology could make a significant contribution to the improvement of the quality of life in European countries.

Though the importance of enzyme engineering and physico-chemical (or biochemical) techniques had long been understood, the originality of the document lay fully in the realization of the importance of genetic engineering and its perspectives:

> "Genetic engineering can be expected to lead, in the very long run, to the creation of new types of microorganisms and of higher plants completely adapted to the requirements of agriculture and of industry. The development of compact and specific molecular converters able to produce, on a large scale and with a minimum of energy losses and of insults to the environment, the range of complex products needed by man should allow the progressive replacement of the presently polluting and high-energy consuming man-made machines. In medicine, the precise knowledge of structures and mechanisms operating at the molecular level should permit, as can be illustrated, for instance, in the research presently carried out for detecting prenatal anomalies or for treating certain forms of inherited anemias, new approaches to early diagnoses and therapies intimately adapted to the molecular basis of the disease or syndrome considered"[33].

---

[33] D. de Nettancourt, A. Goffeau and F. van Hoeck, *Ibid.*, pp. 5–6.

In order that this potential be realized, there had to be a different structure to European research, a structure that implied large multidisciplinary teams and laboratories with the organization needed to coordinate action carefully. A transformation of research structure of this nature was not, and still is not, in the power of small institutions, nor is it always compatible with the traditional isolation and compartmentalization of some of the more important institutes. Furthermore, institutes alone (or in certain cases the smaller countries) just did not have the level of funding needed to initiate and carry through biological research in the three promising domains, in particular, applied molecular biology research. This meant a new coordination of institutes and laboratories' objectives and resources would be needed.

Following their analysis of the various ongoing efforts outside Europe [34], the document's authors pointed out the need for a community R&D program in molecular and cellular biology. This need was particularly clear since, according to the authors, "no attempts have been made to promote the new biological revolution" [35].

It is, of course, true that a certain number of programs had already been developed by the Commission to coordinate and stimulate research as directly applied to agriculture, medicine, life in society, radiation protection, the conservation of the environment and the use of solar power. These activities, although they all had some biological implications and components [36], were essentially carried out as applied research motivated by practical objectives recognized as directly important for the member states. The member states were mainly confronted with the following essential problems [37]:

---

[34] These were the efforts of organizations such as the NIH and the NSF in the United States, the Tallinn Polytechnical Institute in the USSR and the NCRD in Israel, which had launched, or were about to launch, sizable integrated research programs. Particular attention was paid to Japan, where the discoveries in molecular and cell biology had been considered so important for agriculture, industry and medicine that an Office for Life Science Promotion had been set up to plan life science research programs to answer the country's needs and study how to set up a central institute provisionally named the Life Science Research Promotion Center.

[35] D. de Nettancourt, A. Goffeau and F. van Hoeck, *op.cit.*, p. 8.

[36] For instance, on the contribution of research on DNA repair, actively conducted in the Community Radiation Program, to the understanding of gene replication, gene function and gene structure.

[37] D. de Nettancourt, A. Goffeau and F. van Hoeck, *Ibid*, p. 9.

- Food production;
- Energy supply;
- Conservation of the environment;
- Health and man's adaptation to modern society.

These problems corresponded exactly to those that had encouraged certain industrialized nations outside the European Community to set up large applied biology research programs and pipped the nine member states (this was still in 1979) to the post in some of the fields of activity listed in the report. The authors underlined Europe's lagging behind the other nations, focussing on its sizable deficit in trade and comparing the number of patents on biotechnology production methods and the capital allotted. The headway gained by countries such as the United States and Japan was seen as a sign of the deficits. Europe also had a deficit in the level of its production of essential amino acids and proteins for animal feed. By pinpointing these shortcomings, the authors highlighted the urgent need for the member states to pool their resources and define a common biotechnology R&D policy through the development of integrated, coordinated community-wide actions bringing together their contributions to, and participation in, the so-called biological revolution.

This need for a common community R&D policy in biotechnology (genetic engineering, enzyme engineering and molecular pathology) was more than justified because of the triple requirements that:

- European potential and competence be mobilized;
- Activity be planned and rigorously coordinated;
- Support be granted to the sectoral policies of the commission (in particular, the commission's objectives on the optimization of food production, health protection and the improvement of the quality of life).

The authors underlined that:

"Considerable changes in the quality of production and exploitation methods could result from the creation, through genetic engineering or through the domestication of reproductive systems, of novel hybrids combining the genetic features for photo-synthesis, nitrogen uptake, disease resistance or breeding behavior of parental lines from different species, genera or families. Similarly, a completely new approach to health protection may become possible the day molecular biology succeeds to unravel primary pathogenic events in man or to define the principles for manufacturing therapeutically important human proteins in cultures of microorganisms carrying human genes. In the sector dealing with the improvement of life in society and the protection of our environment, very important contributions could be made by the molecular and cellular biologists which may allow, simultaneously, the promotion of new types of bioindustries less detrimental to our environment, the improvement of *in vitro* methods for screening carcinogens or other industrial hazards and the development of micromethods and cytochemical techniques for the prevention and detection of genetically determined conditions"[38].

Three types of action were proposed in this document:

– The development, in the chemical industry, of new bioindustrial methods through the use of biological entities; in particular, this meant focusing on the biosynthetic possibilities that should become available from the use of multi-enzyme complexes or immobilized organelles on a solid support. The goal was to take advantage of the enzyme reactions catalyzed by these systems to synthesize newly-developed products. Another aspect considered was the use of bioreactors in detoxification or energy production processes.
– The transfer of genes to important agricultural species and microorganisms. The aim was to exploit the new techniques of genetic

---

[38] D. de Nettancourt, A. Goffeau and F. van Hoeck, *op.cit.* p. 11.

manipulation to produce new sorts of plants containing genes or groups of genes deliberately selected and plants from other very distant species.
- The clarification of the molecular basis of pathology through the understanding of the intimate mechanisms of disease, which would allow the rational development of preventive or curative treatments through the use of the techniques and knowledge of modern biology.

For the organization and execution of the program, the authors favored the establishment of new research structures (one or more per theme) or preferably, if possible, the conversion of national centers to ensure easy permanent communication between fundamental and applied researchers. These research actions would be better organized at a multinational level. They were to be multidisciplinary and imply a network of organizationally and geographically dispersed groups regulated from a central scientific institution which, for psychological reasons, would have to be highly respected for its science. The project leader would have to come from this institute too.

The societal needs that the authors used to justify the program's creation (food production, energy supply, the conservation of the environment, health and man's adaptation to modern society), are recurrent themes in the history of biotechnology and had already been widely mentioned in reports of the time on the importance of biotechnological science. However, the importance the European project gave to genetic engineering truly put them in advance of national biotechnology policy.

The first document, *Applied molecular and cellular biology (Background note on a possible action of the European Communities for the optimal exploitation of the fundamentals of the new biology)*, took three years to prepare. It was met with great skepticism, especially from the national delegations, some of which (in particular, the French delegation) resented this European initiative, which they saw as a threat to their own embryonic national program. The document was delayed but not dismissed, however.

The First European Biotechnology Program (1982 to 1986) — The Biomolecular Engineering Program

Although its final result was the approval, in 1981, of the first biotechnology program[39], the Biomolecular Engineering Program (BEP), its more immediate consequence was the production of three fundamental detailed reports by D. Thomas[40], of Compiègne University, on enzyme engineering, A. Rörsch[41] on genetic engineering and Christian de Duve[42] on molecular pathology.

The initial 1975 document, these three reports and the risk evaluation work (inspired by the 1979 studies by Ken Sargeant and Charles Evans[43]), the strategic benefit of which could not be ignored given ecological concerns in the member states, in particular Britain and Germany, led to the Biomolecular Engineering Program. The BEP was finally adopted by the council on 7 December 1981 with a budget of 15 MECU for the period 1982–1986 (Fig. 2).

This program covered the following areas:

At the level of contextual measures:

- Bioinformatics: the interface between biotechnology and information technology (data input, databases...);
- The collection of biological material (improvement and integration of existing collections, perfecting techniques).

---

[39] On the next page, you will find the commission's actions and program on biotechnology.

[40] D. Thomas, *Production of biological catalysts, stabilisation and exploitation*, EUR 6079, Office for Official Publications of the European Communities, Luxemburg, 1978.

[41] A. Rörsch, *Genetic manipulations in applied biology: a study of the necessity, content and management principles of a possible action*, EUR 6078, Office for Official Publications of the European Communities, Luxemburg, 1979.

[42] C. de Duve, *Cellular and molecular biology of the pathological state*, EUR 6348, Office for Official Publications of the European Communities, Luxemburg, 1979.

[43] K. Sargeant and C.G.T. Evans, *Hazards involved in the industrial use of micro-organisms; a study of the necessity, content and management principles of a possibly community action*, European Commission EUR 6349 EN, 1979.

2 Political Interpretations of Biotechnology ... 103

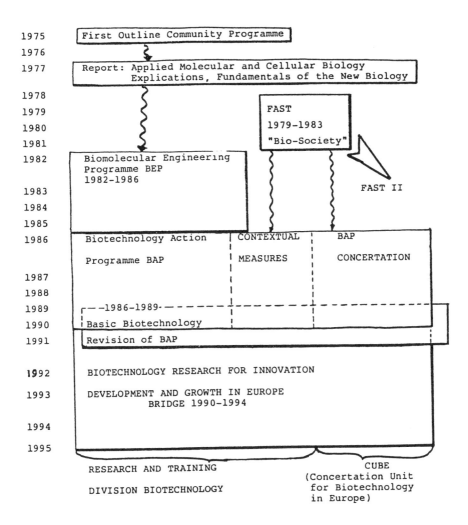

Fig. 2 Community action in the field of biotechnology R&D — Combined organization chart and chronology (source: M. Cantley)

At the fundamental biotechnology level:

- Enzyme engineering: second generation bioreactors, the improvement of immobilized protein stability, the development of new protein activities;
- Genetic engineering: development of production methods for harmless high added-value substances for livestock breeding, veterinary medicine and the agro-food industries. Development of methods aiming for the degradation of plant material, especially lignocellulose, into useful products, and the extension of methods of genetic engineering to plants and microorganisms of agricultural importance;
- Evaluation of possible risks: the development of contamination detection methods and methods for evaluating possible risk associated with the application of biomolecular engineering.

Despite its restricted scope and its very limited budget, which was symptomatic of the commission's prudence in this venture into the field of biotechnology, the first program was undeniably successful. Three main positive characteristics emerged from the establishment of BEP[44]:

- A sizable response from scientific circles[45];
- A high level of transnational cooperation[46];
- The quality and quantity of results allowed the decrease of several barriers between fundamental and applied research.

In addition to these three successes, there was also the training part of the program, which all the national delegations considered fundamentally important and an excellent tool for the transfer of technologies between

---

[44] D. de Nettancourt, quoted in the *Short draft report of the meeting of the CGC, Biotechnology*, Brussels, 7 March 1986, 002/IN2, 12 March 1986, European Commission, Brussels.

[45] 293 research proposals, leading to 103 shared cost contracts, 188 applications for training of which 91 were granted, and the organization of 20 scientific meetings.

[46] 53 formal cooperative efforts established between laboratories in different member states which involved the exchange of equipment and personnel and coordinated experiments being carried out by the contractors, 97 visits between sites of contracting laboratories, 15 meetings of various types and sizes and other international agreements.

countries, the promotion of R&D infrastructure and the increase of general competence in multidisciplinary teams. The level of the scientific output can be seen from the final report of the program[47]. Several world firsts had been achieved such as the cloning of genes from *Streptococcus*, a bacteria used in the cheese industry, and genetic transformations in monocot species from a vector taken from *Agrobacterium*. The characterization and isolation of more than twenty plant genes of major agricultural importance were also original contributions. In total 355 scientific papers were published and 13 patents were granted as a result of the programme.

A comparative bibliometric analysis of publications related to research carried out at community and national levels shows that publications from BEP had a higher number of quotations than average[48] (150%, in fact). More generally, despite its small size, BEP played a catalytic role by bringing an awareness of the European dimension to the research community and through the birth of community research cooperation. The delegations' opinions of BEP[49] were very positive on the whole, but they did point out gaps and weaknesses:

- The weak level of industrial participation. It was, however, recognized that the precompetitive nature of the program, focused as it was on agriculture and not endowed with much of a budget, was not really suitable for funding industry;
- The lack of faith of certain national decision-making centers when the program was first launched;
- The insufficient number of proposals for applications of risk evaluation and the limited results in this field;
- The delays of the payment procedure.

---

[47] *Biomolecular Engineering in the European Community*, E. Magnien, ed. Martinus Nijhoff, 1986.

[48] P. Cunningham, *A bibliometric study of BEP publications (Program of policy research in engineering, science and technology)*, University of Manchester, quoted by D. de Nettancourt "Biotechnologies communautaires: la dixième année", *Biofutur,* April 1991, p. 20.

[49] Meeting of the CGC-Biotechnology, Brussels, 7 March 1986, point 3: "Preliminary evaluation of the achievements of the Biomolecular Engineering Program (BEP) (doc. CGC-IV-86/2). Please refer to the CGC-Biotechnology meeting of 27 November 1986, point 3 of the agenda, "BEP: evaluation of the final report by national delegations", opinions expressed by individual national delegations.

The delegations were aware that the program's impact was limited by the twin factors of time and money. The time factor was due to the particularly unfavorable situation at the program's adoption by the council in 1981 (distrust from many of the national circles, unfinished but apparently competitive national programs, lack of European biotechnology networks, lack of interest from industry, gaps in public information, etc.). This situation was changed by a gradual transition in opinion towards the importance of biotechnology. The financial factor was due to the fact that the contracts were granted insufficient funding and the commission's participation was too small in shared cost activities, most of which were co-funded by other national or international sources.

Despite these shortcomings, the success of this first program opened the way for a reinforcement of community biotechnology. This resulted in the adoption, in March 1985, of a new biotechnology R&D program, the Biotechnology Action Program (BAP), for the period 1984–1989.

These achievements show how much the political context for research within the community had developed since the end of the 1960s.

A Common Research Policy, the Europe+30 Group and the Fast Program

Indeed, in 1974, after many studies and a lot of debate, the community decided to lean towards a common research policy. In a meeting in Paris, on 14 January 1974, the council adopted four resolutions on the development of a common science and technology policy, including the coordination of national policy and the coordinated implementation of projects of community-wide interest[50].

The German, Ralf Dahrendorf, was the first Commissioner for Community Policy on Research and Development (1973–1976). In 1974, he set up a group of experts called "Europe+30", under the chairmanship of

---

[50] European Commission, "Common policy for science and technology", *Bulletin of the European Communities*, supplement 3/77, 1977.

MEP Wayland Young (later Lord Kennet), to consider in what way the long-term (30-year) perspectives could influence the community's choice of R&D targets. At this time, long term studies and predictions tended to be handled by the Rand Corporation, the Hudson Institute and the Club of Rome, famous for its aforementioned predictions on the limits of growth.

The "Europe+30" report[51] was ambitious. It proposed the creation of a new community institution for prospective studies. The final implementation was more modest. It consisted of the council adoption, in 1978, of a community program of Forecasting and Assessment in Science and Technology (FAST) with a budget of 4.4 MECU for 1979–1983. DG XII took the new unit under its wing and contributed to the creation of a new biotechnology initiative.

FAST's mandate was very broad. It included the highlighting of potential perspectives, problems and conflicts that could affect community research and the definition of alternative directions for community R&D action that could resolve or attain them[52]. With only a few staff (six people, all university graduates and exceptionally highly qualified), the group split up its field of study into three themes:

- The relations between technology, work and employment;
- The information society;
- The bio-society.

The theme of bio-society was taken up by two researchers, neither of whom was a biologist: Mark Cantley, a Scottish mathematician who had worked on global systems at the International Institute for Applied Systems Analysis in Vienna, and Ken Sargeant, an English chemist from the Microbiological Research Establishment at Porton Down, after its transferal to the jurisdiction of the Department of Health. Both Cantley and Sargeant thought highly of biotechnology's potential.

---

[51] Wayland Kennet, *The Futures of Europe*, Cambridge University Press, 1976.

[52] *Council Decision of 25 July 1978 on a research program of the European Economic Community on forecasting and assessment in the field of science and technology*, Official Journal of the European Communities, no. L225/40 16 August 1978.

108 *From Biotechnology to Genomes: A Meaning for the Double Helix*

The work on bio-society included a series of long-term perspective studies, colloquia and consultations, particularly with the European Federation of Biotechnology. The results were used in the bio-society section of the first general report of the FAST program, in September 1982 [53]. With very little modification, this report became the first version of a "Community Strategy for Biotechnology in Europe [54]". Reports with that title were published both by DECHEMA (on behalf of the EFB) [55], in 1983, and by the commission, as a FAST Occasional Paper, in March 1983 [56]. The FAST report, including the sections on bio-society, was to be published later under different names [57].

From 1981 to 1984, the commissioner for the research and development Directorate-General (DG XII) and industrial affairs (DG III) was the Vice-President Etienne Davignon of Belgium. In 1982, he was the political force behind the establishment of ESPRIT, a strategic European R&D program on information technology, which, with its 750 MECU budget, was a very large program for the time. When, in February 1983, during his speech on the annual program to the European Parliament, the commission's president Gaston Thorn announced that the commission would follow the ESPRIT approach for biotechnology, his declaration was considered highly significant [58].

---

[53] FAST, European Commission, *The FAST program: Vol. 1. Results and Recommendations*, 1982.

[54] FAST, *A Community Strategy for Biotechnology in Europe*, FAST Occasional Papers, no. 62, European Commission, 1983.

[55] DECHEMA, *Biotechnology in Europe — A Community Strategy for European Biotechnology: Report to the FAST bio-society project of the Commission of the European Communities*, DECHEMA, Frankfurt, on behalf of the European Federation of Biotechnology, 1983.

[56] FAST, *A Community Strategy for Biotechnology in Europe*, FAST Occasional Papers, no. 62, European Commission, 1983.

[57] FAST, Commission of the European Communities, *Europe 1995: Mutations technologiques et enjeux sociaux*, rapport FAST, Futurible, Paris, 1983.

FAST, European Commission, *Eurofutures: the Challenges of Innovation* (the FAST report), Butterworths in association with the Journal Futures, London, 1984.

FAST-Gruppe, *Kommission der europäischen Gemeinschaften, Die Zukunft Europas: Gestaltung durch innovation*, Springer-Verlag, Berlin, 1987.

[58] Quoted by Mark Cantley, *op. cit.*

As already mentioned, there was a lack of any formal commitment in community R&D policy until 1974. By the beginning of the 1980s, the relative weakness of industrial performance compared to Japan and the United States had become more and more obvious.

The Adoption of the First Framework Program

As Commissioner for Research and Industry, Davignon's reaction to the unsystematic approach to community R&D policy was to propose the multi-annual systematic Framework Program, with specifically mentioned targets for the R&D programs. His response to the international economic challenge was to try to increase the amount of expenditure on industrial competition by the end of the first Framework Program (1984–1987). This effort was preceded by a dozen documents, each preparing a "plan by objective". Most of these were drawn up by outside experts, but the one on biotechnology was the work of the FAST group members[59] themselves.

Like the FAST report, this document on biotechnology followed Davignon's philosophy of setting up R&D activity in the context of a wider strategy and objectives. In a resolution of 25 July 1983, the council endorsed the framework program concept and approved the development of biotechnology as a part of the general aim to "promote industrial competition".

President Thorn's speech, in February 1983, was the first mention of biotechnology at a high political level in the commission, and it reflected a perspective far larger than mere genetic engineering.

The 1983 Communication of the European Commission to the Council on the Role of the European Community in Biotechnology

Based on the recommendations of the FAST report, and with contributions from DG III, particularly from its food and pharmaceuticals division, the

---

[59] M.F. Cantley, *Plan by objective: Biotechnology*, XII-37/83/EN, Commission of the European Communities, 1983.

European Commission submitted an initial Communication to the Council at the Stuttgart Summit in June, 1983: "Biotechnology, the Community's Role".[60] A parallel note reported on national support initiatives in biotechnology, in Europe, the United States and Japan[61].

This communication, like the FAST report, underlined the significant progress made in the life sciences in the previous few years and the great diversity of possible applications. It also gave a pragmatic assessment of the large scale of future markets and the power of both Japan and the United States. This was backed up by pointed quotations from contemporary reports, in particular from the American Office of Technology Assessment (OTA)[62]. The relative weakness of biotechnology, in spite of the considerable scientific resources in Europe, was attributed to the lack of a coherent R&D policy and the absence of community-wide structures. There was the BEP program, which had only started the previous year, but given its modest size, it was not considered as more than a starting point for a far more ambitious program under the Framework Program for Community R&D (1984–86). The report particularly stressed the importance of the training and mobility of scientists and technicians, communication with the public, reinforcement of fundamental biotechnology through projects half-way between research and application and improvement of research infrastructure including databases and culture collections.

Even with this support and training activity, biotechnology was seen to be incapable of developing in the community unless a favorable environment was set up to encourage it. Three factors were pinpointed as prerequisites:

– Access to raw materials of agricultural origin (under the same conditions as the Community's competitors);

---

[60] European Commission *Biotechnology, the Community's role*, Communication from the Commission to the Council, Com (83) 328, vol. 8, June 1983.

[61] National Initiatives, Com (83) 328, European Commission, 1983.

[62] OTA, *The Impacts of Applied Genetics: Applications to Microorganisms, Animals and Plants*, OTA, Washington DC, 1981, and OTA, *Commercial Biotechnology: An International Analysis*, Washington DC, 1984. Other reports from the Department of Trade, or prepared for the OSTP were also quoted.

- Adaptation of systems of industrial, commercial and intellectual property;
- Rules and regulations.

With regard to rules and regulations, the communication underlined the need for the harmonization of the internal market, particularly in the health industries, and above all, the need to prevent the appearance of specifically national norms.

The communication was welcomed by the council, and the Commission, very much encouraged, put forward a complete program of actions over the months that followed. The following communication[63], in October of that same year, noted American disbelief in European competitive strength in biotechnology and formed an indispensable strategic action to meet the challenges and opportunities of biotechnology in the following fields:

1) Research and training:
- Horizontal actions concentrated on the problem of removing the bottleneck for the application of biochemical and modern genetic methods in industry and agriculture;
- Actions concentrated on informatics infrastructure;
- Actions on logistical support for R&D in the life sciences through databases, culture collections and networks of information and communication as well as technologies of data input;
- Actions to stimulate certain specific developments in clearly-defined sectors of biotechnology which could contribute to the solution of problems in agriculture policy and the health sector.

(The Biotechnology Action Program implementing the above topics was proposed the following year).

2) Concertation of biotechnology policies:
A central activity of interservice concertation, international and between the community and the member states, with an active follow-up and an

---

[63] European Commission *Biotechnology in the Community,* Com. (83) 672, Commission of the European Communities, October 1983.

evaluation of strengths, weaknesses and opportunities as well as emerging challenges;

3) New regimes on agricultural products for industrial use:
The Commission announced its intention to propose to the council new regimes for sugar and starch for industrial use;

4) European approaches to regulations on biotechnology;

5) European approaches to intellectual property rights in biotechnology;

6) Demonstration projects to ease the transition between research development and exploitation at various levels on a commercial basis.

The Communication recognized that several Directorates General (DGs) should be involved in this six-point strategy. Following an interservice meeting in December 1983 and discussions in several Commissioners' cabinets, Davignon with the help of Commissioners Dalsager of DG VI, Agriculture, and Narjes of the other half of DG III, Internal Market, drew up a document on the external coordination of biotechnology-linked policy. This was accepted on 21 February 1984 and led to the creation of the Biotechnology Steering Committee (BSC)[64] to be chaired by the Director General of DG XII (Science, Research and Development) and open to other services as appropriate to their interests. A secretariat, the Concertation Unit for Biotechnology in Europe (CUBE) (Fig. 3) was set up, and it inherited the staff from the biotechnology section of the FAST program. It was to follow the development and evolution of biotechnology, distribute information linked to biotechnology to the services concerned, and support these services and the Biotechnology Steering Committee (BSC) in implementing the priority actions set up in the October 1983 communication.

The establishment of the BSC indicated that the Commission had recognized the need for a coordinated strategic approach. The location of its

---

[64] On the history and role of the BSC, refer to M. Cantley's *The regulation of modern biotechnology, a historical and European perspective, op.cit.*

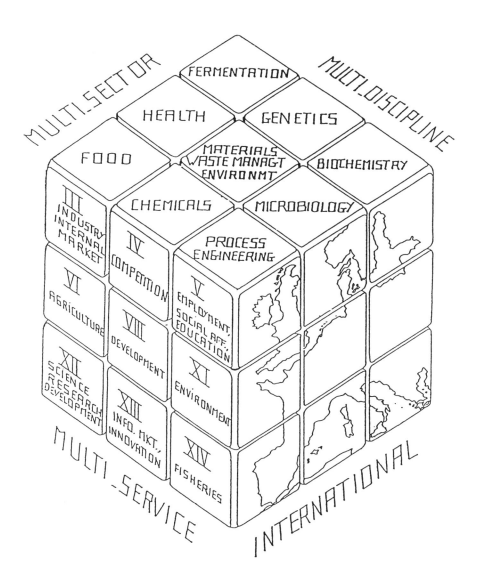

Fig. 3 The concertation unit for biotechnology in Europe (source: M. Cantley)

chairmanship and secretariat in DG XII, as underlined by Mark Cantley, was a reflection of its history and the cause of its subsequent failure to actually assume its role.

The creation of the BSC, nevertheless, represented a plausible solution in the mid-1980s. The questions and problems seemed satisfactorily shared out amongst the DGs. DG XII promoted a new awareness of the fundamental importance of biotechnology in the Commission, drawing its attention in particular to the competitive challenge set by the United States. DG XIII, responsible for the development of the information technology market, was persuaded in March 1982 to set up with DG XII a Task Force for Biotechnology Information, sustained by the Director General, Ray Appleyard.

Reports from the United States Office of Technology Assessment (OTA) and the Department of Trade underlined the commercial potential of biotechnology. A large amount of publicity accompanied the creation and expansion of the first specialist biotechnology companies such as Cetus (1971), Genentech (1976) and Biogen (1981). With similar perceptions percolating through the national administrations in Europe, a political call for a larger biotechnology R&D program began to make itself felt. After many meetings with the member states' experts, DG XII prepared a proposal for the Biotechnology Action Program (BAP)[65] which the commission presented in April 1984, and which was finally adopted in March 1985.

Biotechnology, or the Return of Hopes in Biological Engineering

While the FAST group report reflected on a 1970s concept of biotechnology and only mentioned genetic engineering in passing, biotechnology in Europe through the first BEP and BAP (for Biotechnology Action Program) programs came to mean far more than advanced zymotechnology. It has often been said that history is an eternal repetition of itself, and in the case of biotechnology, this is entirely true. In 1936, Huxley used the term biotechnology to describe his vision of human biological engineering. By

---

[65] See Chapter 5 for an analysis of the contents of this program and its development.

the end of the 1970s and early 1980s, the concepts of biological and genetic engineering had become associated with biotechnology in the context of the emergence of biological engineering of almost unlimited potential. This great potential was often quoted to politically justify increased budget requests for research in the field of genetic engineering and to support the large genome sequencing programs that, in the mid-1980s, were beginning to start up in many nations and at the European level.

It is undeniable, as we will see in the next chapter, that the techniques that appeared in the 1970s and the early 1980s offered new and revolutionary perspectives, especially those of recombinant DNA and DNA sequencing. It was hoped these would lead to the discovery of solutions to the political and economic problems of western society and also to contribute to the improvement of the quality of life of most people living outside developed countries.

Some of the hopes and prophesies on biotechnology of the first leaders of the science, Rudolf Goldscheid, Julian Huxley, John Burdon Sanderson Haldane, Patrick Geddes, Harold Hartley and Lewis Mumford come back time and again. From 1960 onwards, there is an expression of a new era in genetic engineering, an era to be typified by our ability to manipulate large quantities of biological material and elucidate the nature of biological systems.

Already by the 1960s, there had been talk of sequencing genetic material: "Once this knowledge is available, the possibility that the sequence and, hence, the genetic properties of biological systems can be altered offers itself. The commercial and social implications of the availability of methods for such alterations are tremendous"[66].

---

[66] H.M. Truchya and K.H. Keller, "Bioengineering — A new Era Beginning?", *Chemical Engineering Progress*, vol. 61, May 1965, pp. 60–62.

3
The Foundations of the Heralded Revolution

Today, the term "biotechnology" usually calls to mind techniques of genetic manipulation. This interpretation of biotechnology as being the techniques of genetic improvement has overtaken the initial interpretation originating from forty years of traditional biotechnology. It conceals the various economic, political and scientific hopes inspired by what biotechnology was prior to the genetic revolution. There were many prophecies of a new biotechnological era proclaiming a new economic age. However, despite the convictions of its champions, the revolution and the profit it promised was continuously postponed. With the appearance of new techniques of genetic manipulation in the 1970s, the situation deteriorated to such an extent that the history of biotechnology became identified with that of genetics and genetic engineering, so that the biotechnological revolution came to mean no more nor less than the genetic revolution. The hopes, the prophecies and the technological promise of that tradition were misappropriated by the initiators of the genetic revolution.

3.1 From the Frontiers of Genetics to the Birth of Molecular Biology

Classical Genetics

These days, students have relatively easy access to the basic scientific information and laboratory techniques of the "new genetics". This knowledge and methods were open to only the most advanced experts of their parents' generation.

Classical genetics is based on studies of the segregation of traits, which in turn relies on rigorous statistical analysis. It is this statistical analysis which led to our ideas on mutation frequency and genetic evolution in populations. Classical genetics prevailed until 1953 as a well-established science. Its results were already influencing other fields of knowledge, for example, in the battle against the ideas of Lamark and the pretensions of Trofim Denisovich Lyssenko to assert a clear distinction between what is inherited and what is not, between genotype and phenotype. Despite the work of Archibald Garrod, and that of George Beadle and Edward Tatum in 1941 on the fungus *Neurospora crassa*, that led in time to the union of biology and chemistry in the new fields of biochemistry and molecular genetics[1], these two sciences looked poles apart by the end of the first half of this century.

Man has long been interested in genes. In the 19th Century, for example, gardeners and animal breeders developed the art of selection, conservation and crossing and improving commercially-desirable inherited traits in plants and animals to a degree previously unattained. Crossing and selection techniques brought many improvements to root stock, but they belonged to an art which never became a science, because the principles behind successful work could not be scientifically justified. Without rigorous guidance, their power would always be limited.

In the early 19th Century, the landowner-breeders of Austro-Hungarian Moravia, who had learnt of British successes, tried not only to emulate but also improve them. In order to do so, they attempted to integrate their primitive biological engineering techniques with truly scientific study. This led them to appeal to F.C. Knapp, an abbot in Brnö. Although asking an abbot for help might seem strange to us, at the time, this was perfectly

---

[1] In the same way, when Karl Landsteiner and P. Levine discovered the blood groups A, B and O, the genetic mechanism of which would later be demonstrated by F. Bernstein, they opened the huge chapter of immunogenetics that Jean Dausset considerably enriched with his discovery of the HLA system.

logical. Knapp, as the abbot of a large monastery, controlled many farms and vast tracts of land. He was also the director of a large training center that provided him with the expertise of some of the best thinkers of the country. In addition, Knapp could provide access to the genius of Gregor Mendel.

The story of the monk Johann Gregor Mendel is so well known that it need not be more than sketched here. By patiently experimenting on the growth of various species of peas, he demonstrated that the transmission of their external characteristics from generation to generation could be predicted by simple statistical analysis. He noted that certain shape and color traits were expressed dominantly whereas others were recessive. He speculated that the form in which inheritance is transmitted is not as a global representation of the individual, nor as a series of emissions sent from all the parts of the parents' bodies, but as what he called "factors", collections of discrete entities which each governed a characteristic. Each of these "factor" entities could exist in different states, which would determine the different shapes or colours of the corresponding characteristic.

While it cannot be said that the whole science of genetics was born on the particular night in February on which Mendel revealed the results of seven or eight years of research to the Brnö Society of Natural Sciences, it must be admitted that he was the first person to discover the discrete, that is non-continuous, nature of the determinants of heredity and their independence from each other. This was at a time when chromosomes, meiosis and mitosis were still to be discovered and the old concept of heredity, being the result of the mixture of bloods, was still firmly believed. Furthermore, through Mendel's studies, the phenomena of biology suddenly acquired the rigor of mathematics. As François Jacob states in *La logique du vivant*, "It's a whole internal logic which imposes [experimental] methodology, statistical handling [of experimental results] and symbolic representation [that is to say the employment of simple symbolism] by which an unceasing dialogue between experimentation and theory becomes possible on heredity"[2].

---

[2] F. Jacob, *La Logique du vivant, (une histoire de l'hérédité)*, ed. Gallimard, 1970, p. 225.

Although it could have been expected that Mendel's revelations would overturn the practical side of biology, this was not the case. Mendel's work was completely in agreement with the thermodynamic physics of his time, but it had no influence at all on the manner in which his contemporaries did biology. It was the 20th Century that proclaimed Mendel the creator of genetics, and his first paper[3] was considered the birth certificate of genetics. The work was neglected until a British biologist, William Bateson[4], rediscovered the laws of Mendel and made them famous. It was Bateson who coined the term "genetics". At the same time, Hugo de Vries[5], C. Correns[6] and Erich von Tschermark[7] were obtaining results similar to those of Mendel.

The Search for the Essence of Heredity

At about the same time in 1869, almost by accident, a Swiss chemist by the name of Johann Friedrich Meischer[8] discovered a phosphorus-rich acid. Its position in the cell nucleus earned it the name "nucleic acid" even before its precise function was discovered. In 1888, the "filaments", distinguished by Walter Flemming in the coloured nucleus of cells with the limited

---

[3] G. Mendel, Versuche über Pflanzen-hybriden, *Verhaldung des naturforschenden Vereines in Brünn,* vol. 4., 1865, pp. 3–47, trad. fr. Recherches sur des hybrides végétaux, *Bulletin Scientifique,* vol. 41., 1907, pp. 371–419, republished in *La découverte des lois de l'hérédité, une anthologie (1862-1900),* Press Pocket, Paris, pp. 54–102.

[4] W. Bateson, *A Defence of Mendel's Principle of Heredity,* Cambridge University Press, Cambridge, 1902.

[5] H. de Vries, *Sur la loi de disjonction des hybrides,* C.R. Acad. Sc. vol. 130., Paris, 1900, pp. 845-847. H. de Vries, *Espèces et variétés,* trad. fr. ed. Alcan, Paris, 1909.

[6] C. Correns, *G. Mendel's regel über das verhalten der nachkommenschaft der rassen bastarde, Berichte der deutsche botanischen gesellschaft,* vol. 18, 1900, pp. 156–168. Translated into English as *G. Mendel's law concerning the behavior of varietal hybrids* in The Origin of Genetics: A Mendel Book Source (C. Stern adn E.R. Sherwood eds.) San Francisco, Freeman and Co, 1966, pp. 117–132.

[7] E. von Tschermark, *Ueber künstliche kreuzung bei* Pisum Sativum *Berichte deutschen botanischen gesellschaft,* vol. 18, 1900, pp. 232–139. English translation under the title of *Concerning artificial crossing in* Pisum Sativum, in *Genetics,* vol. 35, Suppl. to no. 5, *The birth of genetics,* 2nd part, 1950, pp. 42–47.

[8] F. Miescher, *Uber die chemische zusammenzetzung der Eiterzellen* in *Hoppe Seyler's Medizinisch Chemische Untershungen,* ed. August Hirschwald, Berlin, vol. 4, 1871, p. 441.

microscopy of the time, were named "chromosomes" by Wilhelm Waldeyer, and the bands across them named "chromomeres" by Edouard Balbiani and Edouard van Beneden.

At the beginning of the 20th Century, genetics really took off. A large amount of work was done on the analysis of gene transmission in sexual reproduction in plants and animals, including man. By this time, the mechanism of meiosis had been discovered. In 1902, W.J. Sutton[9] and T. Beveri[10] demonstrated the close parallel between how the determinants of heredity (to be named "genes" by Wilhelm Johanssen a few years later) were transmitted and the behavior of chromosomes as well. August Weismann[11] concluded that the "essence of heredity is the transmission of a nuclear substance of a specific molecular structure".

The research of W. Bateson[12] and R.C. Punnet[13] provided results that seemed to disprove the totally independent behavior of genes. These apparent exceptions to Mendel's laws were explained when Thomas Hunt Morgan[14], who had chosen to study the fruit fly *Drosophila*, put forward the chromosome theory of heredity.

It was still only the beginning of the last century when two fundamental notions were disentangled. The first dealt with the distribution of alleles in a population and became the base of population genetics. In 1908, Godfrey Harold Hardy[15] and Wilhem Weinberg[16] independently formulated what is

---

[9] W.J. Sutton, "On the morphology of the chromosome group of Brachystola magna", *Biol. Bull.* vol. 4, 1902, pp. 24–93.

[10] T. Beveri, *Ergebnisse über die Konstitution des Chromatischen der Zelkerns*, Fischer, Iena, 1904.

[11] A. Weismann, *Essais sur l'hérédité*, trad. fr. Paris, 1892, p. 176.

[12] W. Bateson, *Problems of Genetics*, University Press, Yale, 1892, p. 176.

[13] R.C. Punnet, *Mimicry in Butterflies*, Cambridge University Press, Cambridge, 1915.

[14] T.H. Morgan, Science, vol. 12, 1910. p. 120. T.H. Morgan, A.H. Sturtevant, H.J. Muller, C.B. Bridges, *The mechanism of mendelian heredity*, Henry Holt, New York, 1915. Trad. fr Le mécanisme de l'hérédité mendelienne, Malertin, Bruxelles, 1923. T.H. Morgan, *"Sex limited inheritance in Drosophila"* Science, vol. 32, 1910, p. 120.

[15] G.H. Hardy, Mendelian proportions in a mixed population, *Science*, vol. 28, 1908, pp. 49–50.

[16] W. Weinberg, *"Ueber den nachweiss der vererbung beim menschen"*, *Jahresschriften des vereins für vaterländische naturkunde in Württemburg*, vol. 64, 1908, pp. 368–382.

now known as the Hardy-Weinberg law. In a population where crosses are happening by chance, are in equilibrium, have no particular selections or mutations, and are high in number, the proportion of genes and genotypes is absolutely constant from one generation to another. The second notion came from British medical doctor Archibald Garrod's 1909 study[17] of the innate metabolic faults in human beings, notably alcaptonuria. During his study, Garrod concluded that what was missing in subjects with this benign affliction could be the enzyme necessary to convert homogentistic acid because homozygous people excrete it in massive quantities. Thus, it was made thoroughly clear that genes governed the synthesis and activity of enzymes. This was also proved by Lucien Cuenot at about the same time. It was only about 30 years later that this notion was to really come into its own.

In the meantime, the work of M. Ouslow and John Burdon Sanderson Haldane on the synthesis of floral pigments and Boris Ephrussi and George W. Beadle's study[18] on eye color mutants in the fruit fly showed that a particular gene controls a particular reaction. The need for closer analysis led Ephrussi[19] to yeast, which brought him to clarify the phenomena of non-chromosomal mitochondrial heredity. Beadle went on to work with Edward Tatum on the ascomycete fungus *Neurospora*[20]. From the "one gene-one reaction" principle it became easier to progress to the idea of "one gene-one enzyme" by analyzing the crossing of mutants.

Between 1945 and 1955, the notion of gene itself was modified. Both Lewis J. Stadler[21] and Guido Pontecorvo[22] had separately drawn biologists' attention to the ambiguity of attributing functionally distinct properties to

---

[17] A. Garrod, *Inborn Errors of Metabolism*, Oxford University Press, 1909.

[18] G. Beadle, Différenciation de la couleur cinabar chez la drosophile (*Drosophila melanogaster*), C.R. Acad. Sci. Paris, vol. 201, 1935 p. 620.

[19] B. Ephrussi, *Quart. Rev. Biol.,* vol. 17., 1942, p. 327.

[20] G. Beadle and E.L. Tatum, "Genetic control of biochemical reactions in Neurospora", *Proc. Nat. Acad. Sci.*, USA, vol. 27, 1941, p. 499.

[21] J.L. Stadler, *Science*, vol. 120, 1954, p. 81.

[22] G. Pontecorvo, *Adv. in Enzymol.*, vol. 13, 1952, p. 331.

the same entity. At the time, the gene was considered a recombination unit, a functional unit and a mutational unit. Research on the fruit fly showed that at a certain level of analysis these three definitions were identical. It was up to Seymour Benzer[23], several years later, to specify the different corresponding entities: the unit of mutation is the nucleic base, the unit of recombination is also the nucleic base (recombination being possible between two adjacent bases), but the functional unit is the gene itself. S. Benzer's success is, in great part, due to the nature of what he was working on, the bacteria *E. coli* and its parasite viruses. These bacteriophages are an excellent model that proliferates rapidly and from which many sorts of mutants can be obtained, the properties of which had already been well studied and exploited by Salvador Luria, Max Delbrück, Alfred Hershey, Joshua Lederberg, William Hayes, André Lwoff, Jacques Monod, François Jacob and Elisabeth and Eugène Wollmann. It was this same model organism that allowed F. Jacob and J. Monod[24] to discover the control elements for gene activity, in 1961.

Throughout this period, scientists were gathering a considerable mass of data from studies of families with prevailing mental or physical illnesses. Up until 1953, the study of genetics by statistical analysis of hereditary disease had been greatly distorted by the popular bias towards eugenics. The more heredity was understood, the more it seemed to make sense to try to improve a given population by selecting for the "normal" or "ideal" person. This implied the selective breeding of human beings considered to be superior and the suppression of inferior beings. Only the horrible brutality of the era of national socialism in Hitler's Germany showed the world what the ultimate destination and extreme use of planned eugenics could be.

---

[23] S. Benzer, "Fine structure of a genetic region in bacteriophage", *Proc. Nat. Acad. Sci.*, USA, vol. 41, 1955, p. 344.

[24] F. Jacob, J. Monod, "Genetic regulatory mechanism in the synthesis of proteins", *J. Mol. Biol.*, vol. 3, 1961, pp. 318–356.

Nucleic Acids and the Mystery of Life

Although Robert Feulgen [25] demonstrated, in 1924, that there was DNA in chromosomes, twenty years passed before the role of this substance as the carrier of genetic information was made clear. In 1944, Oswald T. Avery, Colin M. McLeod and Maclyn McCarty [26] showed that bacterial DNA from *Pneumococcus* was able to induce the transformation of other genetically different *pneumococci* and be replicated. Their work was based on Frederich Griffith's 1922 discovery [27] that there were two strains of streptococcus bacteria, the R and S forms and on his further discovery, in 1928 [28], that the S (for smooth) strain of the bacteria *Streptococcus pneumoniae* (then known as *Diplococcus*) caused fatal septicemia in mice, whereas the R (for rough) strain [29] had no effect on them at all. Griffith showed that S bacteria could spontaneously mutate and give rise to R bacteria. He further established that if one injected mice with a mixture of live R bacteria and S bacteria killed by heat treatment, the mice died of septicemia, and that it was possible to isolate live S bacteria in their blood. Something from the dead S bacteria had turned the R bacteria into S bacteria. All that had to be done was to find out what.

The answer was provided by Oswald T. Avery at the Rockefeller Institute in New York. When he read the accounts of Griffith's experiments, he was skeptical as the experiments seemed to disprove the notion of the stability of species. However, after having repeated some of Griffith's experiments,

---

[25] R. Feulgen and H. Rossenbeck, *J. Physiol.*, vol. 135, 1924, p. 203.

[26] O.T. Avery, L.M. McLeod and M. McCarty, "Studies on the chemical nature of the substance inducing transformation of pneumococcal types. Induction of transformation by a deoxyribonucleic acid fraction isolated from pneumococcus type III", *T. Exp. Med.*, vol. 79, 1 February 1944, pp. 137-158.

[27] F. Griffith, "The influence of immune serum on the biological properties of Pneumococci", *Reports on public health and medical subjects,* n° 18, His Majesty's Stationery Office, London, 1923, pp. 1-13.

[28] F. Griffith, "The significance of pneumococcal types", *Journal of Hygiene,* vol. 27, January 1928, pp. 141-144.

[29] The S strain has a polysaccharide capsule that makes its colonies look smooth when they are grown on a solid base in a Petri dish. The R strain does not have this capsule and this makes its colonies look granular or rough.

he realized that the phenomenon was actually happening and decided to look for the chemical identity of the "transforming principle". It took Avery and his colleagues Colin McLeod and Maclyn McCarty fourteen years of frenzied research, using test tube versions of Griffith's experiments, to gather convincing proof that the transforming principle was DNA. When they had removed all traces of protein or other contaminants which could have been responsible for the transformation, they established that DNA isolated from the S strain was capable of transforming unencapsulated R bacteria into completely encapsulated S bacteria. None of the other cellular contents (RNA, lipids, proteins or sugars) seemed to be able to carry out the transformation. The transforming activity of DNA could also be prevented by using an enzyme that hydrolyses DNA, deoxyribonuclease (DNAse), but not with an enzyme that degrades RNA, ribonuclease (RNAse). These results strongly suggested that DNA was what they were looking for. The hypothesis put forward in their article suggested that small quantities of other substances absorbed into the latter, (or so intimately associated with it that they escape detection) might be responsible for the transformation. However, Avery added, if the results of his study were to be confirmed, nucleic acid would have to be considered as having a biological specificity. The work of Avery and his colleagues was, however, thrown into doubt because the isolated nucleic acid was not absolutely pure; in particular, it contained extraneous proteins.

Although proteins were then very much the fashion, the role of DNA was becoming more important due to the work of Avery, Rollin Hotchkiss and Ephrussi-Taylor on *Pneumococcus*, and thanks to the French scientists André Boivin, Roger Vendrely and Colette Vendrely [30], who in 1948, showed that the DNA content of cells was directly related to the number of chromosomes they contained. They had also looked at the problem of DNA function [31]. The definitive proof that DNA was the carrier of genetic material

---

[30] A. Boivin et al., L.R. Acad. Sci., vol. 26, 1948, p. 1061 and R. Vendrely, C. Vendrely, *Experientia*, vol. 4, 1948, p. 434.

[31] A. Boivin et al., "L'acide thymonucléique polymérisé, principe paraissant susceptible de déterminer la spécificité sérologique et l'équipement enzymatique des bactéries: signification pour la biochimie de l'hérédité", *Experientia*, vol. 1, 1945, pp. 334–335.

came with other experimentation such as that carried out by Alfred Day Hershey and Marta Chase[32]. It was mainly Avery's work that, in 1950, inspired Erwin Chargaff[33] to undertake the analysis of a non-degraded DNA preparation of high molecular weight, an analysis which led him to discover that the total quantity of purine bases always equals that of the pyrimidine bases. More specifically, he discovered that the quantity of adenine is equal to that of thymine (A=T), and that the quantity of guanine is equal to that of cytosine (G=C).

The Cambridge mathematician, John Griffith, made some calculations on the electrical charges of each base[34]. He independently discovered that adenine attracts thymine and guanine attracts cytosine. In 1952, Alexander Todd[35] established the manner in which the chemical components of DNA were linked by showing that there is a skeleton made of alternating phosphoryl and pentoxyl residues, and with each phosphate, a group of oxygen and phophorus atoms linked to the next one by a sugar. There are many different types of sugar, but in DNA's case the sugar is a deoxyribose. In addition, one of the four A T C G bases is attached to each sugar on the opposite side to the phosphate link. But this description of a series of bases held together with sugars and phosphates was only a very rough pattern; it left the underlying three-dimensional structure completely unexplained.

The X-ray diffraction spectra of DNA strands analyzed by Rosalind Franklin, Maurice H.F. Wilkins and others provided more clues. The profiles obtained showed that DNA consisted of at least two chains, twined around each other in a helix. Although in the early 1950s, DNA had been clearly designated the genetic material and clinical analysis was generating some

---

[32] A.D. Hershey and M. Chase, "Independent functions of viral protein and nucleic acid in growth of bacteriophage", *J. Gen. Physiol.*, vol. 36, 1952, pp. 39–56.

[33] E. Chargaff, "Chemical specificity of nucleic acids and mechanism of their enzymatic degradation", *Experientia*, vol. 6, 1950, pp. 201–209.

   E. Chargaff, "Structure and function of nucleic acids as cell constituents", *Fed. proc.*, vol. 1à, 1951, pp. 654–659.

[34] Olby "Francis Crick, DNA and the Central Dogma", in *Daedalus,* Fall 1970, pp. 956–957.

[35] Alexander Todd and D.M. Brown, "Nucleotides: part X, some observations on the structure and chemical behaviour of the nucleic acids", *Journal of the Chemical Society*, 1952, pp. 52–58.

information on its composition and structure, the chemical and physical facts still needed to be brought together into a proposal for a symmetrical three-dimensional structure compatible with experimental results. It would also have to have the properties one might expect of genetic material, which were that it would allow:

- Genetic material to contain the information necessary for the structure, function and stability of cell reproduction. (As Erwin Schrödinger supposed, this information would have to be coded in the sequence of the base elements that make up genetic material);
- Genetic material to replicate with accuracy so that the same genetic information would be present in daughter cells over successive generations;
- Encoded information to be decoded to produce the molecules necessary for the structure and function of cells;
- Genetic material to be capable of rare variation, since mutation and recombination of genetic material are the source of the evolutionary process.

This is what Francis Harry Compton Crick and James Deway Watson managed to do, in 1953, in the course of their discovery of the "secret of life". This discovery was so important that it marked a decisive turning point in the evolution of biology and heralded the birth of the new molecular biology, opening the way for a new approach to living beings and processes: a molecular one.

3.2 The Secret of Life: DNA

1953 is to genetic science what 1492 is to geography. Until the middle of the 20th Century, only pioneers had explored living cells. Physiological genetics, which appeared in the 1930s, showed that genes affect enzymes and allowed considerable progress to be made in our understanding of the mechanism the biochemical action of genes. From this data on physiological genetics, a relatively concrete representation of the gene's role was attained.

Research had progressed from the notion of macroscopic and, therefore, immediately recognizable characteristics to that of molecules, enzymes and even macromolecules, the structure [36] and the composition of which were gradually being elucidated. The work of T.H. Morgan on topological genetics allowed the first chromosome maps to be drawn up; that is to say, Morgan's work provided a way to determine in what order the genes were laid out, and how far [37] they were from each other.

However, as F. Jacob says, topological and physiological genetics had their limits:

> "At the beginning of this century, the so-called "black box" method had allowed us to grant heredity a form, to represent it with a system of simple signs and submit it to mathematical analysis. But as it pays no attention to the mechanism, it leaves a gap in understanding between the gene and the characteristic. Genetics draws an increasingly abstract picture of the organism with all these symbols and formulae. The gene is a logical entity which has no body, no presence. We must substitute this abstract concept with a chemical reality. (...) The goals of geneticists towards the middle of this century will be to find the nature of this substance, explain the modus operandi of the genes and fill the gap between character, phenotype and gene" [38].

---

[36] This is, in particular, thanks to X-ray diffraction, which allows us to see secondary and tertiary structures in proteins. For an analysis of the importance of crystallographic study in the history of modern biology, please consult the following works:
  a) Claude Debru, *L'esprit des protéines*, ed. Hermann, 1983.
  b) Horace Freeland Judson, *The Eighth Day of Creation,* Penguin Books, 1995. First published by J. Cape Ltd. 1979.

[37] The unit of measurement is the centimorgan (cM). This unit was created by J.B.S. Haldane in 1919, in J.B.S. Haldane, "The combination of linkage values, and the calculation of distances between loci of linked factors", *Journal of Genetics,* vol. 8, 1919, pp. 299–309. This article was the first attempt to develop an estimate of distance between genetic loci on a map based on frequency of recombination between mutants.

[38] F. Jacob, *La Logique du Vivant*, Collection Tel. Gallimard, 1970, p. 246.

However, neither the tools nor the concepts of biotechnology were suitable for such an analysis. To uncover the mechanisms and structures that drive heredity, there had to be cooperation between genetics, chemistry and physics. As it could not afford to miss out on new techniques of physics for its molecular comprehension of reproduction, genetics had to follow in the footsteps of biochemistry, which had adopted physics techniques in its investigation of the structures of macromolecules. At the beginning of the last century, and even up until 1953, the object of heredity itself was not defined. Even the gene, the unit of inheritance, looked as if it would be a structure in the cell nucleus, a structure of challenging complexity with no toe-holds for experimentation.

Despite the knowledge emerging from physiological genetics, which confirmed the link between gene and protein, and the new hypothesis that DNA was responsible for the transmission of hereditary characteristics, the general opinion prevailing in the 1940s was that genes would be molecules of a special type of protein. Although the chemistry of nucleic acids was relatively well-understood between the two world wars, there was no concern for the discovery of their role or structure.

Watson recalls that in 1950 chapters on DNA in textbooks tended to be so short and boring that he forgot the names of the famous bases. No-one was telling the story of the phagist[39] team and experience[40]. The DNA molecule was considered lacking any fundamental interest because it was monotonous and "stupid". The orthodox image it had among biochemists between 1930 and 1950 was as a "structure composed of a repetitive sequence of four bases (these being of equal quantity), the function of which is to constitute the skeleton of the chromosome[41]". But Watson, who had worked on bacteriophages with Salvador Luria, was convinced of the fundamental importance of DNA.

---

[39] This was a group mainly composed of physicists turned biologists who showed through their study of small viruses that attack bacteria, phages, that phage DNA was the active agent that modified these virus-parasited bacteria.

[40] J.D. Watson's *The Double Helix*, Penguin Books, first published by Weidenfeld and Nicolson.

[41] An extract from the courses given by Professor Jean Gayon at the Ecole Normale de Dijon, in 1987, called *La biologie est-elle méchaniste*.

3 The Foundations of the Heralded Revolution 129

In his book, *"What is Life?* [42]*"*, Erwin Schrödinger had already emphasized that the molecular level, however small it was, could be considered a suitable level for analysis, even for the organization of heredity. Watson, like Schrödinger, was certain that the study of DNA would reveal the secret of life and lead to an understanding of the mechanisms that direct reproduction.

This interest in the mechanism of the duplication of genes and the idea that this problem would be solved with structural chemistry brought Watson to Cambridge, in 1951, to learn about X-ray crystallography in the laboratories of Sir Lawrence Bragg. John Kendrew was already there working on myoglobin, as was Max Perutz on hemoglobin [43]. Amongst the group that Perutz had gathered around him was Francis Crick. A physicist by training, he was working at applying physics and chemistry models to biology, and he was acquainted with crystallography from working with Perutz on the structure of hemoglobin. Crick shared Watson's conviction that DNA was more important than proteins and it was this common interest that brought them together.

> "From my first day in the lab I knew I would not leave Cambridge for a long time. Departing would be idiocy, for I had immediately discovered the fun of talking to Francis Crick. Finding someone in Max's lab who knew that DNA was more important than proteins was real luck" [44].

Inspired by Linus Pauling's success in protein structure research, Watson and Crick decided to apply Pauling's methods. "Within a few days after my arrival, we knew what to do: imitate Pauling and beat him at his own game [45]". This is why they applied considerations of a chemical (an estimation

---

[42] E. Schrödinger, *What is life?* Cambridge University Press, Cambridge, Britain. Translated into French by Léon Keffler, *Qu'est-ce que la vie (L'aspect physique de la cellule vivante)*, ed. Christian Bourgois Editeur 1986.

[43] C. Debru, *op. cit.*

[44] J.D. Watson, *op. cit.* p. 79.

[45] *Ibid*, p. 46.

of the force between atoms) and stereochemical nature when they analyzed the DNA crystal X-ray diffraction photographs that showed a helicoidal structure "provided" by Maurice Wilkins and Rosalind Franklin[46]. In addition to this, Crick and Watson paid attention to Chargaff and Griffith's observations on the relationships between the purine and pyramidine bases and managed to deduce the structure of DNA. In doing so, they beat several well renowned scientists and their teams[47] to the winning post in this scientific race.

The Crick and Watson Model

The initial article on DNA structure and its role, the *"princeps* publication" that would found molecular genetics, appeared as a "note" to the scientific journal *Nature* on 25 April 1953[48]. The article also mentioned Pauling and Corey's hypothesis[49] of a three-stranded interlaced structure and Fraser's theory of another three-stranded structure, two hypotheses that the authors refuted. The fundamental description of the structure as discovered was as follows:

> "We wish to suggest a structure for the salt of deoxyribose nucleic acid (DNA). This structure has novel features which are

---

[46] Regarding Rosalind Franklin's importance in the "Double Helix Adventure," please consult Anne Sayre's *Rosalind Franklin and DNA*, ed. Norton, New York, 1975, and Klug's *article* "Rosalind Franklin and the discovering of the structure of DNA", *Nature*, vol 219, 24 August 1968, pp. 843–844.

[47] For a detailed account of this scientific adventure, refer to the following works: J.D. Watson's *The Double Helix*, Penguin Books, first published by Weidenfeld and Nicolson 1968, trad.fr. *La Double Hélice*, Collection Pluriel, ed. Laffont, 1984, and for a less one-sided account, Horace Freeland Judson, *The Eighth Day of Creation*, ed. Penguin Books, 1995. First published by J. Cape Ltd. 1979.

[48] J.D. Watson and F.H.C. Crick, "A structure for deoxyribose nucleic acid", Nature, vol. 171, 1953, reprinted in some editions of Watson's *The Double Helix*. In the same edition of *Nature* this article was followed by two others, one by M.H.F Wilkins, Alexander R. Stokes and H.R. Wilson, "Molecular structure of deoxypentose nucleic acids", pp. 738–740, and the other by R. E. Franklin and Raymond Gosling, "Molecular configuration in Sodium Thymonucleate", pp. 740–741. These two articles confirmed the model proposed by Crick and Watson so well that the latter published a new article in *Nature* on 30th May 1953, pp. 964–976 the "Genetical implications of the structure of deoxyribonucleic acid" which explored the genetic consequences of the proposed structure.

[49] L. Pauling and Robert Corey, "A proposed structure for the nucleic acids", *Proc. N.A.S.* vol. 39, pp. 96–97.

of considerable biological interest (...) This structure has two helical chains each coiled round the same axis (...). Both chains follow right-handed helices (but not their bases) and are related by a dyad perpendicular to the fibre axis. Both chains follow right-handed helices, but owing to the dyad, the sequences of the atoms in the two chains run in opposite directions. (...) the bases are on the inside of the helix and the phosphates on the outside. (...). The novel feature of the structure is the manner in which the two chains are held together by the purine and the pyrimidine bases. The planes of the bases are perpendicular to the fibre axis. They are joined together in pairs, a single base from one chain being hydrogen-bonded to a single base from the other chain, so that the two lie side by side with identical z-co-ordinates. One of the pair must be a purine and the other a pyrimidine for bonding to occur. The hydrogen bonds are made as follows: purine position 1 to pyrimidine position 1; purine position 6 to pyrimidine position 6.

If it is assumed that the bases only occur in the structure in the most plausible tautomeric forms (that is, with the keto rather than the enol configurations), it is found that only specific pairs of bases can bond together. These pairs are: adenine (purine) with thymine (pyrimidine), and guanine (purine) with cytosine (pyrimidine). (...) The sequence of bases on a single chain does not appear to be restricted in any way. However, if only specific pairs of bases can be formed, it follows that if the sequence of bases on one chain is given, then the sequence on the other chain is automatically determined. (...) The previously published X-ray data on deoxyribose nucleic acid are insufficient for a rigorous test of our structure. So far as we can tell, it is roughly compatible with the experimental data"[50].

---

[50] J.D. Watson and F.H.C. Crick, "A structure for deoxyribose nucleic acid", *Nature*, vol. 171, 1953, reprinted in some editions of Watson's *The Double Helix*.

"It has not escaped our notice that the specific pairing we have postulated immediately suggests a possible copying mechanism for the genetic material"[51].

Each of the strands acts as a template for the production of two double helixes identical to the first, since once the base sequence is defined on one chain, the sequence of the other chain is automatically determined. It is easy to see this as a model of an autocatalytic molecule. Crick and Watson's proposed structure carried within itself the means for replication. Indeed, another fundamental characteristic of the model proposed by Crick and Watson is the discovery of a structure that directly reveals its own function: the faithful replication of a message. The three-dimensional structure of DNA is the base for something far more fundamental the one-dimensional structure of the irregular sequence of bases.

DNA's capacity to reproduce with total accuracy, an accuracy so absolute that the two daughter helixes of DNA are utterly identical, goes against the law of information degradation put forward by Claude E. Shannon and Léon Brillouin, a law which can be resumed as, "in any transmission of information, there is degradation of the information".

The symmetry of DNA is forever reconstituted due to the nature of base pairing. In showing that the symmetry of DNA, this enormous irregular crystal to use Schrödinger's words, underlies its invariance, Crick and Watson took a vital step in further understanding the most remarkable property of biological systems.

One of the consequences of Crick and Watson's discovery, which had great importance in the development of molecular biology and biotechnology, was that it finally gave a plausible material solution to the old genetic notion of mutation. Hermann Muller had received a Nobel prize for showing that it was possible to induce mutation with X-rays[52], but no-one had, as yet, suggested what was modified. If the essential factor in DNA was the linear

---

[51] *Ibid*, p. 237.

[52] Please see Herman Muller's work, *Studies in Genetics: The Selected Papers of H.J. Muller*, Indian University Press, Bloomington, 1962, quoted by Horace Freeland Judson, *The Eighth Day of Creation*, *op. cit.*, p. 626.

3 The Foundations of the Heralded Revolution 133

sequence of coded information, a mutation would alter in that sequence. The law of complementarity provided a very simple explanation: during the reproduction of hereditary material, when a DNA molecule is replicating into two daughter molecules, there are sometimes errors. These result in the wrong base being paired on one of the two DNA strands. In the following generation, one of the strands which is acting as a template will already have an error, and it will be faithfully copied, perpetuating the partial alteration of a nucleotide.

Crick and Watson's model thus confirmed Schrödinger's hypotheses. DNA is, indeed, an irregular autocatalytic crystal, but most of all, it is the base for genetic information encoded in the linear alignment of the bases. DNA is clearly the physical base for the hereditary message. With this fundamental discovery, hereditary mechanisms can now be explained in terms of the information transmitted.

With the advent of cybernetics, in 1943[53], and the mathematical theory of communication[54], information was becoming one of the key concepts in the interpretation of phenomena linked to telecommunications[55]. In 1953, information also became the concept that helped in understanding the mechanisms at work in life sciences, thus confirming Schrödinger's intuition that the apparently strictly deterministic deciding factor in the future development of a living thing is a code that is passed on from generation to generation, a code that would contain the entire pattern for the future development of the individual and the way it would behave as an adult[56].

---

[53] Norbet Wiener, Arturo Rosenblueth and Julian Bigelow, "Behaviour, Purpose and Teleology", *Philosophy and Science*, 1943, pp. 18–24, Warren S. McCullock and Walter Pitts, "A logical calculus of the ideas present in nervous activity", *Bulletin of mathematic biophysics*, vol. 5, Dec 1943, pp. 115–133, Norbet Wiener, *Cybernetics or control and communication in the animal and the machine*, Librairie Hermann et Cie, Paris, 1948, later editions: MIT Press, Cambridge, Mass.

[54] Warren Weaver and Claude E. Shannon, *The mathematical theory of communication*, the Board of Trustees of the University of Illinois, trad. fr. J. Cosnier, G. Dahan and S. Economides, *Théorie mathématique de la communication*, Retz-CEPL, Paris 1975.

[55] For the history of cybernetics, information theory and the emergence of the communications interpretations of the natural and social phenomena, refer to Philippe Goujon's *Les voies de l'information, de la communication à la complexité*, Doctoral thesis defended in Dijon (France), June 1993.

[56] E. Schrödinger, *op. cit.*, p. 71.

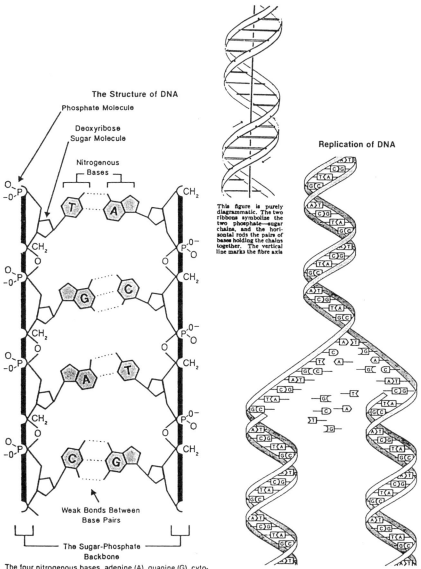

Fig. 4 The DNA model as proposed by Crick and Watson

The importance of the concept of information was reinforced when the British biochemist Frederick Sanger first sequenced a protein — the insulin molecule.

3.3 The First Sequencing of a Protein: Insulin

In 1945, Frederick Sanger began to determine the free amine groupings of bovine insulin, the amino acid composition of which had already been well analyzed[57]. Sanger chose insulin because it was the only protein which he could buy in large and relatively pure quantities. It was a lucky choice because insulin later turned out to be one of the smaller proteins and was therefore, easier to sequence. Nevertheless, it took Sanger a decade to fragment the insulin molecule, separate out the resulting chemical components and rearrange them in their correct order.

Sanger was armed for his task with the analytical methods of organic chemistry, the decomposition or hydrolysis of molecules and the study of the nature of decomposition products. But this work also needed the new technique of silica gel column chromatography, which separated the derivatives of amino acids in accordance with their degree of sharing, with two solvents, one of which was absorbed on the silica gel[58]. It also led to the development of a new method for the identification and quantitative estimation of the free amine groups by the formation of derivatives of dinitrofluorobenzene (DNFB)[59]. Sanger treated the protein with DNFB, hydrolyzed it and determined the nature of the dinitrophenyl amino acid he had obtained. Hypothesizing that insulin had a minimal molecular weight of 12,000 (which would later be proved to be two times too high), Sanger

---

[57] See J.S. Fruton, *Molecules and Life*, New York, 1972, p. 172.

[58] Chromatography on paper; then the use of columns of ion-exchanging resins have since replaced this technique.

[59] Amino acids, in fact, form very stable dinitrophenyl derivates when treated with this chemical which fixes on the free amine functions. The derivates are coloured, while DNFB derivates fixed with phenolic hydroxyl or histadine imidazole do not have any color.

showed that there were two terminal glycine and phenylanine residues as well as residues of two dinitrophenyl derivates on the ε-aminate group of the lysine[60]. He concluded that insulin must be formed by four polypeptide chains linked by disulfide bridges that had already been discovered. The oxidation of the disulfide bridges should lead to two types of free chain. Sanger supposed that they would be identical to each other. He hydrolysed their dinitrophenyl derivative and studied the sequence of the tetrapeptide carriers for the NH_2 terminal extremities (the ones for the tetrapeptides that were carrying lysine in the β chain[61]). From his estimation of sulfur residues in the α and β chains, he proposed several structural hypotheses for the position of the disulfide bridges, one of which was correct. It proposed a structure with a molecular weight of 6,000 which consisted of an α-chain and a β-chain linked by two disulfide bridges, and further on the α chain, an interchain disulfide linkage point[62].

In 1950, the French biochemist Claude Fromageot used the reduction of free carboxylic groups to study the nature of amino acids at the C-terminal ends of insulin. He identified alanine and glycine (albeit in the latter case incorrectly).

In 1951, Sanger looked at the β-chain sequence, which carries a N-terminal phenylalanine residue, and the C-terminal sequence, which was Phe-Val-Asp-Glu. The β-chain was hydrolyzed with concentrated hydrochloric acid. Sanger used electrophoresis, the absorption of ions on exchange resins, and paper chromatography to separate the various peptides. It was then possible to identify the peptides and try to reconstruct the whole sequence. However, this could not be done just through acid hydrolysis, which was a method of some ambiguity. Sanger also had to use proteolytic enzymes (trypsin, chymotrypsin) which produced longer peptides. The combination of the two methods allowed Sanger to establish the sequence of the β-chain

---

[60] F. Sanger, "The free amino group of insulin", *Biochemical Journal*, vol. 39, 1945, p. 514.

[61] F. Sanger, "Some chemical investigation on the structure of insulin", *Cold Spring Harbor Symposia*, vol. 14, 1949, pp. 155–156.

[62] *Ibid*, p. 158.

of insulin, a thirty-residue chain, without any ambiguities [63]. In 1953 [64], he determined the sequence of the α-chain, which had 21 amino acids, and in 1955, he found the exact position of the disulfide bridges.

Sanger's results were of extreme importance for protein chemistry [65]. Even if the techniques he was using could not be applied to much larger molecules than insulin, which only has 51 amino acids, their results supplied an additional argument for the chemical specificity of proteins. This chemical specificity is typified by the uniqueness of the sequence of amino acids in the polypeptide chains. Revealing this sequence was not beyond the reach of chemists, even for complex molecules. Furthermore, it was possible that the sequence was a translation of the nucleotide sequence in deoxyribonucleic acid, which had by then been recognized as the carrier of genetic characteristics. Revealing the mechanisms and the code used in translation became a real challenge to biologists.

Following the discovery of the double helix, the genetic code was finally elucidated, as was the principle for decoding it and the transcription of genetic information. The cybernetic mechanisms that micro-organism genes obey; i.e., the means by which unicellular organisms adjust the workings of the expression of their genome to the environmental conditions, had been revealed. This was a revolutionary clarification of the behavior of creatures that had, until then, been considered inferior — bacteria. It revealed that the bacterial cell is a genetic factory that can adapt itself to many situations. Thanks to its "programming", the genetic information and the regulations that control the working of structural genes, a bacteria can deploy finely tuned strategies to deal with all sorts of modifications in its environment.

Molecular biology had delivered up the most fundamental aspects of heredity: the information encoded in the double helix, the mechanisms of transfer of genetic information and its translation into proteins. It thereby provided a key to the molecular ontogenesis of a large number of complex

---

[63] F. Sanger, "The arrangement of amino-acids", *Protein Chemistry*, vol. 7, 1952, pp. 55–57.

[64] F. Sanger and E.O.P. Thompson, "The amino-acid sequence in the glycyl chain of insulin", 1 and 2 *Biochemical Journal*, vol. 53, 1953, pp. 353–374.

[65] The sequencing of insulin earned Sanger his first Nobel Prize in 1958.

biological structures[66]. Three great principles had been established by the end of the 1960s: the universality of the genetic code, the central dogma[67] and colinearity[68]. These clearly symbolize the journey and successes of molecular biology since the discovery of the double helix. The details of the history of this journey and the discoveries made along the way has been well described since its importance attracts science historians. It is therefore not necessary to go over it again.

The discovery of the structure of DNA by Crick and Watson had an impact as important as that of atomic theory in physics. It was one of the founding elements of molecular biology and genetics. The elucidation of the three-dimensional structure of hereditary material was a considerable step forward. Avery's work had built the bridge firmly between biology and chemistry, and now there was a bridge between biology and physics. It was at the level of the tell-tale macromolecules that the great phenomena of heredity could be analyzed. Everything was in place for the development of molecular biology, which was to progress so rapidly in the course of the following twenty years that it became the focus of most, if not all, biological research.

A talented biologist, who was a convinced opponent of molecular biology, E. Chargaff, had nevertheless, an eminent role in the clarification of the structure of DNA. However, several years after Crick and Watson's publication, he was still to be heard saying that all this molecular biology was not going to further our knowledge of medical matters, or indeed in any applied sector at all, "by an inch". Twenty years passed before the first applications of molecular biology were obtained, thanks to a whole set of innovations. These included the development of methods for the determination

---

[66] For the story of these discoveries, consult Horace Freeland Judson, *op. cit.*, and François Gros, *Les secrets du gène*, ed. du Seuil, Paris, 1986.

[67] This central dogma can be stated as follows: "In biological systems, genetic information always travels from the genes to the ribonucleic acid messengers and from these mRNAs towards the proteins." It summarizes fifteen years of research on the mechanisms of gene expression.

[68] The notion of colinearity is the idea that, in a cell, one might always expect to find a colinear and vectoral adjustment of the alignment of the DNA codes and the alignment of amino acids in the protein coded by that DNA.

of the sequence of nucleotides of nucleic acids and for recombinant DNA technology (techniques of DNA manipulation allowing the isolation of a gene, the insertion of an isolated gene and, when necessary, the modification of the gene in cells of an organism different from that from which it came in order to make it produce a substance that is foreign to it).

3.4 Techniques of DNA Sequencing

Methods for sequencing DNA appeared in the early of the 1970s as extensions of the protein sequencing work that had been carried out since the mid-1940s. As mentioned earlier, in 1953 F. Sanger managed to determine the order of amino acids of the insulin protein, which is used for the treatment of diabetics. Protein sequencing was given a major additional push in 1950, when Pehr Edman, of Lund University in Sweden, discovered how to take one amino acid off the end of a chain at a time[69]. By persistently removing one amino acid at a time from the end of a protein, the sequence could literally be picked out. Previous methods required the fragmentation of proteins into little pieces, the analysis of the order of amino acids in each fragment by various methods and the re-ordering of the sequences obtained in order to find the molecule's entire sequence. Edman's method was not only easier to understand, it was also eminently adaptable to automation. By 1967, tools for determining the sequence of protein amino acids were coming onto on the market[70]. Over the years, these have developed into fast and trustworthy protein sequencing instruments[71].

---

[69] Pehr Edman, "A method for the determination of the amino acid sequence in peptides" *Acta Chemica Scandinavica*, vol. 4, 1950, pp. 283–293.

[70] The various contemporary methods are described in detail by B. Durand in "Séquençage de l'ADN", *Le Technoscope de Biofutur*, no. 23, October 1988, pp. 3–13, and by L. M. Smith in "DNA Sequence Analysis, Past Present and Future", *I.B.L.*, October 1989, pp. 8–19.

[71] M.W. Hunkapiller and Leroy Hood, "A gas-liquid solid phase peptide and protein sequenator", *Journal of Biological Chemistry*, vol. 256, 1981, pp. 7, 990–7, 997. Also M.W. Hukapiller and L.E. Hood, "Protein Sequence Analysis: automated microsequencing", *Science*, vol. 219, 1983, pp. 650–659.

The next step was to progress from sequencing proteins to sequencing RNA[72]. It took Robert W. Holley[73] and his colleagues at Cornell University seven years to determine the order of the 77 bases that constitute one of the forms of RNA. Like the first methods of sequencing proteins, they broke the RNA molecule into little fragments and reconstituted the order of the original molecule. For several years, DNA sequencing was carried out by transcribing it into RNA and then fragmenting the RNA and sequencing it from the small chunks. This is how in 1971 the first DNA sequences were published, namely the complete sequence of the nucleotides of the cohesive ends of the DNA of bacteriophage lambda (λ)[74].

Sanger understood that all these methods were too slow and delicate for use on large expanses of DNA , as he also understood the need to develop new methods of DNA sequencing. He had worked on the protein product of a gene, so it does, in retrospect, seem logical that he should progress to sequencing the gene. Sanger himself was not so sure. He was still very interested in proteins, and he continued to dedicate his time to them. In 1962, he went to work in a new molecular biology laboratory that had just been built in the suburbs of Cambridge, England, where he joined such famous biologists as Perutz, Crick and Brenner. This was a decisive move. As he recalls, with researchers such as Francis Crick around him, it was difficult to ignore the nucleic acids or not to realize the importance of sequencing. The chemical composition of DNA, the four bases that are the bricks of its construction, its structure and the code linking the nucleic bases to the amino acids (3 bases coding for one amino acid) had all been elucidated. But although we understood the genetic alphabet, we could not, as yet, easily read the order of the letters formed in DNA.

---

[72] RNA: ribonucleic acid: a macromolecule that resembles DNA and is involved in the decoding of genes into proteins. Its base sugar is ribose. There are three large categories of RNA: ribosomic RNA, transfer RNA and messenger RNA. The latter is the true matrix for protein formation.

[73] R.W. Holley *et al.*, "Structure of ribonucleic acid", *Science,* vol. 147, 1965, pp. 1, 462–1, 465 and R.W. Holley *et al.*, "Nucleotide sequence in the yeast alanine transfer ribonucleic acid", *Journal of Biological Chemistry*, vol. 240, 1965, pp. 2, 122–2, 128.

[74] R. Wu and E. Taylor, "Nucleotide sequence analysis of DNA II. Complete nucleotide sequence of the cohesive ends of bacteriophage Lambda DNA", *Molecular Biology*, vol. 57, 1971, pp. 491–511.

In order to bring out the sequence of the nucleic bases, a faster means of reading the order of the bases along the DNA needed to be developed. It was not an easy task, and the tools derived from protein sequencing were far too slow. The main problem with DNA was its enormous length; the smallest pure DNA then available was that of the bacteriophage genome, which was 5,000 nucleotides long. Since sequencing, at that time, was at the rate of a few dozen nucleotides per year, it was clear that a new method had to be found. Sanger turned the problem upside down. If the method of breaking DNA up into little pieces and then putting them back together until the entire sequence was clearly not appropriate, why not start at the bases, and for a given strand one wants to sequence, reconstruct the DNA from zero. The tool he chose for this task was an enzyme called DNA polymerase. DNA polymerase (discovered during Nobel prize-winning work by Arthur Kornberg[75] at Stanford in 1955), given a DNA matrix as a model and some deoxynucleotides, will synthesize additional strands of DNA onto the matrix from the four usual deoxynucleotides (dATP, dCTP, dTTP, dGTP). Sanger decided that if he could interrupt polymerase in its work as a catalyst in the formation of the 3'—>5' phosphodiester bridges[76] between the molecules, he could use the information to decode the sequence of that strand.

Sanger's initial method[77] was to provide all the components, but limit

---

[75] A. Kornberg, "Biologic synthesis of deoxyribonucleic acid", *Science,* vol. 131, 1960, pp. 1,503–1,508 and A. Kornberg, *DNA Replication,* W.H. Freeman, New York, 1980.

[76] The enzyme DNA polymerase I is not just a polymerase. It also has exonuclease activities 5'—>3' and 3'—>5'. The enzyme catalyses a strand replacement reaction in which the exonuclease function 5'—>3' degrades the 'non-pattern' strand as the polymerase synthesizes the new copy. The DNA polymerase I 5'—>3' exonuclease function can be removed by cleaving the enzyme to produce what has been named the Klenow fragment. This only leaves the polymerase and 3'—>5' exonuclease functions. As synthesis proceeds in the 5'—>3' direction, each subsequent addition of nucleotides needs a free 3'-OH group to form a phosphodiester bridge. This requirement also means that a short double-strand region with a free 3'-OH group, called a primer region, is needed for the synthesis to begin.

[77] On the evolution of Sanger's method, refer to the following articles:
  - F. Sanger and A.R. Coulson, "Rapid method for determining sequences in DNA by primed synthesis with DNA-polymerase", *Journal of Molecular Biology,* vol. 94, 1975, pp. 441–478.
  - F. Sanger, the Croonian lecture, 1975, "Nucleotide sequences in DNA", *Proceedings of the Royal Society of London,* B191, 1975, pp. 317–333.
  - F. Sanger, S. Nilken and A.R. Coulson, "DNA sequencing with chain-terminating inhibitors", *Proceedings of the National Academy of Sciences* (USA) vol. 74, 1977, pp. 5,463–5,468.

the reaction of one of the four bases needed for the synthesis of DNA. The synthesis would continue until all the deoxynuclotide 5'-triphosphates, of which there was a limited supply, had been used up[78].

It was a good idea, but in practice the method did not work very efficiently. The separation of the various fragments was delicate and the results not very satisfactory. Furthermore, the rate at which the sequences were obtained was desperately slow. It was at this point that Sanger had the brainstorm that, according to his memoirs, constituted the high point of his career as a researcher. For each of the four dNTPs, he added a series of four reactions together, each one containing the four normal dNTPs (one being radioactive) and for each of the reactions, a specific dideoxynucleotide (ddATP, dCTP, ddTTP or ddGTP)[79] (Fig. 5). The incorporation of ddNTPs stopped the lengthening. When the reaction was completed, each test tube contained a set of fragments of varying lengths ending with the same dideoxynucleotide that had been added in the reaction buffer. The size of the fragments showed the position of the ddNTP in the lengthening chain. The newly-synthesized DNA fragments were then ranked by size through electrophoresis for the four sets.

Using this technique, Sanger and his team managed to sequence the first genome, the bacteriophage phi X174. Its sequence was published in *Nature* on 24 February 1977 [80] (Fig. 6), and was greeted with enthusiasm and surprise by the world of molecular biology. Sanger had sequenced the first genome, and in doing so, paved the way for the assault that molecular biology made

---

[78] For example, the quantity of dTTP is limiting in the ACGTCGGGTGC sequence, leaving ACG and ACGTCGGG. The T in the longer of these two fragments is produced just before the reaction runs out of dTTP. The shortest fragment lacks any T. When Sanger lined up the product molecules in order of length, it was clear that the 4th and 9th positions should be a T because he obtained chains of 3 and 8 bases. If one limits the quantity of dGTP in the example just given one ends up with AC, ACGTC, ACGTCG and ACGTCGGT, which means that G is in positions 3, 6, 7 and 9. By limiting each of the four 5'-triphosphate deoxyribonucleotides in one of four experiments, it is possible to determine the position of each nucleotide by putting the fragments from each experiment in length order, and through this, in principle, determining the entire sequence.

[79] Initially, the radioactive dNTP was tagged with ^{32}P radioactive phosphorus. Today, we generally use α-S^{35}-dCTP or α-P^{33}-dCTP; but you should note that only one dNTP is marked, the other three are not.

[80] F. Sanger *et al.* "Nucleotide sequences of bacteriophage phi-X174", *Nature*, vol. 265, 1977, pp. 685–687.

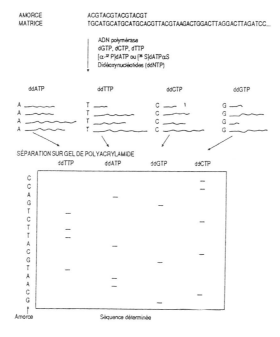

Fig. 5 Sanger's sequencing technique
(After A.D. Woodhead and B.J. Banhart, *Basic Life Science: Biotechnology and the human genome*, Plenum Press, New York, 1988, p. 112)

Sanger's original method is based on the hybridization of a trigger DNA chunk of 15 to 20 nucleotides to the single strand DNA fragment that you want to sequence. Then you synthesize a second strand from the trigger in polymerase and deoxynucleotide triphosphates (dNTPs). Dideoxynucleotide triphosphates (ddNTPs) are then added during the elongation reaction. As they don't have a free 3'-hydroxyle group, they do not allow the elongation of the nucleotide chain, which is stopped as of when one of the ddNTPs is incorporated. Thus, if didesoxyadenosines (ddATPs) are added during the elongation reaction in small quantities, you get a series of complementary fragments of the matrix of different sizes, each ending with an adenosine (A). Fragments ending in cytosine, guanine and thymine (C, G and T) are also obtained by adding, respectively, ddCTPs, ddGTPs and ddTTPs. Part of the deoxynucleotides incorporated are marked with ^{35}S or ^{32}P or ^{33}P isotopes that emit beta particles that can be detected on X-ray film. When the reading procedure is automated with a continuous laser beam, these isotopes are replaced with fluorochromes.

The fragments that come out of the elongation reactions are freed from the matrix by high-temperature denaturation and separated by gel electrophoresis on polyacrylamide gel, allowing for the bands to separate as far as a base. Once the fragments have migrated through the gel far enough, it is fixed, dried and revealed on X-ray film. After the film is developed, you can read the sequence off it from top to bottom (from the shortest fragment to the longest). The last base of each of those fragments can be identified by the reaction from which it comes.

Fig. 6 The sequence for the bacteriophage ΦX 174

This large aperiodic crystal is the sequence of bases for the one chromosome of the bacteriophage ΦX 174. It is the first complete genome ever to be decoded, and it would take another 2,000 of these headsplitting pages to represent the genomes for *E. coli*, and about a million pages to show the DNA base sequences for a human being.

on biological and medical problems during the two decades that followed, not to mention the new hopes he raised for biotechnology. For this work, Sanger was awarded a second Nobel Prize in 1980, admitting him to a very exclusive club[81].

---

[81] Before Sanger, Marie Curie had received a Nobel Prize for physics in 1903 and one for chemistry in 1911. The phyicist John Bardeen, who invented the transistor and the theory of superconductivity, was honored with the Nobel Prize for physics in 1956 and 1972. His compatriot, Linus Pauling, received the Nobel Prize for chemistry in 1954 and the Nobel Peace Prize in 1962, which was a remarkable distinction for a scientist.

Despite these two prestigious awards, Sanger is very modest about his achievement. In an autobiographical article published in 1988 [82], he pointed out that of the three main activities needed for scientific research — thinking, speaking and doing — he, by far, preferred the last because he thought he was better at it. Affirming that he was neither academically brilliant nor maniacally competitive, he added that he thought he would not have managed to get into the prestigious Cambridge University if his family had not been well off. Sanger is, perhaps in his own eyes, just a man who does experiments, but he did some of the most brilliant experiments of the twentieth century.

Several thousand miles away in another Cambridge, in America, Allan Maxam and Walter Gilbert of Harvard were developing an entirely different method of sequencing at the same time. Gilbert was a theoretical physicist, but he turned to biology in 1960, at the age of 28. Within a few years, his work earned him the American Cancer Society–sponsored tenure at Harvard University. Gilbert's first major contribution to molecular biology was the discovery of a protein repressor [83], a biological entity predicted by Jacques Monod and François Jacob [84] to be involved in the inhibition of gene expression when the gene's role in protein synthesis was not required by a cell. The characterization of the identity of this genetic repressor was very difficult, and Gilbert's success established his reputation as an accomplished researcher.

Along with Maxam, Gilbert looked at the regulatory region of the lactose operon, mainly the site recognized by the repressor protein. They isolated a DNA fragment from this region and showed that a small part of the DNA

---

[82] F. Sanger, "Sequences, sequences and sequences", *Annual Reviews of Biochemistry*, vol. 57, 1988, pp. 1–28.

[83] Walter Gilbert and Benno Müller-Hill, "Isolation of the lac repressor", *Proc. N.A.S.*, vol. 56, 1966, pp. 1, 891–1, 889.

[84] F. Jacob, "Le temps des modèles: la régulation binaire", in A. Lwoff and A. Ullman, *Les origines de la biologie moléculaire,* ed. Etudes Vivantes, Col. Academic Press, 1980.

J. Monod and F. Jacob, "Teleonomic mechanism in cellular metabolism, growth and differentiation", *Cold Spring Harbor Symposia*, vol. 26, 1961, pp. 389–401.

J. Monod and F. Jacob, "Genetic regulatory mechanisms in the synthesis of proteins , *J. Biol. Chem.* no. 3, 1961, pp. 318–356.

was protected from enzyme degradation when the repressor was present. The protein repressor was linked to the DNA, and it protected it from digestive enzymes. They transcribed this short chunk into RNA and spent two years laboriously finding the sequence of the short strand's 24-base-pair[85] DNA region using the methods of fragmentation and reconstruction that had proved so useful in the first protein sequencing efforts.

During this research, they were visited twice by a Soviet researcher, Andrei Mirzabekov. His first visit in the mid-70s was very brief. He was trying to find a way to break DNA at predetermined base pairs by selectively adding methyl groups to specific DNA bases. Mirzabekov had discovered that dimethyl sulfate destabilizes DNA and leads to a breakage specifically at the adenine (A) and guanine (G) bases. Mirzabekov, Gilbert, Maxam and a postgraduate student, Jay Gralla, discussed a possible use for this type of specific DNA fragmentation reaction. They already knew the sequence of the *lac* operator to which the repressor binds but wanted to know exactly where the protein link was. The idea was that the protein link did not just block the methylation enzymes but also blocked the chemical methylation, that is, the addition of methyl groups (CH_3) to the DNA bases. If they compared the DNA fragments with and without the repressor, there should be a chunk of DNA which would break easily without a repressor but which would be protected by its presence. The first experiment had no decisive results, but later experiments were more successful and were reported at a Danish symposium in the summer of 1975[86].

These results were an independent verification of the digestion experiments, but the direct chemical fragmentation of DNA produced a far more specific profile of the link. The new method (Fig. 7) had an extremely interesting and seductive characteristic. If one could discover the reaction conditions for the selective fracture of DNA at each of the four bases from

---

[85] W. Gilbert and A. Maxam, "The nucleotide sequence of the *lac* operator", *Proceedings of the National Academy of Science* (USA), 70(12), 1973, pp. 3, 581–3, 584.

[86] W. Gilbert, A. Maxam and A. Mirzabekov, "Contact between the *lac* repressor and DNA revealed by methylation", in *Vinth Alfred Benzon Symposium, Copenhagen: Control of ribozome synthesis*, ed. Nokjeldgaard and O. Maaloe, Academic Press, New York, 1976, pp. 139–143.

which all DNA is made, one could read the DNA sequence directly just by separating the fragments according to length. Maxam adjusted the reaction conditions until he could fragment DNA only at the G base or at the A and G bases at the same time [87]. If a fragment appeared in both reactions, it was a G. If it appeared in the A+G reaction but not in the G reaction, it was an A. Maxam then discovered a similar method for breaking DNA at the cytosine (C) and thymine (T) bases. The base at any position in DNA could then be inferred from the fragmentation in four separate chemical reactions. A sequencing method had arisen [88].

Late in the summer of 1975, while Gilbert was traveling in the Soviet Union, Maxam gave a speech to the annual New Hampshire Gordon Conference on nucleic acids. At this meeting, he distributed a protocol for "Maxam-Gilbert" sequencing. Their method for sequencing DNA wasn't published until 1977 [89].

Sanger was not very happy at the appearance of a competing sequencing method [90], though it was later seen to be complementary in power and efficiency. The approach that people use varies according to what is being sequenced and the way in which the DNA has been prepared, although it could be argued that today Sanger's technique is more widely used. Sanger's initial method has, however, been greatly improved in the intervening years as it was excessively onerous for very long DNA fragments (2,000 bp and more).

Other methods have also been described that allow the subcloning stages to be cut down by using the same cloning vector to get two bits of information, by tinkering with the type of vector or on the trigger oligonucleotides used and by modifying the initial cloning methodology. One of these adaptations

---

[87] The reaction chemicals for the first and second reactions and dimethyl sulfate and piperidine for the purines (A and G) and hydrazine and piperidine for pyrimidines (T and C).

[88] G.B. Kolata, "The 1980 Nobel Prize in Chemistry", *Science*, vol. 210, 1980, pp. 887–889.

[89] A.M. Maxam and W. Gilbert, "A new method for sequencing DNA", *Proc. Nat. Acad. Sci,* vol. 74, USA, 1977, pp. 560–564. This sequencing method was to earn Walter Gilbert the Nobel Prize, but we must also associate Maxmam with this method, because although he did not also receive the Nobel Prize, at the time he was a technician in Gilbert's laboratory and played a great part in the work.

[90] F. Sanger, "Sequences, sequences and sequences", *ref. cit.*

148 From Biotechnology to Genomes: A Meaning for the Double Helix

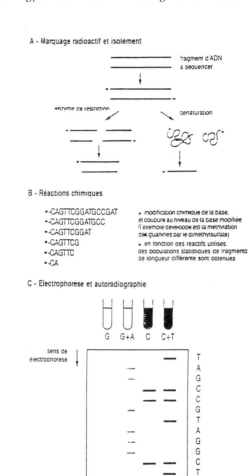

Fig. 7 The sequencing principle according to the Maxam and Gilbert method

The Maxam and Gilbert method, like that of Sanger, is based on getting fragments whose size allows the siting of one of the four bases. Single or double-stranded DNA fragments are tagged with a radioactive marker at one of their ends and then partially and specifically cleaved at the base level in four different reaction buffers. The fragments are then separated according to size in a polyacrylamide gel, and you can read the sequence immediately off the gel.

of Sanger's technique for sequencing very long genes is called "shotgun" sequencing. Fragments of the gene are cloned in the M13 phage at random without any information on their organization in its genome. The fragments are then sequenced and ordered while they are being matched up with the help of a computer. This method gives good results for very long fragments, but usually about 20%, which has not been matched up properly, needs to be resequenced directly.

Given the difficulty of getting hold of missing sequences, more powerful adaptations of the technique were developed, such as a method of sequencing by cloning and enzyme deletions, and another of sequencing by dilution caused by a transposon. A further method appeared known as Church's multiplex method[91] (Fig. 8). This adaptation of Maxam and Gilbert's specific chemical hydrolysis method can also be applied to Sanger's method. It has the advantage of allowing the simultaneous analysis of about one hundred DNA fragments with only one electrophoresis gel.

Other methods have been developed since early 1990, and the automatic methods for sequencing are becoming more efficient and trustworthy[92]. In any case, between 1974 and 1976 two independent DNA sequencing techniques were developed, both elegant solutions to a central methodological problem. The ability to sequence DNA opened enormous fields for experimentation and completed the other major technical triumph of molecular biology at this time: the techniques of recombinant DNA. In fact, before scientists could turn sequencing techniques into common practical procedures, researchers would have to amplify DNA so that there was enough

---

[91] G.M. Church and S. Kieffer-Higgins, "Multiplex DNA sequencing", *Science*, vol. 240, 1988, pp. 185–188.

[92] Another interesting method developed by Pohl implies that the traditional form of sequencing by the Sanger method be followed with the direct hybridization of the fragments obtained by the separation of the gel on a membrane. A system of enzyme elucidation avoids the need for radioactive tags. The advantage of the system is that it is possible to obtain long sequences with a higher resolution than with self X-rays. However, transferring the fragments to the membrane is a very delicate operation. Please refer to the following articles for a panorama of the evolution of sequencing techniques:

- S. Hohan and F. Galibert, "L'évolution des techniques de séquençage", *Biofutur*, Genome Edition, no. 94, October 1990, pp. 33–37.

- M. Parenty, "A l'ère de l'automatisation", *Biofutur*, Genome Edition, no. 146, June 1995, pp. 34–38.

150 *From Biotechnology to Genomes: A Meaning for the Double Helix*

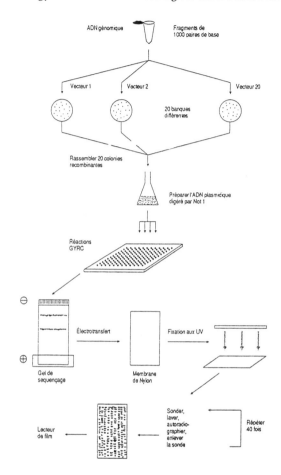

Fig. 8 The stages of multiplex sequencing
(From *Science*, 240, 1988, p. 185)

Several genomic libraries are built in plasmidic vectors, each containing a specific nucleotide sequence as a target sequence. Groups of 20 colonies from each of the 20 libraries are then formed and the plasmids of each group are purified. Enzymatic digestion then allows the inserts containing the genomic DNA fragments, flanked by their two target sequences, to be excised. The sequencing operation is then carried out according to the Maxam and Gilbert method and the fragments obtained are separated out by electrophoresis on a polyacrylamide gel. A probe looking for the characteristic sequence of the vector is used to pinpoint the fragment from that particular clone. It is then removed, and the operation is carried out again using the other probes for the other vectors until each clone is identified. At least 20 reactions can, therefore, be carried out simultaneously, one gel being used for each separation. In fact this technique has not been used and has never worked.

of it to be sequenced. This was solved by Stanley Cohen and Herbert Boyer. However, before examining this revolutionary new technique, restriction enzymes[93], or endonucleases, need to be discussed.

During the 1960s, Salvador Luria[94] and his team discovered that when a particular strain of *E. coli* B/4 was infected with the T2 phage, the phage's descendants could no longer reproduce inside the original normal *E. coli* host strain. They deduced that the phages released by the B/4 strain were modified in such a way that their growth was impossible inside the normal host. They called this phenomenon "controlled and modified restriction" by the host bacteria. By 1960, W. Arber and D. Dussoix[95] had shed some light on the phenomenon of restriction. By growing the λ-phage on *E. coli* and using the descendants to infect *E. coli* B, they showed that most of the phage DNA had been broken up. The DNA was being restricted. The decisive breakthrough of identifying the restriction enzymes that play a role in restriction modification[96] was made in 1970. In 1969, Boyer was studying restriction enzymes at the University of California (San Francisco). The ones he was interested in came from *E. coli* and had the special property that when they were used as biochemical scissors to cut the double strand,

---

[93] We can define a restriction enzyme as follows: an enzyme which recognizes a specific DNA sequence and cuts the DNA either in or close to this sequence, depending on the class of enzyme.

[94] S.E. Luria, "Host induced modficiation of viruses", in *Cold Spring Harbor Symposia Quant. Biol.*, vol. 18, 1953, pp. 237–244.

[95] W. Arber and D. Dussoix, "Host specificity for DNA produced by *E. coli* I. Host controlled modification of bacteriophage lambda", *J. Mol. Biol.*, vol. 5, 1962, pp. 18–36.

[96] W. Arber, "DNA modification and restriction", *Prog. Nucl. Acid. Res. Mol. Biol.*, vol. 14, 1974, pp. 1–37. These restriction enzymes, or endonucleases, recognize a specific DNA sequence and cut the two strands at that point. This is linked to the restriction in the following way: a bacteria has two enzymes (or groups of two enzymes, depending on the number of modifications of which it is capable), which both recognizes a particular DNA sequence. One of them, the modification enzyme, catalyses the addition of a group (which is often a methyl) to one or several of the nucleotides of the sequence. The second, or restriction enzyme, only recognizes the unmodified nucleotide sequence and is used to digest all exogenous DNA unless it has also been modified like that of the cell itself. More than 200 restriction enzymes have already been found to date. All these enzymes have been found in prokaryotes. Restriction enzymes can be divided into two categories according to their mechanism for cleaving DNA. The first group of enzymes recognize a specific sequence of nucleotide pairs and cleave the DNA at a non-specific point beyond the point of recognition. The enzymes in the second group cleave the DNA at the recognition site. The recognition sequences are at the center of a symmetry (palindrome).

they did not cut it neatly. They produced sideways-cut ends, one strand continuing further than the other. The scientists called these "sticky ends" for the very good reason that the protruding strand had exposed bases and these, of course, attracted their other complementary bases. These sticky ends are a bit like the specific hooks on Velcro that tend to join up with complementary ends if you give them half a chance (Figs. 9A and 9B).

Stanley Cohen immediately grasped the potential of these biochemical scissors when he heard Boyer mention them at a conference in Hawaii in November 1972. At that time, Cohen was working on plasmids[97], microscopic rings of DNA that float outside the cell's chromosomes and replicate autonomously, that is to say, independently of the chromosomes of the cell but in a coordinated fashion. As he listened to Boyer, Cohen realized that any piece of DNA obtained with the enzymes that Boyer was talking about could be inserted into a plasmid that had also been cut with the same enzyme, regardless of species. It was theoretically possible to extract a gene from one organism, perhaps a mammal, and insert it, for example, into a fly plasmid (DNA vector) because Boyer's enzymes produced these sticky ends, (for the method, see Fig. 10). Cohen immediately organized a meeting with Boyer, who agreed in principle on a cooperation which would have profound consequences, not only for fundamental science, but equally for the worlds of medicine and international finance.

What Cohen and Boyer were thinking of, and were to carry off, was quite revolutionary as well as remarkable in its theoretical simplicity. They created a composite plasmid from two different types of bacteria and inserted it intact into an *E. coli* bacteria. The new plasmid replicated many times inside each bacterial cell (the bacteria itself dividing every 20 minutes). In this way, they could make copies of the new plasmid genome in a very short time. The report of this historic experiment was published in November

---

[97] Plasmids were discovered by Joshua Lederberg during his work on bacterial genetics. Plasmid DNA is circular and double-stranded. It carries the genes necessary for its own reproduction and perhaps some others. The plasmids most often used in genetic recombination experiments generally carry genes for antibiotic resistance, which makes it easier to identify and select recombinant DNA.

Enzyme	Recognition sequence	Cutting Sites	Ends
BamHI	5'-GGATTC-3'	G↓G A T C C C C T A G ↑G	5'
EcoRI	5'-GAATTC-3'	G↓A A T T C C T T A A ↑G	54
HaeIII	5'-GGCC-3'	G G↓C C C C↑G G	Blunt
HpaI	5'-GTTAAC-3'	G T T↓A A C C A A↑T T G	Blunt
PstI	5'-CTGCAG-3'	C T G C A↓G G↑A C G T C	3'
Sau3A	5'-GATC-3'	↓G A T C C T A G↑	5'
SmaI	5'-CCCGGG-3'	C C C↓G G G G G G↑C C C	Blunt
SstI	5'-GAGCTC-3'	G A G C T↓C C↑T C G A G	3'
XmaI	5'-CCCGGG-3'	C↓C C G G G G G G C C↑C	5'

Fig. 9A Recognition sequences and cutting sites of some of the restriction endonucleases

The recognition sequences are given in single-strand form, written 5'—> 3'. The cutting sites are given as double strand from to illustrate the type of ending provided by a particular enzyme.

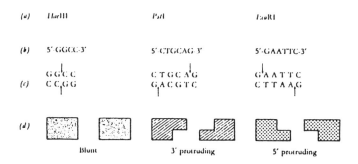

Fig. 9B Ends provided by some endonucleases

The enzymes are listed in a) with their recognition sequences and their cutting sites in b) and c) respectively. In d) there is a schematic representation of the endings generated.

1973 in the *Proceedings of the National Academy of Sciences*[98]. The age of cloning had begun[99].

This ability to obtain limitless examples of genetic fragments, coupled with the rapid DNA sequencing techniques developed by Sanger and Gilbert, allowed molecular biologists to isolate fragments of DNA, multiply them and read their sequences. The gene, a discrete, invisible entity, which had only been postulated a little over a century before by G. Mendel, was now exposed to the explorations of molecular biologists. The black box of our heredity was open, and it would not be long before the scientists started to list the contents. The time of the great sequencing programmes was fast approaching.

Fresh from his success in the cloning experiments with Cohen, Boyer was approached by a 28-year-old financier by the name of Robert Swanson. Swanson, a trained biochemist, turned ambitious venture capitalist, in 1974, when he joined the team of Kleiner and Perkins[100], a small venture capital firm. He was convinced, unlike many businessmen at the time, that considerable fortunes could be made from molecular biology. Swanson began by making a list of the most famous researchers in the field, including Paul Berg (of Stanford University) and Herbert Boyer. Swanson contacted them and proposed a three-way partnership in a firm that would use the new techniques of molecular biology. Paul Berg refused, but Herbert Boyer did not. Swanson asked him whether he thought it would be possible to use the new technique not only to create identical copies of DNA fragments, but also to obtain the products of the genes.

---

[98] S.N. Cohen, A.L.Y. Chang, H.W. Boyer and Helling, "Construction of biologically functional bacterial plasmids *in vitro*", *Proc. Nat. Acad. Sci.*, USA, vol. 70, 1973, pp. 3, 240–3, 244.

[99] We should be very careful not to misunderstand the term "clone". In its original sense, this term implied the creation of a whole organism from a sample of its cell tissue, a concept which fascinated many science-fiction writers (for example A. Huxley in *Brave New World*). However, the definition that became common after Boyer and Cohen's experiment was simply the use of living organisms for the indefinite reproduction of DNA fragments. The process is a sort of biological photocopier. Written into a bacteria, the plasmid that carries a foreign gene can create a thousand copies of itself and that foreign gene (Fig. 9).

[100] Kleiner and Perkins had just put some money into Cetus, a small biotechnology firm in Berkeley, which was exploiting the discoveries in molecular genetics in a commercial fashion.

3 The Foundations of the Heralded Revolution 155

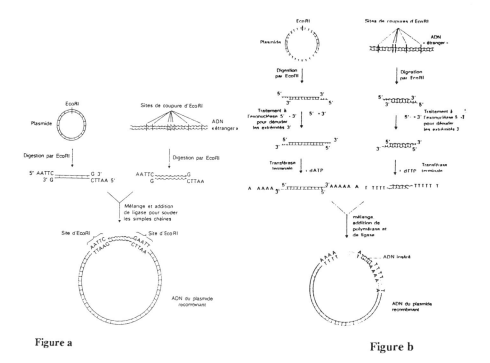

Figure a   Figure b

Fig. 10 Making a recombinant plasmid

We know that some endonucleases cut the two DNA strands in a zigzag fashion at the point of the specific recognition sequence. Foreign DNA can also be cut in segments by a restriction enzyme that will also open the plasmid DNA (for example, pBT322, the most frequently used laboratory plasmid), which opens up into a straight line. Fig A. above illustrates this process for the enzyme EcoRI. It cuts DNA at a particular nucleotide sequence. The single-strand ends of the plasmid and the fragments of foreign DNA are found to be complementary. In a solution, these two types of DNA molecule will meet and pair up to form a circular DNA molecule of greater diameter than the original plasmid, held together by hydrogen bridges across the complementary extremities. When polynucleotide ligase is added, the gaps in the sugar-phosphate structure between the single strands are soldered together with covalent liaisons (four in all) which stabilizes the structure. This is a recombinant DNA molecule. As the ends of the foreign DNA fragment obtained by digestion with *Eco*RI are all the same, these fragments can slot in one direction or the other at random. Their direction makes a difference when the DNA fragment is transcribed (it will contain one or some or parts of genes according to the frequency of *Eco*RI the cut sites) since the initiation of transcription depends on the presence of a promoter sequence and the plasmid's own control sites. In fact, the length of the single-strand ends formed on a DNA molecule by the action of a zigzag cleaving restriction enzyme is quite small, and the probability that the complementary extremities will meet in a solution is also quite small. In addition, some restriction enzymes do not cut a zigzag, resulting in DNA fragments that are double-stranded all the way to the end of the chain (blind-end DNA). In these cases, one of the strands of the DNA molecule is lengthened with another enzyme called terminal polynucleotide transferase. For example, in the presence of ATP, this enzyme catalyses the formation of a poly(dA) on each 3' end of the DNA. So that this will help with the insertion of a fragment of DNA into a plasmid, tails of poly(dT) of the same length have to be polymerized on the 3' ends of the linear plasmid DNA. (Fig. B). Then the two types of DNA are mixed in a solution, the polynucleotide ligase is added, and a circular recombinant DNA molecule is formed.

In other terms, could genetically modified bacteria produce foreign proteins; in particular, could they produce human proteins? Boyer was confronted with general skepticism from his colleagues, but he gave Swanson a positive answer because he was convinced that a plasmid could code for a protein and let the bacteria's internal gene expression machinery produce it. Boyer's "Yes" set in motion forces that were to reawaken hopes in biotechnology, which was going to be so profoundly transformed by the development of recombinant DNA techniques that its whole history would be overshadowed. It was biotechnology's "renaissance".

Boyer borrowed $500 and joined Swanson in the creation of a company to exploit the new science and technologies. They called it Genentech for GENetic ENgineering TECHnology. Genentech was born in April 1974, with Robert Swanson as its only employee. Herbert Boyer was only an associate, remaining at the University of San Francisco. The start-up capital of $200,000 was supplied by Kleiner and Perkins, Swanson's old company, but it was not enough, so Swanson went in search of more money. In April 1977, he scraped together a million dollars from the Mayfield Fund, International Nickel, Innova (in which Monsanto has a shareholding) and Sofinova (a French finance company). At the same time, on Boyer's advice, Genentech signed two contracts with other California research centers, one at the University of San Francisco, the other at the City of Hope, near Los Angeles. Of course, if there were any commercial benefits, the universities would get their share of the royalties. The company's first target was modest: insulin, which may seem a strange choice today. Since Frederick Banting and Charles Best of the University of Toronto had shown, in 1922, that the absence of this critical protein was the cause of diabetes, insulin extracted from cows, pigs and other animals had been used to prevent the fatal accumulation of glucose in diabetics. Since millions of lives had been saved in this way, why was a new source of insulin necessary? There were three answers: first, 10% of diabetics were having secondary reactions to bovine or porcine insulin; second, doctors showed that there was a clear rise in the incidence of diabetes that would continue as more people worldwide adopted western lifestyles. (Some analysts agreed that insulin would become a precious substance, and Genentech thought it could fill that gap or at least prevent shortages); third, and most important, the company wanted to produce a protein with a recognizable name.

They decided to have a first attempt on another gene, coding for the protein somatostatine[101]. In order to do this, they called in two other researchers, Arthur Riggs and Keiichi Itakura, the first scientists to be hired by Genentech. Riggs and Itakura were experts on somatostatine, which at the time was being extracted in very small quantities at very high cost from the brains of thousands of sheep with freshly cut throats. Genentech realized that with recombination techniques it could do better. On 15 August 1977, the scientists managed to insert the gene for somatostatine into the genome of some bacteria that were then "persuaded" to produce somatostatine under its guidance. This success was a big step forward since it was the first time that a human protein had been produced by a living creature outside the body of a human being. From then on, everything moved faster and faster. In April 1978, five venture capital firms (International Nickel, Mayfield Fund, Innova, Sofinova and Hillman Corp.) poured another million dollars into the coffers of Genentech, which was then able to expand to some thirty staff, three-quarters of whom were researchers. In May, it was Wilmington Securities Inc.'s turn when it paid $500,000 for 250,000 shares. At the same

---

[101] Somatostatine is a protein discovered in 1973 by Roger Guillemin's team at the Salk Institute, near San Francisco. It is involved in a series of biochemical activities in the body. Its precise function is to inhibit the production of another natural protein, human growth hormone or HGH. If a person's body does not produce enough somatostatine, it will produce unlimited quantities of HGH, leading to the rare condition of gigantism. Somatostatine is a small protein of 14 amino acids.

time, Genentech signed a new contract with a national medicine research center at the City of Hope. This was only the beginning for them. Insulin remained the strategic objective, and the success they had had with somatostatine had not altered Genetech scientists' determination to be the first to clone the gene for insulin, making it possible to produce proteins in commercial quantities. They were also aware that researchers in other departments of the University of California (San Francisco) were also interested in insulin, as was a group of Gilbert's at Harvard. The race turned out to be intense and exciting.

On 9 June 1978, Gilbert revealed that his group had cloned and expressed insulin in a bacteria. This was a terrible disappointment for Genentech, but they were soon to realize that the race was far from over. The insulin that Gilbert and his team had cloned was only rat insulin and, in addition, the insulin race was not Gilbert's only focus of attention at the time. In 1977, scientists had discovered that in a gene, not all the DNA codes for a protein. There are non-coding sequences throughout the gene, and these are removed during transcription into RNA. These non-coding sequences make a gene a lot longer than necessary for coding for the protein, and to this day, apparently have no function, constituting a frustrating mystery for molecular biologists. Gilbert named these non-coding sequences "introns" and coding sequences "exons". He even speculated very elegantly in *Nature* on the evolutionary origin of these intergene segments [102].

In the end it was Genentech that won the race. In the early hours of 24 August 1978, they managed to mix balanced quantities of proteins from the α and β chains of insulin. The two proteins had been produced separately in *E. coli* (Fig. 11). The researchers used radioactive tags to show that a small quantity of human insulin had been produced. Two weeks later, the results of this historic experience were announced along with the news that Genentech had signed a long-term contract with Eli Lilly, the pharmaceutical giant that, at the time, controlled 80% of the $137 million annual insulin sales market. This was three years after Swanson had approached Boyer, an amazingly short time, certainly a lot shorter than had been predicted by

---

[102] W. Gilbert, "Why genes in pieces?", *Nature*, vol. 271, 1978, p. 501.

other molecular biologists. The beginning of their research was followed with unprecedented speed by the industrial manufacture of this first human hormone produced through genetic engineering in a shining demonstration of how ready the technology was for application and a clear confirmation of Boyer's intuition.

During this same month of August 1978, Genentech signed another long-term contract with Kabigen AB, a daughter company of the California firm AB Kabi, a firm of Swedish origin that was the main world producer of growth hormones. The contract was for the production of the same growth hormone through genetic engineering. This was achieved by Genentech with *E. coli* bacteria, in 1979. Soon after this contract was signed, Genentech signed another contract with Hoffman La Roche for the highly-esteemed and potential "goldmine" protein interferon. In parallel, other research continued, and additional funds were being poured into research coffers. In July 1979, Swanson's company managed to produce thymosin alpha-1, a hormone secreted by the thymus and which is thought to stimulate the immune system.

All this success and development encouraged other companies to invest in Genentech. In September 1979, the industrial chemicals firm Lubrizol purchased a million dollar shares in Genentech for 10 million dollars in cash. This huge transaction was indirect proof that Genentech was looking to diversify and not solely concentrate on its pharmaceutical sector research. By the early 1980s, Swanson's company had more than 120 staff members and announced its new product, proinsulin, the precursor for insulin. In June 1980, Genentech revealed the results of its contract with Hoffman-La Roche: the production of the first micrograms of two types of interferon, leukocyte interferon and fibroblastoid interferon. Genentech, with the contribution of university research centers and the pharmaceutical industry, could boast no less than seven products developed in their four years of existence, in addition to the 200 pending patent requests.

A horde of new biotechnology companies, mainly based in the USA, were springing up. The expertise of these commercial research organizations lay entirely in the genetic modification skills of their employees, thanks to which it rapidly became possible to insert a growing number of human

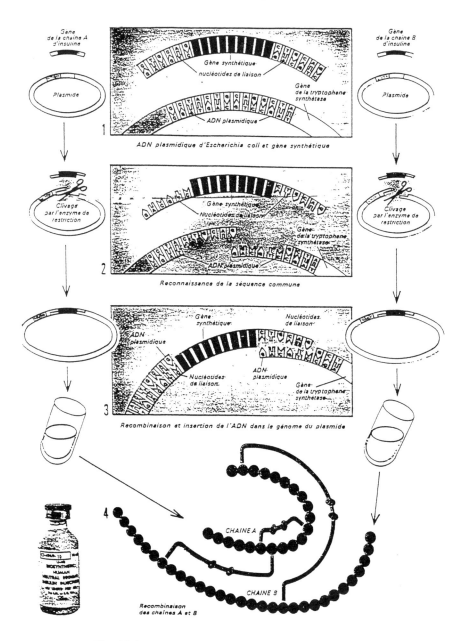

Fig. 11 Production of insulin through genetic engineering

genes into bacteria and other simple organisms. Dozens of medically useful and commercially viable substances have since been created with recombinant DNA techniques.

One of the first companies to take advantage of the new technology was Biogen, co-founded in 1978 by Walter Gilbert. Gilbert had lost the insulin production race, but with Biogen, he fell to work on other possible products including proteins such as the various forms of interferon, which was thought to have anti-cancer and anti-virus properties. Other companies created at this time were Genex (1977), Centocor (1979), Amgen and Chiron, who had their sights set on products such as growth hormones to treat dwarfism, blood clotting Factor IX to treat hemophilia, the tissue activator of plasminogen to fight infarction, the tumor necrosis factor for anti-cancer treatments and erythropoietin (red blood cell regenerator), all of which were produced during the course of the 1980s.

Genetic engineering in the late 1970s was not just about research in laboratories but also implied industrial involvement and profit. During 1980, the world of finance grasped the nature of this high technology industry with real enthusiasm, as clearly seen from the rise of Genentech share prices on the New York Stock Exchange, which on 24 October 1980, rose from $35 to $89 in twenty minutes. *Time* magazine was so impressed by these "technological pyrotechnics" that it put Boyer's photo on its 9 March 1981 cover, a rare distinction for a scientist.

The main shareholders of Genentech made colossal profits. Boyer's 925,000 shares reached about $86 million, and although the share price eventually stabilized at about $70 per share, Boyer had become a very rich man, at least on paper. Many other shareholders in Genentech such as Kleid and Goeddel had been given thousands of shares on their arrival at the company, and these had turned into small fortunes. Never had scientific success been so well rewarded.

3.5 Gene Money, or the Miracles Expected of Biotechnology

Genentech's story has been quoted often to justify the hope that biotechnology will turn DNA into gold. The arsenal of techniques available to biologists led to a belief in a radiant future, a future where biotechnology's often-mentioned potential would finally flower.

With recombinant DNA technology, the techniques of sequencing and Kary Mullis's PCR technique [103] as well as the progress made in gene mapping [104] and physical mapping [105], biological and genetic techniques were opening the gates to new dimensions in the life sciences. They provided a glimpse of how to make yesterday's, today's and tomorrow's medicines at lower cost, developed new diagnostic methods and heralded new therapeutic techniques. In agriculture, which was thought to offer the most potential in socioeconomic benefits, there was very early talk of creating hybrid species

---

[103] Kary Mullis "L'amplification des gènes", *Pour la science, dossier La génétique humaine*, April 1994, pp. 18–25. One night in April 1983, Kary Mullis invented a technique that allowed the synthesis of an unlimited number of copies of genes: this is called the gene amplification method of Polymerase Chain Reaction (PCR). This technique, which earned him the Nobel Prize for Chemistry, works in the following way: first you separate the two strands of target DNA by heating it. Then you pair the target DNA strands with intact triggers while they're cooling down. Secondly, you add dNTPs and polymerase DNA, which adds bases until the complementary DNA sequence is complete; this will keep going indefinitely. The reiteration causes the production of exponential amounts of copies of the target DNA. Adding ddNTP and polymerase DNA allows you to sequence the target DNA. If you use polymerase DNA from *Thermus aquaticus*, a bacteria that lives in hot springs, it resists the heat and greatly improves the method. This technique, which was the subject of a patent request from Cetus in 1985, which was granted in 1987 and 1989, has caused many international legal conflicts on the validity of patents granted to Cetus and then to Hoffman-La Roche. Cetus merged with Chiron in 1991 and sold to Hoffman-La Roche all its patent rights to PCR for $300 million (FRF 1.7 billion). For more on this issue, read Jocelyne Rajnchapel-Messaï's article, "PCR, Il faut payer pour amplifier", *Biofutur*, March 1994, pp. 53–56.

[104] Gene mapping is used to determine the position of a certain number of hereditary characteristics with respect to each other. It is based on the study of the transmission of characteristics and examines the possible associations in their transmission. Two elements are indispensable for this sort of analysis: polymorphism, that is to say different versions of the same characteristic and the gene determining it, and access to the study of as widely extended families as possible in which family relationships are quite certain. The essence of gene mapping is now based on new techniques such as Restriction Fragment Length Polymorphism (RFLP), which is based on variations in the placing of restriction sites limiting a particular section of DNA, such as the fact that the fragments defined by restriction sites can be of different lengths in different individuals. To draw a genetic map, you have to determine the order of a sufficient number of RFLPs along the genome to knit as tight a mesh as possible, and to do this you should isolate the probes defining the RFLPs. For the techniques for drawing genetic maps, please read chapter four of this work.

[105] The physical map is a closer description of molecular reality, and it implicitly refers to the DNA. It determines the position of the genes on the chromosomes and the distance between them along the DNA molecule in nucleotides. The physical map is actually a total DNA sequencing of the region being studied, but more rapid methods allow rougher maps to be made.

3 The Foundations of the Heralded Revolution 163

that could not be obtained by traditional breeding methods and obtaining as many plants as one wanted from a few leaf cells. There was also new hope for the recurring biotechnological dream of creating plant species capable of fixing the air's nitrogen themselves, thereby needing hardly any fertilizer for their growth. In the agro-food sector, genetic engineering was still seen as a potential provider of large contributions and improvements. These important sectors, which demonstrate the medical and agricultural hope being invested in genetic engineering, are not the only sectors which might benefit in the long term from biotechnology; others such as the energy, biofuels and biomass, water depollution, and food sectors stand to benefit as well. With their ability to improve the performances of micro-organisms or create new ones, the new techniques emerging from research were expected to transform the economies of society.

DNA became an "Open Sesame" for economic rebirth, and biotechnology was seen as a beneficial electric-shock treatment to industry. Political and economic managers, aware of how big possible markets could be, became more interested in these new techniques, giving biotechnology its second wind [106].

The progress of molecular genetics, which started in 1960 and had gone from strength to strength, is very different to that of population genetics. Molecular genetics achieved the status aspired to by the proponents of the synthetic theory [107], which it supplanted. With an arsenal of different techniques for exploring the genome, molecular genetics worked on the

---

[106] A study of the use of the word biotechnology in the scientific articles shows the exponential growth of its use between 1969 and 1984 in articles indexed by the *Chemical Abstract*. Since 1983, the use of the word biotechnology has been absolutely explosive and has since been highlighted in its own biological index, the *Biosis Previews*. This sharp rise in interest in biotechnology is characteristic of the attitudes of society as a whole and can also be seen in the *National Newspaper Index* and in *Investest*, an investment database. See M. Kennedy, "The evolution of the word biotechnology", *Trends in Biotechnology*, vol. 9, 1991, pp. 218–220.

[107] In the 1950s and 1960s, physiological genetics and population genetics were biology's avant-garde. Together more coherent and structured, they backed up the neo-Darwinian view of evolution as never before. They were brought together in the synthetic theory of evolution, a theory officially supposed to explain the evolution of species. As its components belonged to the main biological disciplines, it tended to dominate all the life sciences as a sort of exoskeleton theory. This did not last for long since with the discovery of the gene, physiological genetics became molecular genetics, which took first place.

individual and its physiology, areas to which population genetics had always been rather indifferent. Molecular genetics' dominance brought its preoccupation with the individual to the fore. The population approach became secondary to the extent that eugenic ideas began to disappear from biology, or so it seemed. In addition, the physico-chemical aspect of molecular genetics made it look like a "hard science", so its importance was affirmed. The naturalist and other disciplines less related to physico-chemistry were displaced. Any approach that did not fit in with the theories of molecular genetics was accused of ideological bias with terms such as Lyssenkism, creationism, vitalism and so forth. In its ultimate refinement by defenders and enthusiasts, molecular genetics was used to attack eugenics by revealing its unscientific and ideological nastiness, denouncing it indignantly. This discrediting of eugenics was accompanied by a change in the theory of evolution, which like population genetics, became of second-rate interest. Evolution was now simply a framework in which molecular genetics worked. Molecular genetics was about the gene program, not evolution's overall "program[108]".

This revolution and domination of the molecular point of view occurred from 1950 to 1970, but eugenic ideas were not entirely eliminated. Although they had been dissipated by the advent of genetics and molecular biology, eugenic concerns resurfaced gradually and insidiously in biologists' and molecular geneticists' discussions. Only slowly did medical, commercial and economic opportunities offered by genetic engineering, molecular genetics and molecular biology become of prime importance. Edward Tatum was one of the first to move from the idea of molecular biology as a science to biological engineering as a technology. In his inspired speech on receiving the Nobel Prize, Tatum drew up the implications of the 1950s biotechnology revolution. He pointed out that genetics could be used to correct genetic defects and asked his audience to think about the perspectives for biological engineering[109]. Ten years later, Tatum was still highlighting these medical perspectives:

---

[108] Molecular genetics leaves the overall plan to the theory of evolution, on which it imposes but one constraint: the programme must be of a "random need" nature.

[109] E.L. Tatum, "A case history in Biological Research", *Nobel Lecture*, 11 December 1958.

"The time, has come, it may be said, to dream of many things, of genes — and life — and human cells, of medicine — and kings"[110].

Despite the tragedy of the Nazi experiments, discussions on eugenics in the 1960s had a strong continuity with the more positive pre-war eugenics ideas such as those of the Huxley family, J.B.S. Haldane, Theodosius Dobzhansky and Hermann Müller. During two meetings between pre-war eugenicists and the new generation of molecular biologists, the fifteen-year silence on eugenics was finally broken.

The first meeting, Man and his Future[111], at the CIBA Foundation in London, was suggested by Gregory Pincus, the father of the contraceptive pill. During the meeting, the generation of reformed eugenicists confronted the new generation of molecular biologists, and a lot was said about the past. Stanford Professor and Nobel prizewinner Joshua Lederberg pointed out how important it was to focus on medical applications; in particular, curing sick people. Although he brought up the possibility of modifying the genotype (which would affect future generations), he emphasized the importance of changing the phenotype after conception:

"Embryology is very much in the situation of atomic physics in 1900: having had an honorable and successful tradition, it is about to begin[112]!"

The theme of a new biological revolution was reiterated by Tatum at the second colloquium, held at Wesleyan University in Ohio. His own speech led beyond his previous call for biological engineering by suggesting the advent of genetic engineering:

---

[110] E.L. Tatum, "Molecular biology, nucleic acids and the future of medicine", *Perspectives in Biology and Medicine*, vol. 10, 1966–67, p. 31.

[111] G. Wolstenholme ed., *Man and his Future*, J.A. Churchill, London, 1963.

[112] J. Lederberg, "Biological future of man", in *Man and his Future*, ed. G. Wolstenholme, J.A. Churchill, London, 1963, pp. 263–273.

"Biological engineering seems to fall naturally into three primary categories of means to modify organisms. These are: 1) the recombination of existing genes or eugenics; 2) the production of new genes by a process of directed mutation, genetic engineering; 3) modification or control of gene expression, or to adopt Lederberg's suggested terminology, euphemic engineering"[113].

Tatum's vision inspired and terrified biologists and the public for more than a quarter of a century. His announcement of these revolutionary perspectives caused concern and warnings. It prompted Rollin Hotchkiss, who had also been at the second colloquium, to alert the scientific community as to their responsibilities. He urged scientists not to move too quickly and to prevent the uncontrolled growth of genetic engineering in his article, *Portents for a genetic engineering*[114]. These warnings were backed up by extraordinary declamations of the future power of biological engineering[115].

Lederberg was an important influence in the development of public awareness of the new genetics. His warnings were about work on humans, but he had no objection to eugenics as applied to non-human species to produce homogenous genetic material or even as applied to man for his own good. Lederberg's approach was centered on medical applications and implications. Having persuaded C. Djerassi, owner of Syntex[116], of the potential of molecular biology, he was provided with a laboratory to explore the consequences of genetic engineering. In particular, he worked on the

---

[113] E.L. Tatum, "Perspective from physiological genetics", in *The Control of Human Heredity and Evolution,* ed. T.M. Sonneborn, Macmillan, New York, 1965, p. 22.

[114] D. Rollin Hotchkiss, "Portents for a genetic engineering", *Journal of Heredity*, vol. 56, 1965, pp. 197–202.

[115] Let us also note that a number of books and programs pointed out the risks of the new technology, such as G. Rattray Taylor's *The Biological Time Bomb*, New American Library, New York, 1968, Michael Crichton's *The Andromeda Strain*, Gentesis Corporation, 1969, trad. fr. *La variété Andromède*, ed. Pocket, 1995. We could also mention Woody Allen's satiric film *Sleeper*, with the cloning of a person from his severed nose, and the cloning of Hitler from surviving cells which was the theme of the book and later the film *The Boys from Brazil*. The fear of strangers and foreign individuals and organisms was widespread. When the television Star Trek first came out, science and science fiction converged.

[116] Carl Djerassi was the first to synthesize progesterone.

3 The Foundations of the Heralded Revolution 167

mechanisms of hormones, nucleic acids and resistance to foreign organisms, and he investigated the physiology of reproduction to develop new methods for more efficient control of fertility in humans, animals and plants. Lederberg was not the only person working on this. In 1967, Brian Richards and Norman Carey, who were working at the British laboratory of G.D. Searle[117], also began to look at the technological implications of biological engineering and the transfer of genetic material between organisms.

Boyer and Cohen's 1973 experiments amplified worries as well as hopes in molecular biology. It should be remembered that it was the inventors and scientists themselves who, aware of the potential risks, decided to meet at Asilomar in February 1973, and again in June 1975 [118], to alert policymakers and the public to the potential risks of the new biotechnological techniques, in particular recombinant DNA techniques. They recommended urgent action for the provision of appropriate regulations. In June 1973, at the annual session of the Gordon Conference on Nucleic Acids in New Hampton, New Hampshire, at which the risks of research using recombinant DNA techniques were revealed, Maxime Singer and Dieter Soll drew up a letter addressed to the National Academy of Sciences and the Institute of Medicine[119]. They requested the formation of a study committee to examine the biological risks of the new techniques and recommend suitable action. As a result, in February 1974, the National Academy of Sciences appointed Paul Berg the chairman of a study committee. The eleven members of the committee, all active in research involving recombinant DNA techniques, were aware of the speed at which research was progressing and worried about accidents. Their report was published in *Science* on 27 July 1974[120], and in *Nature* at about the same time. It recommended, first of all, that research involving the following types of experiment be postponed until the potential hazards of

---

[117] B.M. Richards and N.H. Carey, "Insertion of beneficial genetic information", *Searle Research Laboratories*, 16 January 1967.

[118] P. Berg *et al.*, "Asilomar conference on recombinant DNA molecules", *Science,* vol. 188, 1975, p. 991.

[119] This letter, entitled "DNA hybrid molecules", was published in *Science*, vol. 181, 1973, p. 114.

[120] Paul Berg *et al.*, "Potential biohazards of recombinant DNA molecules", *Science,* vol. 185, 1974, p. 303.

such recombinant DNA molecules have been better evaluated or until adequate methods are developed for preventing their spread throughout the world.

> Type 1: Construction of new, autonomously replicating bacterial plasmids that might result in the introduction of genetic determinants for antibiotic resistance or bacterial toxin formation into bacterial strains that do not at present carry such determinants; or construction of new bacterial plasmids containing combinations of resistance to clinically useful antibiotics unless plasmids containing such combinations of antibiotic resistance determinants already exist in nature.
>
> Type 2: Linkage of all or segments of the DNAs from oncogenic (cancer-inducing) or other animal viruses to autonomously replicating DNA elements such as bacterial plasmids or other viral DNAs. Such recombinant DNA molecules might be more easily disseminated to bacterial populations in human and other species and possibly increase the incidence of cancer or other diseases.

Second, plans for linking animal DNA fragments to plasmids or bacteriophages were to be evaluated with particular attention to the fact that quite a lot of mammal cellular DNA contains sequences common to those of the RNA tumor virus.

Third, the report asked the National Institute of Health (NIH) to immediately consider the creation of an advisory committee with a mandate to:

> give immediate consideration to establishing an advisory committee charged with i) overseeing an experimental program to evaluate the potential biological and ecological hazards of the above types of recombinant DNA molecules; ii) developing procedures which will minimize the spread of such molecules within human and other populations; and iii) devising guidelines to be followed by investigators working with potentially hazardous recombinant DNA molecules.

Fourth, the report foresaw the organization of an international colloquium of scientists involved in research using recombinant DNA techniques to evaluate scientific progress in this field and discuss appropriate measures to handle the potential biological risks of recombinant DNA molecules.

The Asilomar meeting was carefully prepared and received worldwide press coverage, sparking a complicated international debate involving the worlds of science, politics and economics as well as the public as a whole in the delicate problem of regulating modern biotechnology. An unprecedented appeal for a break in research was launched until experimentation could be regulated; to reassure the public. Furthermore, this call was heard after Asilomar. There was a 16-month break in research until the NIH guidelines were finally made available in 1978.

The events of Asilomar have been covered in many books[121]. These perpetuate an image of the event as the beginning of the age of biotechnology, rather than presenting it as the culmination point of two decades of thought on the threat of genetic engineering. It partially reveals the domination of the new generation, but it also underlines the fact that the discussions of Asilomar were very different from those of the 1960s, although their legacy should not be forgotten. Asilomar explicitly excluded human genetic engineering from the discussion. It is no coincidence that from that time, genetic engineering has come to be associated not only with the manipulation of human beings, but of all organisms. While debate was then mainly focused on biological risks and whether there should be extensive formal regulation or a simple code of practice, the discussions only touched on commercial applications as an incidental, at the beginning of the conference. Lederberg was a notable exception with his vigorous underlining of the potential applications and profits, and he tried to combat the pessimists and their underlying fear by highlighting the importance of the new technologies in diagnostics, therapeutic medicine and the pharmaceutical and agro-food industries.

---

[121] To name but a few: Sheldon Krimsky's *Genetic Alchemy*, MIT Press, Cambridge, 1983, J. Daniel Kevles and Leroy Hood's *The Code of Codes*, Harvard University Press, 1992, J.D. Watson and J. Tooze, *The DNA Story, a Documentary History of Gene Cloning*, W.H. Freeman and Company, San Francisco, 1981.

Lederberg may have been an exception at Asilomar, but along with the public's perceptions of molecular biology and biotechnology, it was his vision of the biotechnological industry that directed the development of the debate. Mark Cantley has clearly shown [122] that political and scientific debates on biotechnology regulation are mostly centered on the industrial and economic implications of molecular biology and its techniques. Communications to the general public have gone from a revelation of the risks and a pessimistic view of the future to the highlighting of potential medical applications and the economic interests of biotechnology, in particular, in molecular genetics.

Genes did not just mean dollars. They were also becoming increasingly linked to hopes of screening and preventing hereditary diseases and identifying "genetic predisposition [123]". From the mid-1970s, researchers were promising that genetic engineering would let us have our cake and eat it too. From 1974 onwards, impressive and speculative advances were listed and compared with equally speculative risks. In 1976, after the series of norm guidelines [124] were published by the NIH, little by little, in all the institutions outside the United States, research began again, ending the moratorium. The guidelines adopted did not dissipate the unconditional opposition to genetic engineering

---

[122] Mark F. Cantley, The regulation of modern biotechnology: a historical and European perspective, a case study in how societies cope with new knowledge in the last quarter of the twentieth century, Chapter 18 of Volume 12 entitled *Legal, Economic and Ethical Dimensions*, of the *Biotechnology* paper published by V.C.H. in 1995.

[123] This notion of genetic predisposition was, and still is, of great help in increasing the number of hereditary diseases which may be treated by supposedly omnipotent science and its techniques in the near future, which always seems to be postponed.

[124] The self-limitation rules that biologists are invited to follow are:
- The rule of obligatory declaration. No conducting of experiments using in vitro recombination techniques without referring it to the ethics committees of the competent ministerial authorities or local institutions.
- The rule of minimal containment a rule of a biological nature. One can only modify cells likely to mutate so that their ecological proliferation would be impossible. The containment would also be of a physical nature according to the degree of risk factors attributed to any one experiment. They should be carried out in isolation areas, with containment in proportion to the possible risk, (hence P1, P2, P3 and P4 laboratories, codes which designate laboratories of ever-increasingly severe security measures).

For a detailed study of the installation of these normative measures, please read Mark Cantley's study mentioned earlier.

by those for whom artificial recombination of chromosomes is a violation of nature and her laws. Nevertheless, there was a realization that the supposed danger was not as real as it had seemed, and simultaneously, another realization that there were many potentially positive aspects to genetic engineering.

By the end of 1979, the fear of the anarchic multiplication of bacteria with unknown properties had died out, and genetic engineering was given a more mature image in society's imagination. The great fear was over, leaving behind a more rational attitude that arose from the new bioethics. Scientists' declarations of the promise of genetic engineering for both economic benefit and the prevention and cure of genetic diseases also helped reassure the public. Genetic engineering was endowed with many virtues: a clean, affordable technology (Genentech's insulin example had (for a time) reinforced the myth that one only had to stoop to gather the products of biological manipulations and sell them as dear as gold) and beneficial to man, with its promises of cures and improvements in living conditions. It would, amongst other benefits, roll out the red carpet over the road to energy independence, bring miraculous medicines, full employment, and on a more mundane note, new markets, which captured the attention of the media, technocrats, policymakers and, of course, industrialists. Bio-industry was the "flavor of the month", like genetic engineering, which had put biologists and biology on the front pages.

Trumpeted announcements of the successes of molecular biology and genetic engineering, their popular image, the promises made by grant-hungry biologists and those justifying funds already granted, rapidly made DNA, genes and genetics "hip" vocabulary and undoubtedly served to reduce media anxieties. This rhetoric[125] reinforced the idea that a new world had been

---

[125] See for example, Sharon McAuliffe and Kathleen McAuliffe, *Life for Sale*, Coward, McLann and Geoghegan, New York, 1981 or US Congress, US House of Representatives, Committee on Science and Technology, Subcommittee on science, research and technology, *Science Policy Implications of DNA recombinant molecule research*, 95th Cong., 1st sess., 1977, or US Congress, Office of Technology Assessment, *Impacts of applied genetics: micro-organisms, plants and animals,* OTA, HR-132, US Government printing office Washington DC, USA, 1981. This OTA report gave official support to the movement that aimed to counter fears. In fact, it was limited to practical commercial industrial benefits.

ushered in with the microbiological production of insulin, announced in 1978. With the gene as its standard-bearer, this new world was based, nevertheless, on old dreams, dreamt well before the advent of recombinant DNA technology. The 1980s have shown beyond doubt that molecular biologists, technocrats and industrialists, mainly in the United States, usurped the technological promises of the century-old tradition of biotechnology. Biotechnology and genetic engineering and its techniques were now no longer obscure denominations of a scientific field at the interface between biology and technology, nor a mere political category (as seen in the emergence of biotechnology programs), but now also an economic category that had become a sort of myth.

Hopes for an increase in conventional biochemical engineering products began to decline[126], and new industrial applications gained impetus from scientists rather than engineers. Biotechnology, with its techniques of recombinant DNA and the pharmaceutical techniques of the hydroma, stimulated new faith in therapeutic proteins, the improvement of biological organisms such as yeast or plants for biological pesticides and for the treatment of genetic diseases.

With these commercial perspectives, scientific advances, industrial involvement and official support converged and biotechnology soon became a normal component of the business world. There were predictions for a radiant future. In the early 1980s[127], it was estimated that the world market for bio-industries would rise from $40 billion[128] to $64.8 billion[129]. In fact, despite several great successes, reality was somewhat darker. This did not, however, prevent biotechnology from being held in high esteem, as its

---

[126] With the fall in oil prices, biogas and gasohol became less attractive. Single cell proteins became less interesting as it became clear that markets were difficult to find. Even industrial enzymes, which were the first products from the field in the 1960s, only had a world market of $500 million in the mid-1980s.

[127] The following estimations are mentioned in Albert Sassons "Les biotechnologies de la bioindustrie", *La Recherche*, no. 188, May 1987, p. 727.

[128] Estimation from the American firm of Genex Corp.

[129] Estimation from T.A. Sheets. In 1985, biotechnology product sales were valued at $10 billion, that is to say, three times the investment made in this field between 1980 and 1985. (cf. M. Bernon, *Biofutur*, vol. 50, 1986, p. 47).

importance lay in future applications and perspectives. It was accepted that the anticipated impacts on both economic and everyday life would be delayed. Although perceived in the light of its economic importance, biotechnology was comparable with information technology, the status of which also depends more on future promise than current commercial applications.

The one essential point that separates biotechnology from electronics and information technologies is that biological products are subject to extremely strict regulation, as with medicines. This fact, which greatly increases the time lag before products can be marketed, had clearly not been taken into account in the early 1980s by those investing massively, and perhaps prematurely, in biotechnology firms. This explains the sudden public infatuation for Genentech when it was floated on the market. Investors had forgotten about the labyrinthine power of the American Food and Drug Administration. Most new biotechnology firms in the early 1980s had no product to sell. Unlike the electronics and computer sectors, their future depended on long term results, the road from fundamental research through development to the commercialization of new products being a long and winding one.

The first field in which the potential value of biotechnological products helped firms overcome the costs of innovation was pharmacology, but even here the development was much longer and more expensive that initially foreseen. In 1988, only five proteins produced in genetically modified cells had been approved by the FDA: insulin, human growth hormone, hepatitis B vaccine, alpha interferon (an anti-virus and anti-cancer treatment) and the tissue activator, plasminogen. Interferon, the most promising substance as far as profits were concerned, turned out to be a disappointment at clinical trials, and many firms changed their targets.

Biotechnology is not affordable for everyone, and the stock exchange ratings of several firms rapidly fell, staff being laid off and bankruptcy lurking. Genetic engineering was "settling". Too many firms had been set up by entrepreneurs jumping onto the biotechnological bandwagon, tempted by apparently easy profits. Three or four years passed without any return on investment, which was unbearable for the venture capitalists. Furthermore, money was getting more expensive; as a dollar could bring as much as 16%

per annum safely invested, why risk it in ventures with long term returns, especially in the pharmaceutical field, where security norms and legislation forced further delays. Only the big firms such as Transgène, Genentech, Cetus, Genex and Biogen managed to survive; the proliferation of smaller firms ended. By the mid-1980s, the pattern of little companies run by one scientist was no longer the paradigm for industrial structure in biotechnology. More and more small companies were becoming research boutiques or were bought out by the pharmaceutical companies. The remaining firms were most often either run by experienced financiers rather than scientists, even if close relationships continued and developed between universities and industrial laboratories.

Despite these disillusionments and the end of the small genetic engineering companies, governments continued to see biotechnology as the technology of the future (along with information and materials technologies), especially as related investments in Japan in the early 1980s were considered by most nations, particularly the USA and the European Union, as being a serious threat.

The question of what sort of policy would be appropriate for biotechnology was one of many being asked in a series of reflections on economic survival. For 30 years since the Second World War, the USA had been the leading nation. Reagan's election in 1980 was a reaffirmation of the electorates' conviction that this would continue to be the case. The United States' success in high technology was, for many Americans, the key to remaining a world power. In Europe, the industries, the nations and the European Union were also reflecting on their industrial future. For all of them, Japan was as much a model as a menace.

3.6 The Japanese Threat and the Human Frontier Science Program

In 1981, the Japanese Science Ministry, MITI, identified biotechnology as a key to the future of industry. Such future technologies, like materials and information technologies, benefited from government funding and perhaps

3 The Foundations of the Heralded Revolution 175

more importantly, were highlighted by industry. Ten-year plans were set up. Biotechnology, particularly, benefited from the American enthusiasm for genetic engineering, but the initiative was also based on a decade of research sponsored under the "Life Sciences" project of the Science and Technology Agency (STA). MITI's interest and targets lay in the applications of biotechnology in chemical engineering, including mass production. During the oil crisis of the early 1980s, these plans were aimed more at energy than medical problems. The subject included bioreactors in industrial use, large-scale cell culture and the application of recombinant DNA techniques. A variety of industrial targets was determined for each program and each of these, in turn, represented research subjects. For example, in bio-reactor technology[130], the target was the development of a chemical industry working at normal temperature and pressure, thus saving energy. This orientation translated into the development of synthesis bioreactors. In its turn, each of these defined research subjects was subdivided into other research themes such as the study of micro-organisms or enzymes. Research on each theme was planned throughout the 1980s. The study of micro-organisms that could produce industrial enzymes was the initial phase of the 1981-84 project. The problem of the cultivation of selected organisms was undertaken in a second phase.

In a separate initiative, MITI set up a Biomass Policy Office in May 1980, with a seven-year program on advanced fermentation technology. This became the Bioindustry Office in 1982. The Ministry of Agriculture and Fisheries had set up its own biotechnology program in 1981. MITI also encouraged many companies to follow the politico-scientific trend. Also, in 1981, the Research Association for Biotechnology was set up with fourteen members. While most of these member firms were from traditional fermentation and chemical sectors, other industrial organizations, looking for diversification, were also attracted to biotechnology. They included Daini Seikosha, better known as the manufacturers of Seiko watches, who were

---

[130] For a description of the Japanese bio-reactor program, refer to D. de Nettancourt, A. Goffeau and F. van Hoeck's "Applied molecular and cellular biology: background note" DG XII, DG XII/207/77-E, 15 June 1977, Table 1-2-3.

suffering from competition from cheap Hong Kong products and looking to branch out into high-tech biochemical equipment. MITI considered this foundation period an initial industrialization period which ended in 1985. It was followed by a period in which industrial infrastructure was built, with the implementation of normative measures for the industrial application of recombinant DNA technologies and growing internationalization. By the end of the 1980s, there was considerable acceleration in the development of biotechnology in companies.

After 1989, there was greater insistence on fundamental research, as MITI's New Generation Basic Technology Program had not managed to plan and structure industrial efforts. At the same time, it was being recognized that bioindustry and its technology, products and industrial systems could be beneficial to humans and the environment. This approach to biotechnology also highlighted the wide range of the field, which was widened still further when, in 1987, in response to the American military Strategic Defense Initiative, the Japanese announced the Human Frontier Science Program. The last page of the 87-page document entitled "Current developments in research in relation to the principal biological functions and future focal points", published on 1 April 1987, gives a précis "summary table which pinpoints the important field for the time being among the research subjects of the Human Frontier Program".

This table gave priority to 14 research subjects, divided up into eight main fields:

- Within the field of brain and nervous system (Neurobiology) the important subjects are: *Reception and cognition, Motor and Behavior Control, Memory and Learning including Heredity and Development*;
- Within the field of heredity, priority is given to: *Expression of Genetic information*;
- Within the field of development, growth and differentiation, priority is given to: *Morphogenesis including Cell Differentiation and Cell Migration*;

- Within the field of immunity, priority is given to the study of *Molecular recognition and Responses including Antigen-Antibody Interactions*;
- Within the field of chemical response (Endocrinology), priority is given to the *Molecular Recognition and Responses including Hormone-Receptor Interaction*;
- Within the field of material conversion (Metabolism) priority is again given to the *Molecular Recognition and Responses including Enzymes-Substrate Interaction*;
- The entire field of *energy conversion including movement* is also listed as being important;
- As regards to the development of leading key technologies, the following five topics emerge: *Sequence Analysis of DNA, Three Dimensional Structure Determination of Proteins, Non-Invasive Measurement of Biological System, Ultramicro-Manipulation, Measurement of the Dynamic Structure of Biological System.*

From this document, it was implicit that the research topics were focused on, but not restricted to, the study of the superior human functions; indeed, all relevant models from bacteria to mammals were considered. Also, whenever possible, all levels of structural complexity were studied: molecules, organelles, tissues, organs and individuals.

With the exception of the development of linked technologies, all the areas were of a fundamental nature and did not touch on applied fields such as the development of medicines and vaccines, microbial fermentation and plant biotechnology. The importance of DNA sequence analysis was highlighted as it was specified as a technology of the future. This was backed up when, from 1981 onwards, a program entitled "Extraction, analysis and synthesis of DNA" was set up with special monies from the Japanese Science and Technology Council from the STA, chaired by Akiyoshi Wada. The project had two goals; first, to revitalize biological research; and second, to attract the attention and interest of firms from outside biology, including firms whose fundamental skills were in robotics, information, computing and materials sciences. Wada's strategy was to set up automatic protocols in molecular biology rather than invent new approaches [131]. His project focused on DNA sequencing. The gene gold rush was being accelerated by the fact that genetic information was leaving the laboratories and moving into a time of practical applications bit by bit. Furthermore, people were realizing that with the improvement of physical and genetic maps and the localization of more and more human genes, the economic potential of research results on genes and sequences could be enormous. It was becoming increasingly clear that DNA sequencing was a strategic essential, especially human DNA [132]. Wada managed to convince Seiko, Fuji Photo, Tokyo Soda, Hitachi and Mitsui Knowledge Industries to join the project team.

The first phase of the project, from 1981–1983, was allocated $3.7 million (¥910 million) and produced a micro-chemical robot built by Seiko and a standardized system for gel electrophoresis from Fuji Photo. In 1984, the project was funded again to the tune of $2.05 million by another branch of the STA and re-entitled "Generic Basic Technologies to support Cancer

---

[131] A. Wada and E. Soeda, "Strategy for building an automated and high speed DNA sequencing system", *Integration and Control of Metabolic Processes*, ed. Ol Kon, Cambridge University Press, New York, 1987, pp. 517–532.

A. Wada and E. Soeda, "Strategy for building an automated and high speed DNA sequencing system", *Proceedings of the 4th Congress of the Federation of Asian and Oceanic Biochemists*, Cambridge University Press, London, 1986, pp. 1–16.

A. Wada, "The practicability of and necessity for developing a large-scale DNA-base sequencing system: Towards the establishment of international super sequencer centers", *Biotechnology and the Human Genome: Innovations and Impact*, ed. A.D. Woodhead, B.J. and Barnhart, Plenum, New York, 1988, pp. 119–130.

[132] The information obtained from human genome sequencing can be used in three main practical ways. The first is diagnostics with DNA probes and the possibility of evaluating the severity of a disease or identifying a predisposition to a pathology. The second application is traditional pharmaceutical research through the comprehension of the physical-pathological mechanisms of illnesses from the genes in question, offering an alternative approach to the traditional "trial and error" sifting of millions of chemical molecules. The third application is gene therapy and the production of therapeutic proteins. Regarding non-human model organism genomes, benefits would be the improvement of industrial strains, a better understanding of the various processes that ensure gene expression and also a possibility (since a number of genes have similarities with genes of more complex organisms, even in organisms phylogenetically distinct, where we can expect indicate a similarity of function), that we can comprehend the function of genes in higher organisms. It is unthinkable to knock out a gene in a human being in studies, but it is absolutely appropriate in far simpler creatures such as yeast or bacteria to disrupt similar genes to those in higher organisms, and unravel their function.

Research". Seiko developed a DNA purification system and another microchemical robot. Fuji started large-scale production of its gel, and Hitachi developed a prototype DNA sequencer[133]. The headquarters of the operation was transferred to the RIKEN Institute at the research city Tsukuba, under the new name "Research Project on the Composition of Genes". Officially, RIKEN is the Institute of Chemical and Physical Research, or RIKagaky KENkyusho. In 1985, Wada's project was pulled into the tornado of debate on a possible Japanese Human Genome Project (HGP).

In 1986, Wada went to the United States to look for support (of a moral nature) for an international DNA sequencing effort and to consolidate the international support base for his project. He visited many research centers and met personalities involved in the nascent debate on the Human Genome Project. Wada's vision was of a series of international centers dedicated to quick and cheap sequencing. As for him Japan was one of the world leaders in sequencing technology, he thought it logical that Japan should concentrate on that particular field. He considered large scale DNA sequencing a potential force for rapprochement through collaboration between the USA and Japan. Unfortunately, this idealistic vision was not fulfilled.

Wada's journey was not just to start up new collaborations but also to get political support for his project by generating foreign pressure on the Ministry of Finance bureaucrats back in Japan. He obtained an extension of funding for his RIKEN project although it was a lot less than he had asked for; he had been hoping for considerably more commitment for a large sequencing center. Wada's position was greatly criticized, in particular because he had defined too ambitious an objective and, in doing so, had exacerbated tension between the USA and Japan.

At home, the industrials were gradually dropping out of the project because their products were not having much success. Wada, now the Dean of the Science Faculty at the University of Japan, was becoming more and more involved in other tasks, in particular that of reforming Japanese science.

---

[133] M. Miyahara, *"R&D on Human gene analysis and mapping systems supported by the Science and Technology Agency of Japan"*, Report memorandum no. 135, Tokyo Office of the US National Science Foundation, August 1987.

180 *From Biotechnology to Genomes: A Meaning for the Double Helix*

In 1988, the sequencing project reins were passed over to Koji Ikawa and Fichi Soeda of RIKEN, who re-evaluated the technical objectives and concluded that the estimates of costs of speed were too optimistic[134]. The target was revised from Wada's defined target of a million DNA base pairs per day to 100,000 base pairs per day of raw data in sequencing capacity.

By 1989, the automated DNA sequencing project was already eight years old and had led to the development of a series of machines capable of sequencing about 10,000 base pairs per day. Isuo Endo of RIKEN found himself in charge of new projects involving the automation of molecular biology procedures (in particular automated cloning). Mostly financed by the STA and industrialists Hitachi, Seiko, Cosmic, Mitsui Knowledge Industry, Tosch and Fuji Film, Endo's sequencing programme bore fruit when he reported, in 1991, that he had reached a sequencing speed of 108,000 base pairs per day[135].

These successes encouraged the government to increase its commitment to the field and finance other initiatives. The STA's annual genome budget of about US$1.3 million (¥200 million) in 1989 and 1990 grew to US$8.6 million (¥1.2 billion) in 1991. Also, the Human Frontier Science Program grew considerably to US$25 million per year. Despite Wada's hopes, all this Japanese initiative and success in the genome field was quickly perceived as a sort of biotechnological "Trojan Horse", a premeditated assault on one of the last bastions of American pre-eminence.

It was a time when a real doctrine and ideology of DNA was being developed, when countries, industrialists, economists and scientists were hoping to hit the jackpot with genetic information, opening a new era of diagnostics, medicines and therapy. There was a genuine threat that the Japanese, with their technological approaches and advances, might usurp western expertise and scientific dreams in what was being called the key

---

[134] M. Sun, "Consensus elusive on Japan's genome plans", *Science*, vol. 243, 31 March 1989, pp. 1656–1657.

[135] D. Swinbanks, "Japan's Human Genome Project Takes Shape", *Nature*, vol. 351, 20 June 1991, p. 593.

frontier of biotechnology, the understanding of information contained in genomes — in the human genome in particular. In fact, this understanding would allow focusing on the 4,000 human hereditary defects linked to one lone gene that might affect 1% of the population (and trying to treat them) or finding new therapies with the development of more appropriate chemical molecules. There were even more lucrative possibilities for dealing with cancer in a more efficient fashion, especially since the first systematic mapping efforts for large genomes were beginning to produce their first results. The western nations had to reply to the Japanese threat. Biotechnology, mapping and genome sequencing would increasingly be the center of attention of nations and governmental organizations.

4

Attack on the Genomes: The First Genetic and Physical Maps

During the 1970s the extraordinarily rapid development of molecular genetics allowed the construction of genetic maps, first of all, for viruses and then for yeast, *Caenorhabditis elegans* and *Escherichia coli*.

4.1 The Problem of Gene Localization

Human gene mapping began in 1911. Edmund B. Wilson, a colleague of Thomas Hunt Morgan at Columbia, was inspired by Morgan's demonstration that the gene for eye color in the fruit fly (*Drosophila melanogaster*) was linked to the sexual X chromosome. Wilson deduced that the gene for color blindness must be on the X chromosome as well because of its specific pattern of transmission (fathers do not pass it on to their sons, and it is extremely rare in women [1]).

The first discovery of a pathological trait site on another chromosome was published in 1968 by Roger Donahue [2]. Whilst studying the distribution

---

[1] E.B. Wilson, "The sex chromosomes", *Archiv für mikroscopie und anatomie entwicklungsmech*, vol. 77, 1911, pp. 249–271.

[2] R.P. Donahue, W.B. Bias, J.H. Renwick and V.A. McKusick, "Probable assignment of the Duffy blood group locus to chromosome I in Man", *Proceedings of the National Academy of Sciences*, vol. 61, 1968, pp. 949–955.

of the Duffy blood group within large families, he noted that it was always transmitted along with a peculiarity on chromosome I. He thereby deduced that the blood group gene was on chromosome I. Three years later, during the IVth International Congress on Human Genetics, it was showed that, at the time, only two genes had been located on *autosomes* (the non-sexual-chromosomes), while 80 had already been attributed to the X chromosome. The first genetic link between two autosomic characteristics was established between the Lewis and Secreteur blood groups in 1951 by Jan Mohr[3]. However, at that time, *autosomes* could not be recognized very easily, so he could not find which chromosome hosted the link. *Autosomes* can be distinguished from each other in specific families when a particular chromosome has a peculiar shape, or when cytologically detectable translocations[4] cause easily detectable clinical characteristics. However, vast regions of the chromosomes other than the X and Y chromosomes remained very difficult to map. Another sign of the technical limitations faced at that time was that geneticists still thought, incorrectly, that there were 48 human chromosomes.

In the late 1960s, two technical developments occurred that freed mapping from its dependence on rare chromosomal abnormalities. The hybridization of somatic cells[5] (general body cells as opposed to germ cells) provided a cell called a *synkaryon* which has a single nucleus containing the chromosome sets of two different animals (human and non-human), and which can create a colony through successive mitosis. "Man-other organism" hybrid somatic cells (mice are often used) were useful for human gene mapping for the following reasons:

---

[3] J. Mohr, "Search for linkage between Lutheran blood and other hereditary characters", *Acta Pathologica e Microbiologia Scandinavica*, vol. 28, 1951, pp. 207–210.

[4] Translocations are chromosomal mutations caused by a change in the position of one or more chromosomal segments and, therefore, of the sequence of genes carried on the segment(s).

[5] The method of hybridizing two somatic cells allows genes on human chromosomes to be mapped and consists of fusing a human cell with a cell of an established cellular line, for example, that of a mouse, to obtain a hybrid somatic cell. The result of this is one single cell containing two different nuclei, known as a binucleate *heterokaryon*. Once the two cells have fused, the nuclei themselves can fuse to make a single nucleus containing the chromosome sets of both the original organisms. This sort of cell is called a *synkaryon*.

- Human chromosomes can easily be distinguished from foreign chromosomes because of their shape.
- During the division of hybrid cells, certain chromosomes would be lost, hopefully the human ones. This chromosome loss stops after a while, providing a stable hybrid line with a complete set of foreign chromosomes and some human chromosomes [6], the number and type of which vary from one hybrid cell line to another. If two genes are close to each other on the same chromosome, they are often expressed together in the hybrid lines. By analyzing a large number of hybrid cell lines and selecting only those with the required genes, it became possible to map genes by finding those which were expressed together. It was a laborious way to study the stability of the links between different known genes, and it suggests how close they might be to each other.

Linking the genes to each other does not necessarily mean pinpointing which chromosomes they are on. At the time, the largest and smallest chromosomes could be distinguished from the others, but it was still impossible to distinguish many chromosomes of intermediate size. The geneticists needed a way to distinguish between all chromosomes. In this respect, Copenhagen's Torbjorn O. Caspersson found the way forward with his use of fluorescent colorants and a microscope. Caspersson [7] and other researchers had discovered that chromosomes could be distinguished by coloring them with fluorescent stains that stuck to DNA, basing their ideas on research that dated back to the 1930s [8]. These stains did not color the

---

[6] M. Weiss and H. Green, "Human-mouse hybrid cell lines containing partial complements of human chromosomes and functioning human genes", *Proceedings of the National Academy of Sciences*, (USA), vol. 58, 1967, pp. 1104–1111.

[7] T. Caspersson, C. Lomakka and L. Zech, "Fluorescent banding", *Hereditas*, vol. 67, 1970, pp. 89–102. T. Capersson, L. Zech and C. Johansson, "Differential banding of alkylating fluorochromes in human chromosomes", *Experimental Cell Research*, vol. 60, 1970, pp. 315–319. T. Caspersson, L. Zech, C. Johansson and E.J. Modest, "Quinocrine mustard fluorescent banding", *Chromosoma*, vol. 30, pp. 215–227.

[8] T.O. Caspersson, "The background for the development of the chromosome banding technique", *American Journal of Human Genetics*, vol. 44, April 1989, pp. 441–451.

chromosomes in a uniform fashion but left each chromosome with a distinctive banding pattern. Geneticists could then tell the chromosomes apart easily. The same technique also showed up deletions, rearrangements and chromosome duplication.

Somatic cell hybridization allowed investigations into the expression of gene functions, the coloring technique allowed each chromosome to be positively identified. These two techniques enabled human geneticists to embark on the quest for a complete map of human genes. Although a considerable step forward, they did not yet provide sufficient precision to accurately locate individual genes, except in the very unusual and rare cases when large changes in chromosome structure involving several million base pairs in length could be seen under a microscope. More subtle changes in DNA could not be detected. It was only rarely that a pathological state could be linked to a detectable modification in chromosome structure. Detection techniques were based on large structural modifications and on the expression of the products of known genes. They could generally not be used to locate genes of unknown function or to systematically map new genes as they were found.

4.2 Polymorphic Markers, Gene Mapping and the Great Gene Hunt

The progress that followed was due to the growing power of molecular genetics. As discussed in the previous chapter, the double helix structure of DNA discovered by Watson and Crick immediately clarified how genes could be transmitted from one cell to another, in a completely accurate way, through the faithful copying of DNA. This discovery also opened up a totally new field of study dedicated to understanding how genes guide cell function.

Molecular biology's distinctive signature was (and still is) the search for function through molecular structure. This highly reductionist strategy was borrowed from physics. Initially the most impressively rapid progress was made in the study of bacteriophages, small viruses that infect bacterial cells.

From its first success in 1960, molecular biology gradually invaded field after field, applying its increasingly powerful tools to questions and organisms of ever-growing complexity. For example, in the latter half of the 1970s, molecular genetics was used with remarkable success in cancer studies, leading to the discovery of *oncogenes*[9].

The first illness characterized at the molecular level was sickle-cell anemia. In 1949, genetics studies by James Neel[10] at Michigan University showed that this illness is, in fact, a genetically recessive one. Linus Pauling's biochemical studies at the California Institute of Technology revealed structural changes in hemoglobin. It seemed inevitable that "Neel's gene" would be causally linked to the protein defect[11], and indeed, in the mid-1950s, a change in one of the proteic chains of hemoglobin was discovered, which suggested a mutation in the DNA coding for the protein[12].

Until recently, most of molecular biology's tools have followed this general pattern of the study of individual genes followed by the biochemical analysis of their products. But the application of molecular techniques to chromosome mapping split molecular biology studies into two different disciplines: the mapping of genes at the individual base pair level and the isolation and analysis of DNA fragments millions of base pairs long.

The notion of chromosome mapping opened the way for the discovery of new proteins through heredity studies instead of through the search for genes associated with known proteins, thus completely reversing the traditional gene hunt strategy. Perspectives for finding unknown human genes began to look very good in about 1978, a year when molecular biology reached a level that allowed researchers to efficiently study problems in human genetics. The techniques of molecular genetics developed in the mid-1970s led to a new type of gene map. In 1970, restriction enzymes were

---

[9] N. Angier, *Les gènes du cancer*, ed. Plon, 1989.

[10] J.W. Neel, "The inheritance of sickle cell anaemia", *Science*, vol. 110, 1949, pp. 64–66.

[11] L. Pauling, H.A. Itano, S.J. Singer and I.C. Wells, "Sickle cell anaemia: a molecular disease", *Science*, vol. 110, 1949, pp. 543–548.

[12] V.M. Ingram, "Gene mutation in human haemoglobin: the chemical difference between normal and sickle cell hemoglobin", *Nature*, vol. 180, 1957, pp. 326–328.

4 Attack on the Genomes: The First Genetic and Physical Maps 187

discovered and rapidly became the precision tools needed to inventory and highlight DNA sequences [13].

Another innovation from the mid-1970s was a very trustworthy method allowing the separation of DNA fragments according to their length [14]. Two teams of researchers independently discovered a way to mark the short fragments of DNA with radioactive phosphorus [15] to detect specific DNA sequences, thus providing yet another sort of mapping [16]. In the initial application of this technique, a DNA fragment containing a hemoglobin gene was identified amongst thousands of other DNA fragments. A gene could be "fished" out of a sea of DNA.

The tools were in place for the construction of a human chromosome map, but learning how to combine the different techniques effectively required further advances. A series of experiments on viruses and yeast showed how the mapping of gene linkages could be done for man. Restriction enzymes were first used to find the genes in viruses [17]. The DNA variations, once noted, would be used to show how a family of similar yeast genes sat together, probably in one single chromosomal region [18]. In a series of experiments that were precursors to the American Human Genome Project

---

[13] D. Nathans and H.O. Smith, "Restriction endonucleases in the analysis and restructuring of DNA molecules", *Annual Reviews of Biochemistry*, 1975, pp. 273–293. H.O. Smith, "Nucleotide specificity of restriction endonucleases", *Science*, vol. 205, 1979, pp. 455–462.

[14] E.M. Southern, "Detection of specific sequences among DNA fragments separated by gel electrophoresis", *Journal of Molecular Biology*, vol. 38, 1975, pp. 503–517.

[15] P.W. Rigby, M. Diechmann, C. Rhodes and P. Berg, "Labelling deoxyribonucleic acid to high specific activity *in vitro* by nick translation with DNA polymerase I", *Journal of Molecular Biology*, vol. 113, 15 June 1977, pp. 237–251.
    T. Maniatis, A. Jeffrey and D.G. Kleid, "Nucleotide sequence of the rightward operator of phage Lambda", *Proceedings of the National Academy of Science* (USA), col. 72, 1975, pp. 1184–1188.

[16] A.J. Jeffreys and R.A. Flavell, "A physical map of the DNA regions flanking the rabbit beta globin gene", *Cell*, vol. 12, October 1977, pp. 429–439.

[17] T. Grodzicker, J. Williams, P. Sharp and J. Sambrook, "Physical mapping of temperature-sensitive mutants of Adenoviruses", *Cold Spring Harbor Symposia on Quantitative Biology*, 39, 1974, pp. 439–446.

[18] T.D. Petes and D. Botstein, "Simple Mendelian inheritance of the reiterated ribosomal DNA of yeast", *Proceedings of the National Academy of Sciences* (USA), 74, November 1977, pp. 5091–5095.

(HGP), yeast DNA was sliced into fragments with restriction enzymes. The DNA fragments obtained were then separated according to length and probed with gene sequences to work out which DNA fragments carried those specific genes[19]. Fragments of DNA with two normal genes were compared with a mutant version of the same gene to highlight the change in a single base pair. A gene could thus be located by linkage to polymorphic markers, isolated by mapping the region in great detail and its mutant identified by analyzing the DNA sequence.

As the genetic studies of yeast and other organisms progressed, new techniques gave medical genetics a decisive shot in the arm. The ability to distinguish chromosomes and localize at least some of the genes gave clinical geneticists something to analyze. As cardiologists could look at the heart and neurologists the nervous system, clinical geneticists could now look at the genome.

Armed with these tools, the first *item* on the agenda for clinical geneticists was to identify lesions on DNA and characterize the diseases they caused. In 1978, Yuet Wai Kan and A.M. Dozy[20], at the University of California (San Francisco), discovered that a particular marker was habitually associated with the gene linked to sickle cell anemia in a North-African family. In 87% of cases, if you cut the DNA with a restriction enzyme, you get a fragment of a particular length associated with the gene that causes sickle cell anemia[21].

---

[19] H.M. Goodman, M.V. Olson and B.D. Hall, "Nucleotide sequence of a mutant eukaryotic gene: the yeast tyrosine — inserting ochre suppressor SVP 40", *Proceedings of the National Academy of Sciences* USA, 74, December 1977, pp. 5453–5457.

[20] Y.W. Kan and A.M. Dozy, "Polymorphism of DNA sequence adjacent to human beta-globin structural gene relationship to sickle mutation", *Proceedings of the National Academy of Science* (USA), 75, 1978, pp. 5631–5635.

[21] Sickle cell anaemia is caused by having both copies of a mutant allele of the gene that codes for the β chain of hemoglobin (which is a heterodimer with an "α chain" and a "β chain"). This mutation causes the replacement of the glutamic amino acid in position 6 from the NH_2-terminal end with a neutral amino acid called valine. This happens in an important part of the molecule, and the presence of a neutral amino acid instead of an acidic amino acid causes the region to repel water; otherwise, it would attract it. In consequence, the polypeptidic β chain takes on a shape that keeps the altered region away from a watery environment; this unusual conformation of the haemoglobin molecules causes clumping and the structural collapse of the red blood cells.

The difference in sequence detected by the restriction enzyme is not at the gene level itself; nevertheless, it marks the chromosome. In families of North-African origin, it then became possible to determine whether a child had inherited the chromosome bearing the marker associated with the illness or a chromosome with a normal gene. By looking for the DNA markers defined by Restriction Length Fragment Polymorphism (RFLP), it was possible to indirectly track defective genes in the families in which they were found. The marker, which does not code for any function, was used to create a link with the mutant gene associated with sickle cell anemia. Kan and Dozy noted that such markers could be very useful for determining links with genes.

Two British groups also saw that normal variations in fragment length amongst individuals could be used as markers to set up correlations with genetic diseases. Alec Jeffreys, at the University of Leicester, was very interested by the study of genetic variation in human populations[22], Ellen Solomon and Walter Bodmer at the Imperial Cancer Research Fund in London, carried out a lot of research to locate genes linked to human pathologies. Bodmer and Solomon were, in particular, thinking of RFLPs with respect to anomalies in hemoglobin, or hemoglobinopathies. In 1979, Bodmer and Solomon[23] published an article in *The Lancet* magazine in which they outlined a method by which the RFLPs could be used as markers for genetic links thereby signaling the absence or presence of particular alleles of a gene.

Since humans inherit two copies of a chromosome, one from each parent, a probe[24] for a particular DNA fragment would select two copies from the

---

[22] A.J. Jeffreys, "DNA sequence variants in the G-γ, A-γ, δ and β globin genes of Man", *Cell*, 18, 1979, pp. 1–10.

[23] E. Solomon and W. Bodmer, "Evolution of sickle cell variant genes", *the Lancet,* April 1979, p. 923.

[24] To detect an RFLP, you need another probe of a DNA sequence sited not far from the enzyme restriction site. For this one, DNA segment is often chosen at random from a collection of cloned fragments that cover the whole of the human genome. It is denatured, marked with a radioactive isotope and used to probe restriction fragments in different individuals. When the bands that correspond to the sequences detected by the probe appear in different positions in different individuals, it's because the DNA being probed has polymorphic restriction sites. The probe and the RFLP that it detects then constitute a very efficient system for genetic marking.

DNA of an individual, the maternal and paternal. Often these fragments, which may differ considerably in length, will show up an RFLP. If a geneticist can get his hands on the parents' DNA, he can then work out which parent bears which allele. The pattern of the restriction length fragment is a sort of fingerprint of that particular fragment of DNA, and the geneticist can follow it in the course of its transmission through subsequent generations. If it turns out that a particular RFLP at the DNA level is always accompanied by that of a mutant gene leading to a genetic disease, then the RFLP can be used to detect the presence of the aberrant gene, even if the exact site, identity and nature of the gene remain unknown.

The same thought had also occurred to some American researchers. Historically, we can date one of the great evaluations in molecular biology methods to a seminar on genetics in April 1978 at Utah University, organized by population geneticist Mark Skolnick of the same institution. During the seminar, Skolnick's collaborator at the University of Utah on hemochromatosis, Kravitz, presented some results that showed the statistical associations between the Human Leukocyte Antigen (HLA) types and protein analyses, which served as a marker for hemochromatosis. David Botstein, at the Massachusetts Institute of Technology, and Ronald Davis of Stanford University, were invited as external experts. The group discussed ways of using statistical associations to locate genes. With Skolnick, they suggested using DNA sequences as markers. Botstein realized that correlating genetic differences with an illness, the way Kravitz generally did to track hemochromatosis, could be used more widely and far more powerfully if a direct analysis of DNA variation was carried out using Botstein and Davis' techniques developed for yeast. R.L. Withe, interested by this proposal, decided to test the hypothesis. He tried to produce markers that would reveal a genetic link in some part of the human genome directly from the DNA.

In 1980 Botstein, Withe, Skolnick and Davis published their article describing this approach in the *American Journal of Human Genetics*[25]. The

---

[25] D. Botstein, R.L. Withe, M. Skolnick and R.W. Davis, "Construction of a genetic linkage map in Man using Restriction Fragment Length Polymorphisms", *American Journal of Human Genetics*, 32, 1980, pp. 314–331.

four Americans realized from the beginning that, at least in principle, they should easily be able to detect markers all along the human genome, and thus have a means of mapping human genes in relation to these markers. Furthermore, the fundamental point of these markers was that they did not have to be inside the gene, only sufficiently close to each other on the chromosome so that the marker and the gene were almost always passed on together from parents to their children.

With RFLPs, and the method as outlined by Botstein, White, Skolnick and Davis, researchers finally had a tool for the consideration of variations between individuals by analyzing the DNA itself directly neither by referring to certain obvious characteristics such as biochemical traits (phenotypes), which might not fully reflect the actual genetic makeup (genotype), nor on the basis of the proteins expressed by the genes. By determining the correlations between the transmission of an illness and the transmission of a particular segment of DNA (a marker), it became possible to locate a gene that causes an illness within one or two million base pairs. The genetic markers associated with illness, furthermore, allowed deficient genes to be tracked from generation to generation and easy diagnosis of carriers of the deficient gene, as well as those who would suffer from it in full.

This sort of work, of course, was still based on linkage analysis, a venerable tool from classical genetics. Linkage analysis had become far more powerful when used with molecular biology techniques since the latter provided lots of RFLP markers, that in certain cases, allowed the identification of previously-unknown genes associated with illnesses. If the markers of one region were consistently inherited with the illness, then there would be statistical evidence that the gene would be somewhere in that region. To find a gene with unknown function and location, you could study the markers of a lot of different chromosome regions and pinpoint the regions consistently associated with the illness. With enough markers you could always find one close enough to the gene, and once you had found an approximate site on the chromosome, the DNA of that area could be isolated and studied in greater detail in hope of discovering the mutation itself.

The power of this theory was demonstrated with panache by White and his colleague Arlene Wygman with their isolation of the first RFLP. The probe for this RFLP was named pAW101 (plasmid Arlene Wygman 101 [26]), and it was shown to have almost as much genetic variation as the HLA locus on Man's chromosome VI, which makes it likely that the two chromosomes should be distinguishable in any individual you may care to consider. The parents should also have different markers. This heterogeneity made it an ideal tool to find a gene close to the "tip" of chromosome 14, right where it was finally located by Wygman.

Raymond L. White went to Utah to take advantage of the famous genealogies of the Mormon families since the Howard Hughes Medical Institute had agreed to fund his efforts to draw up an RFLP map. During the years that followed, the Utah group systematically searched for RFLP markers as they improved the technique. White and his group sampled DNA from more than 40 families available for genetic studies to other groups, providing man's genetic map with quite a lot of markers.

With the help of DNA analysis of some Venezuelan families living near Lake Maracaibo, these new techniques of genetic analysis gave hope for the first time that the gene for Huntington's disease could be found. Nancy Wexler and her geneticist colleagues tracked the transmission patterns of Huntington's in the Venezuelan families. They persuaded family members to let them have some blood samples, which were sent to James Gusella's laboratory in Boston at the Massachusetts General Hospital. There Gusella and his team began to search the extracted DNA for a characteristic pattern of restriction fragment lengths present in the DNA of people known to have the symptoms of the disease, but absent in those who were known with certainty not to be affected. In this way, they hoped to identify the genetic markers inherited with the abnormal gene that were absent in the DNA of those who had the normal gene. In 1983, they found such a pattern and finally linked it to chromosome 4 [27]. Mapping the "Huntington's disease gene" was the first great triumph of the gene hunting technique proposed by Botstein, White, Skolnick and Davis.

---

[26] A.R. Wygman and R.L. White, "A highly polymorphic locus in human DNA", *Proceedings of the National Academy of Science*, USA, 77, 1980, pp. 6754–6758.

[27] J.F. Gusella, N.S. Wexler and P.M. Conneally *et al.*, "A polymorphic DNA marker genetically linked to Huntington's Disease", *Nature*, 306, 1983, pp. 234–248.

Once the gene had been located, many researchers thought that its discovery would only take a few years. Despite the efforts of the consortium (Wexler and the Hereditary Disease Foundation) and wide international collaboration, the gene hunt was a lot harder than originally expected. It turned out that the recombination frequency between the gene for Huntington's chorea and the first marker that had been attributed to it was about 5%, which means the marker is about five million base pairs from the gene for the illness. To identify and clone the deficient gene, you need to have a location closer than a million base pairs from the gene. In fact, it took a decade of intensive work before the international effort led by James Gusella's group finally revealed the gene and the nature of the mutation causing *Huntington's* [28].

Other groups had, in the meantime, started their own illness gene hunt races. Besides the genes for illnesses linked to the X chromosome, which were the first to be located by RFLP analysis [29], several teams discovered genetic markers for cystic fibrosis, neurofibromatosis (also called Von Recklinghausen's disease and characterized by light brown spots on the skin and the apparition of bone and nervous tissue tumors) and familial colonic polyposis (characterized by the apparition of colon polyps that degenerate into cancers). There are now encouraging results for the mapping and identification of genes responsible for certain familial forms of Alzheimer's disease and manic-depressive psychoses.

Once markers are discovered sufficiently close to the gene in question (markers whose recombination frequency with the gene is less than 1%, which indicates a distance of a centimorgan), the next step is the search for the deficient gene itself. The tactic generally used today is to sift a section of DNA; that is, to look amongst a collection of segments of cloned chromosomes for a segment recognized by the specific probe of each marker.

---

[28] The Huntington's Disease Collaborative Research Group, "A novel gene containing a trinucleotide repeat that is expanded and unstable on Huntington's Disease chromosomes", *Cell*, 72, 26 March 1993, pp. 971–983.

[29] The first gene was that of Duchenne's muscular dystrophy and probably also Becker's muscular dystrophy: this was the work of Kay Davies of Oxford University and Robert Williamson at St. Mary's Hospital in London.

The segments detected in this way, which should in principle include the marker and the gene in question, are then cut into successively smaller chunks and then cloned. One then tests their biological activity by using them to detect messenger RNA in tissues altered by the disease. These messenger RNA will only be present in the cell if the gene is expressed. If one of the probes detects messenger RNA from the affected tissues, it is because the probe most likely contains a part or all of the deficient gene. Many genes for illnesses have been isolated and cloned in this way, allowing the development of prenatal and postnatal diagnostic tests.

Markers associated with illnesses were not the only focus of attention. There were also attempts to draw up chromosome maps with increasingly precise locations for arbitrary markers and/or ordinary genes.

4.3 Towards a Complete Linkage Map

Other teams dedicated their efforts to looking for genetic linkage markers and the construction of a complete map. Helen Donis-Keller directed a group at Collaborative Research, Inc., a private firm near Boston. From 1979 to 1981, Botstein had quite a few discussions with the staff of the National Institutes of Health (NIH) in the United States about assembling a complete RFLP map for the human genome. He eventually despaired of ever getting the NIH involved in the affair. Nevertheless, he remained convinced that an RFLP map was of primary importance.

Botstein and Ronald Davis belonged to the scientific advisory board for Collaborative Research Inc. White had also been a consultant for the company, as had Skolnik, although for a shorter period of time. The possibility of an RFLP map was often discussed by the firm's scientific advisors. Collaborative Research was looking for new markers and technical innovation and wanted to set up a limited research and development partnership (a legal funding tool by which funds not normally invested in research could be enticed into it)[30].

---

[30] The federal administration allowed an almost complete tax deduction of research funds invested in the R&D partnership procedure by these groups of investors.

Donis-Keller had come to mapping through nuclear biology. She had worked as a graduate student at Harvard with Walter Gilbert and then gone to Harvard Medical School to work on viral molecular biology with Bernard Fields. In 1981, Gilbert contacted her to ask her to work for the new company, Biogen, and she was taken on as its third American employee. Biogen grew rapidly, work conditions there becoming more and more difficult. In the spring of 1983, Donis-Keller left Biogen for Collaborative Research. Nobel prizewinner David Baltimore directed the group of scientific advisors for Collaborative Research and he, Botstein and a few others convinced Donis-Keller to join them to contribute to ongoing strategic planning. Gene linkage mapping with RFLPs became one of the many projects discussed as a future priority for the firm.

Donis-Keller proposed a project for an NIH Small Business Innovation Research Award, but it was not rated as sufficiently innovative. Donis-Keller and James Wimbush then went to Wall Street to look for $50 million. They almost managed to secure venture capital in a limited partnership agreement, but in the long run, this did not work. Following this failure, they presented the project to Johnson & Johnson, Union Carbide and some other large companies. Donis-Keller sketched out her strategy for building a complete genetic map and using it to locate genes, thereby contributing to a new method of detecting genetic pathologies.

The possession of analytical tools allowing genetic links to be detected not only accelerated the development of genetic tests, but also allowed the company to locate miscreant genes with a speed and efficiency that until then had been unthinkable. Other methods could be used to find the gene itself to provide a target for the development of medicines and genetic therapies. Donis-Keller and the geneticists of Collaborative Research tried and failed on Wall Street again revealing American corporations' lack of acumen, which greatly disappointed Donis-Keller. They simply did not seem to believe that genetics could be of any use in cancer, cardio-vascular illness or other major health problems, all of which were feeding the diagnostics and therapeutics markets.

In 1984, Tom Oesterling became the head of Collaborative Research, and the firm decided to go ahead with RFLPs with funds of its own. The

Collaborative Research Mapping Team developed rapidly, expanding from four members in April 1983 to twenty-four by the end of 1984. By summer 1987, events had moved very fast. The Collaborative Research team decided to finish a complete map in time for the September gene mapping conference in Paris. There were rumors that the Utah team were also hoping to publish such a map, and these rumors, of course, provided a spur to the Collaborative Research team, which finally managed to submit their publication to *Cell* magazine just before they left Paris.

Collaborative Research's announcement of a link analysis map caused a certain amount of emotion at the end of 1987[31]. The corporate office decided to hold a press conference during the American Society of Human Genetics Meeting in October, also announcing the imminent publication of a map with all the public domain markers on it as well as those from the company's own collection[32]. Other groups invited to the press conference opened fire, branding the map incomplete and highlighting the fact that the spacing between some of the markers was indeterminate. Collaborative Research's team, however, thought that the Corporate Office's map would permit the location of 95% of any new gene or marker. White reported that the Utah group would publish maps chromosome by chromosome when they no longer had any unmarked areas. Part of the controversy sprang from the fact that a quarter of the markers on the map had been discovered by other teams and that the families and familial information used had been provided by other groups.

Despite professional tensions, the map began to take shape, thanks to the efforts of the Utah and Collaborative Research teams, and was to become a very powerful tool. The speed and scale of the human gene hunt stepped up drastically at the end of the 1980s, the number of genes being located

---

[31] M. Barinaga, "Critics denounce first genome map as premature", *Nature*, 329, 15 October 1991, p. 571.

L. Roberts, "Flap arises over genetic map", *Science*, 238, 6 November 1987, pp. 750–752.

S. Ackerman, "Taking on the human genome", *American Scientist,* 76, January-February, 1988, pp. 17–18.

[32] H. Donis-Keller *et al.* "A genetic linkage map of the human genome", *Cell,* 51, October 1987, pp. 319–337.

keeping pace. With the availability of maps, the gene gold rush really became competitive, an international scientific sport with wide media recognition promised for the winner. Despite all the rivalries and tensions surrounding mapping efforts, the *Centre d'Etude du Polymorphisme Humain* (CEPH — Center for Studies on Human Polymorphism) kept the lines open between the various groups.

The CEPH was founded by Nobel prizewinner Jean Dausset (with money from a scientific award he had been given and some funds from a private French charitable organization) in 1983 to speed up the mapping of the human genome by linkage analysis. The key to the CEPH philosophy was that the genetic map should be efficiently established by collaborative research on DNA coming from the same sample of human families. The CEPH therefore provided the researchers with high-quality cellular DNA from lymphoblastoid cell lines from each member of a reference panel of large families. They also set up a database that the researchers could supply and share. The CEPH collected cell cultures from members of large families from all over the world, providing 40 families for genetic linkage analysis. Collaborating researchers used CEPH panel DNA to test the families for the segregation of RFLPs detected with their own probes, each team returning the results of their tests as their contribution to the CEPH data bank.

Although most of the CEPH families were initially discovered during studies on the transmission of specific diseases like Huntington's disease, this cooperation was not born of a gene hunt. It was intended to pinpoint DNA markers through systematic analysis of the reference families. The families were not used as references because their genealogy showed the hereditary transmission of specific illnesses, but because there was a sufficient number of living members with clearly defined genealogies whose DNA was easily available. Once the markers were mapped in those reference families, the same markers could be used to locate genes in genealogies with any sort of genetic diseases or other traits.

Groups out on the gene hunt were being equipped with large grants, but those who were trying to produce overall gene maps received very little government aid. Wyman and White's first markers were financed by the National Institute of General Medical Sciences, but for the vast and complex efforts necessary to discover enough markers and put them on a human genome map, almost all of the money came from the Howard Hughes Medical Institute and corporate sources. This gap in government policy was a major factor in later debates on the need for a systematic Human Genome Project. The NIH's reluctance to support the map construction effort was mentioned several times during the debate on future NIH policies in other possible map-making adventures. In any case, the winner of the competition between Donis-Keller, White and other mappers was the entire scientific community. When RFLP markers helped locate the genes responsible for Huntington's disease[33] and Duchenne's muscular dystrophy[34] in 1983, human geneticists took note, and some of them shuffled a bit closer to the mappers. These first spectacular successes were followed by other major breakthroughs such as retinoblastoma, and in 1985, cystic fibrosis.

Genetic linkage with RFLP analysis mapping gradually began to be of fundamental importance. Techniques[35] now allowed a progression from the approximate location of a gene by linkage analysis to a marker, to the discovery of the gene itself and the identification of its product. The first such successful gene hunt, that started out from a chromosome location, ended in 1987 with the cloning of a mutant gene that causes *chronic granulomatosis*[36]. Many more successful finds were to come.

Despite the undoubted usefulness of the above mentioned technique as a means of fixing the boundaries of the zone of research, inaccurate RFLP

---

[33] J.F. Gusella, N.S. Wexler, P.M. Conneally *et al.*, "A polymorphic DNA marker genetically linked to Huntington's Disease", *Nature*, 306, 1983, pp. 234–248.

[34] K.E. Davies, P.L. Pearson, P.S. Harper *et al.*, "Linkage analyses of two cloned sequences flanking the Duchenne muscular dystrophy locus on the short arm of the human X chromosome", *Nucleic Acids Research*, 11, 1983, pp. 2302–2312.

[35] These were advances that had been made in the topographic work needed to mark out the zone and prepare for the search for the gene or genes that are known to be in it, that is to say, breaking up the region with restriction enzymes and cloning it. Pulsed field electrophoresis and YAC construction are among them. Cloning and analysis are complementary techniques, and together they allow researchers to "walk" along the DNA in what is known as "chromosome walking", which means that with a DNA probe known to be close to the gene, you can get closer and closer to isolating the fragments covering the region in question and thereby reconstitute the *contig* for that region from the library.

[36] B. Royer, L. Kunkel, A. Monaco *et al.*, "Cloning the gene for an inherited human disorder — chronic granulomatous disease — on the basis of the chromosomal location", *Nature,* 322, 1987, pp. 32–38.

4 Attack on the Genomes: The First Genetic and Physical Maps 199

maps and other limitations caused by using genealogical studies to locate the genes meant that additional strategies would be needed. The approximate locations of the genes for chronic granulomatosis, Duchenne's muscular dystrophy and retinoblastoma were already known from studies of the transmission pattern, hybrid cells, and patients with small chromosome deletions. Although the Huntington's disease gene had been located, the gene itself was only identified in 1993. In contrast, the cystic fibrosis gene was mapped by RFLP techniques in 1985, and the gene itself found in 1989 [37].

Despite these advances, the size of the task ahead for geneticists was considerable. There were inconsistencies in nomenclature, measurements of genetic distance for the same regions could vary remarkably depending on analytical suppositions, and, in 1990 [38], only a small amount of the genome had been mapped in any detail. Despite all efforts, the human genome remained a *terra incognita*.

Once the American Human Genome Project (HGP) had been launched, the years that followed witnessed a major take-off in mapmaking progress. In October 1992, *Science* [39] published a genome linkage map by the NIH/CEPH Collaborative Mapping Group incorporating more than 1,600 markers, including many far more useful markers than those in the map published by the Collaborative Research team in 1987. Four weeks later, *Nature* [40] published another second generation map resulting from the prodigious efforts of the French Généthon collaboration.

The *Généthon*, a French laboratory of a peculiarly modern nature, was the brainchild of two men, Bernard Barataud and Daniel Cohen. Barataud,

---

[37] C. Roberts, "The race for the cystic fibrosis gene", and "The race for the CF gene nears end", *Science*, 8 April and 15 April 1988, pp. 141–144 and pp. 282–285, Jr. Riordan, J.M. Rommens, B.S. Kerem, N. Alon *et al.*, "Identification of the cystic fibrosis gene: cloning and characterization of complementary DNA", *Science*, 245, 8 September 1989, pp. 1066–1072.

[38] J.C. Stephens, M.L. Cavanaugh *et al.*, "Mapping the human genome: current status", *Science,* 250, 12 October 1990, pp. 237–244.

[39] NIH/CEPH Collaborative Mapping Group, "A comprehensive genetic linkage map of the human genome", *Science*, 258, 2 October 1992, pp. 67–68.

[40] J. Weissenbach, G. Gyapay and C. Dip *et al.*, "A second generation linkage map of the human genome", *Nature*, 359, 29 October 1992, pp. 794–801.

whose child had muscular dystrophy, set up the *Association Française contre les Myopathies* (AFM) to accelerate research on genetic diseases. However, as research is hard on the bank balance, in 1987, he set up the first *Téléthon* on the French television station *Antenne 2* (now *France 2)*. It was a great success, providing the organization with FF126 million, and has been repeated each year since. After four years, the *Téléthon* had collected about a billion French francs. The grants go to several causes, but mostly to research programs. This is how the AFM came into contact with Jean Dausset's CEPH, which was being directed by Daniel Cohen, a researcher personally determined to go as fast as possible in human genome mapping. Cohen had no doubts that mapping would end up being automated, and he had been in contact with the famous French hightech firm of Bertin since 1987. But the CEPH's money was just not enough. So, in 1989, Cohen contacted Barataud and let him into his project, a joint venture with the AMF, to build a modern genome research laboratory. The *Téléthon*'s administrative council agreed, and in 1990, the *Généthon* was set up with a FF150 million budget over three years.

Also, at the *Généthon* laboratory, gene location accuracy was greatly increased by the use of a new type of marker, microsatellites, which are a repetition of one to five nucleotides in 10 to 30 copies, spread out across the whole genome in large quantities. This method led, in 1990, to the publication of a new and improved genetic map of the human genome[41] based on 5264 microsatellites distributed to 2,335 positions with an average heterozygosity of 70%. The *Généthon* map had 814 markers that covered about 90% of the genome. Most of these markers were even more useful than those used in 1987.

With the benefit of hindsight, the systematic search for markers and the construction of a linkage map was one of the most significant accomplishments in human genetics of the 1980s and early 1990s. Even though the work suffered from delayed impact, the maps not only helped in

---

[41] C. Dip *et al.*, "A comprehensive genetic map of the human genome based on 5,264 microsatellites", *Nature*, vol. 380, 14 March 1996, pp. 152–154.

The total length of this map is estimated at 3,699 cM. 59% of the map is covered by gaps of a maximum length of 2cM, and 92% by maximum intervals of 5cM.

4 Attack on the Genomes: The First Genetic and Physical Maps

the hunt for genes, but also provided all sorts of ways to trace gene transmission and study how genes in one region affect genes in others. These maps were fantastic tools that helped open new approaches to the genome. Although the mapping efforts were recognized as part of a genome mapping project [42], they did not originate in the Human Genome Project. On the contrary, researchers building linkage maps could claim certain rights over the Human Genome Project. But the bureaucratic structure that bore the name was rooted in several sources, in particular in three separate proposals that hoped to produce a reference DNA sequence of the entire genome.

Genetic linkage mapping was eventually brought into the bosom of the genome project as it evolved, but it was at best a late addition to initial proposals. The focus on DNA sequencing that led to the genome project was far more than a mere difference in approach. Sequencing proposals formulated in 1985 and 1986 came from a different group of researchers, who were not involved in RFLP mapping. Of course, they had common interests, but the real impetus for DNA sequencing came from those interested in the structural studies of DNA and who had very diverse motives, not from the specialists in classical genetics or the study of characteristic inheritance. So, although the confluence of structural genetics and classical genetics was postponed, it was, nonetheless, inevitable.

The path from the approximate location of a gene by RFLP mapping to isolation of the gene itself is arduous. The work was delicate, the methods not always very reliable and often not very adequate at all. The discovery and precise location of genes needed higher-performance methods of DNA studies for a given chromosome region. In particular, to move more rapidly from an approximate location of the gene to the gene in question, a different type of map was needed — a physical map. Physical mapping would fill the gap between genetic linkage and DNA sequencing, providing a fundamental intermediary stage. Physical mapping techniques were initially developed for other organisms and were only applied to the human genome much later.

---

[42] Since 1985, the CEPH had promised to take a coherent approach to the mapping of the human genome, an approach which was called the Human Genome Mapping Project from that date onwards.
J.L. Marx, "Putting the human genome on a map", *Science*, 229, 12 July 1985, pp. 150–151.

4.4 Physical Genome Mapping: The Reconstruction of a Complicated Puzzle

Between the 1950s and the 1980s, molecular biology applied its sophisticated analytical techniques to the study of organisms of ever-increasing complexity. The methods developed and tested on viruses, bacteria and yeast were then applied to the study of other organisms such as mammals, including man. The transition was not, however, as linear and simple as one might at first imagine. It did not just start with modest creatures, finish them off and then move on to more complex organisms. Molecular research had been applied to man, the mouse and other complex organisms almost from the very dawn of molecular biology. Originally, molecular biology's main objective was understanding the fundamental relationship between DNA, RNA and proteins. Most of the time, viruses and bacteria were used to reveal these relationships and processes. The underlying philosophy for this research was to focus on systems that could be understood mechanically. The strategy was quite openly based on the premise that vital processes could be reduced to molecular mechanisms. During the three decades that followed, the tools of molecular biology were used more and more often to dissect the biology and genomes of organisms of growing complexity and size, from viruses to bacteria, from bacteria to yeasts and then on to multicellular organisms.

Structural studies began with proteins and genes. Since proteins are coded into the genome at the gene level, and because some of the most powerful of the new techniques required the manipulation of gene fragments, DNA studies rapidly became much more important. In a progressively widening group of research fields, more and more academic articles mentioned handling DNA at various stages of experimental protocols, and recombinant DNA techniques provided entirely new ways of probing Mother Nature. Reductionist thinking was pushed as far as it would go in the study of certain bacteria. The systematic description of entire genomes began in the early 1970s and accelerated throughout the 1980s. First of all, the DNA structure of small viruses was described; then bacteria were mapped. As the 1980s progressed, a debate raged about how to map the genomes of more complicated organisms, mainly man.

Linkage analysis mapping had greatly progressed, helping determine the approximate location of a gene, but often it was not sufficient to pinpoint the gene itself. The first steps in precise pinpointing would be to fragment the DNA of the region in question, make copies of the DNA segments obtained and then reorganize them so that the DNA could be studied directly. To achieve that, some cloning would have to be done by cutting DNA into segments with a restriction enzyme, which also opens the DNA plasmid, most often used as a cloning vector[43].

A recombinant plasmid can be used to transform a strain of *E. coli* or another bacteria. The bacteria grow and the plasmids replicate under the control of their own genes, thereby copying the gene in question. A team at Caltech and Harvard, led by Tom Maniatis, thought of cloning the DNA of an animal's entire genome by breaking it into small fragments that could be cloned individually[44]. Millions of distinct bacterial colonies would then contain different DNA inserts of 15,000 to 20,000 base pairs in length. Colony collections like this are called libraries or banks[45]. Because bacteria can be handled in their millions, it is possible to isolate the DNA of an organism, break it into fragments small enough to be inserted into judiciously chosen vectors and copy them. The enormous advantage offered by access to such DNA bank clones was that the DNA was also easily available for later analysis since it could be copied in the bacteria. By setting up a clone

---

[43] Although plasmids are the most often used vectors, bacteriophages or bacteriophage derivates are also used as clone vectors, such as artificial mutants of the lambda phage. The point of the lambda phage is to be able to clone large fragments of DNA, between 14 and 19 kpb. There is a problem with this method, which is that if the genome sequence you are looking for contains one or more cleavage sites for the enzyme you are using, the sequence will be cloned as two or more fragments. Another problem is the average size of fragments obtained after digestion by a 6-base pair restriction enzyme, for example, which can be as small as 40 bp. A representative bank for the entire genome would be made up of a large quantity of recombinant phages, which would make hybridization sifting even more difficult. Other methods (for example, the mechanical fragmentation of DNA, the use of particular restriction enzymes specific clone vectors) can help circumvent these problems.

[44] T. Maniatis, R.C. Hardison, E. Lacy *et al.*, "The isolation of structural genes from libraries of Eukaryotic DNA", *Cell*, 15th October 1978, pp. 687–701.

[45] A genomic library or bank is a collection of clones that represent the whole of the genome.

bank of human genes, Maniatis' team gave a fantastic research tool to medical research and many other fields.

With all that cloned DNA, how would you find a specific gene? It is easy to find the clones that contain an interesting gene when something particular is known about the gene. Maniatis and his team had some DNA probes for detecting rabbit hemoglobin. They used the probes to "light up", and thus identify, colonies containing clones with fragments of the gene that codes for hemoglobin proteins. Then they "fished" out those clones that contained pieces of the gene. By analyzing the fragments in great detail, they were able to find out how the DNA fragments overlapped, and from this data build a map. It was not a linkage map, but a map that reconstituted the entire genome (or in this case gene) from the fragments, that is to say, a physical map.

In a genetic map, the distance between two genes (two markers) is calculated according to the respective proportions of parental combinations and recombinations [46]. The distance on the physical map, however, is measured in base pairs; that is to say, it is the actual physical distance between two genes or gene fragments. Although the order of genes or other markers is always the same on both linkage and physical maps, relative distances can vary considerably.

The two types of map were, and still are, equally important, as they serve different purposes. The genetic linkage map provides a bridge between the study of the transmission of a characteristic and the location of the genes associated with that characteristic. Physical mapping is a way to catalogue DNA directly by region. The problem of how to build a genetic linkage map for man had, in principle, been solved by the discovery of RFLP markers. Another problem was how to make a physical map of the human genome and other genomes of interest. When, in 1985, discussions began on a hypothetical Human Genome Project in the United States, physical mapping of large genomes had only been done for yeast and nematode worms.

---

[46] A distance of 1 centiMorgan between two genes indicates that they have a 1% probability of not being inherited at the same time. In man, this 1% probability means more or less some one million base pairs.

4.5 The First Physical Maps of Large Genomes

Two different groups, quite independently from each other, began to apply the cloning and ordering strategies to draw up physical maps of yeast and nematodes. At the University of Washington at St Louis, Missouri, USA, Maynard Olson and his team had begun to map the chromosomes of baker's yeast *Saccharomyces cerevisiae* (also used in the fermentation of wine and beer). At the Medical Research Council Laboratory in Cambridge, Great Britain, John Sulston and Alan Coulson (later joined by Robert Waterston of the University of Washington) were working on a physical map of *Caenorhabditis elegans,* a nematode worm about one millimeter long. Physical mapping of the genomes of the two organisms began in the early 1980s, producing promising results as early as 1986. These two projects involved well-investigated model organisms and were very important for the understanding of the biology of those organisms and of biology in general. The genetic map of the yeast genome was already well advanced,[47] and yeast progressively emerged as a model for eucaryotic genetics, that is, genetics in organisms with chromosomes in their nuclei. The 12.5 million base pairs initially estimated as the genome of *S. cerevisiae* were a logical first target for physical mapping. There was also an ordered set of genomic DNA clones for yeast[48], which made an extraordinarily powerful addition to the formidable weaponry already massed to conquer the genetics of yeast.

Botstein and Gerald Fink of the Whitehead Institute for Biomedical Research pointed out several characteristics that make yeast a good model

---

[47] C.C. Lindergren, G. Lindergren, E.E. Shult and S. Desborough, "Chromosome maps of *Saccharomyces*", *Nature*, London 183, 1959, pp. 800–802, R.K. Mortimer and D.C. Hawthorne, "Genetic mapping in *Saccharomyces*", *Genetics*, 53, 1966, pp. 165–173, R.K. Mortimer and D.C. Hawthorne, "Genetic mapping in yeast", *Methods. Cell. Biol.,* 11, 1975, pp. 221–233, R.K. Mortimer and D.C. Hawthorne, "Genetic map of *Saccharomyces cerevisiae*", *Microbiol. Rev.,* 44, 1980, pp. 519–571, R.K. Mortimer and D. Schild, "Genetic map of *Saccharomyces cerevisiae*", pp. 224–233, in ed. S.J. O'Brien, *Genetic Maps, 1984,* Cold Spring Harbor Laboratory, Cold Spring Harbor, N.Y., 1984.

[48] M.V. Olsen, J.E. Dutchik, M.Y. Graham *et al.*, "Random-clone strategy for genomic restriction mapping in yeast", *Proceedings of the National Academy of Sciences*, USA, 83, October 1986, pp. 7826–7830, A. Coulson, J. Sulston, S. Brenner and J. Karn, "Towards a physical map of the gemome of the nematode *Caenorhabditis elegans*", *Proceedings of the National Academy of Sciences*, USA, 83, October 1986, pp. 7821–7825.

organism, predominantly the ease with which the relationship between gene structure and protein function can be established[49]. The mass of information on yeast mutants emerging from classical genetics (mutants generated mostly by the selection of yeast able to survive in stressful conditions) combined with yeast's ability to carry out exactly the same efficient and precise genetic exchanges repeatedly, means researchers can induce mutations in known genes. The effects of these mutations can easily be evaluated because yeast grows and reproduces so fast. By inducing mutations, it is generally possible to identify the genes in the genome. If one studies what that mutation does to the gene product and correlates it to the way in which the biology of the organism is modified, it is possible to progress from the structure of genes to protein function and then to physiology.

The power, the extent and the precision of yeast biology hastened attempts to bring together the gene-protein-function triangle, with the consequence that proteins originally discovered in other organisms, but also present in yeast, were easier to study there instead. The fact that the community of yeast scientists was highly interconnected was also important. All the laboratories working on yeast, the *"levuristes"* (from the French *levure* for yeast) already routinely exchanged not only scientific information (through publications and other conventional means), but also strains of yeast, protocols for experiments and other ideas well before any publication. Yeast genetics was sufficiently mature for a structural approach, and as stocks of clones for the entire genome had been promised, it looked like a chance for the *levuriste* community to get their hands on an exceptionally useful tool. Olson's proposal to target a physical map was, therefore generally welcomed by his approving peers.

Olson began his mapping enterprise by building a DNA library, a collection of clones containing chunks of the whole genome, and then ordered all the segments he had obtained by looking for overlapping clones. By finding the adjacent overlapping clone for each segment, he found the order of all the segments and assigned them back to their original chromosomal

---

[49] D. Botstein and G.R. Fink, "Yeast, an experimental organism for modern biology", *Science*, 240, 10 June 1988, pp. 1439–1443.

region. The ordered lambda bank that Olson and his colleagues put together of *S. cerevisiae* strain S 288C had about 5,000 clones, with an average insert size of 15 kb[50].

From 1987–88, several other teams launched fine physical mapping efforts on the yeast chromosomes. For three of these chromosomes (I, II and III), the work to cover their entire lengths by overlapping clones and comparing their physical and genetic maps was already under way[51].

Yeasts are marvellous experimental models for many aspects of eukaryotic genetics. However, these unicellular organisms are not quite so well-adapted for studies on the genetics of the innumerable interactions in more complex creatures. Yeasts are already quite complex, but multicellular organisms are immensely more so, in certain cases with billions of cells interconnected and coordinated as a whole. As a model capable of handling such questions, the ideal creature turned out to be the nematode worm *Caenorhabditis elegans*, mainly because of its short growth and reproduction period and its tendency to self-fertilize, providing automatic sets of identical chromosome pairs. Both these characteristics made it an excellent choice for genetic research[52].

Like many of the great advances in fundamental molecular biology research, the foundation work on *C. elegans* biology mainly took place at the Medical Research Council (MRC) Laboratory, Cambridge, England, under the auspices of Sydney Brenner and John Sulston. Brenner had already, in the 1960s, chosen *C. elegans* as a model for studying multicellular phenomena, in particular those of the nervous system[53]. He had wanted an organism that was as small as possible, but still sufficiently complex to show him the effects of gene mutation, allow him to isolate mutations

---

[50] M. Olson *et al.*, *ref. cit.*

[51] H. Feldman and H.Y. Steensman, chap II of "A detailed strategy for the construction of organized libraries for mapping and sequencing the genome of *Saccharomyces cerevisae* and other industrial microbes", in *Sequencing the Yeast Genome, a Detailed Assessment: A possible area for the future Biotechnology research programme of the European Communities*, ed. A. Goffeau, Commission of the European Communities, Directorate General Science, Research and Technology, Directorate for Biology, Biotechnology Division, BAP, 1988–89.

[52] L.J. Kenyan, "The nematode *Caenorhabditis elegans*", *Science*, 240, 10 June 1988, pp. 1448–1453.

[53] S. Brenner, "Genetics of *Caenorhabditis elegans*", *Genetics*, 77 (1), 1974, pp. 71–94.

affecting the animal's behavior, and observe modifications in the nervous system[54].

In October 1963, Brenner put a proposal to the MRC containing his first ideas for a research plan. He intended to follow a line of research similar to that which had been chosen with such success in molecular genetics, using a very simple organism that could be handled in large quantities. He would attack the problem of cellular development by choosing the simplest differentiated animal possible and subjecting it to the analytical methods of microbial genetics. Brenner quite liked the virtues of *C. briggsiae*, but it was supplanted by *C. elegans*. Brenner's logic in this was similar to that of Seymour Benzer at Caltech in his own struggle to understand the neural functions and other complex phenomena in the *Drosophila* fruit fly. To keep to Brenner's schedule, the project would need mutant nematodes with visible differences and an enormous quantity of structural information on all cell development in the nematode body as well as the connections between them. Several laboratories at the MRC institution in Cambridge found themselves roped into the task.

Through monumental efforts, John Sulston retraced the development of more than 900 somatic cells in nematodes by studying the worm's development under a microscope with a special eyepiece that let him watch each and every cell in the nematode's translucent body[55]. He established a genealogy for all the somatic cells. This was a Herculean task that provided all the basic information needed to follow what was happening when the development of specific cells was interrupted. With this information, he was able to directly observe how the death of a cell affected the nematode's behavior, its chances of survival and how it adapted to its environment. Another *tour de force* was the reconstruction, by the MRC team of Sydney

---

[54] S. Brenner, "Genetics of behaviour", *British Medical Bulletin*, 29, 1973, pp. 269–271.

[55] J.E. Sulston and H.R. Horvitz, "Post-embryonic cell lineages of the nematode *C. elegans*", *Developmental Biology*, 56, 1977, pp. 110–156, J. Sulston, "Neuronal cell lineages in the nematode *C. elegans*", *Cold Spring Harbor Symposia on Quantitative Biology,* 48, 1983, pp. 443–452, J. Sulston, E. Schierenberg, J.G. White and J.N. Thompson, "The embryonic cell lineage of the nematode *C. elegans*", *Developmental Biology*, 100, 1983, pp. 64–119.

Brenner, John White, Eileen Southgate and Nichol Thompson, of the way *C. elegans'* nervous system connections are wired up, by analyzing some 20,000 electron microscope photographs[56]. These two efforts were a considerable triumph for the reductionist approach. It was thought that these fundamental tools would provide an understanding of the whole of the organisms' biology and all its mechanical details. It was not that biology could be explained totally by structural details, but rather that understanding structure was the best way of grasping the genetic and environmental factors influencing the nematode's growth.

The cell lines and neural connection charts were the starting point for the study of what happens when the normal structure of a nematode is disturbed. Structural genetics was the crucial missing link. Since work started on *C. elegans*, the aim had been to correlate behaviour with structure. At the beginning of their research, Sulston and Brenner tried to work out how big the genome was to get an idea of the difficulties ahead. They, therefore, squashed the nematodes and estimated the amount of DNA contained in the cells; then they measured the amount of time the separated DNA strands needed to reassemble themselves into a double helix. (The longer the genome is, the more time this takes). These methods suggested that the genome had about 80 million base pairs[57], making it the smallest genome of all multicellular animals, but this was revised to 100 million base pairs at the end of the 1980s. We now know that the *C. elegans* genome is divided into six chromosomes, which were genetically mapped in 1988[58].

The next critical step was a physical map of the nematode genome. Sulston and Coulson set out on the fantastic job of drawing just such a map. They increased the size of the inserts to make the process of physical mapping simpler. Dozens of laboratories worked at improving the cloning vectors until they could contain DNA inserts of 30 to 40 thousand base pairs, a

---

[56] J.G. White, E. Southgate, J.N. Thompson and S. Brenner, "The structure of the nervous system of the nematode *C. elegans*", *Philosophical Transactions of the Royal Society of London*, Series B vol. 314, 1986.

[57] J.E. Sulston and S. Brenner, "The DNA of *C. elegans*", *Genetics,* 77 (1), 1974, pp. 95–104.

[58] C.J. Kenyon, "The nematode *C. elegans*", *Science* 240, 10 June 1988, pp. 1448–1453.

length which rapidly became the norm. This progress reduced the number of DNA fragments needed to constitute the library, making it easier to slice the DNA into segments, clone them and put them back into order.

Now, a complete map of the *C. elegans* genome could be assembled. Such a map had a great advantage, as chromosomal DNA could not only be mapped, but also cloned and kept available in freezers for later analysis. Sulston and Coulson worked for several years to bring together collections of *C. elegans* DNA fragments and identify them[59]. The DNA fragment patterns were brought together in a computer which sought other clones with a similar pattern indicating that there might be an overlap. By 1986, Sulston and Coulson had already managed to cover 80% of the *C. elegans* genome in 16,000 clones, ordered in contigs[60] and divided into 700 groups. This work was a considerable advance for the nematode specialist community, but there remained some gaps to be filled in. David Burke, Georges Carle and Maynard Olson of the University of Washington, St Louis, contributed to the resolution of this problem through an idea that came from yeast research.

In 1980, Louise Clarke and John Carbon, at the University of Santa Barbara, cloned a yeast centromere[61]. Sequences from this region have a remarkable effect on the plasmids they are inserted into, encouraging correct segregation of the chromatids in 99% of cellular divisions. In 1988, the centromeres of 12 of the 16 yeast chromosomes[62] had been cloned, and it was seen that their sequences were strangely similar. The telomeres[63] were

---

[59] A. Coulson, J. Sulston, S. Brenner and J. Karn, "Towards a Physical Map of the Genome of the nematode C. elegans", *Proceedings of the National Academy of Sciences USA*, October 1986, pp. 7821–7825.

[60] A *contig* is a series of clones where the overlaps are known, and therefore, so is the order.

[61] The *centromere* is a specialized region of the chromosome, which in some species can be distinguished under the microscope by a localized thickening. The region has an important role in the distribution of chromosomes during cell division.

[62] The yeasts are the only organisms where the *centromeres*, the telomeres and the origins of replication have been correctly and clearly defined.

[63] The telomeres, which are sited at each end of the chromosomes, are made up of repetitions of a particular DNA sequence. They appear to protect the end genes from damage.

cloned last of all because their position at the ends of the chromosomes made them difficult to handle by conventional genetic engineering means.

In 1982, Jack Szostak and Elisabeth Blackburn, of Berkeley University, used the telomeres to build a DNA molecule that replicated in yeast cells but stayed linear. A little later, they used the same methods to clone yeast telomeres. At the same time, quite independently, Ginger Dani and Virginia Zakin of the Fred Hutchinson Research Center, Seattle obtained the same result. By using these telomeres that allow the chromosomes to keep their linear integrity (and now that all the elements for replication and chromosomal segregation were available) Andrew Murray and Jack Szostak built a completely artificial yeast chromosome. It was about 20 times smaller than the smallest natural yeast chromosome since it only had 11,000 base pairs. The same team built a second artificial chromosome 55,000 base pairs long using DNA from a bacterial virus. The increase in the length of these artificial chromosomes showed a clear reduction in the frequency of segregation errors.

Burke, then a student in Olson's laboratory, became interested in chromosome structure. With Carle and Olson he managed to use these artificial chromosomes to protect and multiply important fragments of exogenous DNA, which could then be transferred to other cells for experiments. Burke managed to build artificial chromosomes containing inserts of up to several hundreds of millions of base pairs. These molecular works of art came to be known as Yeast Artificial Chromosomes, or YACs for short[64].

The sizable length of cloned DNA fragments in YACs solved several problems at once. A far smaller number of clones were necessary for one chromosomal region. Secondly, the problems of cloning certain genes in bacteria could be circumvented by using yeast, the biology of which is sufficiently similar to that of more complex organisms. Lastly, the larger inserts improved chances of detecting overlaps between fragments, making

---

[64] D.J. Burker, G.F. Carle and M.V. Olson, "Cloning of large segments of exogenous DNA into yeast by means of artificial chromosome vectors", *Science*, vol. 236, 15 May 1987, pp. 806–812, Andrew Murray and Jack Szostak, "Les Chromosomes Artificiels" *Pour la science*, Janvier, 1988, pp. 60–71.

it easier to pick out adjacent and consecutive clones. A smaller fraction of clones was needed to detect the overlaps, and this considerably increased the speed at which entire physical maps could be constructed[65]. The yeast and artificial chromosome host/vector system was clearly the second generation system that genetic engineering had been needing.

Using YACs, the MRC and University of Washington (St Louis) teams managed to bridge the many gaps in the map and reduced the number of *contigs* from 700 to 365 in seven months[66]. At the end of 1989, there were only 190 *contigs* left in the *C. elegans* genome, and by 1992, 90 million base pairs of the genome had been physically mapped, leaving only 40 gaps over the whole of the map[67]. YACs made mapping a lot easier as it helped to fill the gaps.

Physical mapping was also greatly helped by a new way to separate far longer fragments in laboratories. In 1984, David Schwartz and Charles Cantor, then at the University of Columbia, developed a DNA separation technique which they called pulsed field gel electrophoresis[68], a considerable step forward. It overcame some of the limitations of classical electrophoresis, which separates a mix of DNA fragments according to size.

Under the old system, the fragments migrate under the effects of a difference in electrical potential inside a porous matrix of *agarose* or *polyacrylamide* gel. The molecules smaller than the matrix pores are barely hindered and migrate rapidly, whereas the larger ones are slowed by their friction against the pores. This sifting of the fragments allows them to be separated according to their length. However, this method has its limits, because when the DNA molecules in question are too big (a lot longer than

---

[65] E.S. Lander and M.S. Waterman, "Genomic mapping by fingerprinting random clones: a mathematical analysis", *Genomic 2*, 1988, pp. 231–239.

[66] A. Coulson, R. Waterston, J. Kiff, J. Sulston and Y. Kohara, "Genome linking with yeast artificial chromosomes", *Nature*, 335, 8 September 1988, pp. 184–186.

[67] J. Sulston, Z. Du, K. Thomas, R. Wilson *et al.*, "The *C. elegans* genome sequencing project: a beginning", *Nature,* 356, 5 March 1992, pp. 37–41.

[68] D.C. Schwartz and C.R. Cantor, "Separation of yeast chromosome — sized DNAs by pulsed field gel electrophoresis", *Cell*, 37, 1984, pp. 67–75.

the pores) they come up against several pores at once and are slowed down in proportion to their length. The force pulling them is proportional to their electric potential as determined by their length, so after a certain size, they all migrate at the same speed. There is no resolution after the fragments get any bigger than 30 or 40 kilobase pairs, even in the most diluted agarose gels.

The new technique of pulsed field gel electrophoresis overcame this limitation. By occasionally changing the direction of the electrical field, the molecules would be reoriented. A very long molecule (1,000 kilobase pairs, for example) would spend most of its time changing direction. A smaller molecule of 100 kilobase pairs would spend most of its time migrating. With this new technique, it became possible to separate fragments of up to 1,000, 2,000 and even 10,000 kilobase pairs. The easier separations led to the easier construction of physical maps of gene regions for which probes were available. With clones the size of the YACs, this new technique was needed to separate the large fragments. The new method became a fundamental element of physical mapping involving large fragments until it was finally supplanted by even faster and more efficient techniques.

In addition to the physical maps for yeast and *C. elegans* genomes that were progressing at quite a speed, in 1987, two full physical maps for *E. coli* were published — by Ynji Kohara[69] and by C.L. Smith, J.G. Econome, A. Schutt, S. Kico and Charles Cantor[70].

The physical maps for yeast and the nematode worm were turning out to be remarkably useful tools for genetics and the advance of biotechnology. They saved each researcher the work of laboriously developing a clone library and having to sift it independently. The maps were put on computer and made available to the entire scientific community. Researchers discovering new genes or mutant organisms sent their findings back to the system, placing the genetic and physical maps in a dynamic network of cooperating

---

[69] Y. Kohara *et al.*, "The physical map of the whole *E. coli* genome", *Cell*, 50, 1987, pp. 495–508.
[70] C.L. Smith, J.G. Econome, A. Schutt, S. Kico and C. Cantor, "A physical map of the *Escherichia coli* K12 genome", *Science*, vol. 236, 12 June 1987, pp. 1448–1453.

laboratories. These constantly maintained tools — physical maps, genetic linkage maps and cell line maps for the nematode and mutant strains, contributed to unprecedented growth in our understanding of these animals. The quantity of structural detail, the quality of research and the constant quest for improved techniques confirmed Brenner's predictions.

Physical maps have also made the different mapped regions more widely accessible since they are cut up into fragments and each fragment can exist in a freely available clone. If one knows that a given unknown gene involved in some illness is between A and B, and if a clone contig has been established between A and B, then one of the clones on the *contig* has to contain the gene. Therefore, by studying the clones in detail one after another and comparing healthy and affected individuals, one should be able to find the gene in question.

This sort of work obviously has its own problems of collection, validation and dissemination of results, as well as of the availability of libraries and clones. Problems like these definitely have to be dealt with to avoid useless duplication of effort, the non-exploitation of resources once obtained and even errors. This meant that a research, communication and informatics infrastructure had to be set up almost from scratch. M. Olson's 1989 idea to mark each contig map of the human genome with a series of unique sequence tagged sites (STSs[71]) was a step in the right direction. Each STS marker would be a simple chunk of DNA between 100 and 200 base pairs long with a unique fundamental sequence. It could then be recognized as unique and amplified with PCR. This uniqueness of the STSs meant that if two different clones from two different clone collections had the same STS, they had to be from the same region. STSs have also become very useful in genetic linkage mapping[72].

---

[71] M. Olson *et al.*, "A common language for physical mapping of the human genome", *Science*, 245, 1989, p. 1434.

[72] It is, in fact, enough to generate a STS for each polymorphic DNA marker by sequencing the marker and then developing a PCR to amplify a unique sequence in the marker. Polymorphic STSs have proved themselves particularly useful as they can be used both on the physical and genetic linkage maps. They are also points for the alignment of the different scales of the two maps.

Finally, the advent of microsatellite and minisatellite markers completed the arsenal of new markers for the human genome and further improved the chances of identifying specific fragments and locating them precisely[73].

4.6 Towards a Physical Map of the Human Genome

The task of physically mapping human chromosomes was underestimated. It had been promised that the nematode and yeast maps would soon be finished, and that the same techniques could be rapidly and easily applied to the human genome, a far larger and more complicated proposition. Estimate of achieving a complete map of all the human chromosomes within two or three years turned out to be too optimistic. Nevertheless, the first physical maps of human chromosomes were published in 1992, when David Page and his team at the Whitehead Institute produced a map of the Y chromosome[74], and a group in France led by Daniel Cohen put together one of chromosome 20 [75].

The last successes of the 1980s, mapping the *E. coli*, yeast and *C. elegans* genomes, were only a beginning. All these technical innovations and new genome study approaches were heralds of new prospects. In the mid-1980s, while mapping was being developed at an unprecedented rate by using the new techniques, the idea of systematically mapping the human genome began to be taken seriously. Furthermore, the physical maps of the yeast and *C. elegans* genomes were almost finished; the next step would be to sequence them completely. The mid-1980s, in this way, saw the birth of

---

[73] Restriction sites are clearly insufficient as markers. If you have a DNA sample and you want to know whether it contains the site *Eco*R1 2708, you can't.

[74] D. Vollrath, S. Foote, A. Hilton, L.G. Brown, P. Romero, J.S. Bogan and D.C. Page, "The human Y chromosome: A 43 interval map based on naturally occurring deletions", *Science*, 258, 29 October 1992, pp. 52–59, S. Foote, D. Vollrath, A. Hilton and D.C. Page, "The human Y chromosome: overlapping DNA clones spanning the euchromatic region", *Science*, 258, 2 October 1992, pp. 60–66.

[75] I. Little, "Mapping the way ahead", *Nature*, 359, 1 October 1992, pp. 367–368, I. Chumakov, P. Rigault, S. Guillou, S. Foote *et al.*, "Continuum of overlapping clones spanning the entire human chromosome 21q", *Nature*, 359, 1 October 1992, pp. 380–387.

a many projects aiming to map different genomes at the molecular level with a view to sequencing the entire DNA of large genomes, in particular, the Human Genome Project to map and sequence the human genome. The projects size and ambition, and the many reactions and interactions they engendered, opened up the era of systematic genome sequencing.

5

The Human Genome Project and the International Sequencing Programs

As we saw in the last chapter, the evolution of molecular genetics in the 1970s allowed the production of genome maps for organisms of ever-increasing complexity. In 1981 [1], when the first catalogue of human proteins was assembled, it stimulated the idea of mapping the whole of the human genome, an idea probably first mooted at the European Molecular Biology Laboratory (EMBL) [2]. Such a map would provide the information needed to locate defective genes as well as predispositions to a number of pathologies [3]. The sequencing revolution, which in 1987 [4] allowed an increase in sequencing speed from 50 nucleotides per year to several thousand nucleotides per day, brought the tools that would be needed to make the dream of cataloging man come true.

The realization of this dream had begun to seem more of a necessity since the initiatives in Japan, and in particular, those Akiyoshi Wada had launched in genome research, were being considered with concern, mainly in the United States, but also in other countries and at the European Community level. An Office of Technology Assessment, USA (OTA) report

---

[1] "Human protein catalogue — the idea of a Human Genome Project", *Science,* 2 January 1981.

[2] *Idem.*

[3] Victor A. McKusick, "The morbid anatomy of the human genome", Howard Hughes Medical Institute, *Science,* 14 October 1988, p. 229, *Genetic Engineering News,* July/August 1989, p. 30.

[4] *New Scientist,* 21 May 1987, p. 97.

on the field of biotechnology entitled *"Commercial Biotechnology: An International Analysis"*[5] concluded that Japan was the United States' most serious competitor in the commercialization of biotechnology.

The Human Genome Project (HGP) was an attempt to bring together the American focus on fundamental biomedical research and the Big Science mode of thought invented by the Americans themselves during the military and spatial eras. The Big Science concept had been translated to commercial success by the Japanese. At the same time, the Human Genome Project provided a fantastic opportunity to redirect the activities of the big military research laboratories and give a major push to American biological research. It also encouraged worldwide biological research through the reactions and initiatives engendered by the HGP debate at national and international levels.

5.1 The Ultimate Challenge: The Human Genome Project

Not all scientists in the field of biology and molecular genetics are philosophers or prophets. However, some of the leaders in this field can safely be said to have qualities as visionaries, whilst some of them are also entrepreneurs and public figures. A few of these scientists with considerable ambition and vision devised a plan for a future human genome initiative that would mobilize human and financial resources. Their aim was to identify, sequence and analyze all the genes of the DNA molecules that constitute the chromosomes in each human cell.

The aim, easily expressed, but with extraordinarily complex implications, was essentially defined in a US government publication of the time[6] in two terms: sequencing and mapping. The HGP considered work on these maps and the development of new technologies as the two priority objectives until about 1995. However, it was foreseen that the project would continue at

---

[5] US Congress, *Commercial Biotechnology, an International Analysis,* OTA, BA 218, Government Printing Office, OTA, January 1984.

[6] *Understanding our Genetic Inheritance,* US. Department of Health and Human Services, and Department of Energy, NIH Publication no. 50-1590, Washington, 1990, p. 9.

least until 2005. There were predictions that the information generated by the Human Genome Project would be the reference work of biomedical science for the 21st century[7]. Several renowned scientists, such as Renatto Dulbecco[8], Alan Wilson[9], Leroy Hood and Lloyd Smith[10], pointed out the essential importance of the project for the understanding of the origins of man, of human biology in general and for the exploration of the mysteries of human development and illnesses. For Walter Gilbert, sequencing the human genome was a quest for the Holy Grail[11]. The speed with which the HGP went from being a "What if we did?" concept, from 1985–86, to a realistic and implemented program by the end of 1988, is really surprising, reflecting the unprecedented changes that were to affect genetics itself.

A revisionist view of the HGP's history placing its origin in the tradition of medical and human genetics must be refuted. Specialists in these disciplines have tried to assert that "the" project was "their" project. In fact, not one of the three instigators of the HGP was a human geneticist. It is true that each of them used genetic techniques and was involved in biomedical research, but none was really involved in the study of human genetic disease or human heredity. The HGP sprang from a political-economic technological vision, following on from the idea of systematically using the tools of molecular genetics on all genomes. Human geneticists re-appropriated some of the ideas that had been linked to the HGP by redefining its goal, but its roots are clearly not in the great current of human genetic science; although progress in analytical methods for DNA structure allowed the power of genetics to be applied to the study of illnesses and normal physiology in man.

---

[7] *Understanding our Genetic Inheritance*, US. Department of Health and Human Services, and Department of Energy, NIH Publication no. 50-1590, Washington, 1990, p. 9.

[8] Renato Dulbecco, Salk Institute, interview with OTA staff member, January 1987.

[9] Alan Wilson, University of California, Berkeley, at a Symposium for the Director, National Institutes of Health, 3 November, 1987, p. 37.

[10] Leroy Hood and Lloyd Smith, California Institute of Technology, in *Issues in Science and Technology*, 3, 1987, p. 37.

[11] Walter Gilbert, Harvard University, at several national meetings, March 1986 to August 1987.

The real roots of the Human Genome Project can be found in the genetic studies of bacteria, yeast and the nematode worm. Although pioneers had used molecular methods to dissect human pathologies with some success, the emergence of the new genetic technologies, that had been so useful in the study of other organisms, made the study of the human genome easier. These unprecedented technical developments between 1974 and 1986 in genetic linkage mapping, physical mapping and DNA sequencing were the background for the science policy debate that concluded with the idea of a concerted Human Genome Project. The technological capability was available, but for the HGP this new technology would have to be deployed in a scientific project that had a budget and an administrative and institutional structure for its implementation. In 1985, well before the project emerged from the technological "primal soup", several enthusiastic visionaries, quite independently from each other, were already putting forward original and revolutionary ideas for a new great biological enterprise.

The first meeting related to human genome sequencing took place in 1985. It was hosted in the United States by Robert Sinsheimer[12], of the University of California at Santa Cruz (UCSC). Although the genome project did not come of this meeting, it was a decisive moment in the emergence of the whole idea. Proposals for the meeting had begun in October 1984, when Sinsheimer got in touch with biologists Robert Edgar, Harry Noller and Robert Ludwig, asking whether they would care to come to a meeting in his offices. At the time, Sinsheimer held the position of Chancellor of the UCSC, and as such, he had taken part in several major planning efforts in science policy. These had involved the three national laboratories managed for the Department of Energy (DOE) by the University of California (the national laboratories of Los Alamos, Lawrence Berkeley and Lawrence Livermore), the California state proposal to build the Superconducting Super Collider and, more particularly, the Lick Observatory. The UCSC's astronomy faculty enjoyed an international reputation. Sinsheimer, as a biologist, wanted the biology faculty to attain a similar status. He wanted to "put Santa Cruz on the map".

---

[12] R.L. Sinsheimer, "The Santa Cruz Workshop", *Genomics*, 5, 1989, pp. 954–956.

Sinsheimer was not the first to have these thoughts. Others had hatched large scale plans for mapping and technological development before him, but they had never turned them into a genome project. In 1980, the EMBL had seriously considered sequencing the 4,700,000 bp genomes of the bacteria *E. coli*, but the project had been considered technologically premature[13]. Norman Anderson had worked in several DOE-financed laboratories and had made a career of developing molecular biology tools. At the end of the 1970s, with his son Leigh, he campaigned for a concerted effort to catalogue genes and blood proteins. In the early 1980s, Senator Alan Cranston lent his support for their proposal for an early program dedicated to that objective to the tune of $360 million. Discussions on the need to collect DNA sequence results were already taking place. Father and son continued to point out how urgent it was that their program be adopted by the national laboratory system and the DOE. Administrators were aware of their efforts, and this may have contributed to the preparation of the genome project, but it was not the main trigger.

Sinsheimer's inspiration for a DNA sequencing project came from a telescope[14]. A group of University of California astronomers were lobbying for the construction of the world's largest telescope. They were eventually successful with the construction of the Keck Telescope on Mauna Kea in Hawaii, which conquered numerous obstacles to open on 24 November 1984. In 1984, the cost of enlarging Mount Palomar or building a new telescope of similar size was estimated at $500 million, much of this funding being needed to build a huge mirror. Jerry Nelson, of the Lawrence Berkeley Laboratory, had the bright idea of using 36 hexagonal mirrors instead of one big one, thereby cutting the estimated costs by over 80%. Computer adjustment of the mirrors and their alignment could allow this group of smaller and cheaper mirrors to provide the same information as one huge one. An article on the telescope appeared in the *San Jose Mercury*, and soon

---

[13] T.F. Smith, "The history of the genetic sequences databases", *Genomics*, 6, April 1990, pp. 701–707, N. Wada, "The complete index to man", *Science*, 211, 2 January 1981, pp. 33–35.

[14] S.S. Hall, "Genesis, the sequel", *California*, July 1988, pp. 62–69, R. Sinsheimer, "The Santa Cruz Workshop", *Genomics*, 5, May 1985, pp. 954–956.

after, a certain Mr. Kane called the laboratory. He said he knew of a benefactor, Max Hoffman's widow, who might be interested in financing the telescope. Max Hoffman had made a fortune as the Volkswagen and BMW importer for the United States and had left an inheritance of several million dollars. The Hoffman Foundation was being run by three people, including his widow, and to cut a long story short, on the day before her death, Mrs Hoffman signed most of the documents of a $36 million donation for the Hoffman telescope project.

It was the biggest donation in the history of the University of California. Generous as it was, it still was not enough to fund the telescope. Another $30 to $40 million from other benefactors would be needed, and benefactors like that are not easy to find. The University of California had so much trouble finding interested benefactors that they asked the private university Caltech for help. Caltech found other financial partners, but the Keck Foundation to agree to provide all the funding for the telescope, on condition that it be named the Keck Telescope.

As the Hoffman Foundation no longer had the glory of being the main donor, and with the loss of the most enthusiastic trustee, the foundation no longer wished to build a smaller telescope or allow the funds to be used for alternative purposes by the university. Sinsheimer wondered whether an attractive proposal in the field of biology would recapture the attention of the Hoffman foundation. He asked himself whether opportunities were not being missed in biology due to biologists' tendency to think on a small scale unlike astronomers and high energy physicists, who did not hesitate to demand considerable sums to finance programs they insisted were essential to the development of their science[15].

Sinsheimer's laboratory had purified, characterized and genetically mapped a bacterial virus by the name of phi-X-174[16], the same 5,386 bp virus that Sanger had sequenced in 1978[17]. Sinsheimer had been paying

---

[15] Tom Wilkie, *Perilous Knowledge (The Human Genome Project and its Implications)*, Faber and Faber, London, Boston, 1994, p. 76.

[16] R. Sinsheimer, "The Santa Cruz Workshop", *Genomics*, 5, 1989, pp. 954–956.

[17] F. Sanger, "Sequences, sequences and sequences", *Annual Reviews of Biochemistry*, 57, 1988, pp. 1–28.

close attention to the gradual increase in complexity of model organisms being sequenced. Now that he was looking for a large-scale biological target, he thought of sequencing the human genome, 10,000 times longer than the longest viral genome being sequenced at that time. He cast around his UCSC colleagues for advice on the construction of an institute to sequence the human genome, and so it was, in October 1984, that he called Harry Noller, Robert Edgar and Robert Ludwig into his office. Initially, they were startled by Sinsheimer's nerve, but after a bit of thought they approved his idea, judging that the objective and expected results would be profitable to the development of biology. The sequencing process in particular would include physical mapping, of high value in itself. Edgar and Noller prepared a letter of support for Sinsheimer around Halloween 1984, and this became the basis of the letter that Sinsheimer sent to the president of the UCSC, David Gardner, on 19 November 1984. In this letter, he encouraged Gardner to approach the administrators of the Hoffman Foundation with this new idea, pointing out that, "It is an opportunity to play a major role in a historically unique event — the sequencing of the human genome... It can be done. We would need a building in which to house the institute formed to carry out the project (cost of approximately $25 million), and we would need an operating budget of some $5 million per year. Not at all extraordinary... It will be done, once and for all time, providing a permanent and priceless addition to our knowledge"[18].

In March 1985, Sinsheimer also discussed his project with James Wyngaarden, then Director of the National Institutes of Health (NIH). He noted that Wyngaarden seemed very taken with the idea. The latter advised Sinsheimer to approach the National Institute of General Medical Science if the meeting, now foreseen for May, led to a consensus on the project's feasibility. Sinsheimer realized he was being told to find funding, and to do this, he needed the support of some world-famous scientists to give his project some credibility.

The next step was to contact and bring together the experts. H. Noller wrote to Frederick Sanger, with whom he had worked for several years, and

---

[18] R. Sinsheimer, *Letter to David Gardner*, University of California at Santa Cruz, 19 November 1984.

got an encouraging answer. R. Edgar, H. Noller and R. Ludwig set the meeting for the 24th and 25th of May 1985, and invited an eclectic panel of experts. Bart Barrel was Sanger's successor as head of the MRC's large scale sequencing efforts. Walter Gilbert was there to represent the Maxam-Gilbert approach to sequencing. Lee Hood and George Church were working on the improvement of sequencing techniques. Genetic linkage map specialists were also invited: David Botstein, Ronald Davis and Helen Donis-Keller. John Sulston and Robert Waterston were invited to talk about their efforts on the physical map of the *C. elegans* genome. Also recruited were Leonard Lerman, a very technologically-minded biologist, and David Schwartz, who had invented some techniques for the manipulation and separation of DNA fragments several million bases long. Finally, Michael Waterman, of the University of California, was also asked to come because of his expertise in mathematical analysis of sequences and data bases.

During the discussions, the group of experts decided that the development of a genetic linkage map, a physical map and further tools for large scale DNA sequencing seemed realistic goals. Initial sequencing efforts should be part of the development of faster and more automatic techniques that would lessen the costs of sequencing. The gathered experts concluded that it was impossible to sequence the whole of the human genome because such an enterprise would imply great technological progress, as yet unattained. In the meantime, it was felt reasonable to concentrate on sequencing the 1% of regions of great interest (polymorphic regions, functional genes...)[19].

Thus, the idea of systematically sequencing the human genome was laid aside right at the beginning of the meeting. Following the meeting, Sinsheimer wrote to several potential sources of funding, including the Howard Hughes Medical Institute (HHMI) and the Arnold and Mabel Beckman Foundation, enclosing a *résumé* of the discussion. They did not seem very interested. He was not allowed to contact the Hoffman Foundation directly, since the Office of the President of the University of California was now handling links with that foundation. The NIH route was also blocked because considerable funding

---

[19] Human Genome Workshop, "Notes and conclusions from the human genome workshop", 24–26 May 1985, reprinted in Sinsheimer's "The Santa Cruz Workshop", *Genomics,* 5, 1989, pp. 954–956.

would have to be requested to build the laboratory to carry out the work; the NIH had programs to finance the research work itself but not for funding construction of the buildings. These were major obstacles. Sinsheimer concluded that the only solution was, first of all, to find a private benefactor for the building, but his access to serious sources of funding was through the Office of the President of the University of California. The problem was insoluble for him.

Sinsheimer was thinking of going straight to Congress. He discussed the idea of an institute with his congressman Leon Panetta, who was enthusiastic but pointed out that a proposal of that size would absolutely have to come through the hierarchical route, in other words, through the Office of the President of the University of California again. Sinsheimer became more and more frustrated in his efforts to awaken some interest in Gardner's office. He was due to retire, and the prospects of having the Institute of Human Genome Sequencing at Santa Cruz were gradually fading away.

In the meantime, the idea of sequencing the human genome had come to life in its own right and filtered through to other areas.

Walter Gilbert and Biology's Holy Grail

Sinsheimer passed the torch to Walter Gilbert, Nobel prizewinner for his work on sequencing techniques and one of the founders, in 1978, of the biotechnology firm Biogen. Gilbert had finally left Biogen in 1984 to return to Harvard as a professor of the Biology Department, and in 1988, he was named Loeb University Professor. After leaving Biogen, Gilbert had spent some time traveling in the South Pacific, and the Santa Cruz workshop group had great trouble finding him. Robert Edgar finally caught up with him in March, and Gilbert accepted the invitation to the Santa Cruz workshop, to which he made a highly significant contribution. After the workshop, Gilbert became one of the main supporters of the Human Genome Project for most of what turned out to be a critical year. His rational vision and enthusiasm infected other molecular biologists and the general public. Two days after the workshop, he translated the Santa Cruz group's ideas into operational plans in a memo to Edgar. It included a strategy for Sinsheimer's institute, although deep down, Gilbert himself was not convinced that it should be built at Santa Cruz.

For Gilbert, mapping the entire sequence of the human genome was the Holy Grail of human genetics. Through this sequencing effort, all possible information on human structures would be revealed. It would be an incomparable tool for research into each aspect of human functions. Gilbert's Holy Grail metaphor turned out to be a significant rhetorical contribution to the genome debate. It came, perhaps, to mean more than Gilbert actually wanted to say. The Holy Grail myth calls to mind a story which, all things considered, was a perfect match: each of the Knights of the Round Table on quest for an object of indeterminate form, whose history was obscure and whose use was debatable (except that it would bring health and strength to the king and prosperity to the land). Each knight had a different path and lived through his own specific adventure. Gilbert spread the Santa Cruz idea into the mainstream of molecular biology. He gave some informal speeches at a Gordon Conference in the summer of 1985 and at the first conferences on genes and computers in August 1986 [20]. Gilbert's contact network was extremely extensive and he infected several of his colleagues with his ideas and enthusiasm, including James Watson. He helped bring the genome project a lot more publicity than it would have had without him. Several publications echoed his ideas and his role, including the *US News and World Report, Newsweek, Boston Magazine, Business Week, Insight* and the *New York Times Magazine*[21].

Gilbert, Watson, Hood, Bodmer and others gradually became the video documentary stars of the Genome Project. Gilbert and Hood wrote enthusiastic

---

[20] K.D. Gruskin and T.F. Smith, "Molecular Genetics and Computer Analyses", *CABIOS, 3*, no. 3, 1987, pp. 167–170.

[21] K. McAuliffe, "Reading the Human Blueprint", *U.S. News and World Report,* 28 December 1987 and 14 January 1988, pp. 92–93, S. Begley, S.E. Katz and L. Drew, "The Genome Initiative", *Newsweek,* 21 August 1987, pp. 58–60, G. del Guercio, "Designer genes", *Boston Magazine,* August 1987, pp. 79–87, A. Beam and J.O. Hamilton, "A Grand Plan to Map the Gene Code", *Business Week,* 27 April 1987, pp. 116–117, D. Holzmann, "Mapping the genes, Inside and Out", *Insight,* 11 May 1987, pp. 52–54, R. Kanigel, "The Genome Project", *New York Times Magazine,* 13 December 1987, p. 44, pp. 98–101, p. 106.

articles for a special edition of *Issues in Science and Technology*[22], published by the National Academy of Sciences. Along with Bodmer, Gilbert promoted the genome project in *The Scientist* editorials[23]. The idea of a genome project was thus widely diffused, all still in the spirit of Santa Cruz.

Gilbert, however, caused a major controversy when he tried to privatize the genome project. Initially, in 1986, he thought he might set up a genome institute himself. In January 1987, Michael Witunski, the President of the James McDonnel Foundation, contacted Gilbert to suggest that the foundation might be interested in funding the creation of such an institute. This idea fizzled out when the foundation funded a study to evaluate a genome project for the National Research Council (NRC) and the National Academy of Sciences. Gilbert took part in a series of meetings to debate the genome project at the end of 1986, and he became a member of the NRC committee, but in the spring of 1987, he resigned that position. He decided to direct his activities towards more commercial concerns and announced his plans to set up a genome corporation. Gilbert's idea for GenCorp was to build a physical map of the genome, run systematic sequencing and set up a data base. Financial objectives would be the sale of clones, sequencing services and user rights for the data base. Prospective clients would be industrial and academic laboratories and industrial firms such as the pharmaceutical companies, who might need goods and services from GenCorp. The aim was not so much to do what others could not, but to do it more efficiently so that laboratories could purchase goods and services more cheaply than if they had done the work themselves. GenCorp would allow biologists to concentrate on biology rather than waste time making the tools needed for their experiments.

Gilbert, armed with this idea, began a hunt for venture capital at the end of 1987. Wall Street's enthusiasm for biotechnology had gradually soured to skepticism, and the October crash put paid to the funding of GenCorp.

---

[22] W. Gilbert, "Genome Sequencing: creating a new biology for the twenty-first century", *Issues in Science and Technology*, 3, 1987, pp. 26–35, L. Hood and L.S. Smith, "Genome sequencing, how to proceed", *Issues in Science and Technology*, 3, 1987, pp. 36–46.

[23] W. Gilbert "Two cheers for genome sequencing", *The Scientist*, 20 October 1986, p. 11, W.F. Bodmer, "Two cheers for genome sequencing", *The Scientist*, 20 October 1986, pp. 11–12.

Efforts to initiate a genome project at the federal governmental level, widely pushed to the public through heavy publicity, also made investors wary of competing with a possible public sector activity. Although Gilbert could not get hold of the necessary funding, he remained one of the great defenders and enthusiasts of a genome project and optimistic as to technological developments. His enthusiasm was such that it worried younger researchers who shrank from prospects which seemed to them too ambitious. In particular, they hated his monomania for systematic sequencing, and they rejected his view of genome research as assembly-line work. They complained bitterly that they could be asked to attain impossible objectives by policymakers who listened to Gilbert.

Gilbert was censured for setting up excessively ambitious targets, but he stayed in the firing line with the rest of the "genome researchers". He was not just all rhetoric either, his laboratory being one of the first to get involved in large scale DNA sequencing. In 1990, he proposed to sequence one of the smallest genomes of a free-living organism, a small bacteria found on sheep called *Mycoplasma capricolum*[24]. This project was one of the large sequencing projects, then in embryo, that would try to take systematic genome sequencing from a theoretical possibility to a new approach to life[25]. The genetics of *M. capricolum* was not as well known as that of many other organisms. Gilbert's idea had been to determine the entire sequence of the bacteria's genome, which meant sequencing 800,000 base pairs, thought at the time to constitute about 500 genes. He then hoped to reconstruct the organism's biology from its DNA sequence. It was not, for him, that the sequencing itself would allow biologists to handle all the questions of biological interest, but that starting with the sequences would get them there faster. Gilbert's *M. capricolum* project joined other pilot projects to sequence model organisms that were spin-offs at the genesis of the human genome project. They were among the projects financed by the NIH's National Center for Human Genome Research[26] during the first year of its operation.

---

[24] L. Roberts, "Large scale sequencing trials begin", *Science*, 250, 7 December 1990, pp. 1336–1338.

[25] W. Gilbert, "Towards a paradigm shift in biology", *Nature,* vol. 349, 10 January 1991, p. 99.

[26] L. Roberts, "Large scale sequencing trials begin", *Science*, 250, 9 November 1990, pp. 756–757.

Furthermore, Gilbert developed an even more incisive concept of the role of molecular biology, the importance of sequencing enterprises and, in particular, the genome project. He foresaw, for biology, an upheaval of the type Thomas Kuhn called a paradigm shift [27] — an upheaval typified apparently by an increase of theory in science. According to Gilbert, molecular biologists would only do experiments to test theories that came from analysing the masses of information held in computers. Cloning and sequencing, which had cost many a doctoral student and researcher sleepless nights, would, in future, be the work of robots and specialized commercial services.

"To use this flood of knowledge, which will pour across the computer networks of the world, biologists must not only become computer-literate, but also change their approach to the problem of understanding life (...) The view that the genome project is breaking the rice bowl of the individual biologist confuses the pattern of experiments done today with the essential questions of science. Many of those who complain about the genome project are really manifesting fears of technological unemployment".[28]

A genome program strong enough to uphold such a vision would need administrative structure. Building that structure was at least as arduous as building the science of sequencing in the first place. At the beginning of 1987, when Gilbert was plotting GenCorp, there was no center to support these efforts or other similar enterprises in the field of genome sequencing and mapping. Although Gilbert despaired of federal leadership for the genome project. Two federal agencies finally set it up. By the end of the 1980s, the Department of Energy and the NIH had genome programs boasting a total budget of almost $84 million, and genome programs were also running in Great Britain, Italy, the Soviet Union, Japan, France and at the European Community level. This administrative transformation had begun back in 1985.

---

[27] W. Gilbert, "Towards a paradigm shift in biology", *Nature*, vol. 349, 10 January 1991, p. 99.
[28] *Ibid.*

5.2 The Department of Energy Initiative

Independently of Sinsheimer's initiative, the DOE began to envisage a genome program. It might seem a little strange that it was the Department of Energy, but the DOE had been deeply interested in questions of human genetics and mutations for a long time because of its military and civilian nuclear power programs. Several of the roots of human genome research programs do, in fact, originate from the end of the Second World War and strategic defense initiatives.

After the war, Congress and the American government decided not to leave the production of nuclear weapons to the military at the Pentagon (the Department of Defense), but to assign it to the civilian Atomic Energy Commission (AEC). This commission was not only responsible for the development and production of nuclear weapons, but also for the development of civilian nuclear power. In this second role, it was responsible both for promoting civilian nuclear power and for checking safety at nuclear power stations.

Towards the mid-1970s, it was felt that these two roles were not easily concilable, and eventually, the commission was split in two. A Nuclear Regulatory Commission, a control body, was set up to check nuclear sites, and the new Energy Research and Development Administration (ERDA) received a wide mandate including a mission to look for alternative energy sources. The ERDA became the Department of Energy in 1977, but kept its role as the developer and producer of nuclear weapons as well as keeping the control and security of reactors and other procedures involved in arms production. The first biological project linked to nuclear fission was set up by Arthur Holly Compton at the University of Chicago in 1942, the site of the first nuclear chain reaction. During the years that were to follow, the mandate for the Biological Research Program widened considerably to include the study of many biological effects of energy production as well as problems caused by radiation.

In the immediate post-war period, the AEC was a major supporter of genetic research. It had a relatively large budget at a time when the National Science Foundation was only beginning and the NIH was still only embryonic.

The part of the AEC's budget passed on to genetic research put the other genetic programs in the shade, leaving the AEC-funded national laboratories at the forefront of research in the field. This all changed as the NIH's budget swelled over the next three decades. The National Institute of General Medical Sciences (NIGMS) became the main source of funding for fundamental genetics.

One of the main techniques used in these first studies on the effects of radioactivity in humans, and on their genes, was the examination of chromosomes to try to detect the anomalies that would, in theory, be caused by radiation. In 1983, the two main nuclear weapons laboratories at Los Alamos and Lawrence Livermore began to put together a project for the construction of a gene bank. At the time, the laboratories were involved in new separation and classification techniques for chromosomes, particularly one known as *Flow Cytogenetic Analysis*[29]. In 1986, this approach allowed the distinction and classification of all human chromosomes except 10 and 11. By February 1986, the national laboratories had also set up a library of human DNA fragments.

The DOE had an Office of Health and Environment Research (OHER) responsible for funding all life science research linked to environmental questions. In 1985, Charles DeLisi became its director. He began to promote the idea that the DOE should have a larger role in the modern approach to human genetics, in particular, through the new molecular biology. He recognized that sequencing the human genome was a fundamental project and underlined the fact that the DOE, with its two nuclear weapons laboratories, was already used to managing large-scale projects.

DeLisi's idea for a DOE genome project came from attempts to study the modifications of DNA in the cells of survivors of the atomic bomb dropped by the American military over Hiroshima and Nagasaki in 1945, survivors known in Japan as *hibakusha*. This effort was an attempt to understand the consequences, and more precisely, to estimate the frequency

---

[29] This approach involves mixing chromosomes with fluorescent markers. Since different chromosomes are linked to different types of markers, you can tell them apart and rank them in order by projecting laser beams at them and measuring the different amounts of each marker.

of hereditary mutation caused by radiation exposure. The *hibakusha* were severely damaged in the post-war years. They were studied intensively for decades in one of the largest and most complicated epidemiology studies ever undertaken.

In 1947, the US National Academy of Sciences set up the Atomic Bomb Casualty Commission (with AEC funding) to study the biological effects of the bombs. The ABCC was established to gather as much data as possible, but not to provide treatment for the symptoms they found. After this commission had been set up, and also in view of the fact that medical care was not being provided, the Japanese government set up special health programs. In 1975, the ABCC became the Radiation Effects Research Foundation (RERF), which was set up in Hiroshima and Nagasaki and jointly funded by the American and Japanese governments. The RERF continued with the epidemiological investigations and other similar research.

The *hibakusha* were not well accepted in Japanese society, partly because it was believed that they were carriers of hereditary mutations due to their exposure to radiation. In the early post-war years, the extent of mutational damage to A-bomb survivors was the subject of great controversy. Hermann Joseph Muller, who had just been given the Nobel Prize for his discovery that radiation could indeed cause mutation, lent his name to sound the alarm. He suggested that if the *hibakusha* could see the consequences over a thousand years, they might regret that the bomb had not killed them outright[30]. Alfred Sturtevant had an even more apocalyptic view of radiation exposure, that he outlined in a letter to *Science*[31], warning that the A-bombs that had been dropped would, over a very long time, produce many defective individuals, if the human race even survived. Such predictions from one of the most famous geneticists of the time fed the collective fear of radiation, a fear which had existed before the bombs, but intensified because of the mystery surrounding the Manhattan Project and its terrifying offspring. However, the fear was caused more by speculation than fact.

---

[30] Quoted by J. Beatty, "Genetics in the Atomic Age: The Atomic Bomb Casualty Commission, 1947–1956", in *The Expansion of American Biology*, ed. K. Benson *et al.*, Rutgers University Press, 1 November 1991, pp. 284–324.

[31] A. Sturtevant, "Social implications of the genetics of man", *Science*, 120, 1947, p. 407.

This speculation was not purely fabrication, as it was based on animal studies; but in this case, it was not sufficient merely to project findings from studies on other organisms to estimate results on humans. Three decades of medical studies on the *hibakusha* also provided conflicting results. There were not enough facts to allow the decision makers to make the political choices needed. Although radiation clearly did induce an increase in mutation, it was not possible to determine how much of an increase quantitatively, nor to specify what its consequences would be for man. The limited facts available were open to a wide variety of interpretations[32]. Geneticists continued to be haunted by the idea that the *hibakusha* might be carriers of hereditary mutation, and it was hoped that the studies funded by the ABCC could prove that this was not the case.

James V. Neal and other researchers dedicated their careers to this study of the effects of radiation on the genes of the *hibakusha* and their descendants. Neal set up a department of human genetics at the University of Michigan, some of the funding being allocated to studies on the genetic effects of mutations. In the mid-1980s, one group had the idea of trying to apply the new techniques of molecular biology to the study of quantitative measurements of hereditary mutation in human beings.

In March 1984, the RERF held a conference on genetic studies in Hiroshima. The researchers it gathered together recommended that *hibakusha* cell lines be set up and that students of DNA should start using the new methods for direct DNA studies as soon as possible. This recommendation could be interpreted in several ways; therefore, the International Commission for Protection against Environmental Mutagens and Carcinogens decided to hold a meeting dedicated to discussing the new techniques for analysis. This DOE-funded meeting was organized by Ray White, of the Lawrence Livermore National Laboratory at the request of Mortimer Mendelsohn.

White invited an extraordinary group of molecular and human geneticists (many of whom had never met before) to the meeting which took place in

---

[32] As pointed out in the *Le Monde* articles of 28 and 29 April 1996, "Des mutations héréditaires affectent les victimes de Tchernobyl", there is still some uncertainty regarding the transmission of mutations, on the long-term effects of exposure to nuclear waste and weaker radioactive doses, the relationship between the mutation rate and the quantity of exposure the victim received.

December 1984 at Alta, Utah in the USA. Botstein and Davis, who struck upon the idea of systematic RFLP mapping, attended as well as George Church, who was inspired to later improve the Maxam-Gilbert sequencing technique. Most of the younger biologists had never met James V. Neal, and some had not even heard of him. During the meeting, Maynard Olson, later to be a major protagonist in the history of the genome project (the human genome project as well as the yeast *Saccharomyces cerevisiae* genome project), was deeply inspired by Neal's commitment. Olson had just begun to get results from his yeast physical mapping project. In addition, Charles Cantor presented some results produced by the method that he had used with David Schwartz to separate DNA fragments several million base pairs long. White had got further than anyone else, as he had already discovered the first RFLP. As the discussion intensified, the meeting became a crucible, boiling with ideas. The conclusion was that contemporary methods of direct DNA analysis were inadequate and would not detect the rapid growth in mutation frequency expected in *hibakusha* after exposure to A-bomb radiation. The conclusion appears disappointing, but the meeting helped consolidate a group of ideas that developed into what we might call the DOE's genome project.

The links between this meeting and the DOE's project can be found in a congressional report by Charles DeLisi, then a new figure at the DOE. The Congressional Office of Technology (OTA) was, at the time, putting together a report on technologies for measuring hereditary mutation in humans. The congressional committee had considered problems of exposure to Agent Orange, toxins and radiation to be problems of public policy. Mike Gough, then the director of an OTA project, attended the Utah meeting and analyzed the different techniques for a report to the DOE. It was published in 1986 under the title *Technology for detecting heritable mutation in human beings*[33].

Whilst reading the October 1985 preliminary draft report for the OTA[34], DeLisi thought up a project dedicated to DNA sequencing, structural analysis

---

[33] US Congress, *Technologies for Detecting Heritable Mutations in Human Beings* (OTA H 298), Washington DC Government Printing Office, OTA, Washington, 1986.

[34] C. DeLisi, "The Human Genome Project", *American Scientist*, 76, 1988, pp. 488–493.

and computational biology. In his position as the new Director of the OHER, whilst examining a report prepared by experts and external advisors, he reflected on the programs in his charge. It was a surprising time for Delisi to make such a suggestion, as the cold war was getting colder and the United States was building and testing a large number of nuclear weapons. Furthermore, President Ronald Reagan had proposed a space defense project against potential nuclear attack (the "Star Wars" defense initiative), which implied heavy involvement of the big laboratories in the development of new weapons. However, the Strategic Defense Initiative had few concrete effects, and events in Europe and the east gradually wore down a lot of the justification for this policy of intensive development of the defense and nuclear weapons program. Even if the Soviet Union had not imploded, it is unlikely that Reagan's agenda would have been followed. In the 1990s, the weapons laboratories found themselves frenziedly defending their existence by diversifying their goals. It is possible that DeLisi understood that, despite appearances at the beginning of the 1980s, the party was over for nuclear science and technology, and that the time had come to move on to molecular biology.

DeLisi was very enthusiastic. Working with David Smith to develop a major new scientific initiative, they asked the "biology group" at the Los Alamos National Laboratory for their ideas on a genome project. The scientists responded with a very dense and enthusiastic five-page memo, mainly written by the physicist Mark Bitensky[35], which centered heavily on the potential scientific, medical and technical benefits that would be possible using a structural approach to biology. It focused mainly on DNA sequencing and only mentioned physical or genetic mapping in passing. Another argument for them in favor of the project was that a genome project was seen as an opportunity for real international cooperation that might lessen political tensions. The Los Alamos group even checked whether Yale's Frank Ruddle would testify before Congress if he was summoned.

---

[35] M. Bitensky, C. Burks, W. Good, J. Ficket, E. Hildebrand, R. Moyzis, L. Deaven, S. Cramm and G. Bell, Memo to Charles DeLisi, DOE, LS-DO-85-1, 23 December 1986, Los Alamos National Laboratory.

Encouraged by this initial support, DeLisi and Smith began to nag at the various levels of Washington bureaucracy. DeLisi roughed out the policy strategy that would get them the support of the scientific community, their bosses at the DOE and in congress. Smith answered this proposal in his note on rumors of previous discussions at a Gordon Conference and at the University of California the previous summer, but he pointed out that he did not know what had become of them. Smith warned DeLisi that the DOE's proposal would certainly meet with many objections, amongst which were that it was not science but merely technique, politically planned research would be less effective than leaving small groups of researchers free to decide on what was really important and work should concentrate on genes of interest instead of on an overall sequencing effort. DeLisi answered that, of course, there would be complaints about this long-term monotonous job. The point was whether the monotony should span thirty years or should be concentrated into ten. He foresaw that the budget would have to be 100 or 150 million US$ per year for more than a decade and maintained that the political effort should not be focused on whether such a project would leach funding from other work, but rather on how to obtain totally new funds. In an attempt to reach the scientific community, DeLisi and Smith asked the Los Alamos authorities to hold a workshop in order to find out if there was consensus on whether such a project was possible, and if it should be set up, to try and determine the potential scientific and medical benefits, express a political strategy and discuss possible international cooperation, in particular with the Soviet Union.

A group met in Los Alamos on 6 January 1986 to begin planning the workshop, which took place in Santa Fe (New Mexico) on 3 and 4 March 1986[36]. Frank Ruddle chaired the meeting (where DeLisi's idea was met with enthusiasm) and the importance of integrating linkage mapping and starting a systematic physical map were reinforced. There was consensus on

---

[36] Benjamin J. Barnhart, *The Human Genome Project: A DOE Perspective*, OHER DOE Washington DC, 1988, p. 4. This report was presented during the course of a European Commission/DOE meeting in the summer of 1988. *The Human Genome Project — Deciphering the blueprint of heredity*, ed. Necia Grant Cooper, section "What is the Genome Project?", University Science Books, Mill Valley, 1994, pp. 71–84.

the benefits of a genome program, but making the project a reality remained a problem. Anthony Carrano and Elbert Branscombe, of the Lawrence Livermore National Laboratory, pointed out the importance of a clone map and that a program focused only on sequencing could cause needless fears in the scientific community that funding would be hijacked and research be reduced to simple technical activity. Later events proved them right on this. Fear of a monotonous sequencing operation became the main threat to the development of the genome program.

In addition to the strictly scientific interest of the program, the decision makers at the DOE had a hidden agenda for backing such a project: they wanted to take advantage of the considerable scientific and technical resources of the national laboratories. The genome project was a perfect subject for this, the national laboratories' experience in large scale multidisciplinary projects and history of success in applying technique to medical science being its selling points. Large sections of the project could not be easily undertaken by the universities and so this second justification became as important as the first[37]. DeLisi's initial idea was very much generated by a scientific need — the study of mutations — but it was transformed by the inclusion of the more important political goal of saving the national laboratories. DeLisi spoke of the genome project to his immediate boss, Alvin Trivelpiece, who supported him and asked the Health and Environmental Research Advisory Committee (HERAC) to report on such a project. Trivelpiece, as the Director of the Office of Energy Research, reported directly to the Secretary for Energy, then John Hermington, who in his turn spoke to the US President.

On 6 May 1986, six months after his original idea, DeLisi wrote an internal report asking for a new budget line. The report reached Alvin Trivelpiece and continued on to the upper strata of the bureaucracy. DeLisi had roughed out a two-phase program, Phase I of which consisted of the physical mapping of human chromosomes (the central element lasting five to six years), the development of mapping technologies and the development

---

[37] Benjamin J. Barnhart, *The Human Genome Project: A DOE Perspective*, OHER, DOE, Washington DC, 1988, p. 2.

of sequencing technologies. As the physical mapping progressed, parallel work would prepare Phase II, sequencing the whole genome. High speed automated DNA sequencing and computer analysis of the resulting sequences would both be essential for the transition between Phases I and II.

DeLisi spoke of the project as an analogue of the space program, except that it would involve many agencies and require a more widely-distributed work structure, with another agency coordinating. DeLisi thought that this coordinating agency would have to be the DOE. Fiscal years 1987–1991 were to have a five-year budget growing from five through 16, 19, 22 and then 22 million US$. In July 1986, a series of meetings began between Judy Bostock, in charge of the life science budget, and her superior Thomas Palmieri, in the Presidential Office of Management and Budget (the OMB), which prepares the president's budget request to put before congress. Bostock accepted DeLisi's plans, which already eliminated one of the major obstacles on the road to congressional approval. As far as the scientists were concerned, the OHER and the HERAC had approved the plans for a DOE genome initiative in a report of its special *ad hoc* subcommittee[38], which recommended that the initiative be assigned a budget of some $200 million per annum[39]. The HERAC report was published in April 1987, four months after the multiannual budget was agreed to by the OMB and the DOE. The gist of the report was that it favored the DOE as coordinating agency[40].

Despite the green light the DOE had received from its OMB bosses and from the scientific community represented by the HERAC, DeLisi's work was far from over. Before a federal agency can fully implement a major new initiative, Congress must authorize it, and in a separate action, allocate funds for its use. For the launch of the genome program, DeLisi took US$5.5 million from the current budget for 1987 and diverted it. The Committee of

---

[38] Subcommittee on the Human Genome, Health and Environmental Research Advisory Committee, *Report on the Human Genome Initiative, Prepared for the office of Health and Environmental Research*, Office of Energy Research, Department of Energy, Germantown, MD, DOE, April 1987.

[39] For a synthesis of the different budget estimates for the HGP, see *Genome Projects: How Big, How Fast?* OTA – BA – 373, Government Printing Office, Washington DC, April 1988, Appendix B: Estimated costs of Human Genome Projects, pp. 180–186.

[40] Benjamin J. Barnhart, *The Human Genome Project: A DOE Perspective*, p. 2.

Allocations and Authorizations allowed this sort of money shuffling, within reason. For 1988 and later budgets, an action in Congress was required. Finally, after all the political obstacles had been overcome, in particular the difficult process of obtaining congressional authorization and having the necessary funds allocated, the DOE budget for the genome project for fiscal years 1988 and 1989 were in line with the agreement with the OMB, and for 1987 and 1988 respectively, added up to US$12 and 18 million[41]. The DOE's genome project was well on its way, but many obstacles and problems would still have to be overcome, especially that of deciding which agency would direct and coordinate the project[42].

During 1986, as the DOE's project was being born, the situation in the USA was very complex. Of course, the information available on a map of the human genome was priceless, but a lot of attention was being paid to the cost and the monotony of sequencing work. Furthermore, two different scientific communities were getting involved. On one hand, the molecular biologists in the universities and other life sciences research institutes were naturally looking to the NIH for funding, as it was handling most of the federal money available for biomedical research. The NIH seemed a little hesitant, and even reticent[43], to invest in such a project. On the other hand, DOE scientists were hoping to diversify their interests and activities.

In the latter half of the 1980s, there was some real tension between the DOE and the NIGMS (the NIH Institute responsible for funding fundamental research) despite encouragement towards cooperation from DeLisi (of the DOE) and Kirschstein (director of the NIGMS) to the various teams in an effort to avoid direct conflict. The DOE, however, had in its favor well-

---

[41] For fiscal year 1989, the April 1987 HERAC report called for $40 million with a constant increase over a 5-year period to reach $200 million per annum. In fiscal year 1990, the DOE was financing genome research with US$27 million.

[42] Leslie Roberts "Agencies vie over Human Genome Project", *Science,* vol. 237, 31 July 1987, pp. 486–488, J.D. Watson, "The Human Genome Project: Past Present and Future", *Science,* vol. 248, pp. 44–48, *Mapping over Genes: Genome Project: How Big, How Fast?* as above.

[43] According to Ruth Kirchstein at the NIH, the agency had decided that the best approach would be to continue its own substantial program of investment in "initiate-research" in genetics. In 1987, the NIH spent US$ 300 million on research linked to mapping and sequencing, US$ 100 million on human genome research and US$ 200 million on research on other organisms.

equipped and financed laboratories looking for new perspectives that saw the genome project as a chance to diversify. Some of the more eminent biologists opposed the idea of systematically sequencing the human genome. They were suspicious about letting the DOE take charge of the project. During 1986 and 1987, the debate got very heated. The Whitehead Institute in Massachusetts, which had hosted most of the work on genome mapping, became a hotbed of opposition to the sequencing project. Nobel prizewinner David Baltimore, director of the Whitehead Institute, wrote in *Science* that "the idea was taking shape, and he was shivering just thinking of what was to come".[44] Also at the Whitehead, David Botstein, one of the co-authors of the original 1980 article that set up the bases of the human genome mapping strategy, added his voice to those of the skeptics. By June 1986, about three months after the DOE's meeting in Santa Fe, the Cold Spring Harbor Laboratory organized a symposium on the molecular biology of *Homo sapiens*. During this symposium, Botstein pointed out that, "Sequencing the genome now is like Lewis and Clark, going to the Pacific one millimeter at a time. If they had done that, they would still be looking"[45]. Another of the Whitehead scientists, Robert Weinberg, told the *New Scientist*, "I'm surprised consenting adults have been caught in public taking about it (sequencing the genome). It makes no sense"[46]. Such opposition was not confined to the east coast of the USA. James Walsh of the University of Arizona and John Marks of the University of California wrote in *Nature* that, "Sequencing the human genome would be about as useful as translating the complete works of Shakespeare into cuneiform, but not quite as feasible or as easy to interpret"[47].

Skeptical scientists were mainly concerned about the size and cost of the enterprise of sequencing the human genome. Few considered the long-term usefulness of a map or the sequence of the human genome as a progression

---

[44] As quoted in Roger Lewin's "Proposal to sequence the human genome stirs debate", *Science*, 232, 27 June 1986, p. 1600.

[45] Quoted in *Mapping our Genes, Genome Projects: How Big, How Fast? ref. cit.*, p. 126.

[46] *Ibid.*

[47] *Ibid.*

in the life sciences, but many were worried about the best way to reach any intended goal. *Science*'s editor Daniel Koshland defended the genome project in very simple terms in an editorial: "The main reason that research in other species is so strongly supported by Congress is its applicability to human beings. Therefore, the obvious answer as to whether the human genome should be sequenced is 'yes'. Why do you ask?"[48]

But there was no tradition of "Big Science" in biology, at least not the sort of Big Science that physicists and astronomers were used to. Although instrumentation and equipment had become much more sophisticated and expensive over the years, molecular biology was basically unchanged[49]. The large institutes had lots of little groups working in their own laboratories on their own research projects mostly quite independently from each other. Biology had never known an undertaking such as the Manhattan Project, which gathered together thousands of engineers and scientists in an industrially-structured organization, nor was it accustomed to the sharing of work as practiced by astronomers and physicists. Molecular biology still remained a science of individual research or research in small groups, and many scientists feared the loss of this characteristic of their science, which they were used to. There was little doubt for them that the job of sequencing the human genome would take a terrific amount of work, and most of it would be persnickety and repetitive. While the mechanical work of sequencing was being completed, who would have the time and energy to interpret the biological significance of the results? Another more concrete worry was about the money that would be needed. It would be extremely expensive to sequence the human genome, and the funding for such a project would come from the same place as the funding for the rest of molecular biology. Walter Gilbert thought it might cost one dollar to sequence one base. This meant spending US$3 billion over 15 years to sequence the approximately three billion base pairs of the human genome. The sequencing project, therefore,

---

[48] *Ibid.*

[49] D. Baltimore, "Genome Sequencing: a small science approach", *Issues in Science and Technology,* Spring 1987, pp. 48–49. Also the OTA report *Genome Projects: Mapping our Genes: How Big, How Fast?* as already mentioned, pp. 125–128.

would need separate financing, or current research would suffer from the competition for funding. *Science* published an editorial suggesting that the DOE should lead the project since this would prevent draining funds from the NIH and its scientists.

In an introductory commentary to *Science* on 7 March 1986[50], Renato Dulbecco, Nobel Prize winner and president of the Salk Institute, made the sensational remark that progress in the fight against cancer would be greatly accelerated if geneticists took on sequencing the human genome. Dulbecco's article in *Science* revealed the existence of the human genome sequencing project to many biologists for the first time. It stirred up debates in laboratories, universities and research centers all over the world. Like Sinsheimer, Dulbecco had succumbed to the idea of the human genome project because of a tendency to think large-scale. It was during a gala organized at the Kennedy Center on Columbus Day, 1985 by the Italian embassy in Washington (on the occasion of a scientific meeting) that Dulbecco presented his idea to the public. The meeting included a section on USA-Italian cooperation in the sciences. Dulbecco had been invited to give a presentation since he was one of the most eminent Italian biologists on the scientific scene, both in the United States and in Italy. He decided to talk about the genetic approach to cancer and thought that it would be a propitious occasion to broach a daring and innovative subject. Speaking of the past and future of cancer research, he thought it could be greatly enriched by the elucidation of the sequence of the human genome. This meeting in Washington marks the beginning of the Italian genome program.

Dulbecco again made sequencing the theme of his speech on 5 September, on the occasion of the inauguration of the Sandbrook Laboratory at Cold Spring Harbor (Long Island, New York)[51]. He spoke of a transition in cancer biology. "It seems we are at a turning point in the study of tumor virology and oncogenes". The well-known fact that cancers of certain cell types

---

[50] Renato Dulbecco, "A turning point in cancer research: sequencing the human genome", *Science*, 7 March 1986, pp. 1055–1056.

[51] R. Dulbecco, "A turning point in cancer research: sequencing the human genome" in *Viruses and Human Cancer*, ed. Alan R. Liss, 1987, pp. 1–17.

behave differently in different species means that if the primary goal is to understand human cancers, they should be studied in human cells. Dulbecco thought that the DNA sequence could turn out to be the central focus of future cancer research. This vision was more a firm intuition of his than a developed and proven argument, but despite this, and his apologies for this intuitive aspect, he remained convinced of his main conclusion — that the DNA sequence would be a fundamental tool for the clarification of major problems in biology.

There was a lot of negative reaction to Dulbecco's article, but it catalyzed subsequent discussions and along with Sinsheimer's colloquium and the Santa Fe meeting, it contributed to the wave that carried the idea of the genome project through the scientific community. People who had attended the Santa Fe workshop had been talking about it, in particular at a symposium organized at Cold Spring Harbor[52], during which Walter Bodmer and Victor McKusick pointed out the importance of a dedicated attempt to sequence and map the human genome. Also at this symposium, another Nobel prizewinner, Paul Berg, who was not aware of the discussions at Santa Cruz, Santa Fe or at the DOE level, read Dulbecco's article and suggested to Watson that it might be a good idea to have an informal discussion on a sequencing initiative. Watson's faithful network had kept him informed of the Santa Fe and Santa Cruz meetings, and he had already spoken to Dulbecco and Gilbert. He convinced Gilbert to co-chair a meeting on the genome project with Berg, who arrived at Cold Spring Harbor to co-direct a session on cataloguing genome project proposals. Gilbert gave a brief account of the meetings in Santa Cruz and Santa Fe. He pointed out that known DNA sequences were growing at a rate of more than two million base pairs per year and concluded with the importance of starting a genome project. During this 3 June session, the enthusiasm of Berg, Gilbert and others came up against Botstein's skepticism.

It looked as if the Cold Spring Harbor symposium would result in a delay in moves towards the massive sequencing effort. The DOE's efforts,

---

[52] Cold Spring Harbor Symposia on Quantitative Biology, The molecular biology of *Homo sapiens*, 51, 1986, R. Lewin, *Science*, 232, 1986.

in particular, came under attack. With hindsight, the symposium was clearly the trigger event for the international discussion that led to a restructured project benefitting of world consensus. It was a transition from an insistence that from the human genome be sequenced to a plan that gave greater importance to genetic and physical mapping and the study of non-human organisms. The objectives of Sinsheimer, Dulbecco and Gilbert were simple and clear — a reference sequence of the DNA of the 24 human chromosomes. DeLisi's program was fundamentally justified as being the first step to this goal.

The genome project that began to emerge after the Cold Spring Harbor symposium, however, had a different goal — a useful group of maps of human chromosomes, but also some of non-human model organisms too. DNA sequencing, particularly the technology to carry it out faster and more accurately, was still important but no longer a dominant concern. In this redefined genome project, the goal was to apply the new molecular biology techniques on a massive scale for a genetic approach to man similar to that which had long been the case for yeasts, nematodes, and the fruitfly, amongst others.

If systematic sequencing remained a target for controversy, the need for a genome project was becoming more and more obvious to the main leaders and founders of molecular biology. During 1986, Watson had become even more convinced that an attempt on the human genome was desirable and possible. He was also convinced that the project should not be left in the bureaucratic hands of the DOE[53] but should be run by scientists[54] and guided by real scientific needs. This meant that the NIH would have to be involved.

---

[53] J.D. Watson, "The human genome project: past, present and future", *Science*, vol. 248, 6 April 1990, p. 44.

[54] "I don't think the coordination can be done by bureaucrats. It must be done by scientists". J.D. Watson, quoted in Leslie Roberts "Agencies vie over Human Genome Project", *Science,* vol. 237, 31 July 1987, pp. 486–488.

5.3 The NIH Genome Project

At this point, discussion and debate had reached their highest pitch, and some research had to be carried out to clarify the issue. In a politically important initiative, the OTA set up a task force to examine the question[55]. Other analyses of the situation were ordered by the DOE[56] and the Howard Hughes Medical Institute (HHMI)[57]. A crucial step was taken when the National Research Council (NRC) also decided to look at the problem[58]. As the executive branch of the American National Academies of Science and Engineering, the NRC spoke for the scientific community with more authority than any other organization in the USA. It held a meeting at Woods Hole (Massachusetts) in August 1986[59], and afterwards, set up a committee to examine the pros and cons. The proposed projects were studied, and the importance of a genetic atlas was agreed upon.

Mapping the human genome was thought to be more urgent than sequencing work, and the whole project was estimated to require some US$200 million annually over fifteen years. The report also mentioned the need for studies of non-human genomes to give the results obtained in man scientific significance — nobody could seriously imagine genetically altering several base pairs of a human gene just to see what effect such a modification might have, but there is no moral difficulty in carrying out the same sort of genetic experiments on bacteria, for example. The NRC committee found that to obtain major benefit from the human genome, there would have to be a large data base of DNA sequences from the mouse and simpler organisms

---

[55] *Mapping our Genes — Genome Projects: How Big, How Fast?* as mentioned.

[56] Subcommittee on the human genome, *Report on the Human Genome Initiative,* Health and Environmental Advisory Committee for the office of Health and Environmental Research, DOE, April 1987.

[57] M. Pines, "Shall we grasp the opportunity to map and sequence all human genes and create a human dictionary?" Bethesda, MD, HHMI, prepared for a meeting of the trustees of the HHMI, 1986, M. Pines, Mapping the Human Genome (Occasional Paper Number One), HHMI, December 1987.

[58] The NRC advises the National Academy of Sciences.

[59] National Research Council, Mapping and Sequencing the Human genome, National Academy Press, Washington DC, 1988.

like bacteria, yeast, *D. melanogaster* and *C. elegans*. To succeed, the project should not, therefore, be restricted to the human genome but should include intensive analysis of genome sequences from other selected creatures. Watson later claimed to have been the instigator of this widening of the project's field[60].

Whilst the panel of scientists put its report together, the NIH voted on the federal budget for research on the human genome. In December 1987, US$17.2 million was allocated to the NIH's genome project. At about the same time, the DOE had obtained US$12 million for its part of the project, but lost its main champion when Charles DeLisi left his position as Director of the Office of Health and Environmental Research (OHER) in 1987. The NRC and OTA reports were published in 1988. In February 1988, James Wyngaarden decided to organize another meeting[61], held in Reston (Virginia) and chaired by David Baltimore, who had previously been one of the most savage opponents of systematic sequencing of the human genome. However, the NRC's work had diverted the main impetus of the project from sequencing towards mapping, and to the comprehension of the biology through the study of the genomes of other species rather than the brute gathering of data. The Reston meeting transformed the NIH's attitude and set out detailed objectives for the genome project's program to meet.

In early May 1988, Wyngaarden asked Watson to take charge of directing the NIH genome research program. Watson accepted the challenge because, as he wrote later, "He had realized that he would only once have the opportunity to cross the path leading from the double helix to the three

---

[60] The initial DOE project had great weaknesses from this point of view. Very little attention was given to the genetic link mapping and the study of non-human organisms either as a pilot project or in larger studies. DeLisi later explained these weaknesses as the result of a presumption that mapping and studies of other organisms would take place on a separate basis, and that the genome program would simply increase efforts in these closely related but quite distinct fields. This report recommended that the DOE take the leading role in the project.

[61] Refer to J.D. Watson, "The Human Genome Project: Past, present and future", p. 46. The final recommendations of the Ad Hoc Advisory Committee on Complex Genomes that Wyngaarden called together at Reston strongly supported Wyngaarden's proposal to set up an Office of Human Genome Research (OHGR) within the NIH and chaired by an associate director. It was at this meeting in Reston that Watson remarked that the director should be a scientist rather than an administrator.

million base pairs of the human genome"[62]. According to Norman Zinder, Watson's nomination brought the human genome project undoubted credibility. On 10 October 1988, Watson was named Associate Director for Human Genome Research at the NIH, with a 1988-1989 budget of more than US$28.2 million, some $10 million more than the DOE genome research budget for the same year. On the same day, after much negotiation, the DOE and the NIH signed a memorandum[63] on how the two agencies would cooperate on genome research. The genome project was now well under way, and it looked as if the NIH was the leading agency, rather than the DOE[64].

In 1988, Watson's program fell under the aegis of the National Institute of General Medical Sciences (NIGMS) whose Director, Ruth Kirschstein, had previously openly expressed her firm opposition to a program dedicated to human genome analysis. The Office of Genome Research started out with a very small staff, Watson and two others. Within a year, it had developed into a Center for Human Genome Research, independent of the NIGMS and in charge of its own budget. By October 1989, this budget had reached US$60 million, more than twice the US$28 million budget allocated to the DOE genome project, and the center's staff had increased to thirty people. Its 1991 budget amounted to some US$71 million. (Another US$28 million were available for genome research, but the choice of where they went lay solely with the NIH's director).

Watson was characteristically direct in his defense of the project and the importance of entrusting its management to scientists rather than administrators. He caused total consternation when, in 1988, he suggested that the most effective way to meet the project's goals would be to divide the chromosomes up between the countries and the laboratories involved in this sort of research. It is reported that he even said at a press conference

---

[62] J.D. Watson, "The Human Genome Project: Past, present and future", as mentioned above, p. 46.
[63] Charles R. Cantor "Orchestrating the Human Genome Project", *Science*, vol. 248, 6 April 1990, pp. 49–51.
[64] On the question of the NIH's leading role: Jeffrey Mervis "Renowned bioengineer picked to head Lawrence Berkeley Genome Center", section "NIH dominance", *The Scientist*, 3, 11 July 1988.

that because chromosome 1 was the largest in size, and would therefore need the most work for the smallest profit, it should be allocated to the Soviet Union[65].

He caused even worse international tension when he suggested that access to the results of the US genome project should be refused to Japanese scientists if their government did not fund a research program on the human genome of the same size as that set up by the US[66]. Two years later, in an article in *Science*[67], Watson was more circumspect, even though between the lines of his more diplomatic prose one can detect a certain hardness in his proposals. He wrote that, "The idea that the various human chromosomes will be divided amongst various laboratories is far from today's conventional wisdom... Extensive, overall multiple mapping efforts currently only exist for chromosome 21, the smallest human chromosome, which contains, among others, a gene that leads to increased susceptibility to Alzheimer's disease. To my knowledge, the number of investigators wanting to make complete high-resolution physical maps of a specific chromosome is less than ten. Our real problem may be persuading capable teams to focus on those chromosomes that will still have no champions in addition to deciding among alternative proposals for total mapping and sequencing"[68].

In accordance with the recommendations of the NRC, DOE and HERAC reports, the initial objectives supposed an annual funding level of US$ 200 million. In 1995, this level of funding has not yet been reached[69].

---

[65] As quoted in Tom Wilkie's *Perilous Knowledge: The Human Genome Project and its Implications*, Faber and Faber, London, 1994.

[66] For a detailed account of Watson's provocative remarks on Japan, please read Bob Davis's "US scientist eschews tact in pushing genetic project", the *Wall Street Journal Europe*, 18 June 1990.

[67] J.D. Watson, "The Human Genome Project: Past, present and future", as mentioned above, pp. 44–48.

[68] *Ibid*, p. 47.

[69] Diagram 12 A from *Human Genome News*, vol. 7, nos. 3 and 4, September–December 1995, p. 8. Diagram 12b from Francis S. Collins, Ari Patrinos, Elke Jordan, Aravinda Chakravarti, Raymond Gesteland, LeRoy Walters, the members of the DOE and NIH planning groups "New Goals for the US Human Genome Project: 1998–2003".

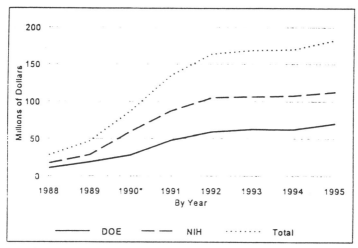

*HGP funding formally began in October 1990.

Fig. 12A Evolution of Human Genome Project finances

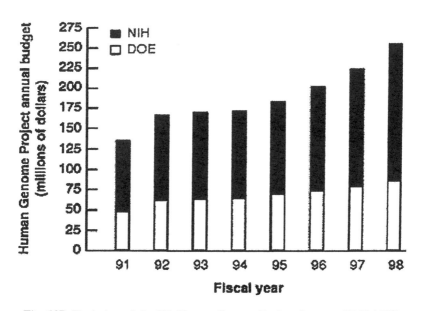

Fig. 12B Evolution of the US Human Genome Project finances (1991-1998)

In other words, there was still pressures on researchers to get involved in the overall mapping of the chromosomes. The international aspects were formulated in more diplomatic tones; nevertheless, they were precisely and rigorously described.

"Early sharing of the human database is much more likely to occur if large-scale mapping and sequencing efforts are undertaken by all those major industrial nations that will want to use this data. It is too early to ask what we should do if we identify one or more countries that have the economic clout to join in the effort, but that apparently do not intend to, hoping instead to take advantage of the information once it becomes publicly available. I do not like to even contemplate such a possibility, since Congress and the public are likely to respond by wanting to move us towards a more nationalistic approach to science. This alternative is counter to the traditions that have allowed me to admire and enjoy the scientific life. The nations of the world must see that the human genome belongs to the world's people, as opposed to the nations"[70].

But the sequence itself remained and remains a long-term objective. After the initial enthusiasm of the 1980s, it became more obvious that the techniques available at the time were too slow and too expensive for an affordable start to sequencing the three million base pairs of the human genome. During the summer of 1989, at a "retreat" at Cold Spring Harbor, representatives of the NIH, DOE and other invited experts set up a five-year program for the American human genome project. The report that emerged, "Understanding our Genetic Inheritance — The US Human Genome Project: the first five years", was submitted to Congress in February 1990. It brought the NRC and OTA evaluations up-to-date and set out objectives to be fulfilled by 1995. Sequencing the human genome was postponed until sequencing costs fell to a minimum of 50 cents per base pair. The scientific objectives for the five year plan were the following:

- To set up a full human genetic map with markers at an average distance of 2 to 5 centimorgans from each other. Each marker if possible to be identified with an STS;

---

[70] *Ibid*, p. 48.

- To assemble an STS map of all the human chromosomes with an interval between markers of about 100,000 base pairs, to generate sets of overlapping clones or closely spaced markers ordered without ambiguity and this for a continuous distance of 2×10^6 bp;
- To improve current methods and/or develop new methods for sequencing DNA to bring sequencing costs to a maximum of 50 cents per base pair (a decision on large-scale sequencing would have to be taken within 4 to 5 years);
- To prepare a genetic map of the mouse genome based on DNA markers. Begin mapping one or two chromosomes.
- To sequence an aggregate of 20×10^6 bp of a variety of organisms, focusing on genomes or fragments of 10^6 bp in length (the model organisms being *E. coli*, the yeast *S. cerevisiae*, *D. melanogaster*, *C. elegans* and the laboratory mouse);
- To constitute a "Joint Informatics Task Force" between the DOE and the NIH for the development of software and databases to support large scale sequencing and mapping projects, to create databases with easy and up to date access to chromosome physical maps and sequences, to develop the algorithmic and analytical tools likely to be of use in the interpretation of genomic information;
- To develop (thanks to a joint DOE-NIH work group) programs for the comprehension and handling of ethical, legal and social problems of the human genome project (through contracts, research grants, colloquia, teaching and educational materials), to identify and define the major problems and develop political scenarios for handling them;
- To support the training of young scientists (pre-doctoral and post-doctoral); to examine the needs for other types of training;
- To support innovative and risky technological development to meet the needs of the genome project;
- To support and facilitate the transfer of technologies and medically important information to interested communities.

From the very start, the idea of the human genome project raised interest and concern from many nations thinking about their own projects, and it very quickly became clear that an international forum would have to be set up.

5.4 HUGO, or the Difficulties of International Coordination

On 29 April 1988, during the first meeting at Cold Spring Harbor, the researchers decided to set up the Human Genome Organization (HUGO) to coordinate the various international efforts and try to minimize useless duplication of effort[71]. HUGO was first chaired by Victor McKusick and then by the British director of the Imperial Cancer Research Fund (ICRF), Sir Walter Bodmer. HUGO's headquarters were in Geneva, but its operational bases were in London, Bethesda and Osaka. HUGO began its work with funds donated by benefactors such as the Wellcome Foundation Trust and other large research organizations. Its praiseworthy objectives were to promote international collaboration, favor the international transmission of information, biological samples and data between researchers through organizing meetings, and more politically, contribute to triggering ethical, economic and legal reflections on the questions posed by analysis of the human genome. Despite this, and the various gifts of funding, this international group of researchers did not really succeed in achieving their goals, "succumbing to the weight of nations"[72].

Despite these difficulties, HUGO took on greater importance in what now appears an attempt at worldwide coordination of work on the human genome, notably through the famous "International Chromosome Workshops" which encouraged collaboration and the sharing of information and resources,

---

[71] "Genome monitoring (the organization to monitor the human genome project should be welcomed)", *Nature*, vol. 335, 22 September 1988, p. 284. On the story of HUGO, please read Victor McKusick's "HUGO news, the human genome organization history, purposes and membership", *Genomics*, 5, 1989, pp. 385–387.

[72] For more information on HUGO's failures, read A. Schoen's article "Les infortunes de HUGO, ou le poids des nations", *Biofutur*, Spécial Génomes, No. 146, Juin 1995, pp. 92–94.

and facilitated the drafting of complete maps of the different chromosomes. HUGO also organizes the Human Genome Meeting (HGM), now held annually in different locations around the world, as well as small specialist workshops on topics such as chromosome mapping, population genetics and genome diversity. HUGO organizes the international Mutation Detection Workshops, which are always oversubscribed.

Concerning the role of the chromosome editors, the following recommendations were expressed:

- After the closing of GDB, editors will no longer be able to edit the database. Chromosome Committees should therefore continue to function.

- Each chromosome committee should work together with the chromosome sequencing center(s) that are engaged in the large-scale sequencing of each chromosome to ensure coordination of the sequencing efforts. This especially pertains to keeping track of regions being sequenced to avoid unnecessary duplication and assuring the goal of a complete finished sequence for the chromosome. In accordance with the anticipated relationship of the chromosome committees and the sequencing centers, the HUGO Genome Mapping Committee (HGMC) will actively recruit new chromosome committee members who are involved with the large-scale sequencing of each chromosome.

- Each chromosome committee will serve as a point of contact for the committee concerning all genomics activities and resources relating to the chromosome. This function will include responsibility and oversight for a HUGO chromosome web page for each chromosome which would provide pointers to all of the appropriate data resources pertaining to all genomics aspects for the particular chromosome.

- Chromosome committees will organize, together with the corresponding chromosome sequencing center(s), meetings of the active chromosome community as necessary to assess and summarize progress in sequencing and annotation of the chromosome. This activity will include finding the necessary funds to support the meeting.

- The HGMC will continue to explore the possibility of editors providing a review of the gene location data in OMIM or perhaps alternative databases.

Unfortunately, the size of possible economic benefits and the growing role of industry in genome research can only damage chances for genuinely free circulation of information on sequences and genes. Regarding this problem, we must stress that at the first (1996) and second (1997) International Strategy Meetings on Human DNA Sequencing, the attendees discussed and strongly endorsed the concept that human genomic DNA sequence data produced by large-scale DNA sequencing centers funded by the HGP should be released as rapidly as possible. In the case of unfinished data, it was suggested that sequence assemblies of 1-2 kb in size be released within 24 hours of generation. Finished data should be submitted to the public sequence databases on a similarly rapid time scale. Most of the funding agencies engaged in supporting the human genome project have adopted policies that reflect the importance of rapid dissemination of genomic sequence data. At the 1998 meeting, the question was raised about similar data dissemination practices for genomic sequence data from other organisms. The attendees unanimously agreed that this was also critical and, as individual scientists, adopted the following statement:

> As extensive determination of the genomic DNA sequence of several organisms proceeds, it is increasingly clear that sequence information has enormous and immediate scientific value, even prior to its final assembly and completion. Delaying the release of either unfinished genomic DNA sequence data serves no useful purpose and actually has the effect of slowing the progress of research. Therefore, the attendees at the third International Strategy Meeting on Human Genome Sequencing (Bermuda, 27–28 February 1998) agreed unanimously to support, as individual scientists, the view that all publicly-funded large-scale DNA sequencing projects, regardless of the organism, should deposit data immediately into the public domain, following the same guidelines that have previously been adopted by this group for human genomic sequence. The scientists attending this

meeting will continue to adhere to these principles and urge all other scientists and policy-making groups involved in large-scale sequencing to adopt them as well.

Unfortunately, this statement does not yet reflect, nor is it binding on, the policy of any funding agency.

In May 1997, HUGO issued a declaration on patenting. It reaffirms the statements on DNA sequence patenting of 1992 and 1995 [73], clarifying the fact that HUGO does not oppose patenting of useful benefits derived from genetics information, but does explicitly oppose the patenting of short sequences from randomly isolated portions of genes encoding proteins of uncertain functions. HUGO regrets the decision of some patent offices, such as the US PTO, to grant patents on ESTs based on their utility 'as probes to identify specific DNA sequences', urging these offices to rescind these decisions and, pending this, to strictly limit their claims to specified uses, since it would be untenable to make all subsequent innovation in which EST sequence would be involved in one way or another dependent on such patents. HUGO urges all large-scale sequencing centers and their funding agencies to adopt the policy of immediate release, without privileged access for any party, of all human genome sequence information in order to secure an optimal functioning of the international network, as well as to avoid unfair distortions of the system. HUGO stresses that only a policy of rapid publication and free availability of human genomic sequence information will secure further international co-operation of large-scale sequencing centers. HUGO emphasizes the differences between the US patent law, which provides for a so-called one-year 'grace period', allowing the authors of published data to subsequently file patent applications for inventions based on such information, and the patent laws of practically all other countries, which do not contain such a provision and where, therefore, no protection for, or based on, published data can be acquired. Finally, HUGO expresses the hope that the free availability of raw sequence data, although forming part of the relevant state of the art, will not unduly prevent the protection

---

[73] Dr. T. Caskey, Pr. R.S. Eisenberg, Dr. E.S. Lander, Pr. J. Strauss, *HUGO Statement on Patenting of DNA Sequences,* ed. Dr. Belinda J.F. Rossiter, HUGO.

of genes as new drug targets, which is essential for securing adequate high-risk investments in biology, and will not result in a shift of activities of the pharmaceutical industry to searching for compounds that give marginal advantages against known targets to taking risks with new targets.

Despite those statements and despite the fact that HUGO now has over 1200 members from more than 62 countries, its effects are still limited even though it contributes to the co-ordination of the global initiative to map, sequence and functionally analyze the human genome, promotes the applications of the findings to the improvement of human health and serves as an interface between the community of genome researchers, the governmental and private funding agencies and the public. In 1986, the HUGO council was:

Table 5 HUGO council 1998

Professor Gert-Jan van Ommen (Netherlands) President
Dr. Anthony Carrano (USA) Vice-President, Americas
Prof. Leena Peltonen (Finland) Vice-President, Europe
Pro. Yoshiyuki Sakaki (Japan) Vice-President, Pacifique
Prof. Andrea Ballabio (Italy)
Dr. C. Thomas Caskey (USA)
Prof. Richard G.H. Cotton (Australia)
Prof. David R. Cox (USA)
Prof. Lev Kisselev (Russia)
Dr. Hans Lehrach (Germany)
Prof. Yusuke Nakamura (Japan)
Dr. Annemarie Poustka (Germany)
Prof. M. Susan Povey (UK)
Dr. David Schlessinger (USA)
Prof. Grant Sutherland (Australia)
Dr. Lap-Chee Tsui (Canada)
Prof. Robert Waterston (USA)
Dr. Jean Weissenbach (France)

The Gulf War, in 1991, clearly demonstrated the fragility of international cooperation and how apparently unconnected political events can affect science. The international database for the genome resides in computers in Baltimore, and many European scientists were using a group of programs called the Wisconsin Package from that database to do *in silico* analysis. The US Department of Trade declared an embargo on access to the computer programs for non-American scientists and organizations, perhaps thinking that the programs could be used by individuals and nations with biological warfare in mind. Although this decision seemed totally without foundation, the NIH was incapable, for quite some time, of having the embargo lifted.

International collaboration was threatened even further when, in the middle of 1991, it was announced that the NIH was filing a patent request for a group of 315 DNA sequences from Craig Venter's group[74] that were for chemicals produced in the brain. The initiative came from Reid Adler, the head of the NIH valorization unit, and was, therefore, not the act of a capitalist in pursuit of profit. An additional complication was that no-one knew from which genes the sequences came, or what their functions were.

In order to label restriction fragments for correct identification in different laboratories, a common reference system was needed. A suggestion from American researchers resolved this need by identifying the genome fragments by sequence-tagged sites (STSs), also known as labels[75]. The idea was, in principle, quite simple: A researcher, isolating what appears to be a new fragment of human DNA, could sequence a sufficiently long portion of it to be identifiable — an STS, or sequence-tagged site. No-one, however, had imagined that someone might, almost at random, extract fragments of DNA from brain cells or from other organs or organisms and then put in a patent request to cover the STSs identified.

The problem stirred up a storm of controversy in the United States, especially when it looked as if the NIH had thrown itself behind the venture without consulting Watson, director of the National Center for Human Genome Research. Watson threw all his weight behind his opposition to the patenting of sequences because he realized that if it were permitted, it would

---

[74] M.D. Adams *et al.*, "Complementary DNA sequencing: expressed sequence tags and the human genome project", *Science*, 252, 1991, pp. 1651–1656. The second patent request in the beginning of 1992 was followed by a description in *Nature* of 2,375 new ESTs (expressed sequence tags), M.D. Adams *et al.* "Sequence identification of 2,375 human brain genes", *Nature,* 355, 1992, pp. 932–634. These ESTs are chunks of mRNA that are taken from complementary DNA banks.

[75] M. Olson, L. Hood, C. Cantor and D. Botstein, "A common language for physical mapping of the human genome", *Science,* vol. 245, 29 September 1989, pp. 1434–1435.

slow down the free transmission of scientific information and thus make international collaboration much more difficult, with a consequential slowing of progress[76]. The worst case scenario was that each nation would have to sequence the entire human genome in secret, hoping that the bits it had identified had not already been covered by a patent requested by another nation.

The problem became one of international politics with particular protestations from Great Britain and France. These protests became louder when it transpired that TIGR (The Institute for Genomic Research, Gaithersburg, Maryland, USA, financed by the firm of Human Genome Science which was in negotiation with SmithKline Beecham), had managed to hire Venter along with his labels (all 60,000 of them by 1995), the release of which had been delayed to the international scientific community[77]. The French thought that human genes should not be patentable. The British sought a compromise since from their point of view, gene sequences with no known use should not be patentable, but sequences of genes with known functions might be patented. The British government added that it was taking a firm stand in that, despite its opposition to the patenting of sequences, it would patent all the sequences already identified if the Americans persisted in their position.

International cooperation was further endangered when, in early 1992, a private American company tried to coax Cambridge's John Sulston across the Atlantic to the USA. One of the most advanced fields in genome

---

[76] We know the saga that this patent request suffered. It was rejected on its first examination by the patent office in August 1992, presented again, and rejected again in August 1993. In early 1994, the NIH's new Director decided to give up.

[77] It was only in the autumn of 1994 that the HGS opened its database of about 300,000 sequences, without a doubt containing 60 or 70 thousand human genes, to academic researchers. Opened is a generous word, since this access was only granted after the researchers signed an extremely limiting contract that gave SmithKline Beecham first refusal of all commercial applications that might result from research carried out with the information in question. The debate on the patenting problems is summarized in L. Roberts "Who owns the human genome?", *Science*, vol. 237, 24 July 1987, pp. 358–370, or T. Damerval and H. Therre, a meeting with J.C. Venter "La conquête du génome", *Biofutur*, 110, April 1992, pp. 18–20. John Carey, "The Gene Kings", *Business Week*, 8 May 1995, pp. 72–78. B. Jordan, "Chercheur et Brevet, la fin d'un blocage", *Biofutur,* June 1995, pp. 86–88. Françoise Harrois-Monin, "Génétique: la mine d'or", *Le vif/L'Express,* 13 October 1995, pp. 79–80. "Actualité des ESTs", *La Lettre du Greg,* no. 6, April 1996, pp. 20–21.

sequencing is that of the nematode worm, which John Sulston was working on in Cambridge, England, as was Robert Waterston of Saint Louis in the United States. They not only developed the sequencing techniques, but also tools for *in silico* analysis and a high-performance work structure. All this progress was made with public money. The private firm, which intended to set up in Seattle, wanted Sulston and Waterston to join them, bringing all their expertise and helping them a private sequencing company. When the genome project started large-scale sequencing, it would have found that this small firm had a near-monopoly on sequencing technologies. Not only international collaboration, but the US domestic program were being jeopardized by this operation. Once more, Watson reacted with vigor and opposed the project. In any case, Sulston and Waterston had already decided to decline the offer.

These incidents, in which Watson authoritatively prevented the commercialization of sequences and sequencing technology, led to a conflict between him and other parts of US government bureaucracy, and in April 1992, the NIH set up an ethical committee to investigate a hypothetical conflict of interest between Watson's position as the director of the National Center for Human Genome Research and his shareholdings in private biotechnology companies. This inquiry was widely perceived in the scientific community as a pretext, and Watson was considered the victim of a personality clash and internal NIH policy[78]. He had, in any case, from the beginning decided to stay with the genome project for four years only in order to be sure it would work[79]. In April 1992, he announced that heading up both the National Center for Human Genome Research and Cold Spring Harbor Laboratory was becoming too hard, and he resigned his position at the NIH to return to Cold Spring Harbor. His departure highlighted certain tensions in the NIH, especially as the director of the NIH accepted his resignation without a single word of regret. His successor had not been

---

[78] C. Anderson "US genome head faces charges of conflict", *Nature*, 356, 9 April 1992, p. 463, P.J. Hilts, "Head of Gene Map Threatens to Quit", *New York Times,* 9 April 1992, p. 26, L. Roberts, "Friends say Jim Watson will resign soon", *Science,* 256, 10 April 1992, p. 171, L. Roberts, "Why Watson quit as project head", *Science*, 256, 17 April 1992, pp. 301–302.

[79] J. Palca, "The genome project: life after Watson", *Science*, 256, 15 May 1992, pp. 956–958.

appointed, and the NIH named an interim director from its own ranks until Francis Collins accepted the post.

Watson had run the human genome project with passion, rigor, energy and charisma. His departure took place at a time when the project seemed in some difficulty, progress continually being hijacked by the debate on access to sequences and gene patenting. This was exacerbated by private sequencing companies (of which there were a growing number) who were putting together systems of paying access to their sequences, thus skirting the issue.

The debate eventually ended with a painfully reached consensus on a formula much like one of the European proposals: *no* to patenting genes as such, but *yes* to those that contain a sequence coding for a protein with a determined function and to which a precise therapeutic or industrial application can be attributed. This rule clearly excludes the patenting of labels. However there remains quite a lot of ambiguity in the term "as such" which promises to be the subject of some juicy legal battles until jurisprudence is established. Watson's successor still faces not only leading the scientific community on this matter, but also developing a new relationship with the world of private enterprise. Watson's hope that the human genome would belong to humanity, and not to the nations, looks as if it will fail because of the huge economic and political potentials linked to the human genome.

Despite the controversy, the genome project was well on its way. Watson had managed to defend the need for a physical map of the genome and for this map to be tackled by large centers rather than by a network of small laboratories[80]. Against the advice of part of the genome research community, who feared that such a megaproject would invite bureaucratic control and lead to a restriction of the freedom of research[81], Watson and the DOE decided to set up large dedicated genome study centers. Today, the American human genome project has 22 research centers specializing in human genome study (Table 6) and is supported by a growing number of national and

---

[80] J. Palca, "The genome project: life after Watson", *Science*, 256, 15 May 1992, p. 957.

[81] For more on these fears, read D. Baltimore's article "Genome sequencing: a small science approach", *Issues in Science and Technology*, Spring 1987, pp. 48–50.

private sector university laboratories. Guidelines on reasonable sharing practices were drawn up during 1992, and the research community responded to these admirably.

Table 6 American research centers dedicated to the Human Genome Project (table from *Human Genome News*, vol. 7, nos. 3 and 4, September–December 1995, p. 9).

(Directors in Italics)

- Affymetrix — *Stephen Fodor*
- Albert Einstein College of Medicine — *Raju Kucherlapati*
- Baylor College of Medicine — *David L. Nelson*
- Children's Hospital of Philadelphia — *Beverly Emanuel*
- Columbia University College of Physicians and Surgeons — *Argiris Efstratiadis*
- Genome Therapeutics Corporation (Collaborative Research Division) Genome Sequencing Center — *Jen-i Mao*
- *Lawrence Berkeley National Laboratory — *Mohandas Narla, Acting Director*
- *Lawrence Livermore National Laboratory — *Anthony Carrano*
- *Los Alamos National Laboratory — *Robert Moyzis*
- Stanford Human Genome Center — *Richard Myers*
- Stanford University DNA Sequence and Technology Center — *Ronald Davis*
- University of California, Berkeley, *Drosophila* Genome Center — *Gerald Rubin*
- University of California, Irvine — *John Wasmuth*
- University of Iowa Cooperative Human Linkage Center — *Jeffrey Murray*
- University of Michigan Medical Center — *Miriam Meisler*
- University of Texas Health Science Center at San Antonio — *Susan Naylor*
- University of Texas Southwestern Medical Center at Dallas — *Glen Evans*
- University of Utah — *Raymond Gesteland, Robert Weiss*
- University of Wisconsin, Madison, *E. coli* Genome Center — *Frederick Blattner*
- Washington University School of Medicine — *David Schlessinger*
- Washington University School of Medicine Genome Sequencing Center — *Robert Waterston*
- Whitehead Institute for Biomedical Research and Massachusetts Institute of Technology — *Eric Lander*

The sharing of information and services has been greatly facilitated by the physicists' gift of the WWW protocols, which have greatly simplified the presentation and transfer of information over the Internet. Furthermore, materials and reagents produced by researchers have been made widely and often commercially available.

As the scientific community hoped, the genome infrastructure considerably accelerated studies on human disease and the identification of genes. It was, notably, the availability of linkage maps of the human genome and increasingly precise physical maps of the chromosomes (Figs. 13A and 13B) that made this possible.

Watson had another influence on the human genome project. It was he who pointed out how important it was to get funding for studies on the ethics of genome research. Initially, he had hoped that around 3% of NIH genome funds could be set aside for ethical questions[82]. His decision to set up a program to anticipate the social implications of genome research was the basis of Ethical, Legal and Social Implications (ELSI), directed courageously by Nancy Wexler, who supported large-scale screening pilot studies.

Watson's motives may not have been as pure as they appear, as ELSI could also act as a socio-political windbreak, that would shield the whole project from potential criticism and fear. Nevertheless, the program was an attempt to articulate the values that should be behind the research and anticipate unwanted social consequences with a view to their prevention. The main characteristic of the ELSI program and its counterparts in the European Union, France, Canada, Germany, Russia and Japan was not their mere existence, that they had been created, but the speed at which the policymakers and politicians accepted them as a norm, despite the lack of any similar program in other scientific fields. No gesture had struck their imaginations and consciences in quite such a way since the debates on experimentation on human subjects and the advent of recombinant DNA techniques that had captured the public's attention twenty years before.

The ELSI research program was a welcome addition to the NIH. ELSI guided the NIH to a more rational research program on screening for cystic fibrosis, but it was soon confronted with deeper questions of social implications well beyond its capacity to answer. Most of these questions and problems related to disparate concepts of political, philosophical, ethical

---

[82] H.M. Schmeck, "DNA pioneer to tackle biggest genome project ever", *New York Times,* 4 October 1988, C1, C16.

5 Human Genome Project and International Sequencing Programs   263

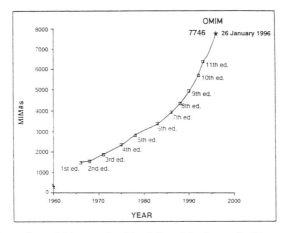

Fig. 13A Total number of hits on the Mendelian Inheritance in Man and on the online version OMIM
(Table from *Human Genome News*, vol. 7, no. 5, January–March 1996, p. 6).

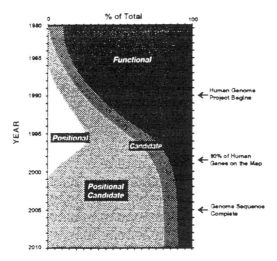

Fig. 13B Tendency of cloning methods for genes linked to human illnesses, 1980–2010

The projections for after 1995 are highly speculative. Based on all the resources and facts on the human genome, the "positional candidate strategy" will allow researchers to combine information on the chromosomal location of a gene with increasingly detailed genetic and physical maps, obtaining an easier identification of the miscreant gene. (Diagram from *Human Genome News,* vol. 6, no. 6, March–April 1995, p. 1).

and legal morals. The genome project attracted the attention of university thinkers outside the molecular biology domain. These thinkers began to examine the new trends in science and, more importantly, the social context under which they must operate.

Although initially the genome project only targeted the sequencing of the human genome, it evolved as its targets became more diverse. There was little disagreement that in the future sequencing techniques would have greatly developed, making sequencing less expensive, faster and more accurate. There was consensus that sequencing model organisms was a good target and would prove very useful. Watson was convinced that model organisms would play a crucial role in the genome project, and in this he concurred with the recommendations of the NRC report[83] that underlined the importance of sequencing pilot projects and programs for the improvement of sequencing technologies. Watson also agreed with the HUGO association that the systematic decryption of the human genome should wait until automated sequencing had attained greater levels of efficiency and accuracy, and that in the meantime, it would be prudent to work on smaller organisms. In 1990, Watson told André Goffeau[84] that he planned to commit a US$20 million per year budget to sequencing small genomes (20% of the total budget that had been announced for the human genome)[85], such as those of yeast, *E. coli*, plants and the mouse. The sequencing of human cDNA had become considerably more popular, but systematic sequencing of model organisms also looked more and more important as an early step towards the major goal of mapping and sequencing the entire human genome[86].

During a symposium on the Human Genome Research Strategies and Priorities in January 1990, Watson stated his intention not to undertake the

---

[83] National Research Council, *Mapping and Sequencing the Human Genome*, National Academy Press, Washington DC, 1988.

[84] This is from the report André Goffeau wrote after his mission from 3–5 September 1990, "Biologie Moléculaire de *Bacillus* à l'Institut Pasteur", 19 November 1990.

[85] R.A. Weinberg, "DNA sequencing: an alternative strategy for the human genome project", *Biofutur*, October 1990, pp. 20–21.

[86] J.D. Watson and R.M. Cook Deegan, "Perspectives on the human genome project", *Biofutur*, October 1990, pp. 65–67.

systematic sequencing of genomes in the USA until the price of sequencing a base pair had fallen to below 50 cents.[87] At the same symposium, Ansorge of the EMBL described his own sequencing system, available commercially from the firm of LKB, which according to Ansorge, should permit 8 kbp to be sequenced per day at a cost that would fall below Watson's target minimum. Ansorge also announced that he was developing a more complete sequencing system in collaboration with the firms of Eppendorff, Amersham and Bertin. This should be viewed in the light of the following remark from Philipson, then Director of the EMBL: "(...) The technique and instrumentation development is probably best carried out on a competitive basis, and doesn't necessarily need international collaboration ..." This wait-and-see attitude of relying on the market to ensure developments in sequencing power is also manifest in the genome center's refusal of the proposal presented by J.A. Hoch of the Scripps Foundation at La Jolla for a project including American participation in sequencing the entire genome of *Bacillus subtilis* set up in the framework of a European program. The proposed project was essentially a collaboration between the United States and Europe, since 10 teams, five American and five European, would participate. This rejection was despite the fact that the proposal had been received and accurately evaluated by the review committee of the Council of the General Medical Institute of the NIH. It was at the level of the Genome Center itself that the decision was made not to accept the American collaboration side of the project. Raymond Dedonder[88] wrote to Watson to support it, and apparently Watson had thought it "appropriate in principle[89]". In his answer, Watson clearly indicated that he had no basic objections to sequencing various different organisms, including bacteria, and that he was favorable to the developments and possibilities of international cooperation. What seems to have been the decisive negative factor was his refusal to start the work before sequencing costs had reached a suitably low figure.

---

[87] Another of Goffeau's reports, on his mission to the Symposium on Human Genome Research Strategies and Priorities, 29–31 January 1990, UNESCO Headquarters, Paris, written 15 February.

[88] R. Dedonder to J. Watson, letter of 15 September 1989.

[89] J. Watson to R. Dedonder (answer to R. Dedonder's letter to J. Watson of 15 September 1989), 9 November 1989.

Despite the NIH National Center for Human Genome Research's (NCHGR)[90] position on this project, several model organism genome sequencing projects did get under way due to Watson's firm approval, often in collaboration with other countries, thus helping start a new era of sequencing genomes of bacteria and higher organisms. This was a sequel to the times when viruses were being sequenced. Because of the simplicity of their genetic material, viruses were, in fact, the first organisms to be completely sequenced. The first DNA genome sequenced was that of the phage *phi*-X174 (over 6,000 bp)[91], followed by total sequences of SV40 (5,000 bp)[92] and the polyomic virus. From 1978 onwards, the hepatitis B virus had been worked on[93]. Other viruses followed, and the list of viruses for which we now have the sequence keeps growing: to name but a few, the adenovirus 2 (36,000 bp), T7 and lambda bacteriophage (some 50,000 bp each)[94], the Epstein-Barr virus[95] and the human cytomegalovirus (230,000 bp)[96].

With the help of reverse transcriptase and molecular cloning, RNA viruses also greatly benefited from systematic analysis. Aside from the sequences of a number of viral oncogenes, the first RNA virus that was completely decoded

---

[90] In January 1997, the NIH National Center for Human Genome Research (NCHGR) was granted institute status and a new name — National Human Genome Research Institute (NHGRI). The new status will facilitate collaboration with other institutes, give NHGRI's director equal standing with other directors and allow NHGRI to operate under the same legislative authorities as other NIH research Institutes. The new name is thought to reflect more accurately the growth and accomplishments of the former NCHGR, which was established seven years ago to carry out the NIH role in the Human Genome Project.

[91] F. Sanger *et al.*, "Nucleotide sequences of bacteriophage phi-X174", *Nature*, 265, 1977, pp. 687–695.

[92] V.B. Reddy *et al.*, "The genome of simian virus 40", *Science*, 200, pp. 495–502 and W. Fiers *et al.*, "Complete nucleotide sequence of SV40 DNA", *Nature,* 273, 1978, pp. 113–120.

[93] F. Galibert *et al.*, "Nucleotide sequence of the Hepatitis B virus genome (subtype ayw) cloned in *E. coli*", *Nature*, 281, 1979, pp. 646–650.

[94] F. Sanger *et al.*, "Nucleotide sequence of bacteriophage lambda DNA", *J.M.B.*, 162, 1982, pp. 729–773.

[95] R. Baer *et al.*, "DNA sequence and expression of the B 95-8 Epstein-Barr virus genome", *Nature*, 310, 1984, pp. 207–211.

[96] M. Chee and B. Burrell, "Herpes virus: A study of parts", *Trends Genet.*, 6, 1990, pp. 86–91.

was that of Rous' Sarcoma (RSV)[97]. Studies on the hepatitis B virus led to work on retroviruses, particularly HIV which is responsible for AIDS[98].

Once started on viruses, the analysis of nucleotide sequences did not stop there. A beginning was made on the analysis of human mitochondrial DNA[99]. The human genome project took the analyses on, working on bacteria and other more complex model organisms.

5.5 The Importance of Model Organisms

E. coli was picked as an official priority in the 1990 plan for the five first years of the HGP, a plan that was revised in 1993[100] due to the slow progress made. In 1991, Blattner's team at the University of Wisconsin was granted US$7.8 million over four years by the NCHGR, of which US$2 million was for costs other than sequencing. Frederick Blattner's laboratory, therefore, became the "*E. coli* Genome Science and Technology Center". *E. coli* was the first and smallest of the model organism sequencing projects to get HGP funding, partially because it was expected that a better understanding of the bacterial genome would solve several scientific conundrums in fields as varied as evolution and cancer.

Despite the team's optimism in 1991, with many researchers feeling that the traditional sequencing would all be over within four years, by 1996 it was still not finished, even with additional efforts in the USA and Japan[101]. The 1993 re-evaluation of the HGP's plans estimated that the work would be finished by 1998. Between 1991 and 1993, Blattner and his group had set themselves the goal of one megabase per year. If the team had been able

---

[97] D.E. Schwartz *et al.*, "Nucleotide sequence of Rous Sarcoma virus", *Cell*, 32, 1983, pp. 9–17.

[98] S. Wain, Holson *et al.*, "Nucleotide sequence of the AIDS virus LAV", *Cell*, 40, 1985, pp. 9–17.

[99] S. Anderson *et al.*, "Sequence and organization of the human mitochondrial genome", *Nature*, 290, 1981, pp. 467–469.

[100] F. Collins and D. Galas, "A new five-year plan for the US human genome project", *Science*, vol. 262, 1 October 1993, p. 43.

[101] Research News, "Getting the bugs worked out", *Science*, vol. 267, 13 January 1995, pp. 172–175.

to keep this up, by 1995 over four megabases, about 80% of the *E. coli* genome, would have been finished. In fact, Blattner and his group had only managed 1.4 megabases, Harvard's G. Church had sequenced 0.1 megabases, and a Japanese effort coordinated by Kiyoshi Mizobuchi at Tokyo University and Katsumi Isono at Kobe University had sequenced 0.19 megabases. 1.6 megabases had previously been sequenced by bacterial geneticists interested in one or another of the 4,000 genes of the single *E. coli* chromosome, but these sequences were full of errors and came from many different strains of the bacteria. As a result, many genome sequencers, including Blattner, thought that those sequences should be checked or even resequenced. Including this check of the 1.6 megabases, at least 60% of the genome remained unsequenced[102] at the beginning of 1996, the delay in great part due to the "archaic" sequencing techniques employed by Blattner. Many critics say that Blattner should have taken up new faster sequencing technologies, namely automated sequencing machines[103].

With NCHGR funding R. Davis, D. Botstein, K. Hennessy, J.T. Mulligan, and G. Harzell had set up a new laboratory to show the feasibility of sequencing the genome of the yeast *Saccharomyces cerevisiae*, at a reasonable cost and rate, as a preliminary stage to a genome center which would sequence the rest of the yeast. Its first task was to evaluate two techniques, automated fluorescence sequencing and multiplex sequencing with computerized film reading. While André Goffeau, now a professor at the University Catholique de Louvain, organized the European network under the European Commission's biotechnology programs and launched the attack on sequencing *S. cerevisiae*, Botstein and Davis' group decided to start with chromosome V and accepted keeping in touch with the Europeans, who were making rapid progress, to make sure that there was no unnecessary duplication of effort. They finally sequenced chromosome V and also worked on the right arm of chromosome VI (600 kb). At another center, the St Louis Center Project in the USA, a group led by Johnston obtained funding to sequence

---

[102] Research News, "Getting the bugs worked out", *Science*, vol. 267, 13 January 1995, pp. 172–175.
[103] *Ditto*.

chromosome VIII of *S. cerevisiae*, beginning in January 1993 and finishing by 1994 [104]. This group also sequenced the right arm of chromosome XII.

Walter Gilbert obtained funding to sequence the small organism *Mycoplasma capricolum*. Gilbert's, Botstein's and Davis and Blattner's projects were joined by other pilot sequencing projects. The American groups working on the nematode *C. elegans* started large-scale sequencing in transatlantic collaboration with John Sulston and Alan Coulson of the MRC Laboratory of Molecular Biology in Cambridge, England and Robert Waterston of the Washington University School of Medicine, St. Louis, USA[105]. A significant part of the genome has now already been elucidated, some 55 megabases (finally reduced down to 36.4 Mb) of the 100 megabase genome. This is the largest quantity of DNA sequences available to date [106] from a simple organism. The *C. elegans* genome work started in 1990 and was finished in 1998. A follow-up project to sequence the closely-related nematode *C. briggsiae* has been announced.

These systematic sequencing projects were, of course, accompanied by mapping projects. A program managed through the laboratories of Dan Hartl and Ian Ducan at the University of Washington (St Louis) mapped large sequences of the *Drosophila* genome in artificial yeast chromosomes. Researchers working on the 120 Mb on the euchromatic part of the *D. melanogaster* genome had sequenced about 2.5 Mb of it by December 1996.

In 1995, researchers put together a genetic map of the mouse genome with about 6,500 microsatellite markers among some 7,300 genetic markers [107], a map which, with the final report on the genetic map for man,

---

[104] M. Johnston *et al.*, "Complete nucleotide sequence of *Saccaromyces cerevisae* chromosome VIII", *Science* 265, 1994, pp. 2077–2082.

[105] J. Sulston *et al.*, "The *C. elegans* genome sequencing project: a beginning", *Nature*, vol. 356, 5 March 1992, pp. 37–41, K.E. Davies, "The worm turns and delivers", *Nature*, vol. 356, 5 March 1992, pp. 14–15, J. Hodgkin *et al.*, "The nematode *C. elegans* and its genome", *Science*, vol. 270, 20 October 1995, pp. 410–414.

[106] *Human Genome News*, vol. 7, nos. 3 and 4, September–December 1995, p. 7.

[107] W.F. Dietrich, J. Muller, R. Steen *et al.*, "A comprehensive genetic map of the mouse genome", *Nature*, vol. 380, 14 March 1996, pp. 142–152.

marks the end of Phase One of the human genome project[108]. Work has already begun on a physical map of the mouse genome. Other researchers are currently working on homologous regions in the mouse and man genomes, such as the region containing the genes for the T cell receptor, which specifies the receptors on the surface of the cells and plays an important role in immune response. Furthermore, the international consortium of Image cDNA had, in 1993, initiated the construction and characterization of the mouse DNA library, integrated and made available facts and begun the construction of clones.

But rather suddenly, the leaders of the scene were taken by surprise by the announcement of the achievement of the sequence of a series of small genomes of less than 2 Mb long.

Craig Venter's team at the TIGR announced in *Science* on 28 July 1995[109] that, thanks to private funding, they had completely sequenced the first genome of a non-viral organism, *Heamophilus influenzae*, a 1,830,137 bp organism responsible for a particularly nasty form of childhood meningitis. Most genetics specialists had previously thought that the first non-viral organism to reveal its intimate secrets would be *S. cerevisiae*. On 9 December 1994, Genome Therapeutics Corp (GTC) of Waltham, Massachusetts issued a press release stating that they had sequenced the genome of *Helicobacter pylori*, which causes ulcers. The firm refused to publish the sequence because they had done the work with private funding and now intended to use the results for research into products for the ulcer medication market, which in 1992, was valued at several billion dollars per year. GTC had sequenced the genome by chance, having sequenced fragments of interest that, coincidentally, probably covered the whole genome. As J.H. Miller of the University of California pointed out on hearing the announcement, those two aspects removed all significance from GTC's claim. TIGR has recently resequenced the *Helicobacter pylori* genome and made it public.

---

[108] "The construction of a detailed map of mouse and man" C. Dib *et al.*, "A comprehensive genetic map of the human genome based on 5,264 microsatellites", *Nature*, vol. 380, 1996, pp. 152–154.

[109] R.D. Fleischmann *et al.*, *Science*, 269, pp. 496–512, K.M. Devine and Ken Wolfe, "Bacterial genomes: A TIGR in the tank", *T.I.G.*, vol 11, no. 11, November 1995, pp. 429–431.

Another successful initiative is the DOE's Microbial Genome Initiative (MGI) which funds TIGR for other small genomes. C. Frazer, Vice-President for Research and Director of cellular and molecular biology at the Institute of Genomic Research led the project, which by running forty automated sequencing machines simultaneously produced, in a few months, the complete sequence of the *Mycoplasma genitalium* genome (580,070 bp)[110], the smallest known genome for a living organism. It was thought that with fewer than the 482 genes this organism has, it is no longer possible for life to exist[111]. What is technically remarkable about this particular program is the procedure of assembling DNA with new and powerful software that searches for overlaps in the sequences submitted to the computer.

In December 1995, TIGR finished sequencing the genome of the bacteria *Methanococcus jannaschi* (2 Mbs)[112], which had been discovered in 1982 living in an extremely hot environment in the ocean depths. TIGR has also sequenced the genome of *Treponema pallidum*[113], the cause of syphilis, in collaboration with molecular biologist G. Weinstock of the University of Texas at Houston and funded by the National Institute for Allergies and Infections Diseases.

There is no end to the development of sequencing projects. R. Weiss of the University of Utah recently put forward a project to sequence the genome of *Pyrococcus furiosus* at the fifth DOE Human Genome Program Contractor-Grantee workshop. At the same workshop, J. Dunn of the

---

[110] C. Fraser *et al.*, "The minimal gene complement of *Mycoplasma genitalium*", *Science*, October 1995, pp. 397–403, D.N. Left, "Massive DNA automation sequences *Mycoplasma*, smallest total genome", *Bioworld Today*, Vol. 6, no. 202, 20 October 1995, p. 1 and 4. (*M. genitalium* develops in the epithelial cells of human respiratory and genital systems but does not cause any real damage).

[111] A. Goffeau, "Life with 482 genes", *Science*, vol. 270, 20 October 1995, pp. 445–446.

[112] Bult *et al.*, *Science*, 273, 1996, pp. 1058–1073. Many other genomes are actually being sequenced by TIGR: *Caulobacter crescentus, Chlamydia pneumoniae, Chlamydia trachomatis mouse pneumonitis, Chlorobium tepidum, Deinococcus radiodurans, Dehalococcoides ethenogenes, Desulfovibrio vulgaris, Enterococcus faecalis, Plasmodium falciparum, Porphyromonas gingivalis, Legionella pneumophila, Mycobacterium avium, Mycobacterium tuberculosis, Neisseria gonorrhoeae, Neisseria meningitidis, Pseudomonas putida, Salmonella typhimurium, Shewanella putrefaciens, Staphylococcus aureus, Thermotoga maritima, Thiobacillus ferroxidans, Treponema denticola* and *Vibrio cholerae*.

[113] Fraser *et al.*, *Science*, 281, 1998, p. 375

Brookhaven National Laboratory and his colleagues announced that they would be sequencing the genome of *B. burdorferi* (935 Mb) that causes Lyme disease[114]. A project put forward in 1989[115] to map and sequence the mustard cress *Arabidopsis thaliana*[116] was financed jointly by the National Science Foundation, the US Department of Agriculture (USDA) and the DOE. The USDA, through a research program for plant genomes opened in 1991 for five years[117], also funds other genome mapping projects for plants and animals of agricultural importance. Its budget is $14.7 million for 1991, $15 million for 1992 and $18 million for 1993, its continuation being dependent on an evaluation carried out in 1996.

Clearly, the human genome project has been the nursemaid for the model organism sequencing programs as they have grown in importance[118] and has also encouraged the development of sequencing operations in other research organizations in the United States. The debate that surrounded its creation in the United States, and the HGP itself, have also contributed to the reactions of the other nations and of Europe as a whole.

---

[114] It is the TIGR which finally sequenced it: Fraser *et al.*, *Nature*, 390, 1997, pp. 580–586.

[115] The project is outlined in "A long-range plan for the multinational coordinated *Arabidopsis thaliana* genome research project", (NSF 90-80), 1990.

[116] Within the framework of an international effort, in which Europe has an important role.

[117] For a presentation of the USDA genome research program, please read S. McCarthy's "USDA's plant genome research", *Agricultural Libraries Information Notes*, Vol. 17, no. 10, October 1991, pp. 1–10, and the article entitled "USDA's high-priority commitment to the plant genome research program", *Probe* (Newsletter for the USDA plant genome research program) vol. 3, no. 1/2, January–June 1993, pp. 1–3.

[118] Given the success of the first five-year plan's execution, and the sequencing of model organism genomes, and given the improvements in sequencing tools and techniques, the NCHGR decided in 1996 to run a three-year pilot study involving six American research centers and costing $60 million to explore the feasibility of large-scale human genome sequencing. The six centers are: TIGR, Baylor College of Medicine, Whitehead Institute for Biomedical Research, Stanford University and two laboratories at the University of Washington.

5.6 The International Dimensions of Genome Research: The First Stirrings in other Countries

The genome debate that began in the mid-1980s and led to the human genome project crossed national borders very quickly due to the little regard the international scientific community has for political frontiers. In many countries, a growing number of researchers were requesting funding from their governments to set up new genome research programs, often meeting with success. The first such program to be set up was that of Charles DeLisi at the DOE, and the second started in Italy.

Italy

The Italian genome program began as a pilot project in 1987 under the aegis of the Italian Consiglio Nazionale della Ricerche (CNR) one year after the first review of the genome program by the DOE. The Italian program sprang from the speech given by Renato Dulbecco in 1985 in Washington DC, during which he presented the idea of a human genome sequencing project as the next battle in the war against cancer. Italian consensus rapidly gathered around Dulbecco's editorial in *Science*[119], and a program was quickly formulated and ratified by the Italian government. At the time, a revitalization of scientific, technological and economic research was being encouraged and biotechnology was seen as one of the priorities. The government announced in April 1987 that it would allocate 209 billion lira (some US$156 million) over five years to a national biotechnology project involving both public research centers and industry[120]. In May 1987, the CNR announced a special biotechnological research program to which it would dedicate about 84 billion lira (about US$63 million) over five years, and this would include a project to sequence the human genome involving contributions

---

[119] Renato Dulbeco "A turning point in cancer research: Sequencing the human genome", *Science*, vol. 231, 7 March 1986, pp. 1055–1056.

[120] G. Oddo, "More business involvement in CNR's finalized programs", *Il Sole 24 Oro*, Milan, 14 July 1987, p. 6.

from all the CNR institutes and laboratories working in biology[121]. This project was seen as a way to bring Italy world recognition in molecular biology.

Dulbecco was asked to coordinate the project with assistance from Paolo Vezzoni. The CNR wanted to finance the program with 20 billion lira (approximately US$15 million) and involve 75 to 100 people every year. For each of the first three years, the budget was US$1.25 million. Right from the very beginning, the strategic approach chosen was, unlike the American approach, that of a network of laboratories cooperating together for a common goal, none of the laboratories in the network having the equipment or the skills in its possession to do all the work. The common goal which provided the project with such unity was the end of the long arm of the X chromosome (Xq28-Xqter). Representatives of the various laboratories meet two or three times a year, promoting active collaboration. Cooperation is also developing with many other European and American laboratories and data bases are being set up.

As well as the human genome, there were Italian contributions to the sequencing of the *B. subtilis* and *Arabidopsis thaliana* genomes in the context of the European Commission initiatives. The sequencing of the *Mycobacterium leprae* and *M. avium* genomes with funding from the WHO was proposed as well as the mapping of *extremophile* genomes. Funding for genome research is conveyed through the Ministries of Agriculture, Health and the Consiglio Nazionale delle Ricerche.

United Kingdom

The next national genome program emerged in the United Kingdom. The British program has its roots deep in the history of molecular biology as a whole. The UK has a long tradition of molecular biology and genetics research and has carried out innovative work in the development of mapping and sequencing techniques. At the time, the Cambridge molecular biology

---

[121] Consiglio Nazionale della Ricerche, "Progetto Stregico CNR: Mappaggio e Sequenzimento del Genoma Humano", 1987.

laboratory hosted the most modern mapping and sequencing technologies and remains, with or without a genome project, one of the great centers of molecular biology in the world. The laboratory's traditions could only be favorable for the birth of a British genome project, although it was not seen as necessary as it was in other countries.

In the United Kingdom, research is mainly funded by the government through the Departments of Education and of Science, the Department of Trade and Industry and the Departments of Health and Social Security. The Department of Education and Science channels research funding to the universities through five research councils. (This structure has been recently modified and the biotechnology and Biological Research Council has been created). Biotechnology is a field that overlaps the mandates of the Science and Engineering Research Council (SERC) and the Medical Research Council (MRC), the two councils whose fields of interest are most linked to genome research. The SERC supports fundamental biomedical research outside the medical domain, as well as financing some research on automated sequencing through its biotechnological directorate, which eases the link between academic research and industrial needs.

The MRC was clearly the main supporter of mapping and sequencing research. As a rough guide, its total budget for genome research in fiscal year 1985–86 was about £4.2 million (or US$7.4 million). The MRC is a body similar to the US NIH in that it supports projects proposed by the researchers themselves, though it also sets up targeted programs in particular domains. It has a long tradition of commitment to molecular biology and is able to set up new units dedicated to specific areas of research when a suitable director and adequate funding are available.

Although the MRC financed a large quantity of research on human genomes and other organisms, and the laboratories it was supporting had the necessary expertise and tools to study hereditary illnesses, the MRC did not, at first, consider a research program to map or sequence the human genome. The situation took a long time to change. Representatives of the Cambridge MRC attended the main meetings for the HGP in the United States; for example, John Sulston was at the very first Santa Cruz meeting. The opinion of British scientists was almost automatically requested in the debates about the HGP because of their contribution to molecular biology. In fact, the UK's scientists had been involved for quite some time in daring projects to push structural biology to its limits. Two figures stand out from the molecular biology community, Walter Bodmer and Sydney Brenner. These two researchers were immediately involved in the maelstrom of the genome debate.

Bodmer was the Director of the Imperial Cancer Research Fund (ICRF) in London, an internationally famous research center, known in particular for its work on the genetic aspects of cancer and other pathologies. He had gained world fame for his studies on genetic variation in human populations. He had been asked by Watson to open the Cold Spring Harbor meeting on molecular biology in June 1986, and he also chaired the HHMI meeting in Bethesda two months later. Sydney Brenner of the MRC laboratory in Cambridge participated in several of the early American meetings and was later invited to sit on the National Academy of Science Committee. Brenner and Bodmer were both close to scientific policymaking circles in the UK, and in the world of molecular biology they acted as British ambassadors. Brenner was very high up at the MRC, taking on several different roles during the genome debate.

The British genome debate began in 1986 when Brenner suggested at an MRC meeting that a unit for molecular genetics should be set up to include genome research. Brenner began the genome program with funds of his own from a science prize he had won from the Louis Jeantet Foundation (about £25,000). The MRC approved and endorsed Brenner's project to map the human genome (mainly with private funding) as long as the research did not cost them a penny. Brenner's proposal was to use the physical mapping techniques developed by Sulston and Coulson for *C. elegans* on the human genome. He thus involved the MRC in the debate on "genome politics and planning".

At Brenner's request, the MRC set up a scientific advisory office jointly headed by MRC Secretary Sir James Cowan and Bodmer, Director of the ICRF. The members of the office reflected the interests of other research councils and contributing charitable organizations. The ICRF and the MRC were to share the financing equally. Coordination was to be ensured by a committee of scientific advisors. In February 1989, the Secretary of State for Education and Science announced that the MRC would receive £11 million over three years for a project to map the human genome. This new office caught the attention of the Cabinet Office and presented a detailed proposal which was approved by the Advisory Council on Science and Technology. Margaret Thatcher gave the genome program the final necessary agreement and thus ensured that funds would be available: £2.3 million for 1989-90 and £4.6 million for 1991-92. The MRC would have to look for alternative funding from non-governmental sources[122] for later years. The British genome program officially began in April 1989 for three years, hoping to obtain an annual stable budget of £4.5 million from 1992 onwards. The scientific strategy was twofold, consisting of the coordination of the work already under way, including the data bases and research linked to the development of techniques, and a boosting of fundamental genome research with additional funding and proclamations of ambitious technical perspectives. The approach to international coordination would be in essence one of "let's get started, and then we will talk about it". Brenner noted that "once a center was set up in the United Kingdom which would be of some value to the research community, then and only then could it be of use in international efforts."[123]

In its first year, the British program focused on automatisation, developing new sequencing and mapping techniques and the mapping of regions of the genome that were of particular interest. It then switched focus to cloning, mapping and a closer inspection of coding regions. The program also included research on the mouse, especially in terms of mapping. The British effort became focused on the regions of the human genome that code for proteins and the development of bio-informatics. The program allowed the importation of the artificial yeast chromosome library from the University of Washington at St Louis, which made the clones more freely available in Europe. A set of DNA probes was gathered together at the ICRF, and the human cell lines stocked at the MRC center at Porton Down were made more widely

---

[122] John Galloway, "Britain and the human genome", *New Scientist*, July 1990, pp. 41-46.

[123] D. Dickson, "Britain launches genome program", *Science*, 245, 31 March 1989, p. 1657, J. Galloway, "Britain and the human genome", *New Scientist*, July 1990, pp. 41-46.

accessible. Furthermore, in 1991, a right of access was acquired to the genome data base at Johns Hopkins, making the United Kingdom the first communications node for that information outside Baltimore.

The project's progress was reported in a newsletter that disseminated information on financial resources of the current program, the various projects, new techniques and summaries of the main colloquia[124]. Tony Vickers was named director of the human genome mapping project in 1990, and later that same year, the Human Genome Resource Center was set up at Northwick Park Hospital, Harrow, also under his aegis, until the end of 1992, when he was succeeded by Keith Gilson.

A new element in the British effort was the establishment of a transatlantic collaboration on the nematode *C. elegans*. The physical mapping effort on the nematode genome, one of the model projects in physical mapping, evolved in the context of this British-American partnership, and so it was that Brenner's favorite organism would once again help to push back borders for biology. The transatlantic collaboration was so successful that it was the basis of a major expansion of genome research efforts during 1993, when John Sulston became genome research director for the new Sanger Center funded by the MRC and the Wellcome Trust[125], located at Hinxton Hall, south of Cambridge. During the two first years of the British genome project, attention in the United Kingdom turned to the organization of the 11th workshop on human genome mapping (HGM 11) to be hosted by the ICRF in London.

British scientists continued to distinguish themselves in international genome politics far beyond what could be expected given the size of their research budgets. Bodmer replaced Victor McKusick at the head of HUGO in December 1989. He presented genome research to the parliamentarians sitting on the Research Committee, a science policy forum, in order to

---

[124] The three first publications were entitled "*G. string*". In later publications it was entitled "*Genome News*". The first publications were mainly Britain-centered, but later editions also covered other European countries.

[125] The Wellcome Trust is by far the largest charitable organization in Great Britain, with an annual budget of £200 million, which is close to that of the MRC.

highlight the solutions in genome research. Cambridge's Smith became the president of the European Community Working Party to formulate a European genome analysis program. Brenner continued to be a major scientific force.

Initial British work on *C. elegans* continued to be a prototype for human genome research. In addition to this program, the MRC financed the drafting of a map of the Fugu fish genome, in the context of the HGP and under the watchful eyes of Brenner and Greg Elgar. Other work in the United Kingdom included technological research on sequencer automation at the University of Manchester's Institute of Science and Technology (UMIST) and research in bioinformatics at the University of Edinburgh. The bioinformatics research unit at Edinburgh had considerable experience in research linked to data bases, and at the beginning of 1989, undertook studies on the software that would be needed to analyze maps and sequencing information. Great Britain was also participating in a number of model organism study projects funded by the European Union such as the mapping and sequencing of *Arabidopsis thaliana*, the PIGMAP project, the EUROFAN project, and studies on cereal and livestock genomes.

In November 1995, the Wellcome Trust announced[126] that it wanted to allocate the Sanger Center sufficient funding to sequence a sixth of the human genome, at least 500 million of the three billion base pairs. During the Budget Speech for 1995-96, the government declared that it would include £9.6 million in its science budget for genome and immunology research. The United Kingdom contributed and continues to contribute to international research efforts such as the EMBL, to which in 1987 the MRC allocated £2.72 million, and certain European genome projects including the sequencing of *C. elegans* and the mapping and sequencing of *Arabidopsis thaliana*.

Now genomics research in the United Kingdom continues to be carried out largely through the agencies of research councils (in particular the MRC, the Biotechnology and Biological Sciences Research Council (BBSRC) and the Wellcome Trust.

---

[126] David Dickson, "Wellcome sets sequencing project in motion", *Nature*, vol. 378, 9 November 1995, p. 120.

The BBSRC funds a large range of genomics projects in animals, microbes and plants through grants to universities, focused initiative funding and direct support to its institutes. The two most important genomics researchers are John Innes Center, which focuses on crop plants, plant species and microorganisms and the Roslin Institute, which focuses on farm animals. The MRC supports research in a similar way through grants to universities and direct support of its units and initiatives. A wide range of programs relevant to genomic research are supported and five of the 40 MRC units carry out significant numbers of programs on genome research. In the most recent allocation of the science budget to research councils, funding of human genome research has emerged as the UK government's major priority over the next three years [127].

The Wellcome Trust has two major institutions for genomics research, the Sanger Center and the Wellcome Trust Center for Human Genetics at the University of Oxford. It also provides support directly to researchers at universities through grants. The Wellcome Trust is committed to funding the sequencing of approximately one third of the human genome and several microbial pathogens through its Beowulf Genomics Initiative. The Wellcome Trust, together with the NIH, is also dedicating new funding to build an Single Nucleotide Polymorphism (SNP) database so that scientists will have access to this information in the public domain. The Wellcome Trust Genome Campus brings together the Sanger Center, The European Bioinformatics Institute and the MRC Human Genome Project Resource Center. Specifically, the genome sequence has been largely concentrated at the Sanger Center and funded by the Wellcome Trust and the MRC. The joint NIH/MRC project sequence of the *C. elegans* genome is now finished.

There is also selected sequencing of the mouse, the Fugu (puffer) fish, *Drosophila* and the rat, and the Beowulf Genomics Initiative involves the sequencing of 14 microbial genomes, a major achievement, in collaboration with the Pasteur Institute (Paris, France), being the publication of the complete

---

[127] Anon, "MRC aand BBSRC win in science allocations", *Research Fortnight*, vol. 5, 28 October 1998.

annotated genome sequence of the cause of tuberculosis, *Mycobacterium tuberculosis*.

The BBSRC supports much of its genomic research on plants through the John Innes Center, which is involved in the European *Arabidopsis* sequencing project. The Roslin Institute also has a coordinating role for EC sequencing projects on farm animals. Expression profiles, libraries and databases are well developed across these different centers.

The Union of Soviet Socialist Republics

While genome projects gestated in Italy and the United Kingdom, a different sort of genome project was beginning in the USSR. Although rooted in rich scientific soil, it was politically very unstable and just when it reached maturity, the USSR disintegrated and scientists found themselves fighting for every rouble.

The initiator of the Soviet genome program was Alexander Drovich Bayev, a man who had survived a Gulag exile from 1937 to 1954, and who had witnessed countless moments of political instability in the late 1980s. The perseverance, vision and dynamism that led him to build a children's hospital in Roulsk during his exile also brought him to create the third largest national genome program in 1988.

Bayev, a talented biologist who had studied under Vladimir Alexandrovitch Engelhardt in Moscow, had lost the allegedly most productive part of any scientific career (from 35 to 50 years of age) during his exile and imprisonment after the Stalinian purges. Bayev remained in voluntary exile for several years after his imprisonment, not wishing to imperil his friends by writing to them. His exile ended in 1954, less than a year after Stalin's death. Engelhardt managed to bring him back to Moscow, where he spent the later part of his career first at the Institute of Biochemistry and then at the Institute of Molecular Biology.

Bayev missed the biological revolution that happened at the beginning of the 1950s. He was in exile when Watson and Crick published their article on the structure of DNA. More importantly, the USSR had few scientists capable of grasping that molecular biology was on its way. Stalin's pet geneticist, Lyssenko, had distorted biological thinking by systematically repressing genetics and molecular biology for two decades. After Stalin's death, Lyssenko gradually began to lose his ideological stranglehold on biology and agriculture. Bayev's mentor, Engelhardt, had been in the vanguard of Lyssenko's opponents. Thanks to him and other dissidents like Andrei Sakharov, Lyssenko was finally publicly ousted in 1964. Englehardt, who had temporarily lost his position in the biology section of the USSR Academy of Sciences because of his anti-Lyssenko stance, became the director of the Institute of Molecular Biology and asked Bayev to come back.

It was at this institute that a young and talented student working for Bayev, Andrei Mirzabekov, distinguished himself as a rising star of Russian biology. He had been born the year Bayev was arrested in 1937. In 1971, Mirzabekov obtained permission to travel to the west to Cambridge, England, where he worked on the crystallization of transfer RNA. He managed to lengthen his stay by six months thanks to support from Nobel prizewinners Aaron Klug, Francis Crick, Frederick Sanger and Max Perutz. In the mid-1970s, Mirzabekov came to represent a link between the USSR and the western world of molecular biology. His knowledge of the power of molecular biology was obtained at its epicenter, the MRC laboratory at Cambridge. Mirzabekov returned to the west to participate in several major events, in particular, the 1975 Asilomar conference. It was during his 1975 visit that he had the now-famous dinner with Walter Gilbert, Allan Maxam and Jay Gusella at Harvard, a meeting important for the subsequent development of a new sequencing technique: the Maxam-Gilbert method.

During the 1970s and 1980s, Mirzabekov and Bayev continued to work at the Moscow Institute of Molecular Biology, now renamed the Engelhardt Institute. With other researchers in Moscow and Leningrad, they contributed to the introduction of new approaches and techniques to Soviet biology. In 1986, nominated by Bayev, as he himself had refused the post, Mirzabekov was appointed the Director of the Engelhardt Institute. Although considerably burdened by his new administrative responsibilities, Mirzabekov remained scientifically active. Several Soviet scientists were very interested in the development of techniques that would make DNA sequencing cheaper, in particular, less hard on reagents and requiring as little robotics as possible.

Mirzabekov's group worked alongside other groups in the USA and Yugoslavia on the improvement of a new sequencing method suggested by Hans Lehrach[128] of the ICRF in London. Bayev and Mirzabekov gradually became the champions of the USSR genome program.

Bayev was let in on the American genome project by Walter Gilbert and James Watson during his 1986 visit to the United States. That same year, in June, Mirzabekov attended the "Molecular Biology of *Homo sapiens*" symposium at Cold Spring Harbor. Back in Moscow, Bayev and Mirzabekov started looking for funding for a Soviet genome program. The search was aided by Mikhail Gorbachov's new policies of *glasnost* and *perestroika*. *Glasnost* made it possible to recognize the damage caused to molecular biology and genetics by Lyssenkism and to authorize repairs to the field. *Perestroika*, as it applied to biology, meant the linking of science and biotechnology to national economic objectives. The first stage was to bring Soviet molecular biology back up to world standard. The Soviets returned to the peer review system and other aspects in the funding and administration of science. Bayev and Mirzabekov contacted their colleagues to build a financial and scientific support base for a Soviet genome project, aiming not to fall further behind the accelerating American movement.

Despite Bayev's conviction that long-term decisions should be made as soon as possible, the genome project met with opposition in 1987 and again in February 1988, when Bayev made a presentation to the USSR General Assembly of Sciences. The opposition was based mainly on its importance relative to other scientific fields. Bayev and Mirzabekov persisted, so convinced of the importance of genome research were they by the movements setting up the American HGP and the British and Italian genome projects. They finally managed to gather colleagues and science policy makers in

---

[128] This technique is based on linking short fragments of DNA of a known sequence some eight base pairs long and the determination of the specific links with one or many fragments of a given DNA. If they link, the sequence is present on the piece of DNA. By linking a large number of fragments like this and identifying the linked fragments, the sequence of targeted fragments could also theoretically be determined. Despite the numerous technical difficulties to be overcome (accuracy of identification, number of fragments needing to be tested), potential advantages such as speed and simplicity make this sequencing technique very attractive.

support of their project. In 1988, a program was approved by the Council of Ministers, allowing the State Committee of Science and Technology to list it as one of 14 priority areas of scientific research. Genetics took on central importance in several other initiatives in biology and agriculture as Lyssenko's shadow vanished in the bright new sunshine of molecular biology.

The Soviet genome program is a good example of *perestroika*. The funding for the project was partially distributed by traditional mechanisms controlled by the directors of the national institutes and by the directors of laboratories within the institutes. Another part of the budget was modeled on the American NIH's specific project funding system. As director of the program, Mirzabekov desperately craved information on the western peer review and administration systems for grants. He hoped to revitalize Soviet molecular biology by applying western scientific management and administration methods, especially those of the NIH. Despite the difficulties of the time, the genome project budget grew from 25 million roubles in 1988 to 32 million in 1989. The existence of a direct telephone line linking the Engelhardt Institute to the west shows how important genome research was considered. The economic problems, debate on decentralization and the stormy transition of a centralized communist economy towards a free market economy imperilled the USSR genome project. However, in 1991, a 40 million rouble genome project was approved in the national budget.

The chaos that followed the failed *coup d'état* against Gorbachov in August 1991 caused a period of great uncertainty. As the USSR imploded, the Russian republic inherited the main scientific institutes hosting most of the former Soviet genome research, but budgets had been savagely cut and the economy and national policy had more pressing problems. Despite this, as Soviet science toppled into crisis in 1992, the genome project kept a fairly large budget going from the Russian Academy of Sciences, perhaps because Bayev and the Engelhardt Institute were never directly associated with the politicians in power under Brezhnev and had long been seen as reformers.

Another influencing factor could have been Bayev and Mirzabekov's use of the promise of new economic development stemming from genome research to justify funding for the project like their counterparts in the USA human genome project. As Russia and the other republics rejected the old order, economic revitalization became the buzzword. The genome project was one of the programs that had already been set up. Although it was now both financially and materially limited, and although conscious of the enormity of the task ahead, the genome scientists stayed enthusiastic and ambitious. Bayev and his colleagues had set up the genome project with a view to the future of biological research, taking into account the importance that structure and function research on genomes (especially the human genome) would have for science, the economy and industry.

Professor Bayev was not only the organizer of the Russian HGP and the head of its Scientific Council, he was its real soul. He continued to head the program until his death in December 1994, at the age of nearly 91. In June 1995, after a period of some uncertainty, Professor Lev Kisselev from the Engelhardt Institute was appointed his successor. Andrei Mirzabekov, now spending part of his time at the Argonne National Laboratory in the USA, is the vice director of the program.

For the first three years (1989–1991), the Russian HGP had rather generous funding (about 30 million roubles plus US$ 10 million per year). However, in 1994 this dropped to US$ 400,000 and in 1995 US$ 600,000. To survive, the Russian HGP drastically reduced the number of grants awarded and adopted a policy of broad international cooperation. The Engelhardt Institute of Molecular Biology distributes funds on behalf of the Ministry of Science. There has been a switch from funding for key institutes (Table 7) to more substantial support for individual chromosome projects and individual research groups. Funds are also available for the purchase of reagents and equipment, support of the Moscow HUGO Office and the Moscow and Novosibirsk database centers, organization of conferences, and for foreign travel.

The program succeeds because of the following three factors:

- The high quality of Russian researchers, their original ideas and approaches;
- Wide international cooperation with the leading laboratories from Europe and the USA; the HUGO Office in Moscow allows stable contacts with other international genome programs;

– Competition for project grants, good organization of research and cooperation between groups.

Table 7 Key institutes involved in the Russian human genome project

Engelhardt Institute of Molecular Biology, Russian Academy of Sciences, Moscow
Institute of Bioorganic Chemistry, Russian Academy of Sciences, Moscow
Nikolai Vavilov Institute of General Genetics, Russian Academy of Sciences, Moscow
Institute of Molecular Genetics, Russian Academy of Sciences, Moscow
Russian State Center for Medical Genetics, Russian Academy of Medical Sciences, Moscow
Institute of Cytology and Genetics, Russian Academy of Sciences, Siberian Branch, Novosibirsk
Institute of Bioorganic Chemistry, Russian Academy of Sciences, Siberaian Branch, Novosibirsk
Moscow Lomonosov State University
Institute of Cytology, Russian Academy of Sciences, St Petersburg.

The main goals of the Russian HGP are: the physical and functional mapping of the human genome, molecular diagnosis of hereditary diseases and malignant tumors, computer analysis of genome structure and discovery of unknown human genes. During five years of the Russian HGP more than 300 papers have been published, many of them in international journals.

The original method of sequencing by hybridization (SHOM) was developed in 1988 by the research group headed by Andrei Mirzabekov. Now, several groups in the world are working on the improvement and usage of different variants of SHOM. In 1994, the Engelhardt Institute and the Argonne National Laboratory (USA) agreed to collaborate in developing further basic and technological aspects of SHOM.

Important progress has been made with the physical and functional mapping of the human chromosome 3 by a team at the Engelhardt Institute, headed by Lev Kisselev and Eugene Zabarovski, in collaboration with the Karolinska Institute (Stockholm, Sweden). More than half of the chromosome is mapped — over 100 chromosome markers have been obtained using high-resolution *in situ* hybridization, and more than 200 markers of *Not*1 restriction sites have been mapped.

An original approach has been developed for the transcriptional mapping of human chromosome 19, and many functional regions of this chromosome have been mapped. This work is headed by Evgeny Sverdlov (Institute of Bioorganic Chemistry) in collaboration with scientists from Germany and the USA.

Professor N. Yankovsky (Institute of General Genetics) is the head of the project mapping human chromosome 13, in close cooperation with several groups in Sweden, Holland and the UK. The project, which involves several Russian institutes, consists of the total cosmid mapping of chromosome 13 and special investigation of the q14 region, where a tumor suppressor gene for human chronic lymphoid leukemia is putatively situated.

Russian researchers have contributed significantly to the development of new computer programs to aid the study of the human genome. These programs are used in Russia and internationally. The informational Human Genome Center founded in Moscow has satellite connections with centers in Europe and in the USA. The Siberian Regional Center for genome information was established in 1995. The information center of the HUGO Moscow Branch is located at the Engelhardt Institute, under the direction of Dr. Y. Lysov. There is a databank, connected with the EMBL DataBank in Heidelberg, and with other information centers.

The Russian HGP pays particular attention to medical-genetical aspects of human genome research. Molecular diagnostic methods for the detection of common human hereditary diseases, in particular prenatal testing, have been developed and transferred to medical practice. Searches for disease genes are performed even though the financial support for this work is much smaller than in many other countries.

The Russian genome project recently started to develop human gene therapy methods, including:

– Introduction of the *dystrophin* gene and/or 'minigenes' into muscle as a possible treatment for Duchenne and Becker myodistrophies;
– Use of antisense genetic constructions as potential tools for anticancer therapy;

- Transfer of the human apolipoprotein gene into liver cells both *in vitro* and *in vivo* as a possible approach to gene therapy for *atherosclerosis*;
- Futher development and application of the ballistic technique for delivery of target genes *in vivo*.

During the first years of the program, there was a separate commission for 'equipment, reagents, probes' which supported the creation and production of new apparatus, reagents and material necessary for human genome investigations. Since 1995, the HGP Scientific Council has spent about 10% of its total budget on purchasing computers, other equipment and reagents distributed subsequently among grant holders.

In Russia, the future was completely uncertain, but the genome project looked as if it would, despite the difficulties, survive to serve as a brick in the construction of the new Russia.

As the genome programs developed in Italy, the United Kingdom and the USSR (and the political entities that followed), parallel developments were occurring elsewhere on the European continent as a response to the USA HGP initiative.

France

French genome research began in several centers that were already deeply involved in human genetics. The CEPH had been contributing to international research on genetic linkage maps for a long time, and it has since extended its efforts to physical mapping and the development of sequencing and mapping technologies. Through the CEPH and several important research teams, the French were playing a major role in the initial genome mapping efforts, participating both scientifically and financially in several European programs as well as in the EUREKA LABIMAP project on sequencing automation which involved the United Kingdom, France, the CEPH and the British firm of Amersham Ltd. Individual funding for genome research at each of the French scientific agencies evolved into a more sizable and targeted program between 1988 and 1990. In 1990, an allocation of funding was made especially to "genome and informatics research". Jacques Chirac, then Prime Minister, made genome research a national priority and, in May

1990, the government announced FF 8 million budget for genome research, distributed through a committee to be chaired by Jean Dausset. It was a start.

In June 1990, the research minister, Hubert Curien, said he intended to set up a more centralized genome research program. In the spring of the same year, he had already asked Philippe Lazar, the Director of the INSERM, to present him with proposals for a French genome program. At the time, the United States had been investing heavily in this sector, and a year earlier Britain had set up its own human gene mapping program. Philippe Lazar, in turn, asked Philippe Kourilsky of the Institut Pasteur to draw up a project proposal after wide consultations with scientists involved in such research. Kourilsky's proposals recommended the establishment of a program that should benefit from new research funding to the tune of FF 100 million per year, with a priority for gene study (and thus that of cDNA), while supporting informatics infrastructure, the study of model organism genomes and technological development. In terms of organization, Kourilsky suggested that a public interest group allows the association of the public and private sectors and pointed out the need for its rapid creation with a director with administrative experience and a flexible attitude to personnel management. On 17 October 1990, a ministerial press conference launched a French genome program that greatly resembled the one suggested by Kourilsky, with fresh funding of FF 50 million in 1991 and FF 100 million in 1992 [129]. Things seemed well on their way. But what happened next dashed all their hopes.

In March 1991, Jacques Hanoune, who had been assigned to explore the terms under which this public interest group could be set up, met with several difficulties. In particular, it seemed that the various potential partners in the public interest group (research organizations, associations and firms) had different ideas as to what the public interest group should be as well as on the sharing of seats on its board. Some did not even see why it should be set up in the first place. Furthermore, even at the ministerial level, support for the project was not unanimous. By August, at the 19th Human Genome

---

[129] B.R. Jordan, "Le Programme Génome Humain Français", *Medecine, Sciences*, 9, 1990, p. 908.

Mapping Workshop in London, there had been no apparent progress. Some French scientists were voicing their concerns in public. A little later, the ministry decided to distribute research contracts for a total of several tens of thousands of French francs in a dubious manner (without a call for proposals) and as soon as possible. This decision caused serious tensions which were worsened by the fact that the credits were not payment credits at all, but only authorizations to begin work on projects. To turn this into real money, the government would have to dip into the finances of the INSERM, the CNRS or the INRA, whose reluctance and disapproval can be imagined.

Despite these difficulties, the public interest group was finally created by a decree of 25 January 1993 [130]. The GREG public interest group (directed by Piotr Slonimsky) was entrusted with the coordination of all genome research until 1995. In that year, it lost all the human genome aspects of research as well as those that involved bioinformatics in the reshuffle announced by François Fillon, on 11 March 1995. GREG was also asked to pay special attention to genomes of animal and vegetable organisms. Apart from the creation of a specific budget line of FF 257 million for 1995 and 1996, the reshuffle put an end to the Directorate General for Research on the Life Sciences. This was the wish of the minister for research and was announced during the presentation of the budget for civilian research and development in 1995 [131]. The Directorate General for Research on the Life Sciences was replaced by 14 scientific and technical committees, each running a coordinated concerted action under one of the seven strategic orientations defined in the plan for Life Sciences[132]. While the Research Minister pointed out that medical and biological research, and genetics research in particular,

---

[130] "Le GIP-GREG, Origine, définition et missions", *la lettre du GREG*, no. 1, April 1994, pp. 1–6.

[131] J. Rajnchapel-Messaï, "le BCRD version 1995", *Biofutur*, 139, 1994, p. 75.

[132] The titles of the seven great strategic orientations were the following: genetics; developmental, reproductive biology and biology of aging, structural biology and pharmachemistry (interface between physics, chemistry and biology and structure of macromolecules and organized macromolecular systems), environmental sciences, physiopathological mechanisms, notably in cardio-vascular and neurological illnesses, bioinformatics and biotechnology.

were priority fields, decisions being made on the organization of genome and health research gave cause for doubt.

Consequently, the GREG saw its role slashed to the structural analysis of animal and vegetable genomes. Its other roles — functional analysis of genes including human ones, human genetics, and bioinformatics were transferred to the control of the three scientific and technical committees (CSTs - *comités scientifiques et techniques*) of the same names, governed by a committee in charge of research activities in genetics and the environment. There was, therefore, an accumulation of structures leading to weighty administration and delays in human genome research. As a budgetary consequence of this reorganization, the GREG saw its finances fall from FF 80 million in 1994 to FF 22 million in 1995. The FF 60 million taken away from the GREG were pooled with the FF 154 million of the 1995 budget and redistributed to the CST covering genetics research. The drastic amputation of state contribution to the CEPH budget (14.5% in 1995) was a strange translation of François Fillon's above-mentioned "strong commitment to genetics research[133]". The fact that the GREG was not given new funding to allow it to support other projects in 1996 made the future of the French genome program look very bleak. In any case, as P. Slonimsky says, "the dislocation that has occurred between research on model organisms and research on the human genome can but damage them both".[134]

Finally, during a meeting of the public interest group's board on 27 June 1996, a decision was made to dissolve the GREG by the representative of the Secretary of State for Research. From a science policy point of view, this decision was surprising, as other nations were investing more than ever in genome research, and other public funding for genome research in France was also becoming hard to come by. In contrast, the CEPH and the private

---

[133] Despite the recommendations that a large sequencing center be set up from a study committee brought together by GREG in 1995, and from the TGS (a committee constituted under B. Bigot's mandate to answer the question of whether a large sequencing center was to be desired), the wait-and-see policy prevailed, since the French scientific community had always had the tendency to be condescending about high-tech activities (*La Lettre du GREG,* Séquençage systematique des génomes, no. 4, July 1995, pp. 18–20.).

[134] P. Slonimsky, as quoted in *Biofutur,* June 1995, p. 23.

funding of the Généthon (raised by the Association Française contre la Myopathie or AMF)[135] had, since 1987, provided between FF 200 and 300 million each year to research; it still afforded a stable basis on which France could hope to maintain its position in the vanguard of genome research, it in particular, in genetic mapping of the human genome[136]. But even this is decreasing since the AMF is gradually withdrawing its financial support.

This unfortunate state of affairs is to be regretted since Watson, when asked what the best genome program was outside the United States during budget hearings in 1993, pointed out that through the Généthon, the French had been the first to aim for super-production. When he was asked which country was the second best in the world, he answered again that it was France[137]. To bring them back up to their former level, a budget of fifty to a hundred million francs per year, the nomination of a director with real power and without too much of an ejector seat and a large but rapid concertation would be needed, to set up a realistic long-term program, for example for five years[138]. In 1997, there was a choice to be made. The scientists were ready, and all that remains is for the political and economic powers to decide. The last hope for the preservation of French genome research, at that time, seemed to be cooperation with European Union programs and some privately funded international initiatives such as that set

---

[135] D. Chen impressed all researchers at the annual Cold Spring Harbor Genome meeting when, in May, 1992 he revealed the results on the physical map of chromosome 21 that showed he was further along than most of his peers had thought. D. Conco, "French find short cut to map of human genome", *New Scientist,* 23 May 1992, p. 5, C. Anderson, "New French genome center aims to prove that bigger really is better", *Nature,* 357, 18 June 1992, pp. 526–527. In *Nature,* 14 March 1996, the Genethon team published a genetic map of the human genome based on 5,264 microsatellites (C. Dib, S. Faure, C. Fizanies *et al.*, "A comprehensive genetic map of the human genome based on 5,264 microsatellites", *Nature,* vol. 280, 14 March 1996, pp. 152–154).

[136] Let us also add to these enterprises and efforts the call for proposals issued by the AFM, the LNC (Ligue National contre le Cancer-National League against Cancer) the *Association Française de Lutte contre la Muscoviscidose* (AFLM) and the Ministers of Research and Health, entitled "Centers of application and development networl of gene therapies" with a total budget of 35 million French francs.

[137] U.S. House of Representatives, Departments of Labor, Health and Human Services, Education and related Agencies Appropriations for 1993, Committee for Appropriations, 25 March 1992, Part 4, pp. 607–608.

[138] B. Jordan, "Chronique d'une mort annoncée?" *la Lettre du GREG,* no. 7, July 1996, p. 20.

up between the Institut Pasteur and the Sanger Center to sequence *Mycobacterium tuberculosis* with funding from the Wellcome Trust.

Nevertheless, we must remain hopeful: there are INRA, CNRS and the new Genoscope.

INRA is responsible for government-funded research for agriculture in France. There are no large scale genome sequencing programs at INRA, rather the focus is on two aspects: specific genome sequences which are being analyzed for genes of agronomic potential relevance and EST sequences which are being produced for *Arabidopsis*, wheat and *Medicago truncatula*. Genome mapping is being undertaken for several species at different INRA research stations. These include two cereals: maize and wheat, two oilseeds: sunflower and rapeseed, fruit species: vine and *Prunus*, forage species: *M. truncatula* and *M. civita*, sugarbeet and six vegetables species: red pepper, melon, rapeseed, cabbage, cauliflower and pea and finally *Arabidopsis*, the plant model species. Several of these mapping projects are being undertaken in collaboration with firms from the seed industry. In terms of the kind of research being undertaken, these include work on Quantitative Trait Loci (QTL) identification in those listed above, positional cloning of genes in *Arabidopsis* and chromosomal organization and functional analysis of *Arabidopsis* by insertional mutagenesis. Most of the research collaborations at INRA in genomics are with other EU-based organizations and funded by the European Commission's Framework program IV. The INRA *Arabidopsis* program is a component of the National French program "GDR *Arabidopsis*". Other outputs of the genomic program at INRA include cDNA and YAC libraries, expression profiles, high density filters for the *Arabidopsis* program and databases (a Synteni database and a database for the CIC YAC library and chromosome III of *Arabidopsis*).

The CNRS genome program provides funds for 30 laboratories in the following:

- Bioinformatics;
- DNA chip development;
- Sequencing and mapping;
- Chromosomal stability;

- Hybrid construction in *Drosophila*;
- *Arabidopsis*.

These approaches are being applied to the majority of model organisms in animals, plants and bacteria as well as humans.

With the establishment of a national sequencing center, France is now set to continue to make a major contribution to the international human genome project.

The sequencing center (Genoscope) was initiated by the French Ministry of Research in January 1997 with funding for 10 years. Genoscope is a joint project between the Ministère de l'Education Nationale, de la Recherche et de la Technologie (MENRT) (50%), the *Centre National de la Recherche Scientifique* (CNRS) (40%) and a private firm, France Innovation net Transfert (10%). The main aim of Genoscope will be to generate a massive amount of high-quality sequencing data which will be released into the public domain as quickly as possible, in line with other publicly-funded sequencing centers around the world.

A mixture of sequencing projects will be undertaken, with at least 50% initiated by the *Centre National de Séquençage* (CNS) and others in collaboration with external centers. Initially, these collaborative projects will be with public-sector labs in France, but international collaboration may be considered in the future. Genoscope will fund all collaborative projects from its own budget and at least one member of the collaborative team will be invited to join the staff of the CNS to coordinate the project. All project proposals will be reviewed by a scientific committee of internationally renowned genome scientists. A strategic committee comprising representatives of French national research institutions will then rank the proposals according to scientific context and national priorities.

Genoscope could also be involved in contracted research for private institutions, subject to approval of the strategic committee, with costs being met by the contracting institution in consultation with other competing private institutions.

As Genoscope matures, its mission will evolve to include dissemination of novel technologies to other French public-sector research groups involved in genome sequencing.

Genoscope is headed by Jean Weissenbach and will employ around 120 people. Staff from Généthon and other labs began moving into the newly built premises in August 1997. Activities were scaled up during the next six months and the CNS hoped to be operating at full capacity by the end of 1998. A number of projects that were started at Généthon will continue at Genoscope, including construction of sequence-ready maps of human chromosomes, international collaboration to sequence the genome of the model plant *Arabidopsis thaliana* and two French national projects to sequence the genomes of *archeabacteria*.

Genoscope is set to perform the random sequencing of ordered BACS from human chromosome XIV. It has also nearly sequenced the Archeon *Pyrococcus abyssi*. A call for proposals has solicited over 50 proposals collaborative projects from which a smaller number will be selected to be performed at Genoscope. The following sequencing projects are in progress:

- *Encephalitozoon cuniculi*, a microspiridium;
- *Ralstonia solanacearum*, a phytopathogenic bacterium;
- Assembly of sequence tag sites (STS) from *Tetraodon fluviatillis*, a fish used in gene mapping;
- A European project on the mapping of mouse ESTs.

A second national center, the National Genotyping Center, has been established with funding from INSERM, CNRS and MERNT. This has the objective of defining genes responsible for common diseases. A further collaborative technology platform has been established for plant genome called Genoplante, to be established alongside Genoscope. Interestingly this is a group of companies (Rhône-Poulenc Rorer, Agro, Biogema and others) conjoined with INRA. The focus will be on the genomes of model species (*Arabidopsis* and rice), mapping in wheat, maize and colza, the three basic themes being genomics, mapping and phylogenetic resources.

Finally, Infobiogen was established in 1995 by the French government (Ministry of Research and Technology) to enhance the use and development of computing in molecular genetics and molecular biology. The primary and current activities of this organization involve sequencing, productivity, quality assurance and control. The aim is that Infobiogen will maintain a catalogue of databanks for the use of French and other laboratories. Other centers are foreseen although details of those projects have yet to be confirmed.

So, after a period of some uncertainty, genome research is now back on track.

Germany

Even though Germany had the most productive economy, and despite its long tradition of biotechnology research, its contribution to human genetics was far less than that of the United Kingdom or France, as a bibiliometric study from 1990 shows[139]. One of the reasons for this was the shadow cast by eugenics and racial hygiene[140] on German genetics and culture. The contribution of German scientists to the ideological foundation of the Nazi racial hygiene program, before and during the Second World War, was only beginning to be discussed and analyzed at the same time as the human genome program was beginning to take shape in other nations. In the eyes of the German people, the science of genetics, especially human genetics, was suspect.

Of course, the "Green" movement was another obstacle to genome research. The Greens were very wary of biotechnology in general and genetic engineering in particular. Towards the end of the 1980s, when genome research was being debated for the first time, the Greens were a rapidly growing political force, remaining so until the elections of 1990. Their influence then declined because they were not prepared to embrace the reunification of the two Germanies with the enthusiasm required. The Greens were worried about the potential dangers of genetics, in particular, the way in which its results could harm people. The AIDS epidemic, along with the growing importance of molecular genetics in the fight against disease, and

---

[139] European Science Foundation, *Report on Genome Research*, ESF, 1991.

[140] This national complex might explain the negative German reaction to the commission's proposal for a research program on predictive medicine (cf. Wolf-Mickael Catenhusen, "Genome research from a politician's perspective", *BFE*, Vol. 7, no. 2, April 1990, p. 138).

5 Human Genome Project and International Sequencing Programs 297

the realization of its positive potential in the fight against certain specific diseases, tempered this wariness and opposition. Despite the relative paucity of German genetic research, German scientists were impatient to join the world genome research effort. They hoped to build a science that would be considered (in Germany) as being of social benefit and not as a menace. This meant they would have to break with genetics' inheritance and give it a new moral image.

The German research council, the Deutsche Forschungsgemeinschaft (DFG) initiated a human genetics program in 1986, entitled Human Genome Analysis with Molecular Methods. Up and running by 1987, it has been financed by the BMBF since 1990. In September 1987, representatives of the DFG rejected a proposal prepared by a group of scientists for a concerted genome project[141]. Funding for individual projects, however, continued through a program called Analysis of the Human Genome by Molecular Biological Methods, which included the input, handling and analysis of results, technological development, fundamental genetics and support for European programs[142]. The budget was extended in 1990 for another six years at a level of DM 5 million a year.

Other proposals for the constitution of genome programs were rejected. A proposal prepared by some scientists for the Bundestag reached the Ministry of Research and Technology, but stopped there[143]. A meeting in June 1988 led to a consensus that the German effort should concentrate on the informatics aspects of genome research under EEC funding. This led to a three-part project under the aegis of the German Cancer Institute to constitute a nodal network database in Heidelberg, develop a genome database integrated

---

[141] US Congress, "Mapping our genes – Genome Projects: How Big? How Fast?" OTA BA 373, Washington D.C. Government Printing Office, 1988, p. 144.

[142] D.J. McLaren, *Human Genome Research: A Review of European and International Contributions*, MRC, United Kingdom, January 1991.

[143] D.J. McLaren, *Human Genome Research: A Review of European and International Contributions*, MRC, United Kingdom, January 1991. For a reflection on the political perception of genome research in Germany in the late 1980s, see W.M. Catenhausen's "Genome research from a politician's perspective", *BFE*, Vol. 7, no. 2, April 1990, pp. 136–139.

with the genome database at Johns Hopkins and start an initiative to identify the Open Reading Frames (ORFs) in DNA sequences.

The reunification of Germany in 1990 brought together two very different scientific research structures. Human genetics in East Germany had been focused on clinical applications. A program started in 1986 had begun to bring techniques of molecular biology to bear on the diagnosis of three of the most common genetic illnesses: cystic fibrosis, Duchenne's muscular dystrophy and phenylketonuria. East German scientists could barely contribute to the advance of genetic science due to their limited access to the necessary technology for research, but they could provide records for family structures that had excellent profiles and clinical histories. When east joined west, it brought with it a social structure that had more than its fair share of scientists.

In 1992, while the genome program was growing in strength, the euphoria of reunification faded in the light of comprehension that the two Germanies had become very different during the four decades of separation. True unification was going to take time. One of the happy results of unification was access to the Max Delbruck Institute for Molecular Biology in Berlin, an institute that had previously been part of East Germany's scientific establishment.

In 1994[144], a brochure entitled Biotechnologie 2000 was published describing the main budget lines of the Bundesministerium fur Forschung und Technologie (BMFT)[145]. A project, "Methods for the sequencing of DNA and DNA technology", had begun in the summer of 1993 with a three-year budget of DM 40 million. The emphasis was on the widest development of new methodologies for the genome program. For the period 1993–1997, DM 23 million was allocated to interinstitutional collaboration in the field of molecular bioinformatics. The BMFT attributed a further special budget for genome research to the DFG. One of the programs funded with this special budget, "Analysis of the human genome with methods of

---

[144] "La recherche sur le génome en Allemagne", *Biofutur,* July–August 1994, pp. 35–37.

[145] The BMFT is the main source of funding for research in Germany. A part of the BMFT's research budget is distributed through the DFG (*Deutsche Forschungs-gemeinschaft*) for basic research, and through the Max Planck Gesellschaft.

molecular biology" was given DM 25 million for 1991–1997, DM 7 million in 1993 being divided amongst 46 projects. The DFG was the main source of funding for the university teams mapping the human chromosomes. The main German research centers, which obtained 90% of their funding from the BMFT (the remaining 10% coming from the Länder Ministries), were also conducting genome research. Most in universities, in particular research linked to human genetics and chromosome mapping, is still being run under the specific programs of the DFG and the BMFT.

It was only towards the end of 1995 that Germany decided to contribute to the international effort to decipher the human genome. Although some human genetics research had been going on, it was being carried out by isolated teams. The new research program announced by the Federal Minister for Research, Jürgen Ruttgers, had a budget of DM 77 million for three years. Jointly committing the DFG and BMFT, it gave priority to the development of therapeutic and diagnostic applications for genetic diseases. A very important branch of the program depended on the concepts supported by Hans Lehrach, namely the massive production and large-scale distribution of cDNA (or genomic banks in the form of high density filters), each bearing tens of thousands of clones regularly dispersed by on-site robots and the centralization of results obtained from the use of these clones. Also foreseen is the establishment of a center for resources to be in charge of these operations, with several dozen researchers and technicians, heavy equipment and appropriate funding. In addition to the DM 77 million that has already been promised, the ministry will contribute to the construction of two resources centers, one in Berlin and one in Heidelberg, with another DM 20 million. The prospect of a proper German human genome project (Fig. 14 [146]), other than various collaborations to EU genome programs, has been met with highly favorable international acclaim.

Within the German human genome project (DHGP), there is a resource center which generates and disseminates genomic and cDNA libraries of a wide range of organisms and tissues. There is also a microsatellite center

---

[146] Source: "Elements of molecular medicine, Human Genome Research", Federal Ministry of Education, Science, Research and Technology, BMFB, p. 20.

300 *From Biotechnology to Genomes: A Meaning for the Double Helix*

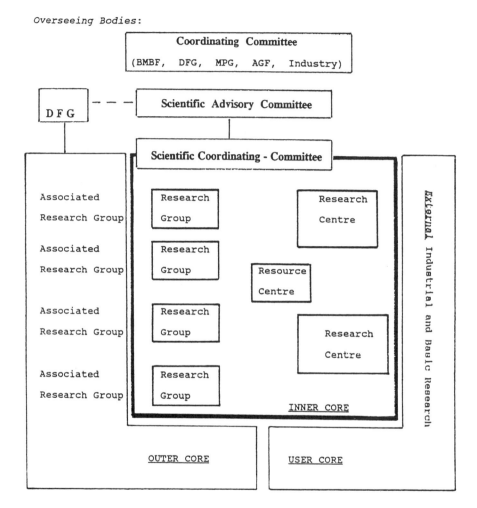

Fig. 14 Diagram of German human genome project organization

which aims to generate markers based on dinucleotide microsatellites at about a 5cM resolution. The range of organisms under study within this program is substantial and includes work on several model organisms including the mouse, rat, zebrafish and *Drosophila*. Between 1996 and 1999, 45 projects are being funded which cover a wide range of genomic research such as mapping, sequencing, expression analysis, functional analysis and bioinformatics in both human and model organisms.

Within the DHGP are a range of funding priorities: one on congenital diseases of man, another on the complete genome sequencing of two small eucaryotes (*Dictyostelium* and *Neurospora*) and another on congenital mutagenesis. A second phase (1999–2002), will put special emphasis on pharmacogenomics, susceptibility genes in diseases and wider studies on the role of genes and gene networks. The particular emphasis is on functional and expression analysis, bioinformatics and the integration of genome research into pharmacology and medicine. Within the DHGP, there are also projects specifically focused on technology development, and in this first period, emphasis has been on development of bioinformatics tools. The second program on technological development is concerned with automation, miniaturisation and procedures, particularly for functional analysis.

Belgium

There are no specific programs for genomics research funded by national organizations in Belgium. However, there are several scientists within the country who are active in the broad area of genomics and receive funding from national or international agencies and institutions.

Laboratories with EU or international funding are involved in sequencing projects on *S. pombe*, *Leishmania*, *Dictyostelium*, and *Arabidopsis*. Libraries are largely limited to small collections of specific genome regions and reside in most laboratories involved in gene mapping and identification. Most of the local databases relating to particular diseases or chromosome regions contain information which is also passed on to Genbank. There are no major programs focusing on post-genome research although many laboratories are running relevant projects.

Denmark

In Denmark, human genetics has had a long history. The Danish medical offices have maintained scrupulous archives of clinical files for decades.

Denmark also has archives of thousands of cell lines available to human genome research. Particular efforts have provided a large collection of archives on healthy families (who carry no known genetic diseases). This collection is constituted of genetic material from reference families and contains samples of red blood cells, serum, plasma, thrombocytes, lymphocytes and skin biopsies. Most of the families, though small, were large enough to be part of the family group of the CEPH. Besides this work, there is strong clinical genetic capability in Denmark.

All genetic illness was referred to one single hospital, the Rigshospitalet in Copenhagen, which drastically simplified the process of constructing a clinical register. While smaller families were less useful for drafting a linkage map, documents from genetic evaluations were a major tool for the discovery of genes responsible for genetic diseases. Danish genome efforts, therefore, continued a traditional interest in clinical genetics. The Danish government and public were more interested in genome research in the context of genetic illnesses than in mapping *per se*. However, encouraged by movements towards genome research in other countries, in 1991 the Danish set up a Human Genome Research Center at the University of Aarhus. This center, which was really the old Bioregulation Research Center with a new name, began work on 1 January 1991 with a mandate to carry out linkage and physical mapping and to characterize the mutations that cause human hereditary illnesses. The mandate also included functional studies on genes. Its annual DKK 10 to 15 million budget depended on government funds allocated to the Danish Research Council, with additional funding from the university and the private sector.

Other than the University of Copenhagen's Institute of Medical Genetics, which has a long tradition and great interest in gene mapping and sequencing, and the Human Genome Center at the University of Aarhus, the government has set up 10 new biotechnology centers and allocated some DKK 500 million to their operational budget for five years. DKK 410 million have been used to set up new research centers in the technical universities and in private firms. Like the USA, Italy, the United Kingdom, Germany and Denmark also eventually set up a national human genome research program.

Finland, Sweden, the Netherlands and Spain

While there is some genome research going on in Finland, a country that also regularly participates in international projects, for a long time Finland could not develop a concerted effort in the field. Nevertheless, there is now a policy of support to biotechnology and molecular biology. Since 1994, the Finnish Academy has implemented a genome research program with estimated funding of about US$4 million per annum over a seven-year period, with the human genome as its main target in the context of hereditary illness, gene therapy and population genetics. The study of smaller genomes within this program is still not confirmed.

Like other European nations, Sweden, the Netherlands and Spain also have developed genuine concerted genome research programs, albeit smaller in size.

Canada

In Canada, the genome debate followed those going on both in Europe and the United States. Canadian genetic science is well respected; the Canadian genetics laboratories were a clear international benchmark with powerful networks in British Columbia (around Toronto) and Quebec. Charles Scriver of McGill University, who had contributed to the involvement of the Howard Hughes Medical Institute (HHMI) in genome research, tried to repeat this success in Canada. Scriver was part of the Medical Council bureau in the HHMI. In 1985, he persuaded the other council members and some HHMI administrators of the importance of genome research and databases. He had been involved in the discussions on the emergent human genome program in the United States. Scriver was not alone in this, since Ronald Worton of Toronto, well known for his highly successful contribution to the search for the gene for Duchenne's muscular dystrophy, also sat on the HUGO council.

In Spring 1989, genome scientists met in Toronto to discuss a possible Canadian genome program. Besides its network of biotech research laboratories funded by the National Research Council (NRC), which focused on protein engineering, from 1989 onwards Canada also had 15–20 university laboratories that had the skills and equipment necessary to contribute to the genome program. The meeting's four organizers, Ford Doolittle, James Friesen, Michael Smith and Ronald Worton, drew up a White Paper that included a large list of scientists lending their support to the project. In October, the White Paper was sent to the government and to the three main funding councils. The Science Minister, in his rapid and encouraging answer, pointed out that the budgetary climate required that the funding councils implement the project with funds they already had. The National Science and Engineering Research Council (NSERC) developed a model for a predefined project to be the subject of a wide call for proposals. In June 1990, the authors of the White Paper rejected this model as being too monolithic preferring a more open project similar to that of the United States. They proposed that the Science Ministry fund it. The Medical Research Council gave its agreement to this alternative, and the NSERC and the Social Sciences and Humanities Research Council formed an Inter-Council Human Genome Advisory Committee, chaired by Charles Scriver.

In early 1991, the committee recommended that a genome program be set up in Canada immediately. "Immediately" proved to be a very relative term. As genome funding entered its third year in the United States and Italy, Canadian scientists were beginning to worry about their ability to contribute to the international genome effort as well as Canada's competitiveness in fields such as medicine, agriculture, the pharmaceutical industry and so forth. The government's delay in responding was only partially due to bureaucracy, as there were other factors to the delay, including a severe economic recession.

Eventually, the government declared genome research a priority field. On 2 June 1992, William Winegard, Minister of Science and Technology, announced the immediate creation of the Canadian genome program at the International Biorecognition conference. Ronald Worton was appointed the director of a five-year program (1992–1997) with a $12 million budget of fresh funding, a $5 million commitment from the National Cancer Institute and $5 million from the Canadian MRC. The aim was to set up a program in the fields of mapping and sequencing the human genome and genomes of model organisms[147], the collection and distribution of data and the evaluation of legal, social and ethical problems linked to genome research. It was hoped that industry might also share the financial burden.

In 1995, the Canadian government and industry put funds into a new genome research center, the Center for Human Genome Research and Molecular Medicine. Hosted by the Samuel Lunenfeld Research Institute at the Toronto Mount Sinai Hospital, it was to work on the identification of genes linked to hereditary disease, the analysis of predisposition to certain pathologies and the development of gene therapy. The main provider of funding from the industrial complex was the American firm Bristol Myers Squibb, which invested CAN$10.3 million over five years in molecular biology research. The center later attracted many other investments from pharmaceutical industrials, to facilitate technology transfer and play a catalytic role in the emergence of spin-off companies.

The Canadian genome program was thus constituted by the joint efforts of the three funding councils and the NCI. It was a new independent initiative, boasting a management committee chaired by a scientist, Worton, that included representatives of the four agencies. This autonomous program was very different from traditional research funding in Canada. Its management was an institutional innovation that fulfilled the needs of multidisciplinary research.

Japan

Genome research in Japan mainly developed from the program set up and managed from 1981 by Akiyoshi Wada, entitled "Extraction, analysis and synthesis of DNA" and supported by special funding from the Council of Science and Technology. This program, which focused on DNA sequencing, was refinanced to the tune of 450 million yen by a new department of the

---

[147] As examples of studies on model organisms under the program, one might note: the *Bacteria Sulfolobus solfataricus*, W. Ford Doolittle, Dalhousie University, the yeast *S. cerevisiae*, H. Bussey, McGill University, *C. elegans*, A.M. Rose, University of British Columbia, *D. melanogaster*, R.B. Hodgetts, University of Alberta, *A. thaliana*, B. Lemieux, York University and mitochondria, B.F. Lang, University of Montreal.

STA under the title "Generic basic technologies to support cancer research". The base of operations was moved to the RIKEN Institute at the science city, Tsukuba.

In 1985, the sequencing program set up by Wada became involved in the growing debate on a future American genome program. Following the disappointing results of the marketing efforts for Fuji sequencers and gels and Wada's appointment as Dean of the Science Faculty at the University of Tokyo, the reins for the sequencing project were passed in 1988 to Keji Ikawa and Fichi Soeda, who re-evaluated its technical objectives. In 1989, the sequencing project had produced a group of machines that allowed a throughput of 10,000 base pairs per day. A new series of projects to automate cloning and other molecular biology processes were launched under the supervision of Isao Endo of RIKEN. Endo's project would need 600 million of the STA's yen over ten years, as well as contributions from large companies. The STA also began to fund university projects, including the mapping of chromosomes 21 and 22 under the direction of Nobuyoshi Shimizu at the University of Keio. The STA genome budget for 1989 and 1990 was approximately 200 million yen and reached 1.2 billion yen in 1991. The 1991 STA program began a formally approved extension of previous pilot projects, including specific projects being led by Eichi Soeda at RIKEN, Masaaki Hori of the National Institute of Radiological Sciences, Isao Endo, also at RIKEN, J.E. Ikeda of Tokai University and Hiroto Okayama.

In January 1990, under the auspices of Y. Murakami of the RIKEN Institute, sequencing work had begun on chromosome VI of the yeast *S. cerevisiae* within the context of the European Commission's project to sequence the entire organism. In the United States, the RIKEN program was brandished as proof of the Japanese technological lead, proof that would serve supporters of the American human genome program.

The genome program became a source of tension between the United States and Japan, tension made worse by Watson's notorious comments that Japan was not contributing enough to the international human genome effort. Ken-Ichi Matsubara tried to identify and enlarge the Japanese genome effort by giving it more of an academic support base. As the director of the Institute for Molecular and Cellular Biology at Osaka University, he was also an advisor to the STA project. He hoped that the Ministry of Science and Culture, the MESC, more commonly known as Monbu-sho, might become considerably more involved than the STA. Monbu-sho financed most of the academic science in Japan, mainly within the nine major universities and the forty smaller universities scattered throughout the prefectures. Matsubara felt that Monbu-sho should make genome research a priority from the 1989 budget onwards. Monbu-sho gave a positive answer, granting genome research 300 million yen per year over two years. This sort of commitment to what was only a pilot project still did not look large enough from overseas, especially to Watson, who threatened the Japanese effort with isolation.

In 1990, genome funding in Japan was estimated at US$ 5 to 7 million. Monbu-sho's proposals for 1991–1996 included funding for post-doc travel and a program that would study the social and ethical consequences of genome research. The primary sources for research funds, namely bureaucrats and industry, were only just becoming aware of the cause and effect link between science and technology, and even in 1991, obtaining research funding was still a problem. The STA and Monbu-sho programs were hoping to grow substantially from 1991 onwards. Monbu-sho's budgetary plans, drawn up by Matsubara and the Committee of the Science Council, were slashed by officials at the end of 1990. The Ministry of Finance, in an effort to eke out government funds, reduced funding even more drastically. Monbu-sho and the Ministry of Health and Welfare finally only obtained 400 million yen each for 1991. As negotiations began for the 1992 budget, the Monbu-sho program looked as if it would stagnate well below researchers' hopes, while the STA's genome budget was increased by 50%.

The first five years, from 1991 through 1995, allowed an appropriate environment to be set up for genome research and some pilot studies to be started. Professor Matsubara's Monbu-sho grant allowed the development of fundamental research on the human genome, and the funding that reached Professor Kamehisa contributed to the development of "genome informatics". STA funding to Dr Masaaki Hori was put to the development of human genome mapping technologies. Further funding was granted to Dr Takeo Sekiya. During this five-year plan, a computer network, "GenomeNet", was set up, and a strategy was drawn up to map expressed human genes. Other major results include the physical analysis of human chromosome 21, a high-density cosmid map of chromosomes 3, 6, and 8 and the identification of several genes linked to hereditary disease.

In addition to the Monbu-sho and STA programs, the Ministry of Health and Welfare set up a program of genome research to concentrate on genes associated with human disease and the technologies to find them. Commitment on questions of an ethical, social or legal nature, although similar to that in the United States and in Europe, was to a lesser degree. The ethical program for Monbu-sho was directed by Norio Fujiki of the Fukui Medical School, but his annual budget was only of 5 million yen; Tadami Chizuka, Professor of European History at the University of Tokyo, had a very similar budget from the Ministry of Health and Welfare, most of which was for the translation of foreign documents on genome research.

In 1991, the Ministry of Agriculture, Forestry and Fisheries announced a genome budget of 621 million yen for the mapping and sequencing of the rice genome. Later, the same ministry reduced the budget to 372 million yen, questioned the commitment of Japanese industry to the project and the quantity of information that should be shared or withheld by the companies involved. In 1992, the rice genome project[148] received a large boost from the Japanese practice of dedicating a quarter of horse race profits to scientific and technological projects. The rice genome work is mainly being carried out in Tsukuba by researchers at the National Institute for Agrobiological Resources (NIAR) and STAFF, the Society for Technology Innovation of Agriculture, Forestry and Fisheries. Many industrial firms have also joined the program, including Mitsui and Mitsubishi. Apart from generous funding, the program can also boast 10 ABI 373 sequencers and four catalyst robots.

While Japanese scientific agencies whined about their poverty and pointed out the future promise of their work, and the bureaucrats of the Ministry of Finance were clipping the wings of their best scientists with their austerity policies, the prefectoral government of Chiba and some private corporations

---

[148] The Rice Genome Project (RGP) currently has three main themes that are at different levels of development: the constitution of a genetic map with the help of RFLP markers, the creation of labels from cDNA and the constitution of a physical map.

began to set up plans for the construction of the Kazusa DNA Institute. In general, advisors for the project were the same as those advising the four governmental agencies, namely Monbu-sho, the STA, Kosei-Cha (the Ministry of Health and Welfare) and MITI. Part of the Chiba project, led by Mitsuru Takanami at the University of Tokyo, was acting as a free DNA sequencing service for Japanese laboratories. Another part worked on the development of technologies, the search for and the analysis of sequences and structural biology. The DNA research institute was to be a central part of the Kazusa Science Park, a sort of magnet to attract other research institutes. Itaru Watanabe and Susuma Tonegawa managed to convince Chiba prefecture officials to support the institute, thus ensuring a 5 billion yen budget. Most of this money came from the prefecture's coffers, but there were also contributions from Nippon Steel, Tokyo Electric Power Company, Tokyo Gas, Hitachi, Mitsui Toatsu Chemicals and local banks. In 1991, the funding pool had reached 9 billion yen. Although to the outside world it appeared quite distinct from the genome program, the work being done seemed to be part of a wider master plan, which in truth didn't exist.

This budgetary saga reached its peak with the 2 billion yen (US$ 15 million) granted by the Japanese government in 1992 for non-agricultural genome research. This is compared to the US$160 million allocated to such research in the United States. It was an increase on the 1991 budget, bringing Japan to a level equivalent to that of the United Kingdom and making the proportion of money spent on genome research per GNP the highest outside the USA. In 1993, the United Kingdom moved to second place due to a large contribution by the Wellcome Trust.

During this time, the mighty MITI was preparing its own strategy and plans, hoping to benefit from interest in the genome project in order to galvanize groups not yet involved in biological research. A MITI official had put forward the idea of a genome program as early as 1987. Sumitomo Electric had heard him and reacted with enthusiasm. Michio Oishi of the University of Tokyo was appointed the academic contact to plan MITI's incursion into genome research, which would allow the participation of companies in other sectors. The academic plans for the projects were drawn up by Oishi while industrial support was organized by the Japan

Biotechnology Association, a private organization acting as an interface between MITI, academia and industry, incubating ideas and contributing through its activities to a consensus on the new initiatives. The main idea of MITI's project was to improve research for new instrumentation, to cultivate interest in fundamental biology and to encourage long-term commitment from a consortium of companies with a mixed balance of government and private financing. MITI gave its blessing to Kazusa's DNA research institute, thus conferring stature and credibility, although no actual MITI funding.

While the institute (Fig. 15 [149] shows the main research centers, institutes and universities involved in genome research) was opening its doors, it seemed possible that MITI would unleash a substantial effort on its own ongoing operations and initiatives over the following years.

The genome project engendered many programs at various ministerial levels which, in general, had separate planning and budgetary processes. Wataru Mori, a member of the Science and Technology Council which advised the Prime Minister, tried to set up a committee coordinating the various programs. He established a genome committee of scientists associated with the genome programs of various agencies, trying to overcome the fragmentary nature of the Japanese effort, directed as it was by bureaucrats with little comprehension of the science underlying genome research or its importance. In April 1995, the Bioscience Subcommittee of the Science Council of Monbu-sho published its plans for the continuation of the Japanese human genome project in the universities over the next five years (1996–2000). The report underlined the importance of large scale cDNA sequencing, functional analysis and bioinformatics. Specific financing will be allocated to research teams and to consolidate work being carried out at research centers.

Monbu-sho has already approved funding for Professor Yoshiyuki Sakaki's group working on the program "Genome Science: New frontiers in biology" which has an annual budget of 600 million yen. Three projects will be

---

[149] Map taken from the article entitled "the human genome project in Japan: the second five years", *Genome Digest*, vol. 2, no. 4, October 1995, p. 6.

5 Human Genome Project and International Sequencing Programs 311

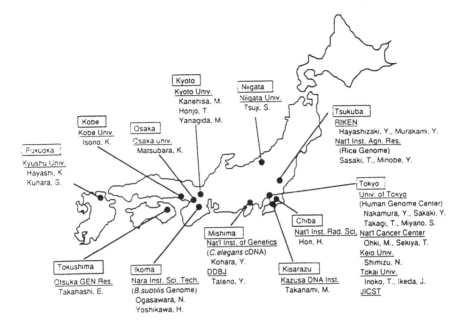

Fig. 15 Map of Japanese major genome research groups

supported: an analysis of the structure of the human genome (Dr Misao Ohki), functional analysis of genomes (Dr Yuji Kohara) and bioinformatics (Professor Minoru Kamehisha).

The "Japan Information Center for Science and Technology" has launched a three-year program for sequencing the human genome with a budget for the first six months of 520 million yen. Projects from the Hidetoshi Inoko, Yisuke Nakamura, Yoshiyuki Sakaki and Nobuyoshi Shimizu groups have been approved. Details on another STA project are forthcoming. The project supported by Kosei-sho will continue under the direction of Dr T. Sekiya, with an annual budget of 200 million yen.

On the basis of the pioneering accomplishments of the previous five-year projects, and taking into consideration the report of the Bioscience Subcommittee of the Science Council of Monbu-sho, the Creative Basic Research "Human Genome Analysis" (representative: Professor Kenichi Matsubara) and the Scientific Research on Priority Area 'Genomic

Information' (representative: Professor Minoru Kanehisa), Monbu-sho has started a new five-year project named 'Genome Science' (1996–2000). The budget for 1996 was 60 million yen (about US$5 million). The 'Genome Science' project consists of the following three research groups pursuing the following new goals:

- Structural analysis of the human genome (group leader: Misao Ohki, Research Institute of National Cancer Center). The tasks of this group are to construct sequence-ready high-resolution physical maps and finally a sequence map of disease-related loci such as those of chromosome 21q, 11q and 1p. As a result, the group aims at discovery of new disease-related genes primarily concerning cancer and neurological disorders. The group will also develop the technology and methodology of genome analysis needed for these purposes.

- Functional analysis of the genome (Group leader: Yuji Kohara, the National Institute of Genetics). This group first aims to develop the methodology and technology needed for exploring the function of many novel genes which will be discovered by genome sequence analysis. Ultimately, it aims to clarify the roles of individual genes in the networks and cascades of genetic information. For this purpose, the group will at first do advance analysis of genomes of experimental organisms such as the *Bacillus subtilis*, yeast and nematodes as well as 'body mapping' of the mammalian genomes.

- Biological knowledge and genome information (Group leader: Minoru Kanehisa, Kyoto University Institute of Chemistry). In view of the coming era of a flood of sequence data, this group aims to develop the data processing technology needed to discover the biological meaning of the vast sequence data, to construct various databases for linkage, integration and systematization of the information obtained through genome analysis with the existing knowledge processing systems for utilization of the databases.

In addition to the above groups' research, several proposals from individual investigators having excellent ideas, materials and methods have been selected. These groups and individual researchers have been coordinated by the coordination group (represented by Yoshiyuki Sakaki). This group is also responsible for the ethical, legal and social issues related to the genome project.

In July 1997, the Science Council of Japan prepared a report on the basic plan for the promotion of life sciences in Japan, in which the various genome studies have been included as the fundamental key for the development of the life sciences in the 21st century in Japan. Based on this report, the ministries and agencies are now reorganizing their activities as indicated in Fig. 16. First, a reorganization of "the Human Committee" to "the Genome Science Committee" should be noted. The new committee, chaired by Dr. Masami Muramatsu, consists of leading genome scientists in Japan and will play a central role in the advancement of genome projects in Japan by establishing a long-term plan for genome science and coordinating the activities of each ministry and agency. Among other new activities, the following development should also be noted.

STA is setting up a new 'genome center' at RIKEN, in which three projects will be conducted: i) the large scale sequencing and comparative analysis of the human genome; ii) the development of a comprehensive 'gene encyclopedia' of the mouse; iii) comprehensive investigation of protein structure.

Monbu-sho has set up a very high-performance supercomputer at the Human Genome Center at the Institute of Medical Science, University of Tokyo. The equipment includes two 128-processor Cray Origin 2000 systems, a Hitachi SR2201 and a Sun Ultra Enterprise 10000. These systems will make the center one of the largest configurations for genome analysis in the world.

- The Ministry of Agriculture, Forestry and Fisheries (MAFF) will start the second phase of the Rice Genome Project, focusing on the sequencing analysis of the genome.
- MITI will start a new program called "Genome Informatics Technology", with the objective of developing informatics and related technologies required for post-sequencing genome study.

314 From Biotechnology to Genomes: A Meaning for the Double Helix

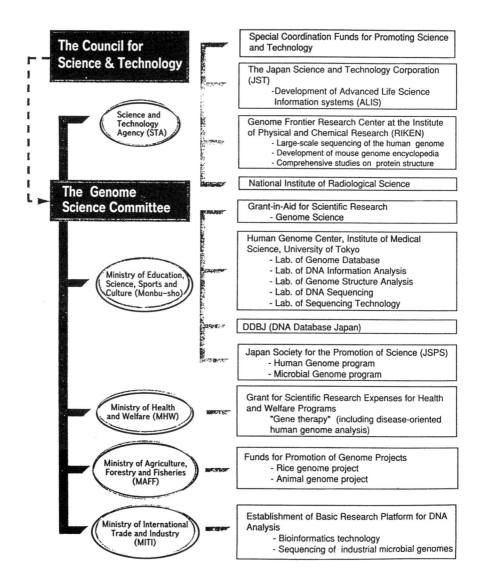

Fig. 16 New organization of Japanese genome projects

In addition to these new developments, some non-governmental organizations such as Kazusa DNA Research Institute, Helix Institute and Genox Institute are focusing on cDNA analysis for drug-discovery purposes.

The Japanese genome project is now at a pivotal stage in the establishment of a solid foundation for the 21st century. Despite the commitment from the Japanese government, the future of genome research remains uncertain in Japan. Will it be dominated by researchers from the academies hoping to raise Japan to the summit of world research or by local governments whose fiscal bureaucracy is opening the door to new Japanese regionalization, or by the Japanese Ministries, which hope to widen their mandates and power by duplicating industrial policy that has succeeded so well in the fields of electronics and the automobile? Only time can tell.

China

The Chinese human genome program seems to be developing slowly, but there is a program on the rice genome that was described in *Probe* magazine[150].

From a US$ 4 million budget for the five first years, with the aim of mapping and sequencing the rice genome program (a program which should be finished in about twelve years) the project should help improve Chinese agriculture. Ongoing research focuses on:

- Localizing, characterizing and sequencing biologically important genes;
- Characterizing and mapping genes important in agriculture and applying these results by growing the product;
- The establishment of an ordered set of DNA clones to use for the construction of physical maps of the chromosomes;
- The elucidation of gene expression mechanisms;
- The establishment of an internationally connected informatics center.

---

[150] "China's rice genome research program", *Probe*, vol. 3, no. 1/2, January–June 1993, p. 7.

316   From Biotechnology to Genomes: A Meaning for the Double Helix

Table 8 shows how this Chinese program is set up[151].

A real human genome project in China was initiated only at the beginning of 1994. The first phase of the project is sponsored by the National Natural Sciences Foundation of China (NSFC). The 19 laboratories constituting the Chinese human genome community are also funded by the State Commission of Education (SCE), the Ministry of Public Health and local governments such as the Shanghai Municipality.

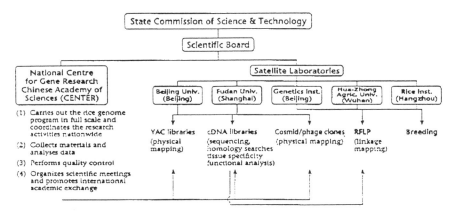

Table 8 Organizational chart of China's rice genome project (from the *Rice Genome Newsletter*, Vol. 1, No. 2, 1993)

This project is supervised by an advisory committee composed of highly respected leaders while an academic committee, chaired by Drs Boqin Qiang (Beijing) and Zhu Chen (Shanghai) is responsible for coordination of nationwide efforts. A secretariat is affiliated to the academic committee to take care of the academic arrangements, document preparations and daily communications.

The first phase (FY 1994–1997) of the HGP in China is composed of three main programs: I. Resource Conservation; II. Technology and

---

[151] Taken from "China's rice genome research program", *Probe,* vol. 3, no. 1/2, January–June 1993, p. 7.

Development; III. Studies on Disease Genes. Through the efforts of the participating labs, some progress has already been achieved over the past four years.

Resource conservation

[P.I.: Prof. Jiayou Chu (Institute of Medical Biology, CAMS, Kunming) and Prof. Ruo-Fu Du (Institute of Genetics, CAS, Beijing)]

China's population of 1.2 billion people represents mankind's largest human gene pool. The Chinese population is composed of 56 ethnic groups, of which the majority is Han. The Han, together with the 55 minority groups and numerous genetic isolates, provides invaluable genetic resources for human genome diversity research and gene cloning. However, due to the increasing mixture of different ethnic groups, there is an urgent need to collect and preserve diverse cells and genetic materials before they blur.

Since 1994, through the joint efforts of researchers at the Institute of Medical Biology, the Chinese Academy of Medical Sciences (CAMS) (Kunming), the Chinese Academy of Medical University, Harbin Medical University (Prof. Chun-Xiang Lui, Harbin), the Institute of Genetics, the Chinese Academy of Sciences (CAS) (Beijing and the Institute of Genetics, the Chinese Academy of Sciences (CAS) (Beijing) and the Institute of Zoology (CAS, Kunming), blood samples have been collected from two Han reference groups and 10 selected minority ethnic groups distributed in Yunnan Province in southwest China and Heilongjiang Province in northeast China. Over 500 EBV-transformed lymphoblastoid cell lines have been established from these samples following internationally standardized methods for both sampling and manipulation. Biodiversity studies have been initiated on selected loci including the mitochondrial DNA, Y chromosomel and other genomic markers. Recently, genetic diversity was also analyzed using microsatellite markers covering each of the 24 human chromosomes; genetic distance was measured according to the allele frequencies. Additionally, special attention has been paid to the collection of large families with a distinct disease or certain phenotypes.

Technology and informatics development

[P.I.: Prof. Zhu Chen (Shanghai Municipal key laboratory of Human Genome Research, Shanghai)]

It is essential for this project to modernize facilities, update techniques for human genome research and expand resource centers in order to make the best use of current limited equipment and experimental resources.

The Shanghai Municipal Key Lab. Of HGP has established a YAC resource center in collaboration with CEPH (France) and with the support of the EU-China Biotechnology Program. Other genomic libraries (YAC, Cosmid), as well as various cDNA libraries, have also been constructed with Chinese materials by researchers from the Institute of Cell Biology (CAS, Shanghai), Hunan Medical University (Changsha), the Institute of Basic Medical Sciences (IBMS) (CAMS, Beijing) and the Shanghai Institute of Cancer Research (SICR, Shanghai).

The development and improvement of tool enzymes such as rare cutters has been undertaken by Prof. Jiang-Gang Yuan's group at IBMS. Techniques to develop STS, EST and other genetic markers have been established by Fudan University, IBMS and SICR. These research groups have also provided GenBank with the first batch of novel markers from China. Additionally, techniques for gene identification such as genotyping, FISH, CGH, microdissection, differential display, subtractive hybridization, exon-trapping and cDNA selection have been successfully used for the other HGP programs in China. The informatics network has been set up by Prof. Runsheng Chen in the Institute of Biophysics (CAS, Beijing) and has been connected to the international information centers.

Studies on Disease Genes

[P.I.s: Prof. Ren-Biao Chen (Shanghai Second Medical University, Shanghai) and Prof. Yan Shen (IBMS, CAMS, Beijing)]

This project consists of two aspects: identification of disease genes and studies on genetic diseases having high frequency in China.

Cloning of cancer-related genes is a major task of this project. Several novel genes or complete/partial cDNA clones have been identified to be associated with esophageal cancer susceptibility by Prof. Min Wu's group at the Institute of Cancer Research (CAMS, Beijing). Several novel genes and a number of ESTs with deletion and/or altered expression have been cloned from hepatic cancer by Prof. Jianren Gu and by Prof. Da-fan Wan's group at SICR (Shanghai).

Molecular cloning of the genes associated with exocytosis has been carried out by Prof. Hanxiang Deng's Lab. in Human Medical University (Changsha). Two fusion genes, PML-PARα and PLZF-PARα, involved in the specific chromosomal translocations t(15;17) and t(11;17), respectively, in acute promyelocytic leukemia, have been published by Shanghai Municipal key lab. of HGP (EMBO J. (1993) 12, 1161 ; PNAS USA (1994) 91, 1178; PNAS USA (1996) 93, 3624). This group has also cloned a series of novel genes named RIGs (retinoic acid inductible genes). Regions known to have genes involved in human cancers such as 17q12-21 and 13p14 have been selected for regional positional cloning and have yielded ESTs and other sequences by Prof. Long Yu's group at Shanghai Fudan University (Shanghai) and Prof. Guoyang Liu's group at IBMS, CAMS (Beijing). In addition, in an attempt to approach complex traits and studies on non-insulin dependent diabetes mellitus have been initiated by the Shanghai Institute of Endocrinology.

With the goals of developing gene diagnosis and/or further gene therapy in China, several known disease loci; i.e., FMR-1, Wilson's disease, β-thalassemia, Marfan syndrome and G-6-PD deficiency, have been selected by IBMS (CAMS, Beijing), the Shanghai Institute of Medical Genetics (SIMG), and the departments of Medical Genetics of Shanghai Second Medical University and of Sun Yat-sen Medical University, respectively, with the purpose of identifying mutation patterns in the Chinese population and exploring possible heterogeneity. Progress has been made in these diseases. In β-thalassemia, Prof. Yi-Tao Zeng's group at SIMG has succeeded in modulating the splicing patterns of β-globin gene with IVS-2-654 G to T mutation to increase the synthesis of normal β-globin gene with hydroxyurea (*Brit. J. Haematol.* (1995) 90, 557).

Prof. Yan Shen's Lab. at IBMS, CAMS (Beijing) has focused on the alternative splicing patterns of FMR-1 gene in different tissues of the brain and identified dozens of STSs or ESTs in this region. Prof. Chuan-Shu Du's group in Yat-sen Medical University has identified seven mutations of G-6-PD gene in the Chinese population.

Issues related to HGP

In addition to the three main programs mentioned above, efforts have also been made to address issues related to HGP.

Programs have been set up to educate the public and train local medical officers and clinicians regarding the significance of genome research as well as to provide knowledge about medical genetics. Different educational methods have been used including new media, lectures and courses organized by scientists from the Chinese genome community, together with biomedical and scientific bodies such as the Chinese Medical Association, the Chinese Genetics Society of China and the Chinese Society of Medical Genetics.

Considering the great impact of HGP on society, the ethical, legal and social issues have been taken very seriously and the guidelines of HUGO, WHO and UNESCO have been followed. More than 36 Chinese scientists have been active as HUGO members, demonstrating that human genome research in China has become an integral part of the international efforts of HGP.

International Organizations and Cooperation

In 1986 and 1987, the genome debate was expanded by the media. It was also widening geographically with a growing number of nations jumping on the bandwagon. The need for international cooperation became more and more obvious. The response to this need was the holding of international conferences, for scientists a knee-jerk reaction to this kind of situation.

The first major international conference on human genome research was organized by Santiago Grisolia at the Institute of Cytology in Valencia,

5 Human Genome Project and International Sequencing Programs 321

Spain. Grisolia, a biochemist intrigued by genome related research, was fascinated by the personalities involved in the genome debate that was going on in the United States. During the summer of 1987, he began to organize a sumptuous conference in Valencia. He had originally planned to invite about 50 scientists in positions of influence with regard to international cooperation. However, the conference took on a life of its own as a growing number of scientists influential in their own countries expressed their wish to attend.

The Workshop on International Cooperation for the Human Genome Project took place from 24 to 26 October 1988. It turned out to be a sort of evaluation of what could really be hoped for from mapping efforts, and more specifically, genome sequencing efforts. It revealed a large number of similar efforts in various nations progressing towards the establishment of large scale genome projects. Most of the participants learnt about the Soviet and Japanese efforts for the first time. The first French program was then but embryonic. The workshop produced the Declaration of Valencia, which encouraged international cooperation, although the precise mechanism for its implementation caused a minor controversy[152]. A first version of the declaration urgently called for HUGO and UNESCO to commit themselves to the cause. Victor McKusick and James Wyngaarden chaired the final plenary meeting, during which the declaration was discussed and the final document drawn up, only mentioning HUGO.

UNESCO had a new director-general, Frederico Mayor, a Spanish biochemist. He wanted to renew UNESCO's commitment to science and restore the reputation of the organization that the Americans and British had left in 1984, complaining about its slowness, its overly complex bureaucracy and its reluctance to accept the new world order of information and communication. Mayor saw that genome research and other scientific efforts were a good opportunity for reconciliation, as scientific programs are a little less sensitive and politicized that many other UNESCO-funded programs. In April 1989, the United States Congress held a hearing on whether the USA

---

[152] L. Roberts, "Carving up the human genome", *Science*, 242, 2 December 1988, pp. 18–19.

should rejoin UNESCO. The genome project was referred to frequently as an example of UNESCO's involvement and was listed as the first candidate amongst the life science disciplines. A year after a US State Department affirmation of its refusal to contribute to UNESCO, Mayor came to Washington hoping that American policy could be changed. UNESCO continued to obtain positive feedback on its reforms, but the difficult environmental climate and an isolationist, nationalistic attitude prevalent at that time in the USA undermined support for an American return to UNESCO.

Following a meeting in February 1989 of the "genome advisors" of Europe, the USA, Japan, the Soviet Union and Australia, Mayor set up a scientific coordination committee to lead the UNESCO genome project. The latter began to take shape during meetings organized in Paris in February 1990 and Moscow in June 1990. The UNESCO program, with a budget of US$ 260,000 over two years, focused on training scientists from countries which would otherwise have been unable to contribute to genome research and on training visits abroad. UNESCO also contributed funding to several international meetings in 1990 and 1991. The UNESCO program provided funding to young scientists from the third world, grants co-funded by the Third World Science Academy, to allow them to travel to Asia, Europe or the USA, and to train in laboratories. The candidates' applications were evaluated after a meeting in Valencia in November 1990. In the first year of the program 16 grants were allocated; this is now an annual event. UNESCO associated itself with the Third World Science Academy to build a directory of centers interested or involved in genome research.

Jorge Allende, an energetic Chilean biochemist, had many connections with scientists in Europe and the United States. He contributed to the adoption of a resolution[153] to promote genome research during a meeting of the Latin American Network of biological science at Quito, Ecuador, in mid-1988. The resolution called on developed countries to ensure that the genome project also included contributions from developing countries such as those of Latin America. Participants at the conference had asked Allende to pursue

---

[153] This effort led to the creation of the Mexican and Brazilian genome programs.

5 Human Genome Project and International Sequencing Programs 323

his proposal for a regional workshop on genome research in Santiago (Chile), and they also requested partial funding from the UNESCO. In June 1990, a workshop entitled 'Human Molecular Genetics and the Human Genome: Perspectives for Latin America' brought together scientists from 12 Latin American nations with scientists from the United States, officially launching the Latin American Human Genome Program.

Allende also brought out a special edition of the *FASEB* journal dedicated to world genome research in which he described Latin-American efforts to promote international training and collaboration[154] with genome efforts emerging in technologically advanced countries[155]. He promised to set up an efficient mechanism for north-south cooperation. The workshop produced another resolution of similar intent and continued requests for UNESCO support. UNESCO was hoping to see the constitution of similar regional networks in Africa, southern Asia and eastern Europe. In order to encourage genome research in developing countries, UNESCO sponsors international meetings and co-funds the north-south conferences. At the second north-south conference in Peking in November 1994, the delegations from Brazil, Kenya and Shanghai made impressive presentations, respectively, on techniques of molecular biology linked to genome research, studies on the genome of the parasite *Theileria parvan* and mapping and cloning the region on which the gene for muscular dystrophy is located.

Third world populations are of central importance in the understanding of man's origins and genetic diversity. Consanguinity levels of sometimes up to 20% are not unusual in some regions, making recessive gene-linked diseases more common and, therefore, easier to detect. Clear records of consanguinity also provide opportunities to map genes. The technique depends on the availability of genetic maps to compare parental chromosome regions. If patients with a disease systematically inherit the same chromosomal region, the gene for the illness logically lies on that portion of the chromosome. A gene can, therefore, theoretically be mapped by studying very few patients.

---

[154] J. Allende, "A view from the south", *FASEB Journal*, 5 January 1991, pp. 6–7.

[155] J. Allende, "Background on the human genome project", *Red Latinoamericano de Ciencias Biologicas,* 28 June–1 July 1988.

Large families for this sort of gene search can most often be found in the third world simply because the populations are higher. As mentioned earlier, the search for the gene linked to Huntington's disease was greatly accelerated by the study of large families in Venezuela. The pathologies of hemoglobin were first studied in those living in the "Malaria Belt" (the Mediterranean basin, Southeast Asia, and a part of the Middle East) or in people whose ancestors had lived there. The call for the involvement of third world countries was therefore not a disinterested gesture. It was also a necessity for those countries where health problems were at their worst and where methods of genetic diagnosis could be of great use, on condition that they were cheap and sufficiently trustworthy. Without coordination, developed countries would ignore efforts to study diseases afflicting people in the third world. Those aiming to develop technologies that would render the tests simpler and less expensive would find their efforts stagnating without the help of more technologically advanced countries.

HUGO provided great hope for a durable support mechanism for vigorous international collaboration flexible enough to avoid being crushed by bureaucracy. But it soon appeared that the international direction hoped for and recommended by HUGO was nothing but a "Christmas wish", as no political act of will answered that call for cooperation.

There is also a human genome diversity project, a complement to the human genome project under the auspices of HUGO, which aims to obtain a more precise definition of the origins of the different world populations by integrating genetic knowledge with historical, anthropological and linguistic studies. It should afford us better understanding of the primordial roots of the human race, the history of man's biology, population movements and sensitivities or resistances to various human diseases.

Since 1994, an international network has been running to sequence the genome of the parasite *Leishmania*, a flagellant protozoa of the Trypanosome family, under the aegis of WHO. Twelve million people worldwide are thought to be suffering from this parasite, with 400,000 new cases per year. This cooperation remains more political than financial, however, since most of the teams involved are operating with national or private funding.

The terms "genome program" and "genome project", often used to designate all the work being done in the field, are in fact quite incorrect. They designate no international budgetary or administrative reality, but are the juxtaposition of national or regional initiatives, like the American Human Genome Project, the British Human Genome Mapping Program, or research being done in France under the aegis of GREG and in the Genome and Health program. These various public programs have been supported by influential groups already well established as institutions including the NIH, the DOE, the Howard Hughes Institute in the USA, the INSERM, CNRS and AFM in France, the MRC, the ICRF and the Wellcome Trust in the United Kingdom, the National Institute for Radiological and Human Sciences, the Genome Center, MITI and Monbu-sho in Japan.

HUGO's apparent failure, or even relative failure, to deliver the international cooperation expected of it reveals the balance between two contradictory arguments in genome research: internationalizing the quest for knowledge (cooperating in the process of research versus national appropriation of economic advantage) and competition for the exploitation of results.

Of course, international scientific cooperation does exist, for example the *S. cerevisiae* and the *C. elegans* projects, LABIMAP, the agreement between Washington University at St Louis, USA, and RIKEN in Japan and the *Drosophila melanogaster* and *Arabidopsis* projects. It is, however, undeniable that there is a preponderance of the national dimension in public policy, revising the concept of molecular biology as a "heavy" science. According the OECD, molecular biology will soon enter a new phase, that of programmed "heavy" science. If you take the exploration of human DNA as an example, a study generally considered to be "heavy" science, you can see that it is not, for all that, too heavy for a country such as the USA to undertake alone. Genome sequencing would then be in a sort of category of its own, a sort of national "heavy" science enterprise. The call for the internationalization that HUGO supported was more directed at the grouping of intellectual means than financial ones.

While more and more nations put genome research[156] into their science policy, the perspectives opened by genome research seemed limitless for medicine and biotechnology as well as for economic and industrial growth. Given the activities being undertaken by the large nations, Europe had to react. By the end of the 1980s, Europe was not really considered a trustworthy or necessary partner in the various projects under way in the United States and Japan[157], where it was eyed with economic wariness. The European Commission, on the initiative of some of its civil servants, put genome research into its agenda for biotechnology; and in the predictive medicine program (studies on the human genome), it set up European networks dedicated to sequencing model organisms. This decision allowed Europe to maintain its position and competition in the world biotechnology race in human and non-human genome studies, the localization and study of genes and in the exploitation of these discoveries.

---

[156] In 1995, 16 countries (not counting the EU), that is to say, Brazil, Canada, China, Denmark, France, Germany, Israel, Italy, Japan, Korea, Mexico, the Netherlands, Russia, Sweden, the United Kingdom and the United States could boast a national genome program.

[157] *First report on the State of Science and Technology in Europe,* Commission of the European Communities, 1989. J.M. Claverie, *Genome mapping and sequencing, and biotechnology databases in the USA, strategic situation and risks,* January 1990, (report drawn up following a journey to the United States in December 1989).

6
European Biotechnology Strategy and Sequencing the Yeast Genome

While systematic genome mapping and sequencing programs were being set up in many countries, the European commission hoped to bring genome research going on in various member states together into a coherent whole. This intention emerged very early during the genome debate as the result of a convergence of interests of the member states, their desire not to fall behind in the genome race and their increasing awareness of the importance of genetics and new techniques of molecular biology.

During 1986 and 1987, several proposals were made for genome research programs by scientists and civil servants of the European Commission. Sydney Brenner had been very much involved in the American genome debate. He had made great contributions to the establishment of the British genome program and was asked to represent the opinions of British mappers and sequencers during the USA NRC study on the need for the human genome project. He had also suggested to the British Medical Research Council that a unit of molecular genetics be set up that would include genome research. In a short proposal received on 10 February 1986[1], he also alerted the commission to the issue.

Around the same time, André Goffeau, now a professor at the Université catholique de Louvain (Belgium) and a science officer at the European

---

[1] Sydney Brenner, *Map of Man* (First draft sent to B. Loder), European Commission, 10 February 1986.

Commission, put together his plan for sequencing the genome of the yeast *Saccharomyces cerevisiae*. Other sequencing projects at the community level were later to see the light of day — projects on the human genome and the genomes of *Bacillus subtilis*, *Drosophila melanogaster*, *Arabidopsis thaliana* and the mouse. These projects received scientific, budgetary and political support through various European biotechnology programs. The Fourth Medical and Health Research Program (MHR4) implemented by the European Commission under the Second Framework Program (1987–1991) stipulated that a new program would be set up for the analysis of the human genome. Following Brenner's proposal, and after much discussion and reformulation of the project, a program for the analysis of the human genome was adopted and launched for an initial phase from 1990–1992.

The time at which these programs were proposed was politically and strategically favorable. First of all, at a scientific level, new techniques were making it possible to consider tackling the large genomes. At a political level, a recognition of the importance of biotechnology in the maintenance of European scientific, industrial and economic competition in the future, prospects opened by genome research for bioindustry, and most of all, the promise of future markets to be won, required rapid reaction at the community level. This reaction took place in a climate of a deliberate political intention to set up an integrated approach to the complex multidimensional domain that biotechnology represents.

6.1 Towards a New European Research Policy for Biotechnology

"The need for an integrated policy" in the field of biotechnology had, in theory, been recognized. In fact, the phrase was the title and the message of a European Parliament proposal adopted in February 1987[2]. This resolution

---

[2] *European Parliament (February 1987) Resolution on biotechnology in Europe and the need for an integrated policy* (the Viehoff Report); Doc. A2-134/86, *Official Journal of the European Communities*, C/76/25-29, 23 March 1987.

6 European Biotechnology Strategy and Sequencing the Yeast Genome 329

and report was the fruit of sixteen months of work, debate and elections involving six different parliamentary committees, respectively those of Research, Industry, Agriculture, Environment, Consumer Protection, Social Affairs and Citizen's Rights. It was an elegant piece of rhetoric. Fulfilling "The need for an integrated policy" it mentioned was a little more difficult.

In the early 1980s, there was fear that Europe would lose its place in international economic, industrial and scientific competition. A group of European Commission officials reacted to this fear, among them Mark Cantley, responsible for the social aspect of the Forcasting and Assessment in Science and Technology (FAST) Group report, who was trying to get this message across. A provisional report entitled "Biotechnology and the European Community: the Strategic Challenges", was drawn up by Cantley, his colleague Ken Sargeant and Ricardo Petrella, the Director of the FAST Program. It was discussed by the community in June 1982. The report suggested that biotechnology had become important for three reasons:

- The development of multi-sectoral applications;
- The easing of resource constraints it implied;
- The global challenges of the USA and Japan with the potential to help the third world.

These three points were considered relevant to the biological revolution of the previous twenty years, a revolution that had involved the domestication of microorganisms, plants and animal cells. A joint committee was set up between DG III (the Directorate General handling the internal market and industrial affairs) and the FAST group, with Cantley as Secretary. On 8 February 1983, a few months after the publication of the final FAST report, Gaston Thorn, who was then President of the European Commission, made a speech including a one-liner announcing that "the Commission would take the same approach for biotechnology as it had done in the so-called ESPRIT program for stimulating information technology". It was this one-liner that guaranteed the necessary official support. Two days later, a meeting was held between DG III and DG XII to negotiate a way to make this come true. Since agricultural applications were considered important, and DG III was only involved in industry, DG XII, with the backing of the FAST program and its neutral position vis-a-vis the great differences between industry and agriculture, was ideally suited to take the lead in the forthcoming initiative.

An integrated approach had already been proposed in the 1983 *Communications on Biotechnology*[3], which strictly reflected the recommendations put together from 1979 to 1983 by the FAST group[4] in collaboration with other units, particularly with the unit handling biological research. The 1983 communications on biotechnology list, six priority courses of action which should involve at least the following five general directorates:

- Research and training (DG XII);
- New regimes on agricultural outputs for industrial use (DGs III, VI);
- A European approach to regulation affecting biotechnology (DGs III, V, VI, XI and XII);
- A European approach to intellectual property rights in biotechnology (DGs III and VI);
- Demonstration projects at the interface between agriculture and industry (DGs III, VI and XII);
- Concertation (of an interservice nature, hosted administratively by DG XII).

To support this six-point strategy, in February 1984, the Commission created an internal structure under the Biotechnology Steering Committee (BSC), which was open to all services in accordance with their interests. The Concertation Unit for Biotechnology, or CUBE, was set up as a secretariat for the BSC by redeploying the "biotechnology" team of the FAST program.

On 12 December 1984, the Commission discussed the need for a lobby group with the 17 leading firms in bio-industry. In addition to Vice-President Etienne Davignon and Directors General Fernand Braun of DG III and

---

[3] *Biotechnology, the Community's Role*, Communication from the Commission to the Council, COM.(83)328, 8 June 1983. *Biotechnology in the Community*, Communication from the Commission to the Council, COM. (83)672, October 1983.

[4] FAST Programme, *Results and Recommendations*, December 1982. FAST Occasional Paper no. 62, *A Community strategy for biotechnology in Europe*, March 1983.

Paolo Fasella of DG XII, the meeting was graced by Sir John Harvey-Jones, President of the Council of the European Chemical Industry and President and Chief Managing Director of ICI, Jacques Solvay of Solvay, Hans-Georg Gareis, Research Director at Hoechst, Callebaut of Amylum, Loudon of Akzo, Hilmer Nielsen of Novo, Alexander Stavropoulos of Vioryl in Athens, John Jackson of Celltech and others. Monsanto Europe did not take part. Dieter Behrens of Dechema was present in recognition of his leading role in the European Federation of Biotechnology and his work with FAST.

Despite a telling silence as to any possible industrial association, it was recognized that the Commission would need an industrial advisor. In June 1985, therefore, at the behest of the The Council of the European Chemical Industry (CEFIC), the European Biotechnology Coordination Group (EBGC) was born of the cooperation of five associations, CEFIC for the chemical industry, the European Federation of Pharmaceutical Industry Associations (EFPIA), the Confederation of Agro-Food Industries (CIAA), the International Association of Agrochemical Industries (GIFAP) and the Association of Microbial Food Enzyme Producers (AMFEP). There was considerable reluctance on the part of some of these associations, in particular the EFPIA, to undertake anything more sophisticated than the EBCG, which had neither a chairman nor a budget.

The EBCG was created as a component of a coherent structure within which a biotechnology R&D subgroup was to be set up under the auspices of CEFIC as part of their already highly active research groups. With regard to regulation, the EFPIA led the debate through a body known as the European Committee on Regulatory Aspects of Biotechnology (ECRAB). Regarding patents, The Union of Industries of the European Community (UNICE), as the general body representative of industry in the Community, already had a working group that could take into account all needs for action by or response from industry.

During the years that followed, sectoral European associations joined the EBCG. One such was FEDESA, the association for the animal health industry, which was set up in March 1987 to provide representation of a sector troubled by the sad tale of steroid, or analogue hormones, which had drawn public and political fire (not appreciated by the Commission at a time when the managers of the Common Agricultural Policy were struggling with growing problems of surplus). The development of large-scale production, and later (but foreseeable) dispersion of genetically modified micro-organisms and plants, fed debate on the nature of contemporary regulatory approaches and structures. These debates took place in the United States and in several European countries amongst groups of national biotechnology experts and amongst the Commission's civil servants.

In all the countries this involved debate cried out for an inter-agency or interministerial debate forum. For the Commission, the forum took the shape of the Biotechnology Steering Committee (BSC). DG XII, with a few additions from other DGs, was the main origin of the Commission *Communications on biotechnology*. The BSC was mainly interested in obtaining funding for R&D programs and managing them efficiently through the development of a systematic approach by the framework programs. The FAST program had produced wise strategic analyses but remained firmly rooted in DG XII. It was not an autonomous Commission institution, as had been advised by the "Europe + 30" group[5]. Their recommendations would not be implemented unless they coincided with other powerful interests in the Commission's services or beyond. The creation of a separate Forward Studies Unit by the President of the Commission and a Science and Technology Options Assessment (STOA) group by the European Parliament made it clear how non-central and non-neutral FAST really was. None of these initiatives managed, during the 1980s and early 1990s, to get close to the central and authoritative position occupied in the United States by the Office of Technology Assessment. DG XII's vast scientific network and its CUBE unit had little weight in the conflict over regulation.

The Directorate General for the Environment, DG XI, began to participate in the BSC from mid-July 1985 onwards and was aware of the discussions that had involved the American Environmental Protection Agency and Department of Agriculture. Technical arguments on details of regulation problems were not appropriate for discussion by directors at the BSC level.

---

[5] Wayland Kennet, *The Future of Europe*, Cambridge University Press, 1976.

6 European Biotechnology Strategy and Sequencing the Yeast Genome 333

In consequence, in July 1985, the BSC set up the Biotechnology Regulation Inter-Service Committee, (BRIC), with the following functions:

- To review the regulations applied to commercial applications of biotechnology;
- To identify existing laws and regulations that may govern commercial applications of biotechnology;
- To review the guidelines for recombinant DNA research;
- To clarify the regulatory path that products must follow;
- To determine whether current regulations adequately deal with the risks that may be introduced by biotechnology and to initiate specific actions where additional regulatory measures are deemed to be necessary;
- To ensure the coherence of the scientific data which will form the basis of risk assessment and, in particular, to avoid unnecessary duplication of testing between various sectors.

The chairmanship was divided between DG III and DG XI. The secretariat was assigned to the CUBE unit of DG XII. DG VI (Agriculture) pointed out that although it accepted this arrangement, it still felt very involved, having already had some experience in the regulation of sectors that would certainly be important spheres of biotechnological application. The representatives of DG V (Employment and Social Affairs) were active participants due to their responsibilities for the safety of employees and public health. The BSC, in theory, remained the governing body for the BRIC to resolve any conflict that might emerge. This did not work in practice for two reasons:

- The frequency of meetings: while the BRIC met each month and had 15 meetings between September 1985 and 15 October 1986, the BSC only met three times in that period;
- The conflict between the DGs could not be resolved at the BSC level under the chairmanship of DG XII; but had to depend on a last-chance arbitration and resolution at the Commissioner level or during a session of the Commission itself. Although the Commission meets every week, its ability to manage technical problems requires a lot

of preparation and considerable detailed briefing, so can only be used in exceptional cases for questions of paramount importance.

The BRIC was, therefore, the active center of debate, where interservice discussion on biotechnology regulation took place between 1985 and 1990. The BSC gradually disappeared, meeting once in 1987 and for the last time in 1988. The briefing documents for the July 1988 meeting underlined the various modifications that had occurred since 1983 in the strategic environment for biotechnology, such as:

- The scientific progress and consequent pervasive significance of biotechnology;
- Industrial development;
- The GATT (Uruguay Round) negotiations, which would liberalize agricultural trade and might, therefore, be expected to alter the attitudes of agricultural ministers to S&T innovation;
- The competitive challenge;
- The Single European Act, adopted July 1987, to become effective July 1989, facilitating (by majority vote on Article 100A) the progress towards a common internal market by 1992, and with an extensive new section (Title VI) devoted to Research and Technological Development and starting with the blunt language of Article 130F: "The community's aim shall be to strengthen the scientific and technological basis of European industry and to encourage it to become more competitive at international level".

The last BSC meeting boasted the presence of an impressive number of European Commission Directors from DGs as diverse as DGI (External Affairs) to DG XII (Science, Research and Development), including DG III (Internal Market and Industrial Affairs), DG V (Employment and Social Affairs), DG VI (Agriculture), DG VIII (Development) and DG XI (Environment). The arguments in support of a strategic re-evaluation were accepted, and it was agreed that all the services should present their opinion on a review of the biotechnology strategy and the preparatory notes in writing. These points of view were gathered together into an internal document entitled "Interim document towards a redefinition of community strategy for biotechnology". But the rapid approach of the end of the Commission's four-year mandate, from 1986–1989, distracted the Commissioners and their close colleagues with thoughts of their own political and personal futures, so their priority at the time was not reforming community biotechnology policy.

When he circulated the document in December 1988, the BSC President, DG XII's Director General Paolo Fasella referred to particular communications that had appeared since July of that year, such as the proposal for a directive on the protection of biotechnological inventions[6]. He also referred to other proposals in gestation such as DG VI's proposal for a Community-wide system for protecting plant varieties, DG XII's proposal for the next biotechnology program (BRIDGE), the mid-term summary of progress in the Uruguay Round of the GATT negotiations, and the Bio-Ethics and Biosafety conference in Mainz, 8/9 September 1988 and Berlin, 27 and 30 November 1988. He made clear his intention to continue preparing a communication on biotechnology for consideration by the new Commission that would take into account the interim document and other questions, with a view to an integration of the whole, in a draft to be submitted to the next BSC meeting. But there never was another BSC meeting. The new Commissioner for Research and Technological Development had other priorities, in particular, the institutional reorganization of DGs XII and XIII, the implementation of the second framework program (1987–1991) and the preparation of its successor, the third framework program (1992–1995), as well as an improvement of procedures for research contract management. Vice-President Pandolfi and his cabinet took no part either in the regulatory debate on the two biotechnology directives taking place in the Parliament and the Council nor in the wider questions of strategic biotechnology policy, apart from dismantling the CUBE unit in DG XII that had been most involved in its management and the indefinite postponement of the community program on human genome R&D in view of the authorization of a deeper examination of its ethical aspects.

---

[6] *Proposal for a Council Directive on the Protection of Biological Inventions,* Communication from the European Commission to the Council, COM(88)496, October 1988.

Amongst the structures that were created between 1984 and 1987 were two specific committees, the CGCs and IRDAC. On 29 June 1984, the council decided to set up the "Management and Coordination Advisory Committees" or CGCs. The mandate of the CGCs was to assist the Commission in its task of defining and preparing research, development and demonstration activities and in its role of management and coordination as it implemented community science and technology strategy. The CGCs would:

- Inform and consult the Commission on scientific and technical questions in their sphere of competence[7];
- Regularly follow and compare national scientific and technological development research programs in the field of interest to the Commission and provide the Commission with information emerging from such comparisons to identify what activities of coordination could be undertaken by the member states;
- On the basis of the scientific and technical objectives set out in the framework program, help the Commission identify and select the themes or actions that could be subjects of community research, development or demonstration activities;
- Contribute to the optimum execution of community research, development and demonstration programs for which the Commission is responsible, in particular, to contribute to the detailed description of projects and their selection, evaluate the results obtained and ensure multiple links between the execution of the programs at community level and the research and development work being carried out in the various member states and under their own responsibility;
- Formulate opinions on scientific and/or technical cooperation foreseen between the European Community, third countries and/or international organizations in the specific areas in question.

Each CGC would be designated by the Commission on the basis of nominations from the member states and would consist of two representatives

---

[7] Although the opinions were to be transmitted formally to the Parliament and the Council, the "conclusions" were only forwarded in reports or minutes.

of each member state (Table 9), elected for four years, who could be assisted and replaced by a maximum of three national experts, and two representatives from the Commission who could be assisted or replaced by other representatives, and a president elected by the representatives of the member states for two years, renewable once only[8, 9]. Each CGC (Table 10) would set up its own procedure and meet at least once a year. The CGCs had the power to set up *ad hoc* working committees for limited periods of time with clearly defined terms of reference.

At a different level, on 29 February 1984, a Commission decision set up an Industrial Research and Development Advisory Committee (IRDAC). This was justified by its need to promote industrial competition (one of the main aims of the framework program), the importance of permanent dialogue between industry and the European Commission and the recognition that industrial R&D had increasing influence over working conditions and employment.

The creation of IRDAC is based on the continued existence of CORDI (Advisory Committee in Industrial Research and Development), set up by a Commission decision of 27 July 1978, and was the adjustment of the consulting system to the new guidelines and political requirements in the field of science and technology. IRDAC's mandate was to advise the Commission either on its own initiative or at the Commission's request on the preparation and implementation of community policy on industrial R&D, including resulting industrial and social impacts. In particular, it could contribute to the analysis of the Commission's needs and opportunities in the field in question and provide the Commission with appropriate information on industrial R&D. At the Commission's request, IRDAC could consult with it in relation to other community initiatives in industrial R&D. To fulfill these requirements, IRDAC could:

- Give its opinions or draw up reports for the Commission on general problems linked to industrial R&D not specific to any one sector;

---

[8] See Table 10 for the various CGCs to be set up.

[9] The initial CGC Biotechnology and IRDAC committees were composed as laid out in the Tables 9 and 11.

338 From Biotechnology to Genomes: A Meaning for the Double Helix

Table 9 Members of the biotechnology CGC
(During the period of the BAP program)

BELGIUM	**ITALY**
R. Bienfet	A. Albertini
J. de Brabandere	M. Moretti
A.M. Prieels	(G. Magni) (b)
G. Thiers (b)	(M. Lener) (b)
FEDERAL REPUBLIC OF GERMANY	**LUXEMBURG**
H. Klein	F. Arendt
R. Wandel	A. Betz
(E. Warmuth)	
N. Binder	**THE NETHERLANDS**
	B.A. Beide
DENMARK	M.C.F. Van den Bosch
K.A. Marcker	R. van der Meer (c)
I. Petersen	(E. Veldkamp) (b)
	(H.J.J. Grande) (b)
GREECE	
C.E. Sekeris	**PORTUGAL**
A.S. Tsaftaris	F-J. A. Carvalho Guerra
A.L. Stavropoulos	A. Xavier
FRANCE	**SPAIN**
M. Lelong	A. Albert
P. Printz	R. Revilla Redreira
(M. Guignard) (b)	
P. Douzou	**THE UNITED KINGDOM**
G. Pelsy	D.G. Lindsay
	A.F. Lott
IRELAND	(H. Pickles) (b)
J. O'Grady	R.H. Aram (a)
J. Ryan	R.A. Jonas (a)
E.P. Cunningham (a)	(F.P. Woodford) (b)
B. Finucane (a)	

a) resigned before the end of the program
b) expert or replacement
c) president

– Exchange information with the Commission on the impact of initiatives at community level affecting industrial R&D.

The IRDAC Committee had 12 members (Table 11); appointed by the Commission, with substantial experience in R&D work in industrial firms, research institutions or other organizations. In order to ensure coherence

6 European Biotechnology Strategy and Sequencing the Yeast Genome 339

Table 10 CGC committees and sectors

SECTOR	CGC COMMITTEEs
Industry	1) industrial technology
	2) scientific and technical norms
	3) biotechnology
Raw and other materials	4) raw and other materials
Energy	5) energy, nuclear fission and control of fissile materials
	6) energy, nuclear fission, fuel cycle handling and waste storage
Aid to development	7) non-nuclear energy
	8) research linked to development
Health and safety	9) medical research and that linked to health
	10) radiation protection
Environment	11) environment and climatology
	12) linguistic problems

with the work of CORDI [10], facilitate the exchange of information and take into account the interests of the relevant European organisations in the field of industrial R&D, such as the Union of Industries of the European Community (UNICE), the European Center for Public Enterprises (ECPE), the Federation of European Industrial Cooperative Research (FEICRO) and the European Trade Union Confederation (ETUC), the committee also included four members appointed by the Commission after consultation with the aforementioned organizations. The list of committee members is published by the Commission in the *Official Journal of the European Commission*. Each member was appointed for a period of three years and could not be re-elected.

At about the same time, the Council, the Commission and the associations representing European industry expressed their recognition of the various challenges set by progress in biotechnology. The establishment of the CGCs underlined the Commission's intention to implement the strategy proposed in 1983 and the importance it allocated to research, development and

---

[10] Advisory Committee on Industrial Research and Development (CORDI) set up by Decision 78/636/EEC published in the O.J. no. L203, 27 July 1978, p. 36.

Table 11 Members of the IRDAC - WP5 committee

M. H. NIELSEN	P. MANGOLD
G. ALLEN	B. MCSWEENEY
K. BAKER	G. MIGNONI
R. BROWN	M. MILLER
W. M. CASTELL	G. NOMINE
G. DELHEYE	K. POWELL
H. DORNAUER	J. P. REYNAUD
P. GEYNET	M. ROSE
D. GUNARY	V. RUBIO
A. HERRERO	E. SHEJBAL
M. HIRSINGER	K. SIMPSON
B. JARRY	A. STAVROPOULOS
R. JEAMBOURQUIN	M. V. D. VAN WEELE
N. KOSSEN	R. VAN DER MEER
B. LE BUANEC	C. VELA
M. LE HODEY	J. VAN VELDHUYSEN
W. LEUCHTENBERGER	D. VON W. ZANTEN

demonstration programs, including their efficient execution and the coordination of research activities at Commission level, whilst not neglecting research activities going on in each of the member states.

The creation of IRDAC was a consequence of European Commission Vice-President Davignon's intention to lead community research towards goals of European Community significance. The European Commission was, at the time, battling to reinforce its industrial competition, and research was clearly seen as one of the necessary ingredients for the revitalization of industrial activity. The Commission, therefore, had a real need for dialogue with industrial experts on a permanent basis. It was not long before biotechnology benefited from the existence of IRDAC (in 1986, it set up a Working Party on Biotechnology (WP5), which began to meet from June 1986 onwards under the chairmanship of Hilmer Nielsen of Novo Industries). As its mandate was limited to R&D, its opinions could not include larger questions of community policy, although several participants were subsequently involved in the 1989 launch of the Senior Advisory Group for Biotechnology (SAGB) under the aegis of the Council for the European Chemical Industry. Since its creation, IRDAC has been very active in providing of opinions and advice to the Commission.

The intention to ensure dialogue between industry and Commission services at the highest level was also notable in the creations of the EBC and the BRIC, initiated by Commission invitation and modeled on two American industrial associations, the Industrial Biotechnology Association founded in 1981 and the Association of Biotechnology Companies formed in 1983 to organize the defense and pursuit of the common interests of American bio-industry[11]. In many countries, similar events were taking place; for example, the UK's Interdepartmental Coordinating Committee on Biotechnology was set up in 1982 following the Spinks report[12].

6.2 The 1980s: An Implementation of the 1983 Strategy?

The years that followed 1983 saw the advent of Commission actions on each of the six priorities listed in the 1983 communication on biotechnology, but this was more a witness to correct analysis and predictions of the inevitable than perfect political coordination.

Demonstration projects

In 1983, it was foreseen that proposals for demonstration projects to promote the transfer of fundamental research results to practical applications should logically follow research programs. The sixth element of the Commission's plan (demonstration projects and concerted pre-demonstration activities) was delayed until the research, training and concertation activities had been

---

[11] These two associations merged in 1984 to become the Bio-Industry Organization.

[12] The Royal Society, ACARD, ABC, *Biotechnology: Report of a Joint Working Party, the "Spinks" Report*, March 1980.

approved[13]. Such projects were to be the focus of distinct subsequent initiatives, logical sequels rather than companions to the programs for research and training and other initiatives.

In addition to the research programs, the Commission had been undergoing a sort of cultural conversion. Since the 1970s, it had tried to shed its role of agricultural controller, which had dominated all other roles. At the 1983 Stuttgart summit, it was decided that the cost of the common agricultural policy could not be allowed to grow indefinitely. In the great agricultural debate, it was recognized that "the agricultural sector could no longer be separate from the rest of the economy[14]". Clearly, there was a need for a reorientation of agricultural policy, especially given the numerous constraints it imposed by limiting the field of action for new policies. Following the Stuttgart summit and the great debate on agriculture, the following priorities were identified after extensive analysis and consultation:

– Gradual reduction in production in surplus-ridden agricultural sectors and a lightening of the consequential burden on the taxpayer;
– Increase of the diversity and improvement of the quality of production with reference to the internal and external markets and consumer requirements;
– More efficient and systematic handling of the problems of small-scale agricultural exploitation;
– Better support for agriculture in domains where it is essential for land use to maintain social balance and protect the environment and the landscape.

In consideration of these priorities, it seemed inevitable that there would be a change in direction for agricultural activities in the establishment of a

---

[13] There had been, since the mid-1980s, a small amount of activity of 1 MECU/year, managed in tandem by the Directorate Generals of Agriculture and Science, Research and Development. It included a set of studies and activities to define new ways of reducing agricultural surpluses by finding other uses for the excess and looking for new products and new processes. The largest project involved the drafting, by the European Federation of Biotechnology, of a decision-making matrix for agricultural research projects.

[14] *Green Paper*, COM(85) 333 Final. Ken Sargeant, of CUBE, at a hearing for the British Parliament, testified that "agriculture has become separated from industry, and we feel that the two worlds should get closer to each other".

new partnership between research, agriculture and industry. Various new directions could be foreseen for alternative use of agricultural land (leisure, sport, tourism etc.). Farmers' priorities would have to change from intensive agriculture aimed at maximum production to goals of lower market cost and higher quality, including quality of life. A complete reorientation of the usage of agricultural products was also foreseen. Industry might constitute a good new market for certain excess agricultural products.[15] In this debate on new directions for European agriculture,[16] biotechnology was expected to play a major role, if not a primary one.[17] Several factors, therefore, made biotechnology demonstration projects appropriate, amongst which were:

- The need to show new possibilities and directions for the further evolution of European agriculture;
- The recognition by investors and individuals of the uncertainties, perspectives and long term promises associated with several sorts of biotechnological innovation;
- The need to show that despite these uncertainties, there is a "New Deal" in European biotechnology that is a consequence of the progress of community research and initiatives for improving conditions in this context.

The 1986 TAMDA document *Towards a Market-Driven Agriculture* and the Commission's *Communication on agro-industrial development* were followed by the proposal for the ECLAIR program (European Collaboration Linkage of Agriculture and Industry through Research[18]) and the FLAIR

---

[15] L. Munck and F. Rexen, *Cereal crops for industrial use in Europe*, EUR 9617 EN, 1987.

[16] *Towards a market-driven agriculture: a consultation document from CUBE*, XII/951/ December 1985.

[17] *Towards a market-driven agriculture (a contribution from the viewpoint of biotechnology)*, a discussion paper from CUBE, DG XII, February 1986. *Biotechnology in the Community: stimulating agro-industrial development*, discussion paper of the commission, COM (86) 221/2, April 1986. K.H. Narjes (November 1985, *The European Commission's strategy for biotechnology in industrial biotechnology in Europe: issues for public policy*, ed. D. Davies CEPS, Brussels, 1987.

[18] *Proposal for a Council Decision to adopt a future multi-annual programme (1988–93) for biotechnology based agro-industrial research and technological development:* ECLAIR, COM (87) 667, December 1987.

program (Food-Linked Agro-Industrial Research). These two programs began in 1989 [19].

ECLAIR, which was granted 80 MECU for its framework program, was aimed at promoting biotechnological innovation at the interface between agriculture and industry at several levels: upstream from the farm (better input, including products and methods less aggressive to the environment), downstream from the farm (new processing procedures) and at the farm itself. In all cases, an integrated perspective was to be encouraged. For example, the inputs could be changed in order to facilitate later steps in the production chain and vice versa. ECLAIR was a program of central importance to consumers at a time when political attention to their interests had greatly increased. From this point of view, the February 1989 creation of a Consumer Policy Service within the Commission, in line with one of the many recommendations of the European Parliament [20], is highly significant. The development of quality and methods of testing was also strongly adapted to the development of the internal market.

The focus was on the development of research programs such as the Biotechnology Action Program (BAP) and its successor, BRIDGE, and on an orientation towards market requirements, even though this was only at a precompetitive level. FLAIR, with its budget of 29 MECU for 1989–1994, combined a continuation and an expansion of the EEC-EFTA COST action into a new "shared-cost" action. The main objective was to apply the new technologies, both biological and others, to the improvement of quality, innocuity and nutritional value in products from the food and drink industries. Other demonstration activities were also undertaken, such as AIDA (Agro-Industrial Demonstration Action).

In parallel with the third biotechnology program, BRIDGE, which had been gestating since the middle of the 1980s, there were also a number of shared-cost research programs linked to biotechnology. As well as the

---

[19] *Proposal for a Council Decision to adopt a multi-annual R&D program in food science and technology, 1989 to mid-1993*, COM (88) 351, June 1988.

[20] European Parliament (April 1989), *Report on consumers and the internal market, 1992 (the Abber Report)*.

European Collaborative Linkage of Agriculture and Industry through Research (ECLAIR), the Food-Linked Agro-Industrial Research (FLAIR) and AIDA activities, there was also the action on Tropical Agriculture and Medicine, Science and Technology for Development (second program), which was allocated 80 MECU over the period. Agricultural Competition and the Use of Resources were allocated 55 MECU. Other programs (Medicine, Non-nuclear Energy, Biomass, Environment and the Science program) also included sectors or activities in relation to biotechnology.

Research and Training: The New Bap Program

The European Community's biotechnology activities did not begin with the 1983 *Communications*. Since the mid-1970s, some of the commission's civil servants, in particular D. de Nettancourt, F. Van Hoeck and A. Goffeau, had been advocating an initiative in genetic engineering and enzymology. Their efforts were modestly and belatedly recompensed by the Biomolecular Engineering Program (BEP). Adopted by the council on 7 December 1981 with an 8 MECU budget[21], it was then revised by the same body on 26 October 1983 and allocated a further 7 MECU[22]. The program ran from 1982 to March 1986.

The 1983 *Communications* on a strategy for biotechnology were well received at meetings of the Council of Industry Ministers (November 1983) and Research Ministers (November 1983, February 1983). In consultation with national and industry representatives, an ambitious proposal for a program was put forward, allowing the content of the program to be better adapted to the medium- and long-term requirements of European industry and agriculture, to be totally compatible with national activities and to remedy European weaknesses in the field of biotechnology. These weaknesses were such that Europe was barely considered a serious rival by its US competitors. The dialogue with industry continued with CEFIC's publication on 24 January 1985 of its opinion of the Commission's proposal.

---

[21] *O.J. no. L375*, 30 December 1981, pp. 1–4.

[22] *O.J. no. L305*, 8 November 1983, pp. 11–13.

346 From Biotechnology to Genomes: A Meaning for the Double Helix

The CEFIC assigned greatest priority to the following sectors:

Subprogram I: Mathematical models
                Data bases
Subprogram II: Physiology and genetics of species important to industry and agriculture
                Technology of cells and tissues cultivated *in vitro*
Subprogram III: Evaluation, follow-up and encouragement of research in the field of safety.

IRDAC also gave its advice and opinion on the following program of biotechnology R&D. The proposal for the research action program on biotechnology was published in April 1984[23]. It clarified the domains for research and training activities in biotechnology, including:

– *Sub-program 1*: Contextual measures (bio-informatics and collections of biotic materials)
– *Sub-program 2*: Basic biotechnology: precompetitive topics in the following areas :
  * Technology of bioreactors;
  * Genetic engineering;
  * Physiology and genetics of species important to industry and agriculture;
  * Technology of cells and tissues cultivated *in vitro*;
  * Screening methods for the evaluation of the toxicological effects and the biological activity of molecules;
  * Assessment of risks.

Although the research and training activities were the main body of the proposal (which also spoke of concertation activities under Action II) and would use up all the 88.5 MECU that had been requested, the April 1984 *Communication* was an opportunity to reformulate the Community's

---

[23] European Commission, *Proposal for a Council Decision adopting a multinational research program of the European Economic Community in the field of biotechnology*, COM (84)230, 1984.

6 European Biotechnology Strategy and Sequencing the Yeast Genome 347

biotechnology strategy. The document opened with unflattering quotations on European capabilities from American documents and reiterated the priority activities from the October 1983 *Communication*. The proposal focused on two priorities, research and training and concertation, but it also considered the activities from four other angles:

- On new regimes on agricultural outputs for industrial use: proposals for revisions to the sugar and starch regimes were announced as imminent for the former and in preparation for the latter;
- On intellectual property rights in biotechnology: the Commission mentioned the establishment of an inter-service working group on biotechnology, identifying needs for improvement which could be used to prepare the community position in the discussions and negotiations likely to follow the OECD inquiry, then in progress, on patenting in biotechnology;
- On demonstrations projects: proposals would follow later, once R&D projects were in progress and targets could be better evaluated;
- On a European approach to regulations affecting biotechnology: the communication presented the Commission's views in some detail. Monitoring of questions of biological safety, in particular the regulation of recombinant DNA (genetic engineering) work and the social dimensions of biotechnology would be included in the "Concertation Action". The document reiterated the general case for a common internal market, then referred to consultations the Commission had undertaken "with committees of government experts, with industry and with its own expert scientific committees in various sectors touched by biotechnology such as pharmaceuticals, chemicals, human nutrition, animal feedstuffs etc".

The conclusion of these consultations was drawn up as follows:

> It would appear that the application of current community regulations in the various fields will meet current regulatory needs, provided there is close cooperation between the competent authorities in the member states and the Commission. Such cooperation can be achieved

by greater recourse to the existing institutional or scientific committees and, as necessary, use of the new information procedure for technical standards and regulations adopted by the Council in its *Directive 83/89/EEC* of 28 March 1983. On the basis of its experience from the use of these various instruments, the Commission will put forward general or specific proposals appropriate to create a regulatory framework suitable for the development of the activities of the bio-industries and for the free circulation of goods produced by biotechnology.

The document then tackled the problems of the pharmaceutical sector and announced the Commission's intention, following consultations with industry and the pharmaceutical committee, to submit legislative proposals to the council in July 1984 to implement a community-wide agreement on medicines produced by or derived from biotechnology.

As for concertation, the communication laid out a strong justification for a network of supervision and information in strict liaison with the biotechnology policymakers in the member states, with industry through industrial associations and other organized interests, and "in conjunction" with the Commission and member state services concerned.

The following list of nine tasks was proposed, and subsequently adopted, as the mandate of the concertation action; the only material amendment between the April 1984 proposal and the March 1985 Council Decision being the inclusion in item 8 (at the request of Parliament) of the words "and risks".

> Concertation activity will be implemented with the objectives of improving the standards and capabilities in the life sciences, and enhancing the strategic effectiveness with which these are applied to the social and economic objectives of the Community and its member states.
>
> In conjunction with the relevant services in the Community and the member states, the following tasks will be executed:
>
> – Monitoring the strategic implications of developments elsewhere in the world for biotechnology-based industry in Europe;

- Working with the services of the Community, member states and other interested parties to identify ways in which the contextual conditions of operations for biotechnology in the Community may be further improved, promote its development in all useful applications and the supporting scientific capabilities;
- Responding to the needs for research and information in support of the specific actions of other services of the Commission;
- Identifying opportunities for enhancing, through concertation and cooperation, the effectiveness of biotechnology-related programs in the member states and promoting collaborative initiatives in biotechnology with and between industry and universities;
- Consideration of how the safe and sustainable exploitation of renewable natural resource systems in Europe may be enhanced by the application of biotechnology;
- Promoting, in co-operation with developing countries and relevant institutions, the pursuit of the same task within their respective regions;
- Monitoring and assessing developments in biotechnology bearing on safety and other aspects of the 'social dimension';
- Disseminating knowledge and increasing public awareness of the nature and potential risks of biotechnology and the life sciences;
- Establishing and *ad hoc* system of collaboration between groups and individuals with interests and capabilities in the life sciences and biotechnology, thus creating networks, as informal and flexible as possible and adapted to the particular problems under study. The networks to have the triple function of providing an active input into the program, encouraging coordination through the exchange of information between the participants and assisting the broader diffusion of information envisaged in the preceding task.

The BAP proposal was adopted by March 1985 with relatively little controversy[24]. The program began on 1 Jan 1985 for a period of five years,

---

[24] *O.J. no. L83*, 25 March 1985, pp. 1–7.

350 *From Biotechnology to Genomes: A Meaning for the Double Helix*

the end of the contracts being foreseen for 31 December 1989[25]. Despite the Commission's request of 88.5 MECU, the program only received 55 MECU. Although the budget limits were fixed by the Council's conclusions of 19 December 1984, the research group of the Council accepted that the budget share of each subprogram be shuffled by the CGC. Following discussions with the CGC, the division for the management of the five-year biotechnology program was as follows:

Action I	Subprogram I	9.5 MECU
	Subprogram II	40.0 MECU
Action II	Concertation	5.5 MECU
TOTAL		55 MECU

Article 3 of the decision foresaw the execution of an evaluation of the program, to be submitted to the council and Parliament. This evaluation would form the basis of a commission submission for the revision of the program in accordance with the appropriate procedure.

The BAP program, fundamentally of a precompetitive nature, showed many new traits, revealing a clear trend in community biotechnology policy.

At the level of its contents, there was clearer recognition than in the first BEP program of the progress made by the new biology, in particular, of genetic engineering techniques and their perspectives for the priorities of the 1983 communication. This recognition was in agreement with the general world trend of a major awakening to biotechnology (in particular, the new molecular technologies) in an increasing number of fields, making old dreams from the beginning of biotechnology's history, already reiterated in the reports of the 1970s and the early 1980s, finally possible.

At the level of its form, the concept of the European Laboratories Without Walls (ELWWs) emerged. The major biotechnological weaknesses at the member-states level were attributed to fragmentation of effort in biotechnology, the lack of scientists and technicians with appropriate advanced training and insufficient context and infrastructure within the Community.

---

[25] For the Commission staff involved in BAP, see Table 12.

Table 12 Table of Commission personnel involved in the implementation of the BAP program[26]

Directorate-General XII Science, Research and Development Director General: P. Fasella		
Cost-shared actions (Directorates C, E, F and FUSION) Deputy Director General: H. Tent		
Directorate F, Biology Director: F. van Hoeck		
Division F-2, Biotechnology Head: D. de Nettancourt		
BAP research actions : 1985–1989		
Bio-informatics	P. Reiniger	(1)
	B. Nieuwenhuis	(2)
Culture collections	P. Reiniger	(1)
	A. Aguilar	(2)
Bioreactors	A. Goffeau	(3)
	B. Nieuwenhuis	(4)
	I. Economidis	(5)
Protein engineering	B. Nieuwenhuis	(10)
	A. Goffeau	(3)
Industrial microorganisms	A. Aguilar	(10)
	A. Goffeau	(3)
Plants and associated Microorganisms	E. Magnien	(10)
	A. Vassaroti	(9)
Animals	P. Larvor	(6)
	H. Bazin	(10)
	A. Klepsch	(7)
In vitro testing methods	P. Larvor	(6)
	A. Klepsch	(7)
Risk assessment	U. Bertazzoni	(10)
	I. Economidis	(5)
Yeast chromosome III	A. Goffeau	(10)
Sequencing pilot project	A. Vassaroti	(9)
BAP training actions: 1985–1989		
all sectors	A. Goffeau	(3)
	D. de Nettancourt	(8)

1) Until 31/03/87 2) from 01/04/87 3) until 31/12/86
4) until 15/09/87 5) from 16/09/87 6) until 31/07/87
7) from 01/08/87 8) from 01/01/87 9) from 01/11/88 10) throughout the entire BAP

---

[26] Table taken from *Biotechnology R D in the E.C., Biotechnology Action Programme (BAP) 1985–1989, vol. I, Catalogue of BAP Achievements,* Edited by A. Vassarotti and E. Magnien, ed. Elsevier for the Commission of the European Communities, Paris, 1990, p. 19.

The point that research was fragmented was not only about the isolation and dispersal of national efforts. Biotechnology is, by nature, a fast-evolving multidisciplinary science with complex solutions. In order to allow this, a critical mass of disciplines must overlap in the handling of many day-to-day practical problems. This means new types of research cooperation between the member states in different sectors of research. This was clearly foreseen by the Commission, and from the beginning, Community initiatives in the field of biotechnology were targeted towards conquering isolation and the relative insignificance of national efforts. BEP had already specified transnationality as the main criterion for accepting proposals. This specification was reinforced in 1985 when a new criterion was adopted for community R&D proposals — the transnational nature of projects financed by the European Community. As more money became available both in national activities in the member states and through the adoption of the BAP Community program, and scientists became more interested, the demand grew at an accelerating rate (300 proposals in BEP to 1,357 in BAP). The time had come for a rationalization of Community research and a deeper definition of its role, to complement, prolong and widen national activity. At various levels, it was thought that the European taxpayer's money would be better spent encouraging a larger number of groups to cooperate over national borders. The transnational criterion was to have significant effects, particularly in biotechnology, a deeply multidisciplinary field more likely to benefit from that sort of teamwork and cooperation[27].

The application of the transnationality criterion was justified in the 1986 establishment and adoption of the idea of the European Laboratories Without Walls, which was initially proposed by André Goffeau. It was a rapidly adopted, highly successful idea[28]. There was a real need for flexible structures that would allow research groups to be held together over a longer period

---

[27] A. Vassarotti and E. Magnien, *Biotechnology R & D in the E.C. (BAP 1985–1989) Vol. 1: Catalogue of achievements,* 241 pages, *Vol. II: detailed final report,* 564 pages, Elsevier, Amsterdam, 1990.

[28] R. van der Meer, E. Magnien, D. de Nettancourt "Laboratoires Européens sans murs: une recherche précompétitive ciblée", *Biofutur,* July–August, 1988, pp. 53–56, E. Magnien, A. Aguilar, P. Wragg, D. de Nettancourt, "Les laboratoires Européens sans murs", *Biofutur,* November 1989, pp. 17–29, L. Flandrog, "Un bilan largement positif", *Biofutur,* November 1989, pp. 30–34.

of time than that required to put a research proposal together. Cooperation at the action level itself was a logical extension of cooperation by exchanging information. Created as multi-partner structures to facilitate interaction and the international division of labor, the ELWWs gradually came into existence, each of them involving between five and 35 university or industrial laboratories[29].

1986 also saw the inception of generalized peer review in the evaluation of proposals submitted to Community biotechnology programs. Each of the BAP sub-sectors were evaluated in this manner. It brought experts of recognized talent and competence (identified by the Commission and the member states) together for three days in Brussels during which they read, evaluated and gave a mark to each of the proposals submitted to them. The number of proposals (all transnational) assigned to each group of evaluators varied between 10 and 25. The experts used, amongst others, the following criteria to rank the proposals:

- The technical competence of the proposers;
- The scientific interest of the proposal, its originality and its relevance to the content of the program;
- The degree of community integration through transnational cooperation (priority being given to joint proposals from laboratories in different member states);
- The degree of industrial involvement;
- The precise evaluation of risks likely to result from the research foreseen.

6.3 BAP's First Year

The 1985 call for proposals drew 1,357 of them, 35% of which had an industrial link and 80% transnational. From these proposals, the Commission

---

[29] E. Magnien, A. Aguilar, P. Wragg and D. de Nettancourt, *"Les laboratoires Européens sans murs"*, Biofutur, November 1989, pp. 22–29.

negotiated 262 shared cost contracts, over 93 projects of between two and four years in length[30]. Despite 169 expressions of interest from industries, only 16 of the contractors were industrial firms. 189 fellowships were also granted, most to relatively junior scientists, pre-doctoral or recently appointed post-doctoral researchers.

By the end of the first year of BAP, a number of conclusions were evident[31]. Generally, the activities taken on since 1985 had been in accordance with the guidelines laid down in the 1984–1987 first framework program of Community research and development activity approved by the Council. The BAP program was an undoubted success in the scientific community, and from its first year on, contributed to the building of a high-performance infrastructure for European biotechnological research, as well as the breakdown of national frontiers between laboratories through the ELWWs. Real enthusiasm had evolved for transnational cooperation, even amongst those who had initially been skeptical.

However, the part of BAP concentrated on R&D infrastructure (research contracts with groups of universities or institutes, training fellowships and contextual measures) was not sufficiently financed to allow it to reach its full potential, as can be seen from the massive response to the call for proposals, of which only 15% could be funded. The precompetitive nature of the research and the small size of the projects, with an average grant of 50,000 ECU per annum, enough to employ one scientist for three to four years, ensured that most of the proposals were submitted by university or institute groups. Amongst these were almost all of the high-level laboratories active in the fields chosen for BAP. BAP's allocated budget could only fund

---

[30] For a statistical analysis of the BAP programme (structure of the BAP projects, distribution of the laboratories according to the size of the project, analysis of the contractors' publications), see A. Vassarotti and E. Magnien, *Biotechnology R & D in the E.C. (BAP 1985–1989) Vol. 1: Catalogue of achievements*, 241 pages, Elsevier, Amsterdam, 1990, p. 4 and pp. 8, 9.

[31] These were mainly the opinions expressed during the CGC meeting on the evaluation of the BAP programme by a group of experts (Charlotte af Malmborg, Pierre Feillet, Fotis Kafatos, Jan Koeman, Pier Paolo Saviotti, Gunther Schmidt-Kastner and Geoff Walker) brought together by the Director General of DG XII as individuals and not as representatives of particular organizations or countries. For a complete evaluation, see *Research evaluation*, report no. 32, EUR 11833 EN, 29/88 DN46, 29 September 1988 rev. 2 February 1989.

262 such grants, several of which would be collected together in one transnational project. This meant that 85% of the proposals had to be rejected. Among these, there were too many excellent proposals that filled the eligibility criteria, in particular those of transnation and industrial relevance. There was concern that such a high rejection rate (due to insufficient budget) would cause a reduction in applications and a slowdown in European biotechnology, in the short term, a demobilization of programs, researchers and industrialists.

To allow BAP's objectives to really be attained, its financial resources had to be brought in line with its potential capabilities. Until the advent of BAP, it had been close to impossible to estimate what size a community biotechnology program concentrating on transnational cooperation within the Community should be. During the preparation of BEP, this had resulted in a Commission unable to protest at the council slashing the budget from 88.5 MECU to 55 MECU. One of the justifications of this reduction was that it was not possible to predict how R&D infrastructure would react to a call for proposals with transnationality as a main criterion. During the first year of the program, this question was answered, fueling the debate over the reopening of discussions on BAP's financing. At the end of this first year, the Commission and the CGC were able to remind the Council of its promise that the CGC could ask for BAP to be revised.

The revision, agreed upon by all the delegates to the CGC during its meeting on 7 March 1986[32], began that same year for the remaining BAP period of 1987–1989. It was a pivotal time, as in the spring of 1987, the Single European Act was adopted opening new directions for Community research with its Articles 130F to 130P[33]. The objective put before the

---

[32] Meeting of the Biotechnology CGC, Brussels, 7 March 1986.

[33] Title VI: Research and technological development: article 130F of the Single European Act (1987): 1: The community's aim shall be to strengthen the scientific and technological basis of European industry and to encourage it to become more competitive at international level. 2: In order to achieve this, it shall encourage undertakings including small and medium-sized undertakings, research centers and universities in their research and technological development activities; it shall support their efforts to cooperate with one another, aiming, in particular, at enabling undertakings to exploit the community's internal market potential to the full, in particular, through the opening up of national public contracts, the definition of common standards and the removal of legal and fiscal barriers to that cooperation.

community in these articles was to reinforce the scientific and technological base of European industry and the development of industrial competition. This was not a contradiction of previous efforts (BEP and BAP), but was an innovation due to the official recognition that the member states were prepared to assign to the fundamental need of Community research for economic development. Research acquired social significance and took on a real economic role.

The translation of this new political vision from words into action was through the assignment of more attention to industrial needs, more systematic consultation with IRDAC and the adoption of a "top-down" approach to definition of future needs in research. An indirect effect was the proliferation of a number of specialized industrial clubs trying to establish dialogue with the Community's institutions. This can be seen, for instance, in the 1987 creation of the Green Industry Biotechnology Industrial Platform, GIBIP, which brought together 21 companies with massive investments in R&D in biotechnology to bring about developments in modern agriculture.

Other major events of 1986–1987 included the enlargement of the Community to include Spain and Portugal, the adoption of the second framework program (1987–1991) which planned to increase efforts in research, training and concertation activities in biotechnology, the latter having been the point of a communication to the Council on 21 May 1986[34], and the preparation of the next biotechnology program, to be called Biotechnology Research for Innovation, Development and Growth, or BRIDGE.

The objectives of the revision were laid out as follows:

- The intensification of efforts in the sector of the program concerning the evaluation of the risks associated with modern biotechnology;
- The increase in the volume of activities being undertaken (visits, electronic networks, colloquia, workshops) aiming at a better and more rapid dissemination of information on the programs and research results;

---

[34] COM (86) 272 Final.

- An increase in European industrial involvement in research activities;
- An encouragement of pilot studies and feasibility projects to prepare community activities; in the field of R&D in biotechnology during 1990–1994. One of the actions suggested was the mapping and sequencing of the yeast genome, livestock animals and certain species of plants;
- The increase in training activities at all levels;
- The adaptation of (personnel) resources relevant to the concertation activities;
- The reinforcement of concertation activities;
- The extension to Spain and Portugal of all the activities foreseen by the program.

This revision, and the objectives laid out, were only accepted by the CGC delegations on condition that a new "BAP 2" program be planned for the future. All the delegations considered the establishment of the training section of the program as a priority task. Furthermore, most of the delegations hoped for a more systematic involvement of industry in the preparation of programs and calls for proposals, although some of them (France and Germany in particular) could see problems for industrial participation in the execution of research projects with very precise targets. A possible compromise would be to establish university-industry tandems with research activities mostly taking place in the universities, but financed by industry and the Commission. Another possibility foreseen was for the Commission to group together industries that would help it to conceptually and financially support precompetitive targeted research in networks of laboratories pinpointed by Commission services[35].

The general objective of the revision was to improve member states' ability to maintain competition *vis-a-vis* the rest of the world in the preparation of agricultural and bioindustrial products. The program also envisaged the improvement of new methods to evaluate biological risk and the uniform and harmonious development of policies and regulations on promoting modern

---

[35] Doc. CGC - IV/86/3. Revision of the Biotechnology Action Program.

biotechnology in the Community. This revision of the BAP program was delayed for two years by the repeated failure of the adoption of the second framework program (1987-1991).

The Council's eventual adoption of the proposal for the revision of BAP on 29 June 1988 increased BAP's budget by 20 MECU and allowed some of the less developed aspects to be reinforced, in particular, the lack of participation from Spanish and Portuguese laboratories (these two countries had joined after the closing date of the first call for proposals). This new funding allowed a further 116 laboratories to be selected for funding from a further 276 proposals received over 1989 and 1990.

In accordance with opinions from the CGC and IRDAC-WP5, this increase would also allow an increase in the number and intensity of activities in the bioinformatics sector, a sector increasingly perceived as a sensitive issue. As F. Van Hoeck pertinently said in the CGC-Biotechnology meeting of 7 March 1986, "any proposal accepted after the second selection round in April 1986 still has to be funded from BAP's initial budget". The revision constituted a new exercise, initiated so that the program could be re-fitted to the objectives it had been assigned. It was only at this stage that the proposals left on the reserve list in November 1985 and April 1986 could be reconsidered for funding[36].

This time of deep transformations also included:

- The development and implementation of a real European strategy in the field of biotechnology;
- The delayed adoption of the second framework program, for which the Commission had requested an overall budget of 7,735 billion ECU over five years and to which 5,677 billion ECU had been assigned. Eventually the position taken by the member states (10 in agreement and 2 reserved positions) was for 5,396 billion ECU. As for the budget line for the "exploitation and valorization of biological resources", the Commission's request for 350 MECU in accordance with the Presidency compromise had turned into a final budget

---

[36] *Report of the CGC-Biotechnology meeting,* Brussels, 7 March 1986.

adoption of 280 MECU, of which 120 MECU only were for biotechnology (105 MECU for the new biotechnology program and 20 MECU as the top-up under the BAP revision), 105 MECU for agro-industrial techniques and 55 MECU for agricultural competition and the management of agricultural resources;
- The revision of the second European biotechnology program, BAP.

In view of these, the European Commission, the CGC, and other committees considered and finally approved several programs linked to fields of genome research. The intention was to keep up in the international race in genome studies, to discover genes of industrial interest, and not to be pre-empted in the new markets that might emerge from these discoveries.

6.4 The Revision of the BAP Program

Although the decision on the second framework program was still pending, the CGC-Biotechnology began to discuss their priorities for the content and implementation of the BAP program revision as well as those of its successor, the BRIDGE program. This was done in agreement with the delegations, who would only permit the revision of BAP if a successor was planned to nurture the seeds sown in the field of academic research, the so-called BAP II program (which became the BRIDGE program)[37]. As R. van der Meer pointed out to the members of the delegations, the extension of BAP I was to be carried out with this second program in mind. In the second phase, "BAP II", the target was no longer only the Europeanization of research, but also attaining an international reputation in the field of applied biotechnology, a task at which Europe had excelled in the sector now called agro-business. To reach an important world position, it was considered vital that this field grew more sophisticated, strongly directed and influenced by research. Biotechnology was seen as capable of contributing to a stable and modern research base, resolving several problems in the sector, including those of

---

[37] CGC-Biotechnology meeting, 7 March 1986.

surpluses and deficits of agricultural products, increasing the added value of agricultural products and contributing to the developments of new products and procedures [38].

This industrial and market-directed orientation of the BAP program revision was supported by some of the member state delegations and IRDAC WP5 and greatly proclaimed during the industrial conference held at Louvain-la-Neuve in Belgium, where possibilities for closer collaboration between industry and the ELWWs were discussed. But the program had to remain precompetitive, that is to say it could not include applied research. If it did, BAP itself would be in competition with programs such as EUREKA. Some delegations, in particular those of West Germany and Great Britain, nevertheless, made clear "the need to act speedily in the preparation of community programs in competition at different levels with the EUREKA initiative and the Japanese Human Frontier Science Program" [39]. The revision of the BAP program provided an opportunity to respond to these scientific initiatives (both seen as competitors to the program), respond that would include increased efforts in the rationalization of infrastructure to sustain the Europeanization of culture collections. The CGC-Biotechnology, hoping for a quick decision from the Council on the next framework program, set the main objectives of the revision in four main fields:

- The extension of all the activities in the ongoing program to Spain and Portugal;
- A doubling of training activities;
- The enlargement and rational organization of risk evaluation and bioinformatics activities, taking into account the interim evaluations available and political needs, answers to the call for expressions of interest to support the TAMDA initiative and the recommendations of the working parties on culture collections;
- An exploration of activities to prepare the forthcoming BRIDGE program.

---

[38] R. van der Meer, Revision of BAP I, preparation of BAP 2, 28 February 1986.

[39] Meeting of CGC-Biotechnology 31 March 1987.

Because of unexpected delays in the adoption of the second framework program, it was felt that the BAP revision had to be implemented rapidly, and that the small window of opportunity available (1988–1989[40]) had to be used to set up activities that would make way for a more ambitious program.

The amplification of the BAP program not only provided 4 MECU to allow Spanish and Portuguese laboratories to join in ongoing actions mid-project, but also authorized several significant developments, such as an increase in work on risk evaluation, (assigned 4.48 MECU) which would eventually involve 58 laboratories across the Community. This increase in risk evaluation was in answer to growing parliamentary and public concerns about genetic engineering and the possible release of organisms genetically modified by such techniques.

The amplification of the BAP program also contributed to the reinforcement of bioinformatics activities in Europe in accordance with FAST recommendations, and following a series of studies and workshops, co-financed by concertation actions and DG XIII under the title of BICEPS (Bio-Informatics Collaboration European Program and Strategy), also engendered the AIM program (Advanced Informatics in Medicine). A Task Force for Biotechnology Information, with the support of DG XIII and input from the concertation actions, started off several important initiatives involving infrastructure. They included support for the BIOREP data base, biological research projects financed with public funds, the European node (CERDIC, NICE) of the Codata data base of hybridomas, and financing, in cooperation with industry a major strategic evaluation of the biotechnology information infrastructure in Europe, following the ASFRA report[41].

BAP funded projects to develop new software for data input, compression, stocking and comparison, handling 3D images and facilitating access to data bases and the use of computer tools. Furthering the Codata task group's 1985 analysis, "Coordination of Protein Sequence Databanks", identifying an urgent need to set up a European data base in a European network, BAP

---

[40] This short period for carrying out the contracts under the BAP revision involved spending a lot of money very fast, which required trustworthy spending mechanisms.

[41] J. Franklin, *The role of information technology and services in the future competitiveness of Europe's bio-industries,* Report for COM. EC, 1988.

also supported the constitution of a protein sequence data base at the Max Plank Institute and the EMBL data base, as well as the research and development of next-generation databases for molecular biology and standards for the international exchange of data.

In the highly competitive field of genome sequencing and analysis, several projects should be pointed out. Firstly, a project on the automation of DNA sequencing involving two groups, one at UMIST in Manchester, UK, directed by W.J. Martin and one at TCCIP in Konstanz, Germany, led by R. Massen. They collaborated on building a new vision-controlled robot that could identify and select plaques and colonies on Petri dishes for automatic sampling, and then transfer candidate samples for further growth in individual containers. Secondly, two other groups, one at the University of Konstanz led by F.M. Pohl and one at the RAL in Didcot, UK, led by J.E. Bateman, achieved the automation of electrophoresis and direct blotting as well as building a gas counter to directly image DNA sequences marked with radioactivity.

The only project really dedicated to systematic sequencing was the BAP funded attempt to sequence yeast chromosome III, taken on in the BAP revision as a feasibility study, adopted with a budget of 2.4 MECU (and the source of the proposal submitted under BRIDGE) and to systematically sequence the entire genome of the yeast *Saccharomyces cerevisiae.*

Preparations for the BRIDGE Program

While the BAP program was being revised, the Commission's civil servants were trying to define the nature and targets of Biotechnology Research for Innovation, Development and Growth in Europe (BRIDGE), which would start at the end of 1990 for four years with a 100 MECU budget (see diagram of consultation process.)

IRDAC's WP5 had limited influence on the revision of the BAP program, and while bioinformatics, risk evaluation and feasibility studies had been determined as priorities, the WP5 decided to concentrate their influence on which areas should be covered, and how they should be implemented, in the BRIDGE program.

Irdac's Opinion on the Future of European Biotechnology Programs [42]

- Concentration in medical biotechnology must be accelerated and increased.
- Training must have high priority.
- Substitutes for imported agricultural products must be a high priority for biotechnology applications. Training and political measures to this end are to be encouraged. Involvement in projects which are inherently uneconomic (such as bioethanol) must be avoided.
- As a prerequisite for applications of biotechnology, the basic biology of crop plants needs more attention. Efforts should be made to increase the efficiency of animal husbandry with biotechnological tools. The Commission should monitor developments in transgenic animals.
- Many key areas of industrial biotechnology outside agriculture need R&D support.
- Consistent European biotechnology strategies must be elaborated.
- Training is fundamental to long term success. Funds must be made available quickly when needed.
- The CEC should support research which serves to increase the long term competition of European industry, for example, by enhancing transnational collaboration.
- The CEC should match funds in most cases of European R&D collaboration. Higher levels should be considered when appropriate.
- Conditions for provision of start-up capital for risk ventures must be encouraged.
- 250 MECU should be made available by the Commission for BRIDGE funding.
- Industrial secrecy cannot be ignored when the CEC urges collaborative ventures.

---

[42] *IRDAC Opinion on Future R&D Programs in the Field of Biotechnology*, final version of the document approved in the last meeting of IRDAC WP5 Biotechnology on 24 November 1987.

- The legislative, regulatory, technological and commercial infrastructure must be engineered to make European biotechnology competitive in all markets.
- In a single European market, a concept strongly supported by industry, there must be a unified European approval procedure for biotechnology-based products.
- Recombinant DNA regulations must be harmonized. Risk assessment in biotechnology must be a high priority area. Patent regulations must be brought into line with those of the USA.
- Bioinformatics is a vital part of the infrastructure for biotechnology; it must be strongly supported.

IRDAC pointed out that the European biotechnology industry, despite its weaknesses, should exploit its leads over the USA and Japan, particularly in agrobiotechnology, biosensors, enzymology and waste treatment. Precautions were to be taken in joint projects that could result in a flow of expertise out of Europe to our competitors' advantage. Outside the agricultural sector and linked activities, IRDAC's WP5 recommended the following fields as being of priority:

- Bioinformatics (a vital part of the infrastructure for biotechnology; it must be strongly supported);
- Protein engineering and related techniques;
- Genetics, physiology and metabolites of industrial micro-organisms such as yeast, bacteria and the filamentous fungi;
- Transformation of mammalian cells by recombinant DNA, cell fusion, etc.;
- Culture and expression of mammalian cells;
- New technology for downstream processing;
- Novel biotransformations;
- Research in fats and oils, production, extraction and transformation;
- Use of enzymes and antibodies in biosensors;
- Automation in biotechnology.

Most of the members of IRDAC's WP5 thought that the Community should not be supporting research, whether fundamental or applied, that

6 European Biotechnology Strategy and Sequencing the Yeast Genome 365

could not be seen to have a clear link with useful end products; that is, even fundamental research should be application-directed.

One of the explanations for the strong political support for the EUREKA program is that political masters are more likely to finance R&D with an expressed objective of increasing industrial output in the member states, as opposed to R&D funded by the Commission, which is by definition precompetitive, with no immediately obvious commercial objective. IRDAC's WP5 understood that in the wake of the Act of Union, this direction and focus would have to change towards more short-term applications of biotechnology.

Although no conclusion was reached during the discussions as to a possible "halfway house" between fundamental and market-oriented research, the discussions did produce the concept of "Feasibility R&D", R&D that would show that a new product and/or process would be possible both from the economic and technical point of view.

The various consultations and recommendations of the IRDAC WP5 group identified BRIDGE's main task as building a link from fundamental research to the expectations of European industry. This would mean reinforcing the ELWW networks already in existence and extending them to new areas considered of great significance for the community.

The BRIDGE program, however, was still of a precompetitive nature and targeted towards medium and long term objectives essential to the strategic reinforcement of European industry. It aimed to advantage and encourage transnational research, the catalytic effects of which would accelerate the production of biological data, materials and processes needed for the exploitation of organisms of industrial interest. It also aimed, through competitive research, to set up the scientific basis for new guidelines to regulate novel economically-important production methods. This program, which implemented the new legal and scientific measures of the Single European Act, in conjunction with the priority assigned to an industrially oriented approach to research[43], added a new criteria for the attribution of

---

[43] *IRDAC Round Table on Bio- and Agro-Industrial technology,* February 1989.

priority to research proposals in the selection process: effective industrial participation, whether financial or practical, in transnational projects.

The challenge presented by BRIDGE was to reconcile the preference for a knowledge-driven approach with the involvement of industrial laboratories, a complex challenge given the conflicts that could emerge from the multidisciplinary and transnational requirements and the constraining conditions related to industrial participation (confidentiality, equality of rights and duties).

Other priorities and advice that emerged from the various consultations included:

- An increase of concertation and training activities;
- An increase of risk-evaluation activities;
- The opening of the biotechnology research program to the EFTA countries where the project aimed to bring the leading teams of those countries into European scientific networks, as they had been under the two COST programs.

Like BAP, the BRIDGE program was divided into two actions: Action I for research and training and Action II for concertation. The research activities were divided into four technical areas that each had sectors defined by the priority needs of biotechnological research and European industry:

- Information infrastructure (processing and analysis of biotechnological data, culture collections);
- Enabling technologies (protein design, macromolecular modeling, biotransformation, genome sequencing);
- Cellular biology (industrial microorganisms, plant and associated organisms, animal cells);
- Pre-normative research (*in vitro* evaluation of the activity of molecules, safety assessments; associated with the release of genetically modified organisms).

These sectors, and the chosen shared-cost research activities, aimed to fulfill European biotechnology R&D needs and the expectations of industry. BRIDGE research activity and the other activities such as training and

6 European Biotechnology Strategy and Sequencing the Yeast Genome 367

concertation were implemented by the Commission with the help of the Committee of an Advisory Nature for BRIDGE (the CAN-BRIDGE), a committee of delegates from each of the member states.

In order to rise to the challenge of linking precompetitive knowledge-driven research with industrial economic needs and the involvement of industrial laboratories, the research activities were one of two formats to be both of which foresaw industrial participation:

The N-projects (N for network) were the integration of research efforts linked to sectors defined below into appropriate community structures (the ELWWs). The main difficulty in these sectors was seen as being the lack of fundamental knowledge and know-how. At this level, research could be of an exploratory nature, not requiring very large quantities of funding and providing opportunities for wide sharing of the financial burden. Mixed consortia of public and private laboratories were set up to cover most areas of the program and promote adequate technology transfer to industry. On average each BRIDGE N-project involved five or six laboratories from three or four different countries.

The T-projects (T for targeted) were directed at removing specific bottlenecks in research that were caused by structural problems or problems of scale. They were a new innovation under BRIDGE. There were only a few pre-selected targets, but unlike the N-projects they were far more heavily funded. In general, they brought together 30 European laboratories and over 100 researchers, with a goal of significantly changing the "state of the art" and competitive position of European research in their own domain within the time frame of the program. For these projects, a totally different proposal submission system was designed. Laboratories were invited to target their contribution to a few specific tasks in the overall project. The Commission actively negotiated the conditions for collaboration and the distribution of roles amongst selected partners.

T-projects generally ended up as three-point structures, the key elements of which were:

- The groups of contracting laboratories would be united under the coordination of a named expert;

- The supervising body, consisting of Commission staff, member state representatives and the coordinator, in charge of controlling the progress and implementation of the project and the circulation and development of information outside the project. The coordinators were responsible for the integration of work and the stimulation of the interface between laboratories participating in the project. Their role, already essential in the preparation and submission of proposals to the Commission, also took on other administrative and scientific duties once the project was selected and being implemented. The coordinator's duties included the regular publication of internal newsletters, the organization of meetings, the publication of lists and catalogues of techniques, protocols, reagents, experimental probes that could be exchanged, the negotiation of functional relationships with the industrial platform, etc.;
- An optional industrial platform, a flexible way to keep industry informed of developments. It was a free organization, independent from and with no contractual arrangement with the Commission. It was a forum for industrialists, big or small, who had not excluded the possibility that the T-projects would, in the distant future, produce something to their benefit and wanted to stay in touch with Community research. They provided a certain flexibility to industrial scientists who did not want to get involved in the research contractually but wanted to get up-to-date information on current research and did not mind paying the small financial contribution to be part of the industrial platform. They also provided a forum for answering scientific questions from users of research results and a channel for feedback from applied biotechnology research to fundamental biotechnology research, feedback being essential for the implementation of Community biotechnology strategy and the objectives of the BRIDGE program.

The following T-projects were set up:
- Characterization of lipases for industrial applications (three-dimensional structure and catalytic mechanisms);

- Improvement and exploitation of lactic acid bacteria for biotechnology purposes;
- High resolution automated microbial identification;
- Animal cell biotechnology;
- Factors regulating growth and differentiation of plant cells.

In addition, the BRIDGE program generated the first European actions in the field of systematic large genome sequencing. One T-project was to lead, by 1996, to the completion of the *Saccharomyces cerevisiae* yeast genome sequence, and another promoted informatics networks. The program, which started with a "feasibility" phase, was a model for later activities in European genome research. The history of this project is particularly interesting, because it was so very successful.

6.5 The Origins and Nature of the Yeast Genome Sequencing Project

André Goffeau was one of the co-authors of the document[44] that unleashed the movement towards the first European biotechnology program, BEP. His role at the commission included the launch of a series of new and innovative actions. It was his idea to prepare for the yeast genome sequencing project over 1986–1988 and then initiate it, at the same time as the Commission was setting up the new biotechnology program, BAP, revising it the following year and preparing for the next program, BRIDGE.

Just how large were the tasks of preparing new programs and managing contracts at the time, both of which were entrusted to Goffeau with the trustful supports of de Nettancourt and Van Hoeck? He had to consult researchers, policymakers at various levels (political, Commission, member states, industrialists), write reports and draft projects, have them considered,

---

[44] D. de Nettancourt, A. Goffeau and F. Van Hoeck, *Applied Molecular and Cellular Biology (Background note on a Possible Action of the European Communities for the Optimal Exploitation of the Fundamentals of the New Biology*, Report DG XII, Commission of the European Communities, XII/207/77-E, 15 June 1977.

and in general generate a wave of opinion to sweep away the inertia characteristic of large institutions. For André Goffeau, also a professor at the University of Louvain, there were visits to the various contractors in their laboratories, the establishment of the training program and evaluation duties on the proposals coming in under that program[45]. Furthermore, at the request of F. Van Hoeck and D. de Nettancourt, Goffeau was taking part in meetings with highest-level decision makers, such as the European Parliament and CREST. In retrospect, André Goffeau recognizes "that all this activity was very constructive, and those years were a fantastic apprenticeship to the processes at work inside the arcane decision-making institutions. It also provided me with a better understanding of the scientific community through the many visits I had to make to contractors and the fact-finding missions overseas for the Commission." But he freely admits that, at the time, as a committed scientist with a second life in the laboratory, he found the commission stuffy, and it was becoming increasingly hard to reconcile the two halves of his double existence. He wanted to leave the Commission and return full time to his first love — scientific research, which always had been, and still is, his passion. De Nettancourt understood this, but asked him to set up a major research program before he left.

It so happened that, on that very morning, André Goffeau had been reading R. Dulbecco's famous article, "A Turning Point in Cancer Research, Sequencing the Human Genome" in *Science*[46]. This article, mentioned earlier, pointed out that a better understanding of cancer could and should be obtained from sequencing the human genome. Another factor at the time was the opinions of a fellow civil servant, Mark Cantley, who predicted that genome research was going to be increasingly important in the future. There was also the structure of the Commission, which had managed, with great difficulty, to put together the first supranational European biotechnology programs. André Goffeau was impressed by Dulbecco's article, but the entire human genome looked far too ambitious. And that is how Goffeau came to propose sequencing the yeast genome.

---

[45] From 1987 onwards, this responsibility was transferred to D. de Nettancourt.

[46] R. Dulbecco, "A Turning Point in Cancer Research, Sequencing the Human Genome", *Science*, 231, 1986, pp. 1055–1056.

Yeast was not a haphazard choice for Goffeau. He had begun his studies with a degree in Agronomical Engineering (specializing in the tropical regions) in 1956 from the Agronomical Sciences Faculty of the catholic University of Louvain, UCL, in Belgium. He obtained his doctorate in Botanical Sciences in 1964, also from the UCL, after a year (1959–60) as a research engineer at the National Institute for Agronomic Studies in Yangambi, in what was then the Belgian Congo. He obtained his professorial degree in 1969, again from the UCL, but this time from the Science Faculty.

Most of this training involved plant physiology, although he had gained pre- and postdoctoral training in enzymology during long research visits. From January 1961 to August 1963, he had been delegated by EURATOM to the INRA (National Institute of Agronomical Research) agronomical research station at Versailles, France, with Dr J. Bové. From then until August 1964, again as a scientist detached from EURATOM, he worked in the Animal Morphology laboratory of the ULB (Free University of Brussels) with Professor J. Brachet. From then until June 1966, he worked as a Commission scientific officer detached to the Institute for the Use of Nuclear Energy in Agriculture at Wageningen in the Netherlands with Dr de Zeeuw. And finally, from February 1967 to August 1969, he was detached by the Commission as a scientific officer to the INA (National Agronomical Institute) in Paris, France, with Professor H. Heslot, where he learnt genetics and at the suggestion of Appleyard (Van Hoeck's new boss and a pioneer of microbial genetics), took some interest in yeast.

From 1969 onwards, as his career as a civil servant took up more and more time, André Goffeau's scientific activities were restricted to the UCL, where he started lecturing and directed the FYSA unit (biochemical physiology) from October 1969, the most distinguished member of which was Doctor Albert Claude, who joined FYSA in 1972 and was awarded the Nobel Prize for Medicine in 1974. Goffeau's scientific interest was concentrated on the mechanism of energy transducing membrane transporters in yeast.

Through his research, André Goffeau had a good understanding of the genetics and physiology of yeast. He had also come to know the yeast scientific community, who describe themselves in French as *les levuristes*, especially Piotr Slonimski, who had encouraged him in 1980 to organise the 10th International Conference on the Genetics and Molecular Biology of Yeast at the UCL in Louvain la Neuve, which brought together 475 participants.

Given this research, these fields of interest, this academic career path and the factors determining his suggestion, it is hardly surprising that Goffeau, the civil servant, proposed a the complete sequencing of yeast as a great research project, especially at a time when European civil servants were under enormous pressure and influence, as they implemented the new European research strategies, and had to think up new programs. Compared to the human genome project, which at the time was still under discussion in the United States, sequencing the yeast genome also seemed a far more accessible proposition.

Initially, Goffeau's Commission bosses, D. de Nettancourt, F. Van Hoeck and the CGC Chairman R.R. Van der Meer, thought the proposal was not appropriate; but rapidly they saw that Goffeau was presenting them with no other option. In the meantime, Goffeau was testing the water with a few phone calls and visits to certain eminent *levuristes* and industrial firms. During these discussions, he obtained the following reactions to his idea of a European yeast genome sequencing project[47].

From the scientists

- P. Slonimski of the Laboratory of Molecular Genetics of the *Centre Nationale de la Recherche Scientifique (CNRS)*, Gif sur Yvette, France, thought that yeast was the best target for sequencing the genome of a eucaryote, but feared that several American researchers were already tackling individual yeast chromosomes.
- M. Olson of the Department of Genetics at the University of Washington, School of Medicine, in St Louis, USA, wrote that no similar project was under way in the United States and that he would make his organized DNA library of 95% of the S288C strain of the genome available to anyone who wanted it.

---

[47] The various reactions have been quoted from A Goffeau's *Sequencing the Yeast Genome: Preliminary draft for assessment and proposal of a new research and development program*, January 1987.

- D. McConnel of the Department of Genetics, Trinity College, Dublin, Ireland, pointed out that from a scientific point of view, the project should be carried out. However, he did not think that enough qualified people would be available and wondered whether it might not be necessary to first promote the development of new sequencing equipment based on other principles.

From the industrialists:

- D. Eyben of the Research Department at Stella Artois Breweries, Leuven, Belgium, thought that the project would be justified in the eyes of the brewing world and should be evaluated by the microbiological research group of the European Brewer Committee (EBC).
- J. van der Platt of the Yeast Research Department at Gist-Brocades, Delft, the Netherlands, thought that industry would probably only be interested in some *S. cerevisae* genes. However, Gist-Brocades might well take part in the project.
- A. Hinnen of the Department of Microbiology at Ciba-Geigy, Basel, in Switzerland, said that from the scientific policy point of view the project was interesting because yeast is a fundamental organism in biotechnology and industrial cloning, and Ciba-Geigy would support the project.
- M. Kielland-Brandt of the Department of Physiology at the Carlsberg Laboratory, Copenhagen, Denmark, pointed out that sequencing a laboratory strain of the yeast genome would provide probe tools that would accelerate studies on industrial strains of yeast used in brewing.

During his visits, André Goffeau received the following reactions:

- A. Hinnen and the management at Ciba-Geigy (Switzerland) confirmed that the project was interesting and that they would provide it with their support.
- Von Wettstein, Kielland-Brandt, Nilsson-Tilgreen and Keiden, and another 30 scientists at Carlsberg in Copenhagen, thought the project

was scientifically interesting but did not think that the brewing industry could pay for it.
- J. van der Platt, de Leeuw, Sanders, van den Berg and G. Groot, and several young scientists at Gist-Brocades in Delft, the Netherlands, said that the project was just as necessary as a dictionary is when you are looking up the meaning of a word, therefore, Gist-Brocades would take part in the project.
- U. Stahl and seven other scientists in the Department of Microbiology at the Institute for Fermentation Technology and Biotechnology of Berlin regretted that their department had no sequencing experience. They thought that five German industrialists would be interesting in supporting the project financially.
- Lastly, Willmitzer of the Schering Free University of Berlin Biotechnology Institute thought the project was interesting because yeast is the only eucaryote where homologous recombination can be used to get null mutants (or others). The project, he thought, would be of some interest to the great chemical industrialists (Schering, Bayer, Hoechst and BASF) and could eventually be extended to *Arabidopsis*, which has five chromosomes totaling 100,000 kb.

Goffeau was convinced of the scientific viability of his project. He drew up a first sketch of the program, entitled *Sequencing the Yeast Genome: preliminary draft for assessment and proposal of a new research and development program*[48].

The First Draft for the Yeast Program

In this first version of the document, in January 1987, André Goffeau justified and developed his original idea. As an introduction, he mentioned the human genome project and the debate around it. He highlighted the difficulties and limitations of the human genome project, such as the high cost (estimated

---

[48] A. Goffeau, *Sequencing the Yeast Genome: Preliminary draft for assessment and proposal of a new research and development program*, January 1987.

at between US$ 0.5 and 3 billion), sequencing technologies that were still too slow, and ethical problems. He also pointed out that it had already been suggested that the genomes of mice, yeast, *Drosophila melanogaster*, *Caenorhabditis elegans*, rice and *Arabidopsis thaliana* be sequenced, all of which have far smaller genomes than man, but which have considerable industrial and scientific use without causing any problems of an ethical nature.

Yeast as a Model Organism

Yeast was chosen as a model organism over other options for several reasons. It is part of human history, through bread and wine, and was intimately involved in the birth of biotechnology. In the 1970s, yeast had progressed from its status as a sort of *Escherichia coli* with a nucleus to that of Universal Cell[49], becoming a choice target for the study of many biological processes shared by all eukaryotes. Because of the similarities of function between simple eucaryotic cells and those of higher organisms, cloning cDNA from a plant or an animal through functional complementation of a yeast mutant is quite easy. Yeast is thus a sort of biological test tube, providing conditions close to those in the original animal for the functional study of animal proteins. It is also an ideal host for high industrial value proteins, since yeast, and this is exceptional in eucaryotes, has its own natural plasmid. This allows genetic amplification systems to be developed very rapidly, and they are the base of the overexpression of heterologous genes coding for proteins of industrial interest. The last argument in favor is the interest that industry has in yeast and the fact that it can be cultivated in affordable conditions. Molecular genetics could improve the performance of yeast in its traditional uses in bakery, brewery, distillation and winemaking.

Considerations of a genomic nature were also part of Goffeau's decision. "The last genetic map[60] of *Saccharomyces cerevisiae* describes the localization of 568 genes distributed over 16 chromosomes of different sizes

---

[49] L. Hartwell, *The Cetus, UCLA Symposium on Yeast Cell Biology*, Keystone, Colorado, 1985, quoted by A. Goffeau in the January 1987 version of his draft.

ranging from 200 to 2,200 Kb (kilobase pairs) plus a single gene located on a 17ch chromosome"[50]. Unlike other genomes of higher organisms, it was thought that the yeast genome has few repeated sequences, mobile elements or introns, and that the coding sequences were gathered into a very compact genome of about 6,500 genes of an average length of 2kb. When compared to the genomes of *Caenorhabditis elegans* (100 million base pairs), *Arabidopsis thaliana* (100 million base pairs), *Drosophila melanogaster* (150 million base pairs), the mouse (3 billion base pairs) and man (3 billion base pairs), the yeast genome is quite modest in terms of size. Estimated at 13 million base pairs, its genome is 200 times smaller than man's, and that meant the cost of the sequencing program would be reduced by the same factor. Furthermore, by 1987, 200 to 300 of the genes had already been sequenced, about 5% of the whole genome.

Goffeau had more arguments in favor of sequencing yeast in particular. Firstly, our understanding of the biochemistry, genetics and biology of yeast was far ahead of our understanding of other organisms. Secondly, sequencing yeast was neither fundamental research nor a pure application of science, so there would not be too much competition between the participating scientists and industrialists. Such a project seemed totally suitable for the development of a world network of cooperation with clearly defined objectives and activities. Thirdly, the ability to isolate chromosomes and the fact that in 1987 an ordered set of genome clones was already available, were undeniable technical advantages. To build such a library for other genomes such as *Arabidopsis thaliana* would need serious work for two to four years. Lastly, Goffeau maintained that the understanding of the fundamental structure of the yeast genome would provide us with new information that could feasibly be applied to larger genomes. For example, about 6,500 gene probes would be available to help hybridization locate similar genes in many of species of direct industrial interest.

It also seemed to André Goffeau that to start a project that could be completed with the technology available at the time was the most efficient way to promote the development of sequencing technologies and to encourage

---

[50] A. Goffeau, in the above mentioned document, January 1987, p. 4.

dissemination of the know-how that would be needed to tackle the larger genomes, in particular, those of plants and man.

Goffeau had also foreseen the objections:

- Although it is a full eucaryote, yeast is still just a single-cell organism and cannot be used as a model to study tissue development and cell differentiation;
- The information expected from sequencing would, in any case, eventually become available in the long term without any specific program by bringing together the results of free competition amongst the 2,500 *levuristes* across the world;
- An organized program would not use time and money efficiently unless the bureaucracy was not suffocating; otherwise, bringing the money together could take longer than actually sequencing the organism itself;
- Traditional yeast-based industries had very little experience in genetics research. The prosperous times that they were enjoying would not encourage them to develop new tools, however useful they might be;
- The money could be diverted to other fields of more direct applicability.

After presenting the pros and cons of the yeast genome sequencing project, André Goffeau considered the technical aspects of its implementation.

Technical Considerations

To sequence a genome, one first isolates the DNA in the form of a complete organized library of DNA fragments. This library has to be characterized as finely as possible to prove, by recovery, that all the bits are still there. In 1987, M. Olson and his colleagues had constructed just such a library with about 5,000 clones of an average insert size of 15 kb from the S288C strain of *Saccharomyces cerevisiae*[51]. A *Hind*III and *Eco*RI restriction map was

---

[51] M. Olson, J. Dutchnik, M. Graham *et al.*, *Proceedings of the National Academy of Sciences*, USA, 1986.

also under way. In addition to this "bottom-up" approach, there was also a "top-down" approach under way: the 16 chromosomes had been isolated and their respective DNA was being characterized with restriction enzymes such as *Not*I and *Sfi*I, which provided a small number of large fragments. By 1987, eight of the chromosomes had been mapped in this manner. The two complementary approaches provided a global restriction map and a limited number of DNA fragments covering the entire yeast genome.

As for organizing the sequencing itself, Goffeau had two options, each with their respective advantages and inconveniences. It could, on the one hand, be run at a large centralized establishment with greater efficiency and making it easier to adapt to the more advanced technologies. The disadvantage would be that a new research site would have to be set up and scientists would have to be hired. On the other hand, the sequencing tasks could be distributed between several separate laboratories. This would involve all the scientific *levuristes*, possibly half in Europe and half in the USA, as well as the industrial yeast community. In order to avoid spreading the work too thin and making it less efficient, the project would have to be restricted to researchers who already had good sequencing experience as well as direct interest in the yeast genome. The advantages seemed numerous; in particular, such a structure would be functional immediately, providing the necessary expertise. No heavy investment would be needed. The disadvantages that Goffeau could see were that dispersed groups would be less open to new technologies than a centralized institute and that tight quality control was necessary. In fact, he envisaged that sequences would have to be done twice[52].

As for sequencing techniques, the choice would have to be left to the researchers since several techniques were already available and new techniques were continually coming on to the market. Goffeau also looked at the new sequencing machines that could greatly accelerate the completion of projects, in particular, an Applied Biosystem sequencing machine which was one of the most powerful, but whose efficiency and accuracy had yet

---

[52] In his 22 May 1987 answer to the questionnaire sent out by Goffeau, F.M. Pohl said that the quality control element could be handled in a more formal manner than by simply resequencing the same piece of genome.

to be proven at that time. It was clear that a centralized research group would be able to adapt more quickly to new techniques. However, a decentralized approach would provide a larger market for new equipment and allow many laboratories to learn new techniques benefiting later multi-faceted projects of similar or greater complexity.

The number of clones needed, counting the checking work, was thought to be 3,000 (with an average size of 15 kb for the 1 clones). If 50 kb could be sequenced per year per laboratory, then the project would require 900 man/years or 112 laboratories employing two people each for four years. This was a worst-case estimation since technical progress was expected to reduce these figures very rapidly.

As for general organization, one single laboratory should be responsible for the constitution of the ordered library. M. Olson of St. Louis was everyone's candidate for this job since he was providing his library to the rest of the world. In addition, a central coordinator was considered essential to carry out the following tasks:

- The organization of the various sequencing groups and ensuring that the clones were distributed. For American and Japanese participants, a central coordination set-up for their continents would probably have to be foreseen;
- The attribution of responsibility for homogenization of software and even hardware, which would be of great importance in exchanging information, comparing results and promoting feed-back;
- The gathering, recording and analysis of resulting sequences. Centralized analysis was thought necessary to limit and locate ambiguity and for large scale searches for specific regulation sites, analysis of replication, recombination and integration *loci*, for structural and functional prediction, for the search for structural similarity and for the constitution of all sorts of statistics;
- Promoting the distribution of results to the scientific and industrial communities.

In his project, A. Goffeau thought it of major importance that the coordination tasks be undertaken by laboratories with a good understanding of yeast and its genome and long-standing interest in it. These laboratories should also be recognized and respected by the *levuristes* so that they could communicate with authority.

As for computer structure and needs[53], H. Werner Mewes of the Max Planck Institut für Biochemie in Martinsreid, Germany, made the original evaluation using, at Goffeau's request, the US project as a model, involving the installation of a central computer installation with a Cray supercomputer that cost US$15 million and US$3 million per year in maintenance, with an ancillary staff of 30 people costing another US$3 million a year. The terminals, the workstations, the servers and the networks were estimated at another US$2 million. If the project lasted for ten years, the computing cost would be three cents per base pair. If the project was decentralized, the central computer for the yeast genome sequence could be a superminicomputer of the Convex type, costing about a twentieth of the Cray, and with a team and maintenance costing only US$300,000 per year. Local computing equipment would be adapted to the length of the sequences. If the project was very dispersed (50kb per lab), small PCs would be enough for all routine data entry and analysis. If there were to be larger sequences per laboratory (300 kb), investment in a powerful workstation with appropriate memory (Microvax, IBM 6150 or Sun Station at US$25,000 each) would be useful. These computers would provide almost 40 times the capacity of normal PCs for only five times the price. The total cost estimated for all the equipment (including personnel and hardware) was US$ 1.6 to 2.85 million.

If the laboratories were decentralized and each received 300 kb to sequence, half the computing needs would be in the laboratories themselves, each with its own powerful workstation. The workstations looked expensive, but they would promote the common evolution of know-how and good technology transfer between industry and the laboratories involved. Furthermore, intelligent workstations and to a lesser degree superminicomputers such as the Convex would, unlike the Cray, provide open structure

---

[53] The advice of Drs. H.W. Mewes and F. Pfeiffer of the Max Plank Institut für Biochemie, Martinsreid, Germany (F.M. Pohl, in his answer to the questionnaire sent out by André Goffeau, 22 May 1987, and H.W. Mewes and F. Pfeiffer, in their answer of 7 June 1987).

6 European Biotechnology Strategy and Sequencing the Yeast Genome 381

without coercive norms and could be used for other more mundane tasks as well. The cost of local computers could also be partially supported by host laboratories.

The way in which the central computer should be linked up to the local stations remained to be seen, but several rather complex options were proposed. This problem finally solved itself by the unexpectedly rapid development of the Internet.

Goffeau thought that an overall budget of US$50 million would be required for a period of five years (one year to set up the structure and the basic techniques and four years for the sequencing work), US$30 million of which would go to the sequencing itself. Auxiliary research to develop and promote the automation of sequencing techniques and the construction of ordered libraries of other species of scientific or industrial interest would have to be encouraged. The work and the budget could, and he thought should, be shared between the European, American and Japanese funding agencies. This would mean that the Commission's budget for sequencing would only be about 3.3 MECU per year over five years. This estimation turned out to be fully correct. If the EC decided to carry out the project on its own, the budget would be 10 MECU per year, which André Goffeau thought a modest investment given the scientific, technical and industrial spin-offs that could be expected, and if one considered the push it would give to European science policy.

Since this project was at the interface between applied and fundamental research, Goffeau did not think that European industry could be expected to make much of a contribution. However, a quick survey carried out by Roy Tubb showed that many industries (brewing, yeast producing, chemical industries and cloning firms) were ready to support the project through membership in a yeast industrial club. Membership fees of about US$10,000 did not seem excessive since the members would have privileged access to information. On this basis, it was hoped that about fifty European industrialists would join, which would cover five to 15% of the costs of the project. This money could be spent exclusively on coordination costs. It was also possible that some industrialists would also take part in the sequencing, obtaining in-house expertise in the field. The chance of selling this information and products such as sequences, probes, and software to those not participating in the program was also taken into account, although it did not seem very likely that the profit to be made would offset much of the total cost.

Backed by the positive results from his preliminary discussions, overall strong support from the scientists he had contacted and the industrialists (even if some of the industrialists were still reserved as to whether they would participate), Goffeau presented his project to the CGC- Biotechnology at its 31 March 1987 meeting during preparations for the BRIDGE program.

The discussion on this presentation was postponed to the next CGC meeting planned for 22 June 1987.

6.6 Critical Discussions and the Adoption of the Yeast Genome Sequencing Project

During the 22 June 1987 CGC meeting, in the course of his detailed presentation of the project intended to provoke the reactions of the various attending parties, Goffeau made several further observations. He reminded the committee of the reasons for choosing yeast above other genomes for the planned sequencing project and then made a quick survey of sequencing technology available at the time, which can be summarized as follows.

In Japan, Tokyo Seiko had built a robot that, according to Wada[54], would be operational by 1990 and could sequence 300 million base pairs a year instead of the 20,000 base pairs being sequenced per year at the time. In the USA, Applied Biosystem's sequencer had come on the market in 1986 for US$100,000, and it could sequence 100,000 base pairs a year. Thirty sequencers had been delivered at the time, three of which had come to Europe. The deliveries had each taken three months. Two scientists had divulged to Goffeau privately that, in practice, many of the users were having problems with the machine[55]. In Europe, Genoscope, a Swiss-built

---

[54] A. Wada, *Nature*, 1987.

[55] A. Goffeau's mission report *Preparation of a Yeast Sequencing Program*, USA, 14–19 June 1987, Brussels, 1987.

machine, was thought to be as fast if not superior to the American machine. Several machines had already been bought by European laboratories. At the European Community level, two of the BAP contractors (Martin[56] at Manchester University and Thomas Pohl[57] at the University of Konstanz) were developing automated sequencing equipment that they told Goffeau could be better than the American and Swiss machines.

Professor Goffeau also pointed out that the community had all the researchers necessary to carry off the project; he had details of 825 senior *levuristes* in the member states (Table 13).

Table 13 Distribution of the 825 senior *levuristes* in the member states (1987)

France	221		The Netherlands	39
Great Britain	216		Ireland	39
Germany	127		Denmark	21
Belgium	81		Spain	20
Italy	55		Portugal	3
Greece	3			

Several leading representatives of this scientific community supported the project with enthusiasm. This convinced Goffeau that human and technical resources were not going to be a major difficulty. However, doubts were being voiced in the USA that the European Commission would be able to handle a project of such size, because of the European-level structure needed for the efficient completion of such a major project. During a fact-finding mission to the United States from 14 to 19 June 1987[58], Goffeau recounts the following conversation with Maynard Olson: "Olson thought that it would be more efficient for the yeast genome to be sequenced by a few laboratories rather than through an extensive network. He did not think such a program could be run efficiently by the EEC. Although at the time, there

---

[56] W.J. Martin and R.W. Davies, *Biotechnology,* 4, 1986, pp. 890–895.

[57] F.M. Pohl and S. Beck, *EMBO J.*, 3, 1984, pp. 2905–2929.

[58] A. Goffeau's mission report *Preparation of a Yeast Sequencing Program,* USA, 14–19 June 1987, Brussels, 1987.

was no budget for a similar project in the USA, he thought things could change rapidly. Olson planned to go to Los Alamos to bring the problem up with the DOE. He was also waiting for the visit of a Japanese delegation from the RIKEN to discuss similar Japanese projects. Although he did not want to be formally included in a possible sequencing project, Olson thought that sequencing the yeast genome, from a purely scientific point of view, was an enormous task and that the information it would generate would be of enormous value, both from the scientific point of view and that of the possible spin-off applications".

The fact that many laboratories might not have enough willpower to press on to the fixed goal in the face of difficulties was another objection to the network approach. Goffeau conceded that it would be tough to start the project with 50 laboratories asking each of them to mobilize four researchers per year, but he pointed out that unlike the United States and Japan, none of the member states was able to embark on such an enterprise on its own.

In short, Goffeau's message to the CGC was that a detailed two-stage feasibility study would have to be undertaken before the project could be expanded under the BRIDGE program. In the first stage of the study, one laboratory (his own, under his control) would begin a pilot trial as soon as possible, in September 1987, to sequence a series of overlapping clones made available by M. Olson, in order to find out what problems might occur and also to be able to draw up a sequencing protocol for a 15-laboratory group to use from the beginning of May 1988. These 15 laboratories would each take on a full-time researcher to carry out sequencing work for two years with the (not yet automated) equipment available at the time, intending to sequence 50,000 base pairs per year over two years. The laboratories would be linked by a computer network, a valid exercise in bioinformatics in itself, which would have the advantage of being based on a clearly defined objective. One of the laboratories would have to coordinate the whole affair. The goal in view was to sequence an entire chromosome.

After these two years, in May 1990, there would be an evaluation of the pilot phase as well as a review for further financing, reports and an analysis of Europe's stance *vis-a-vis* a future enlargement of the initial phase and the budget to the USA, Japan, Canada and Switzerland, allowing more chromosomes to be sequenced. Goffeau's budget proposal was to devote a total of 3.4 MECU to the program from 1987 to 1989.

Goffeau pointed out that now that the era of large-scale DNA sequencing was upon us, Europe needed to begin the project as soon as possible, perhaps on chromosome III; otherwise, it would lose a unique opportunity for a competitive advantage over the USA and Japan. Goffeau reduced the apparent dichotomy between cooperation and competition with the USA with his vision of close association. For him, Europe should first reach a strong position from which to negotiate, after which it could dictate the conditions of cooperation. Benefits to Europe itself could not as yet be predicted very accurately, but they were thought to be considerable and would include:

- The discovery of new genes of as yet unknown function;
- The improvement of sequencing methods and the development of new generations of equipment within Europe;
- The development of new computing and communications tools;
- A benefit for industries that use yeast in the form of a general understanding of the genome likely to be used a great deal, for example, in the development of vectors or probes;
- A preparation for megasequencing projects.

In general, the reactions from the various delegations were favorable. However although he stressed the importance of the project, Goffeau received some criticism from the committee, linked in general to the science policy strategies prevalent in the member states at the time.

For example, the Danes pointed out that although it would be easy to make the decision to sequence, they were worried about the exploitability of the information. Accompanying measures would have to be planned in the field of DNA structure-function studies. It remained to be seen whether the European scientific community was best placed to exploit the results. The Irish asked that the motivations of the trial project be more clearly defined. Was it a pre-run for the analysis of the human genome a tool to be offered to the yeast-based fermentation industries? It was pointed out by the Dutch that the lack of a link between the sequence and the fundamental knowledge needed to handle biological problems in the fermentation industries meant that the project was ill-adapted to the BRIDGE target of building links between education and industry. The answer to this problem would be forthcoming from industry, but this meant that industrial support was even more important. The Italians mentioned that the community's expectations should not be formulated in commercial terms because that was not the point of the exercise.

For Goffeau, the issue was whether Europe was or was not going to be a runner in this particular technological race. He showed that the exploitation of results would be done by particular molecular biologists, half of whom work in Europe anyway. For him, only these specialists could really make use of the results and could probably do so more efficiently and usefully than the Department of Energy (DOE) in the United States. The scientific community had told him they were ready to make a head start on the project. Potential industrial partners could be brought in too, especially from the German chemical industries, brewing and agro-food industries.

France pointed out the many positive infrastructural aspects that would result from such a project, in particular, in instrumentation and computer networks. Britain agreed that the project had many qualities — its great size, strong image, clear impact, clearly defined requirements and easy evaluation in terms of cost and results. It was not certain that the EMBL would be involved. After his investigations, André Goffeau did not think that the EMBL was interested in direct involvement, during the preliminary phase at least, but participation could be foreseen for the interpretation of results.

With regard to American involvement, André Goffeau pointed out that the whole project depended on the goodwill of the aforementioned American, Maynard Olson[59], who would distribute his yeast DNA clones from the three libraries he had built from strain S288C to European laboratories. The first library had 4,000 small clones, and the other two libraries, built within the previous ten years, contained about 50 large non-overlapping clones of about 300 kb in length. Olson had agreed to provide the clones on request. As for the urgency of getting the project under way, it was not a question

---

[59] *Ditto.*

of whether the USA would set up a coordinated yeast sequencing program, but when. Goffeau thought that the project would fall under the auspices of the human genome project as an unavoidable phase of its development.

The German delegation voiced its preference that the coordination be international rather than European and was sure that the United States would be prepared to help with the costs. Britain expressed its reservations as to European ability to be competitive. Italy, however, was less pessimistic and pointed out that EURATOM was an example of a very optimistic initiative at a time of debate over a European framework for the exploitation of atomic energy. It was not Europe's destiny to be outstripped by its competitors as long as the community could take initiatives. Goffeau agreed that as the community would have to begin with self-reliance, it could defer negotiation with the United States to a later stage of the project.

As for the budget needed for the first preliminary phase (3.4 MECU), Goffeau did not expect co-financing from the national laboratories. The Chairman of CGC-Biotechnology, however, eliminated the possibility that a Community program would pay for this project indiscriminately. He expressed the view that a European program is only legitimized when funding priorities are determined by the member states, which was not the case. This legitimization should also in its turn encourage the attribution of funds to the project from the national budgets. Goffeau was later to use this argument to justify his assertion that a good project should not compete with other projects for established funds, but should receive funds according to its quality. The delegations finally reached a general consensus as to the global objectives of BRIDGE and on the need for national and industrial support to be manifested in direct co-financing.

Despite the positive and constructive responses that Goffeau had received from potential partners after his first questionnaire was sent round, many of the delegates remained worried by the nature of the work which was seen, like the human genome project, as being unmotivating and unlikely to reward the effort put into it. Lastly, the choice of yeast, justified by its industrial uses, worried several delegations who feared that the food and brewery industries would not want to participate as much as anticipated. Indeed, yeast fermentation is only one of many bioprocesses used in agriculture and industry. On the other hand, the delegations recognized that yeast was, indisputably, the best model for eucaryotic studies. Various declarations to this effect from Italy, Spain, Germany and Goffeau confirmed that studies on the yeast genome could benefit the analysis of other genomes such as those of *Streptomyces*, an antibiotic-producing bacteria, plants and of course that of man.

Dreux de Nettancourt, backed by the German delegation, wound up the meeting by concluding that the precompetitive nature and long-term industrial interest of the project and the fact that none of the member states could hope to undertake the task in question on its own, qualified the project as a European one *per se*. The budget for the pilot phase was considered excessive by some of the delegations and the chairman of the CGC. However, Italy and Britain thought the budget was not excessive, and that exploratory studies or perspective analyses that would only need a small amount of money were quite different from a pilot sequencing trial with a heavier cost burden, but just as justifiable, for its ability to produce quantitative information that would allow objective evaluation. This presentation to the delegations of CGC-Biotechnology, and their reactions, were encouraging, although many issues remained pending.

Goffeau continued to collect opinions and comments from his colleagues and industrial correspondents (Tables 14, 15 and 16) for the next meeting of the CGC on 9 September 1987. More specifically, he was informed by L. Philipson[60] that the EMBL was not interested in participating in the yeast project. The reasons for this reticence were the following:

- The EMBL was, at the time, planning to experiment with sequencing part of the human genome with the sequencing machine it had developed;
- The EMBL was planning to extend its database;
- Discussions deep inside the EMBL indicated a major yeast genome sequencing program was being prepared in the USA, and this would be difficult to compete with within a concerted European effort.

---

[60] L. Philipson in his letter to A. Goffeau, 28 July 1987.

Most of the answers to this later round of investigations by Goffeau were enthusiastic and positive, many of them pointing out how important the project was from the political, scientific, technical and industrial points of view as well as for the analysis of higher organism genomes, in particular, that of plants and man, while highlighting the need for functional analysis and the identification of cell products. However, there were still some which were far more critical. As Goffeau had foreseen, the main objection was with regard to cost. M.S. Rose[61] of the ICI Corporate Bioscience Group and A.J. Kingman[62] of St Catherine's College pointed out that in the prevailing climate of severe restrictions in science funding, it was madness to squander the little money and staff available on a project with such an unattractive cost-benefit ratio and involving such a boring activity. H. Nielsen and Dr Niels-Fill[63] of Novo Nordisk had grasped the long-term benefits for European science and industry in participating in megasequencing projects. They conceded that such programs could stimulate the equipment and chemical industries in yeast-related fields and those touching on the computing aspects of manipulation and analysis of projects. For this reason, they wondered whether it would be appropriate for BRITE or ESPRIT (other non-life science European Community science programs) to co-finance the project. They could only give a qualified support to the yeast genome sequencing program if it were to be considered a conceptual exercise rather than an industrial project. Overall, they thought that the idea of the European Laboratories Without Walls would not adapt to confidential industrial research. They did not think more money should be spent on the yeast project. On the industrial level, a frequent criticism was that the yeast genome results would have to be disseminated widely[64].

---

[61] M.S. Rose, in his answer of 3 June 1987 to the questionnaire sent out by A. Goffeau.

[62] A.J. Kingman, in his answer of 19 August 1987 to the questionnaire sent out by A. Goffeau.

[63] A. Goffeau, mission report of 30 September 1987 on the preparation of "Sequencing the Yeast Genome", a discussion at Novo Nordisk with Messrs. H. Nielsen and Niels-Fill at Copenhagen on 20 August 1987.

[64] For example, take S.I. Lesaffre's answer of 7 May 1987. They recognized that such a project was interesting, but the fact that the results would be widely disseminated was one of Lesaffre's main reasons for refusing to support the project financially.

Goffeau received expressions of interest from non-European nations. D.Y. Thomas, Professor at McGill University, Canada, had divulged that he hoped to participate in the yeast genome sequencing project since he had recently been enlightened as to the utility of such research. His group had already, in fact, discovered a few genes in a single 8 kb clone[65]. P. Nuesch, president of the Schweiz-Koordinationsstelle für Biotechnologie (SKB), Basle, Switzerland, and Ciba-Geigy, also of Basle, thought highly of the yeast genome sequencing project and mentioned the contacts already in place between the European Commission and Ciba-Geigy's Basle R&D department[67]. Japan's Professor Okamoto also supported the project.

From then on, Goffeau saw that the sequencing work could be shared. Answering his letter[68], Olson let Goffeau know what clones were available for chromosomes III and VII and the related mapping information. Although the maps for these two chromosomes were still incomplete at the time, Olson was confident that they could be rapidly finished so that clone availability did not delay the sequencing work. Showing his goodwill, he asked Goffeau to let him know as soon as possible what his priority regions for sequencing were on those two chromosomes so that his groups could redirect their mapping work accordingly. He continued discussions with the RIKEN group from Japan for a pilot project sequencing chromosome VI. Decisions on future directions would depend on its results. Olson wished to be on good terms with the Europeans to avoid unnecessary duplication of effort[69].

In his letter to the European Commission[70] in response to a Japanese document on the Japanese rice genome sequencing program, Ron Davies of the United States underlined the importance of coordinating European and

---

[65] Letter from D.Y. Thomas to A. Goffeau of 13 July 1987.

[67] D. de Nettancourt's mission report "Discussion with the Office Fédéral Suisse de l'Education et de la Science on the possible participation of Swiss laboratories to Community R&D Biotechnology Programs", 29 June 1987.

[68] A. Goffeau's letter to Olson of 11 August 1987.

[69] M. Olson's letter to A. Goffeau of 7 October 1987.

[70] R. Davies to E. Magnien, 6 May 1987.

6 European Biotechnology Strategy and Sequencing the Yeast Genome

Table 14 Organisms, numbers and different answers obtained

	+	+/	−
CGC	9	2	1
Scientists	18	3	0
Industries	5	3	6

Table 15 Answers of the industries

Industries:		
+	+/−	−
Gist Brocades	Petrofina	Vioryl
Artois	Solvay	Henkel
Le Petit	Lesaffre	Limagrain
Ciba-Geigy		Nivkerson
Intelligenetics		ICI
		Novo

Table 16 Scientists who answered positively

Scientists:
Aigle Michel (Bordeau, F.)
McConnel David (Dublin, Irl.)
Felmann Horst (Munich, G)
Fukuhara Hiroshi (Orsay, F)
Goffeau André (Louvain-la-Neuve, B)
Heslot Henri (Paris, F)
Hollenberg Cornelius P. (Dusseldorf, G)
Jacquet Michel (Orsay, F)
Oliver Steve (Manchester, GB)
Planta Rudi (Amsterdam, NL)
Pohl Fritz M. (Konstanz, G)
Sentenac André (Saclay, F)
Slonimski Piotr (Gif-sur-Yvette, Paris, F)
Steensma H. Yde (Delft, NL)
Thireos George (Gr)
Thomas Daniel (F)
Wettstein (Von) Diter (Copenhague, D)
Zimmermann K.F. (Darmstadt, G)

Answers To Goffeau's Questionary (September 1987)[66]

---

[66] Provided by A. Goffeau.

American efforts in the megasequencing field in order to avoid duplications with the Japanese program. Julian Davies[71], director of the unit of microbial engineering at the Pasteur Institute, was also strongly convinced that an international agreement was an absolute necessity and pointed out that the coordinator of the entire project would have to be a good manager.

André Goffeau had left nothing to chance in his preparation and inquiries. Each field (computing, database, sequencing techniques, functional search, bioinformatics, communications problems, strains etc.) was probed and explored in his efforts to build an optimally efficient organization of laboratories and researchers to start sequencing the yeast genome as soon as possible with one of the chromosomes as a first phase. Preliminary studies had already been foreseen, especially an evaluation of the proposal on the point of view of the impact on industries using yeast[72]. Several scientists and industrialists including H. Werner Mewes[73], Steve Oliver[74], A.J. Kingsman[75], Clara Frontali[76] and Piotr Slonimski highlighted how important it was to do more than just sequence and were pressing Goffeau to complete the yeast genome sequencing program with a functional study of the Open Reading Frames (ORFs) discovered through identifying the gene products, etc. They pointed out that problems of inter-correlation, regulation and expression should be considered to reach a total structural and functional definition of the yeast genome, a definition only possible if an appropriate database were set up. H.W. Mewes, in his 7 June 1987 answer to Goffeau's questionnaire, said that as well as collecting these results, a large part of the task of such a database would be:

---

[71] Letter from Prof. J. Davies to A. Goffeau of 22 July 1987.

[72] Roy Tubb in his answer to Mark Cantley, who had sent him the file of preliminary studies. Tubb expressed his interest in doing an evaluation of the impact on the yeast (using industries) and put forward a study plan entitled *An evaluation of the long-term benefits of genome sequencing data to the yeast dependant industries.*

[73] H. Mewes in his 7 June 1987 letter to A. Goffeau.

[74] Dr S. Oliver to A. Goffeau, in answer to the questionnaire.

[75] A.J. Kingsman to Dr. M. Probert of the British MRC, in a letter of 19 August 1987.

[76] Dr. Clara Frontali in her letter to Mark Cantley of 8 July 1987.

6 European Biotechnology Strategy and Sequencing the Yeast Genome 393

- The identification of ORFs through comparisons with other databases;
- The advanced analysis of the protein-sequence results to identify structure and function properties of unknown genes;
- The organization of the database to ease access and communication.

The importance of a combined protein/nucleic acid approach with high-level computer analysis techniques and the identification of proteins on two-dimensional gels were cited as being a considerable step forward in the analysis of the yeast genome, in particular, for a better understanding of the dynamic processes of the cell cycle and interactions between the organism and its environment.

Backed by this new information and different points of view[77], Goffeau made another presentation of his project to the CGC-Biotechnology meeting of 9 September 1987[78]. His conclusions were:

- That a professional computational infrastructure would be needed;
- That the current databases were not appropriately equipped;
- That a new infrastructure for the organization and administration of the project would have to be foreseen and that some preliminary technical solutions were already on offer;
- That the project would be a unique contribution to the development of computer network systems;
- That an immediate start (2nd phase of BAP) of a pilot project was needed involving 10 to 20 participants working on the small chromosome III.

As for the chromosomes, the distribution was foreseen as follows (Table 17):

---

[77] Tables 14, 15 and 16 represent the results of his preliminary enquiries at this stage.

[78] The detailed draft agenda from the CGC-Biotechnology Meeting of 9 September 1987 at the European Commission in Brussels.

Table 17 Distribution foreseen for the chromosomes

Chromosomes	Size/Remarks	Scientist to sequence
XII	2,200 kb (contains about 1,000 kb of ribosomal repeats)	R. Planta (Amsterdam)
VII	1,300 kb	A. Goffeau (Louvain la Neuve)
II	800 kb – good restriction map from an industrial strain and isolated clones for 600 kb were available	H. Feldman (Munich)
III	380 kb – Olson strain (S288C) was used for a detailed map of a 180 kb segment made by Caroll Newlon) near 40 kb already sequenced in Oliver's laboratory	S. Oliver (Manchester)
I	250 kb – Olson strain had been used to map 174 kb in detail	I. Steensma (Delft) and R. Kaback (USA)

Chromosome VI was, at that time, the object of an agreement between Olson and a Japanese group[79].

Except for the British, who thought the project still conceptually vague, most of the delegations were not so much discussing the merits of the initiative, as how to implement and pay for it. With regard to means of implementation, the German delegation pointed out that modern management methods would be needed if Europe really wanted to be competitive. Several delegations (Germany, Denmark, Great Britain) remarked that the project coordinator would have to be a Commission official with good experience in the management of modern projects. Particular attention would have to be paid to international coordination beyond the borders of the European Community. Ireland seconded several scientists who had said that the project would be of no use if it only consisted of producing a simple sequence, so it should also develop capabilities of data interpretation (structure-function), a task that would really catch the interest of industrialists.

---

[79] *Washington University Record*, Vol. 12, no. 15, pl. Dec 1987.

As for the financial implications, the clearest reservations were made by the Netherlands and Spain, who pointed out that the sequencing project could hardly be called a "feasibility project", given its size. They thought the project was getting too large and that it might block the way for any other potentially interesting projects that could be submitted for the preparation of BRIDGE.

The CGC's chairman was against the Commission paying for everything. Several delegations were ready to reconsider their decision and increase the support they would allocate to the project if industry could be persuaded to play a more systematic part. The Netherlands hoped that national financing could top up an initial Commission contribution of 1 MECU. Spain and Italy offered to co-finance one or two laboratories participating in the network.

In practice, by the end of this meeting, Germany's proposal to allocate the project 40,000 ECU from the BAP budget was approved by Greece, Italy, Great Britain and Ireland, and then ratified. This allowed a beginning on specific studies on the more delicate elements of the project without, at this very early stage, having to make a full call for proposals. Goffeau was instructed to take into account the results of these studies and the CGC's remarks before finalizing the preparatory papers for his full program. De Nettancourt also reminded him of BAP revision limitations, such as the budgetary ceiling which limited feasibility projects to 2 MECU. If there were no other input from industry or the other nations, this would allow ten laboratories two years' work before an extension under the BRIDGE program would be needed.

These 40,000 ECUs were used to fund the following preliminary studies :

- Steve Oliver, at the Manchester Biotechnology Center of UMIST in Manchester, would obtain the clones from M. Olson and draw up a report proposing "A detailed strategy for the sequencing of chromosome III";
- Horst Feldmann of Munich and H.Y Steensma in Delft would write a report on a "Detailed strategy for the construction of organized libraries for the mapping and sequencing of the genome of *S. cerevisiae* and other industrial microbes";

- H. Werner Mewes and F. Pfeiffer, Martinsried, were to propose a "Detailed assessment on the bioinformatics network required for the collection and processing of the yeast genome sequence data";
- R. Terrat, at the CRIM in Montpellier, would write a paper on the "Computer environment for yeast genome sequencing";
- Fritz M. Pohl would draw up a "Detailed assessment of equipment necessary for sequencing the yeast genome";
- Roy Tubb, who had been sent the Goffeau file by Mark Cantley and had become interested right from the start in studying how to evaluate the project's impact on industry[80], was asked to study the "Long-term benefits of genome sequence data to the yeast-dependent industries";
- Antoine Danchin, from Pasteur in Paris, was asked to give a global report on microbial genemes.

These studies, each costing 5,000 ECU, were to cover in detail the various aspects of setting up the yeast genome sequencing program: chromosome library building, the sequencing itself, computing aspects, seeing what equipment would be needed and disseminating the information to European industrialists using yeast.

F. van Hoeck, in his letter to P. Fasella[81] calculating the budget appropriation for these seven studies within the context of the preparations for BRIDGE, broke them down into the following two stages:

- A so-called "feasibility" stage over two years (1988–1989) during which a 20-laboratory European network would sequence one of yeast's smaller chromosomes (III, which has 300 kb), using microsequencing techniques available at the time;
- A "scaling-up" stage over 5 years (1990–1995) during which the BRIDGE program would sequence all the yeast chromosomes using the sequencing methods being developed at the time in Europe, the

---

[80] R. Tubb, in his letter of 2 September 1987 to Mark Cantley.

[81] F. van Hoeck, Note for the attention of Mr Fasella, "Engagement pour l'exécution d'une étude en vue de la préparation du programme BRIDGE", Brussels, 1 December 1987.

USA and Japan. It looked likely that this work would be shared with the USA, Japan and possibly Canada.

For F. Van Hoeck, these two stages were compulsory stepping stones to sequencing the human genome, which would not be started until after 1995. In the meantime, sequencing the yeast genome and other small eucaryotic genomes would be a phase indispensable for the development of equipment and computer tools. An understanding of the 5,000 to 20,000 new genes that could be found would be a vital basis for the analysis of the human genome. For scientific reasons (the discovery of new genes) as well as for applications, and from the industrial point of view for the automation of biochemical analysis, it was important that Europe began real activities in this field as soon as possible. According to Van Hoeck, each month lost could mean market losses in the next five years, and it was clear that the sequencing work being planned was the ideal framework for Community cooperation because it could only be carried out on an international basis.

The studies were to be finished by 31 January 1988 so that an initiative for sequencing the yeast genome could be set up. They slotted into the global strategy then under discussion for macro-sequencing DNA. The 35,000 ECU to pay for them was rapidly granted. In December 1987, P. Slonimski[82] asked for and received another 5,000 ECU to constitute a European yeast DNA sequence data base that would collect and assemble the yeast sequences to be published before 1988 as well as those sequences published in 1988. This proposal also contained the development of a computer program that would identify overlaps, analogies and motifs of particular interest for yeast genome analysis.

The Opinion of IRDAC's Working Party 5

In the meantime, IRDAC's WP5 had not been able to exert any influence over the deployment of the BAP revision funds. So it decided to concentrate its attentions on the BRIDGE program. During its fourth meeting in Brussels,

---

[82] F. van Hoeck, *Note à l'attention de M. P. Fasella concernant l'engagement pour l'exécution des 7 études en vue de la préparation du programme BRIDGE,* 30 October 1987.

on 24 November 1987, it gave its opinion on the yeast genome sequencing proposal as presented by André Goffeau. During his presentation, Goffeau had laid out the various ways in which industrialists could become involved, other than providing a direct financial contribution:

- The participation in sequencing;
- The development of equipment, software and chemical substances;
- The development of applications for some 5,000 new genes;
- Privileged access to results through the Yeast Industry Club (Platform).

IRDAC's WP5 had already pointed out that fundamental research (such as yeast genome sequencing) should not automatically be financed by the community. They added the following criticisms:

- Yeast is only one of many organisms used in industry, and obtaining the complete sequence of its genome did not look as if it would be of any particular advantage since it would not benefit industries very much. Furthermore, as it is only a single-cell organism, its usefulness as a model was questionable since multicellular organisms have control mechanisms that have no analogue in single-celled models.
- The pharmaceutical companies were much more interested in the human genome. Hoechst, for example, was interested in participating in the human genome program, although not in Germany, where strong ethical lobbyists were a hindrance to research. Hoechst thought they would be able to export the work to the USA or somewhere similar, maybe under the auspices of a body like the WHO. In any case, they were interested in the human genome, not that of yeast.
- The industry could suffer if the yeast genome were studied rather than the human genome, and there were concerns about the size of the budget as other projects would lose funding in consequence.
- The systematic character of the projects was questioned. Would it not be better to try to identify functions of specific industrial interest, rather than sequence 6,000 genes without any clear use?
- Novo's geneticists thought that additional information on the yeast genome would be of little use in, for example, improving the

production of insulin; therefore, they wondered whether industry would be ready to pay such a high price for such a small benefit.

More generally, the project was felt to be of a more academic nature, and given the competitive situation of European industries, not appropriate to industrial (and therefore, commercial) needs, except for structure/function relationships which might be of some use. IRDAC's WP5 confirmed this opinion, already expressed during its meeting of 26 May 1987, that the yeast genome sequencing project should not be elevated to the status of a BRIDGE T-project.

While the various studies were being drawn up, and despite the negative opinion he had received from IRDAC, Goffeau continued to receive indications of interest in his project and meet scientists who could or should play an important role. The round table discussion on the human genome organized by *Biofutur* magazine[83] indicated that French scientists preferred to work on human genes (in particular, those responsible for known hereditary disease) rather than embark on a chromosome in a systematic manner. Systematic sequencing was not a priority, but nobody disagreed with Goffeau's point of view that sequencing the small genomes (*Escherichia coli*, yeast and others) was an unavoidable stepping stone to sequencing the human genome.

André Goffeau participated in a meeting organized by F. Gros and B. Hess[84] to draw up a proposal for a joint effort in the field of genome technology, an effort which, according to its authors, was needed to keep Europe competitive in the field of genome technology, a field in which Japan and the USA already had several projects. The Committee for the European Development of Science and Technology (CODEST) launched an appeal for an *ad hoc* committee to set up a detailed preparatory consultation meeting between scientists, engineers, industrialists and other directors of institutions to help the European Community and Europe in general make these decisions.

---

[83] Round Table on the sequencing of human and microbial genomes, December 1987.

[84] A. Goffeau's mission report, *Réunion de concertation sur le séquençage du Génome Humain avec B. Hess, F. Gros, et G. Valentini*, Paris, 18 November 1987, Brussels.

The yeast genome project, therefore, fitted in perfectly with the vast European movement towards genome research, one of the main motivations of which was not to be left behind the United States and Japan and lose promising markets. The Japanese-initiated Human Frontier Science Program, for example, required a European initiative in response.[85] What the scientists wanted was a mechanism to accelerate European action on initiatives such as the HFSP or similar developments in biosciences. The CODEST study group, a suggestion of Hess, was one way to do this, although the idea was met with skepticism. European programs were generally seen as recommended models of initiatives. It was pointed out that in the USA more investment was being made in genome research and cooperation would be needed on instrumentation, especially instrumentation used on the human genome. The yeast project, from this point of view, could be one of the first European initiatives in the field of genome research, response that would ensure Europe remained competitive in this promising field.

During his visit in August 1987, to the University of Manchester Institute of Science and Technology (UMIST) to meet L. Martin, one of his BAP contractors, Goffeau was able to have a long chat with Steve Oliver[86] about preparing the yeast genome sequencing program. Steve Oliver was the director of the UMIST Biotechnology Center, a teaching and research organization that had grown from a collaboration between three departments namely biochemistry and applied molecular biology, chemical engineering and instrumentation. There were 15 people on the team in five groups: applied molecular biology, headed by Oliver himself; instruments, headed by Martin; protein chemistry, headed by M. Minier; *Streptomyces*, headed by J. Cullum and intermediary metabolism, headed by N. Broda.

At this time, Oliver was one of the few scientists who had mastered continuous fermentation as well as molecular biology. He was also one of the even fewer number of scientists worldwide to have sequenced more than

---

[85] A. Goffeau's mission report, *Preparation Human Frontier*, Oviedo, 21–23 August 1987, Brussels 30 September 1987, p. 3.

[86] A Goffeau's mission report on his 13 August 1987 visit to S. Oliver (at the UMIST, Manchester) during the preparation of "Sequencing the Yeast Genome", Brussels, 30 September 1987.

20 continuous kilobases of the yeast genome. He had already received (from his former mentor Caroll Newlon, USA) a restriction map and two libraries (BamH1 in YIP5 and ECORI in lambda) from a circular fragment of 180kb of chromosome III, of which he had already single-strand sequenced 40 kb[87]. Furthermore, the strain he had used was derived from strain S288C, the same strain that would probably be used in the United States for mapping and sequencing. Chromosome III was one of the smallest yeast chromosomes at 400 kb, only chromosome I being smaller. Even though Goffeau had considered other possibilities, it became clear that chromosome III was probably the most appropriate chromosome to use for the feasibility study foreseen.

There were other possibilities.

Chromosome I - an American called R. Kabback and H.Y. Steensma of Delft were finishing the physical map and the subcloning of this very small chromosome of 150 kb. No systematic sequencing work had been started as yet.

Chromosome II - although larger than chromosome III, it was still a respectable 800 kb. A German scientist, H. Feldmann of Munich, had already done much of the physical mapping and subcloning. However, no extensive sequencing had as yet been carried out and he had used an industrial strain.

Chromosome VII - a physical map of about 200 kb had been produced by an Israeli scientist, J. Margolskee. Goffeau himself had sequenced a continuous fragment of 23 kb, but the whole chromosome was too large at 1,500 kb[88].

Chromosome XI - very long at 2,000 kb. However, much of it, almost 1,000 kb, was part of a repetitive RNA ribosomal unit being studied by a Dutch scientist, Rudi J. Planta.

---

[87] In early 1988, the published chunks of chromosome III added up to 65 kb of sequencing.

[88] In order to check the feasibility of sequencing the entire yeast genome and better understand the difficulties that would be involved, a student of A. Goffeau's named Weiming Chen had sequenced and analysed a region of 17 kb around the centromere of chromosome VII of *S. cerevisiae*. This was during this work which earned Chen his science doctorate.

As for organizing a network of 10 to 20 laboratories working together on the DNA sequence of chromosome III, during their chat, Oliver and Goffeau came to the following conclusions:

- Olson would have to be asked for his chromosome III library as soon as possible to provide S. Oliver with the 120 kb not included in the chromosome ring he had been studying;
- In addition to funding on the basis of the results ($1 per base pair), there would also have to be a basic payment of at least 50,000 ECU/year[89] to allow the laboratories to take on a full-time technician and a post-doc to supervise the work;
- All the members of the network would have to meet often and be linked electronically;
- Specific fragments (for example a contig of 40 kb, comprising 10 overlapping clones) could be worked on by two different groups working as a pair. This would provide training opportunities if the groups were well chosen;
- Once it was obtained from M. Olson, the organized library of the genome could be safeguarded in a plasmid culture collection by one of the BAP contractors who would guarantee attentive maintenance;
- When the fragment to be sequenced covered a gene which had already been published, it would not be necessary to sequence it again, even if the sequence was from another strain. However, an overlap of at least 100 bp would have to be sequenced at both ends.

---

[89] The need for this additional sum of 50,000 ECU had already been pointed out in the remarks made about Goffeau's project on 16 November 1987 by Prof. A. Pierard of the Université Libre de Bruxelles, also Assistant Director of the CERIA-COOVI Research Institute, along with Professor N. Glansdorff of the Vrije Universiteit Brussel, the Director of the CERIA-COOVI Research Institute, and Dr J. de Brabandere of the Belgian Science Policy Unit and a member of CGC-Biotechnology. Glansdorff and Pierard thought that the project was underfunded. Their experience in DNA sequencing told them that 100 kb a year in 15 kb fragments would keep a four-person team fully occupied, which would eat up most of the budget as it stood. An additional sum of 50,000 ECU would be needed. Finally, a payment of 5 ECU/bp was adopted instead. With regard to equipment, it would be impossible to carry out the project without providing the sequencing team with full-time use of a sequencing machine worth 100,000 ECU, an automatic synthesizer of nucleic acids (12,000 ECU) and a computer (10,000 ECU). Letter to J. de Brabandere from A. Pierard and N. Glansdorff, 1987.

After this meeting, the general conclusion between André Goffeau and Steve Oliver was that chromosome III should be used for the feasibility study. A minimum base funding would have to be provided during the feasibility study and Oliver would probably have to take on coordinating the work.

During the same trip, Goffeau managed to meet M. Rose, a member of IRDAC and head of research at ICI, and two of his colleagues. They gave him their opinion on the yeast genome sequencing project, reporting that ICI had no activity linked to yeast whatsoever, being mainly involved in plant and human genomes. They considered it a good idea to avoid developing activities in the megasequencing field by using the American databases when they became generally available. They could not see what interest the yeast project held for industry, but they realized that it was a necessary step on the road to sequencing the human genome and appreciated its clearly defined objective. They also thought that the laboratories that would best be able to carry off the sequencing project would be those producing the equipment (centrifuges, electrophoresis and chemical robots), the necessary products (restriction enzymes, ligases, polymerases, acrylamide, gels, etc.) and using the yeasts (brewers and distillers).

While the yeast genome project was firming up over certain points (choice of target, structure of the network and opinions of the various potential partners), Goffeau saw the CGC-Biotechnology again at their meeting of 16 and 17 February 1988 to present the BAP progress report for 1987 and talk about the progress of the various studies to evaluate the needs and opportunities for yeast genome sequencing that the CGC had requested. He also presented the questionnaire, to be filled in by potential partners, which would have to be returned by 31 March 1988. Two million ECU could be assigned to sequencing under the BAP revision, which would allow the conclusion of 10 associated contracts to sequence chromosome III. Each delegation to the CGC was invited to suggest two or three laboratories in their country that had experience in the field of DNA sequencing, a good understanding of yeast genetics and a good research infrastructure. The new questionnaire was thought to be counter-productive considering all the previous surveys carried out on the same theme, but since the member states' contribution had previously been somewhat patchy, it was finally seen as another chance to suggest new partners.

The choice of yeast as a target organism was still being discussed although its superiority as a candidate for sequencing work, lying in the existence of an ordered DNA library, was recognized. It would take at least two years to build such a library for another organism. *Bacillus subtilis* was presented as another possible target organism for sequencing in a project involving five European laboratories working with another five in the United States. There was also a proposal to map the bovine genome that would involve eleven European laboratories. Few delegations backed the *B. subtilis* proposal, but the CGC decided to postpone its decision until the seven studies on the yeast genome proposal had been delivered. They were sent to the CGC on 16 May 1988, along with a request that they provide the names of the experts attending the working group meeting on yeast genome sequencing to be held on 24 May 1988, which was attended by Commission officials, the CGC-Biotechnology and the experts in question. These studies can be briefly summarized as follows:

Results of the Detailed Evaluations of the Yeast Genome Sequencing Project[90]

H. Feldmann and H.Y. Steensma's study on a detailed strategy for the construction of organized libraries for mapping and sequencing the genome of *Sacharomyces cerevisiae* and the industrial microbes had come to the following conclusions. Recent progress in the field of pulsed field gel electrophoresis would allow the separation of DNA fragments from 10,000 to several million base pairs. It should be possible to prepare libraries of specific chromosomes by cloning the DNA of a single chromosome in cosmid vectors or phages. Several methods could then be brought together to build physical maps and correlate them with genetic maps. This approach could

---

[90] These summaries are taken from "Sequencing the Yeast Genome: A Detailed Assessment" edited by A. Goffeau. *Sequencing the yeast genome, a possible area for the future biotechnology research programme of the European Communities,* 1988.

lead to organized libraries of specific chromosomes. The construction of an organized library from a specific yeast strain could be finished in a relatively short time. Detailed mapping work could be split up between several laboratories. It was thought that one chromosome would take a year if all the modern techniques were used. The sequencing project, however, could begin before the mapping was finished. They believed that the progress of the sequencing project greatly depended on the development of new, avant-garde rapid techniques of sequencing. They advised using the raw material from a well-known strain such as S288C.

Steve Oliver's study, *A detailed strategy for the sequencing of chromosome III*, indicated that the yeast *S. cerevisiae* would be the ideal target organism for a pilot sequencing project. M.V. Olson in St Louis, Missouri had built a large-scale restriction map by using two enzymes — *Sfi*I and *Not*I. The 62 *Sfi*I sites defined 77 fragments that had all been measured, and most of which had been identified, by particular probes. The 53 fragments produced by *Not*I had been partially analyzed. The low resolution map was used to assign contiguous fragments of a phage bank of the whole genome to their respective chromosomes. Thirty four of these 62 *Sfi*I junctions (55%) and 25 of the 36 *Not*I junctions (69%) had been identified in the lambda clones, which were then subjected to high resolution mapping. Two hundred of the 350 contigs had already been assigned to chromosomes, and Olson was hoping to finish the high resolution map in 1988.

In addition to chromosome VI, which was the subject of an agreement between Olson and a Japanese group, Kaback, Steensma and their co-workers had already cloned 75% of the chromosome I and were building a detailed map. However, the ideal target was chromosome III, because:

- The whole ring contig, derived from chromosome III and formed by a crossover between HML near the telomere on the left arm and Mat on the right arm, had been cloned and mapped by Newlon[91]. Newlon and Olson were ready to make their libraries and banks available to the European consortium;

---

[91] This library, however, was still incomplete. In addition, the chromosome III that had been used to clone the *Eco*RI fragments came from a different strain that that of the ring.

- S. Oliver's laboratory had already carried out a lot of sequencing on this chromosome, about 40 kb; generally on one strand;
- The sequences that had already been achieved for this chromosome added up to at least 65 kb;
- Olson had assigned 225 kb of the 335 kb of chromosome III to his lambda clones.

Although the libraries available for chromosome III came from a variety of sources and strains of *S. cerevisiae*, Oliver did not think this was a major problem. Other than at the two main polymorphic sites (LAHS and RAHS), there did not seem to be much variation in sequence.

In conclusion, S. Oliver suggested that each laboratory in the consortium sequence one single BamHi fragment from Newlon's library as their first responsibility, using the Mung Bean/Exo III directed deletion strategy, and that both sequence be checked by two other laboratories, both sequencing the BamH1 junctions on each side by using clone phages and internal primers, referring back to the results obtained in the first sequence. Each laboratory would be involved in producing results and checking the information produced by another laboratory. Since two laboratories would check the work of one of the other laboratories, conflict could be avoided between any two particular laboratories on the quality of the results. This would oblige each laboratory to be in contact with two others, which would ease inter-laboratory communication, essential to a European sequencing consortium.

H.W. Mewes and F. Pfeiffer, in their detailed assessment of the bioinformatics network required for the collection and processing of the yeast genome sequence data, concluded that a data collection center and its connections to the laboratories involved would accomplish some of the more important tasks in the overall yeast genome sequencing project. This center would need major attention and support to help it fulfill its role. Any restriction on the center and its links would generate delays, malfunctions and lead to the production of information of lesser quality, which could have a negative impact on the overall success of the project. This data collection center should have experience in the running of other data bases. Although it should resolve many new problems, it should not be an experimental data collection center. Developing entirely new procedures would be to make the entire project run an intolerable risk. Available software should not be rewritten.

A coordination of the data collection activities for nucleic acid, protein sequences and other genetic information was seen as indispensable. The protein sequences should also be recorded in the data base. All the information should be accessible through a proven relational data base management system. It was impossible to suggest which hardware should be used because of rapid developments in the market. According to the authors, local groups of running work stations and file servers would provide the highest performance and flexible solutions. The VAX/VMS solution was recommended. There were many sequence analysis programs and data base management systems that had already been developed for this sort of hardware configuration.

As for the laboratory sites, they could not provide strict recommendations. If small systems were to be purchased, then the compatibility and availability of appropriate software was of paramount importance. Connections to the network were essential. As for equipment the laboratories already had, it would not be the same in every one. Electronic communications between the center and the laboratories would have to include data exchange, message systems and interactive procedure. The question of data security also had to be considered during the network link-up.

In his study on the computer environment for yeast genome sequencing, R. Terrat concluded that there was no reason that the local workstations should be the same and that laboratories which already had powerful machines would not have to buy new ones. The important thing to have was the computer network with the data base and common utility software. The degree of data and software distribution would be restricted by current limitations on international computer networks. The standardization underway would soon allow better distribution. Finally, it was noted that implementation would require particular development work on the communications software for the workstations, training, maintenance, and other developments to adapt the applications to the expected evolution in computer networking.

F.M. Pohl, in his detailed assessment of equipment necessary for sequencing the yeast genome, concluded that since most of the laboratories participating in the project had the fundamental technology and equipment, sequencing a yeast chromosome between 1988 and 1990 would not require a collective major effort towards a specific method or tool. Coordinated and well-documented use of a wide-ranging set of methods and equipment should lead to the evolution of an optimal solution for sequencing the whole yeast genome. This would allow each laboratory to use technological evolution taking into account its own local circumstances. It was essential that information be freely exchanged between the laboratories. Newly established laboratories should give preference to PCR amplification of DNA and colorimetric or fluorescent detection of sequencing scales. In this way, investment in radioactive or microbiological methods could be kept down as much as possible.

Prudent decisions would have to be made on available financial resources such as whether to spend on equipment or personnel. Buying a fluorescence sequencer, not counting the maintenance costs, would cost as much as five or six years of a technician's salary, for example.

R. Tubb's study on the long-term benefits of genome sequence data to the yeast-dependent industries concluded that a wide range of technical possibilities were available for the genetic manipulation of industrial strains and were being applied not only to yeasts used in brewing, distillation, breadmaking and wine-making, but also to the use of yeast in biotechnological products. The European brewing industry, for example, was in the vanguard in using genetically modified strains in foods and drink.

However, for yeast genetics to reach its full potential in industrial use, a better understanding of the strains in use and fundamental aspects of the physiology and biochemistry of yeast was needed. A commitment to sequence the entire genome of *Saccharomyces cerevisiae* would thus be a convenient exercise that should provide a proven basis for further systematic studies. Such a project, at this stage, could not be justified by its short-term benefits. The strategic importance of obtaining the yeast genome sequence in the long term was, however, recognized by a large group of firms interested in yeast.

Sequencing yeast was certainly an opportunity to develop the interface between academic and commercial activities. A European Yeast Club would provide a mechanism to coordinate such actions. Although such a club could be based on genome sequencing, there were wider perspectives for it. Of the 13 firms that had provided a positive answer (a "Yes" or a "Maybe") to R. Tubb's proposal sent to 40 companies, nine thought they could contribute US$10,000 and four had said that such an amount was too high and would be difficult to justify at this stage. The idea of a Yeast Industry Club that would provide the members with pre-publication access to results was in general, well received.

The Experts' Recommendations on the Yeast Genome Sequencing Project.

With the various detailed evaluation studies on the yeast genome project in hand, the experts designated by the delegations met on 24 May 1988 under the chairmanship of Professor G. Magni (a member of the CGC and a national expert from Italy) in order to make recommendations for the yeast genome sequencing project. The meeting was attended by:

Prof. Michel Aigle (national expert from France)
Dr. N. Gasson (national expert for Great Britain)
Prof. C. Hollenberg (national expert for Germany)
Prof. G. Magni (national expert for Italy)
Prof. D. McConnel (national expert for Ireland)
Prof. R. Planta (national expert for the Netherlands)
Dr. P. Sharp (representing the Irish National Delegate to the CGC)
Prof. A. Tsiftsoglou (national expert for Greece)
Dr. Evelyne Dubois (national expert for Belgium)
Dr. J.P. Garcia-Ballesta (national expert for Spain)
Prof. Piotr Slonimski (expert chosen by the Commission)
Dr. W. Mewes (expert chosen by the Commission)
Prof. S. Oliver (expert chosen by the Commission)
Dr. D. de Nettancourt of the European Commission
Dr. Etienne Magnien of the European Commission
Dr. I. Economidis of the European Commission
Prof. André Goffeau of the European Commission

At the end of the meeting, the experts recommended that:

1) Prof. S. Oliver of UMIST should become the coordinator of molecular resources, with responsibility to:
   1.1 prepare and distribute the accredited clone sets of chromosome III of *S. cerevisae* so that each participating laboratory could begin the sequencing early in January 1989 so that the entire chromosome could be finished before the end of 1990, including the checking of the overlaps from an independent library;
   1.2 assemble and maintain at least two independent DNA libraries for chromosome III;
   1.3 represent a source of primers/vectors;
   1.4 receive and conserve the sub-clones being produced by sequencing laboratories;
   1.5 check sub-clones if there are conflicts;
   1.6 assign and implement resequencing work;
   1.7 advise and train people as necessary.

They requested that a budget of ECU 200,000 be allocated to him to allow him to fulfill these roles.

2) Dr W. Mewes of MIPS should be made coordinator of computer resources, with the following duties:
   2.1 collect the sequencing results;
   2.2 annotate the entries;
   2.3 analyze sequencing results (search for homologies and identification of ORFs);
   2.4 control the quality of information submitted;
   2.5 organize the computer network then needed to exchange information between sequencing laboratories and keep an appropriate on-line system going;
   2.6 support participating laboratories with hardware and software know-how;
   2.7 provide training on the data bases when needed;

6 European Biotechnology Strategy and Sequencing the Yeast Genome 411

2.8 distribute results in due time, in particular, ensure their transfer in an appropriate manner to the EMBL database;

2.9 develop software to ensure transfer, quality control and analysis of the information received.

To allow him to complete all these tasks, a budget of 200,000 ECU was recommended.

3) Professor P. Slonimski of the Center for Molecular Genetics at the CNRS at Gif-sur-Yvette (FR.) coordinator for the search for functions for the 200 unknown ORFs expected to be discovered during the sequencing of chromosome III. He should be appointed to:

3.1 organize a brainstorming session during the first half of 1989 to identify disrupted strains and phenotypes for the characterization of unknown genes;

3.2 prepare a standard protocol to search for mutant phenotypes;

3.3 maintain his group as a host laboratory for training.

These tasks would not be rewarded through a contractual financial contribution but individually and independently through meetings, training grants and other services.

4) As for the sequencing network, each of the laboratories or laboratory consortia was to receive sequencing units from chromosome III of *S cerevisiae* (Table 18) — a unit being the minimum sequence of 8 kb of a *Bam*Hi library cloned into a YIP5 vector with at least three overlaps of 1 kb each around the BamHi site from independent clones with inserts coming from a partial digestion of chromosome III. All the clones would be supplied by the coordinator as well as any restriction maps for *Bam*HI and *Eco*RI.

Everyone was encouraged to look for functions for the unknown ORFs. Each sequencing unit would obtain a total funding of 50,000 ECU of the European Community's money, which would only represent about half of the real costs to the participating laboratory.

Table 18 Scientists recommended by the experts designated by the delegations

Director of laboratory	Localization	Units
Michel Aigle and Marc Crouzet	Bordeau	2
L. Alberghina	Milan	1
Alistair Brown	Glasgow	2
Giovanna Carignani	Padova	1
Bernard Dujon	Paris	1
Horst Feldmann	München	2
Françoise Foury and Guido Volkaert	Louvain-la-Neuve and Leuven	1
Laura Frontali	Roma	1
Hiroshi Fukuhara	Orsay	1
J.P. Garcia-Ballesta	Madrid	1
N. Glansdorff and W. Fiers	Brussels and Gent	2
Leslie Grivell	Amsterdam	2
François Hilger and M. Grenson	Gembloux-Bruxelles	1
C.P. Hollenberg	Düsseldorf	2
P. Jackman	Norwich	1
Giovanna Lucchini	Milan	1
D. Mc Connel	Dublin	1
Steve Oliver	Manchester	2
Peter Philippsen	Giessen	2
Rudi J. Planta	Amsterdam	2
Fritz Pohl and F. Zimmermann	Konstanz and Darmstadt	2
Piotr Slonimski and J.M. Bulher and Claude Jacq and Michel Jacquet	Gif-sur-Yvette and Saclay and Paris and Orsay	5
H. Yde Steensma	Leiden	1
George Thireos	Heraklio	1
Dieter Von Wettstein	Copenhagen	1
The following reserve list was approved :		
J. Conde	Sevilla	1
F. Foury and Guido Volkaert	Louvain-la-Neuve and Leuven	1
François Hilger and M. Grenson	Gembloux and Bruxelles	1
A. Jimenez	Madrid	1

From BAP to BRIDGE — An Extension of the Pilot Project?

The yeast genome sequencing pilot project seemed to be well on its way. However, it was only a preliminary phase to a far wider sequencing program to be introduced under BRIDGE that would constitute Europe's answer to the large-scale sequencing programs underway in the States and Japan.

Although overall, the CGC members recognized the need for European action, those of IRDAC did not entirely agree. In order to organize a common contribution to the preparation of BRIDGE, R. R. van de Meer, Chairman of CGC-Biotechnology and H. Nielsen, Chairman of IRDAC WP5, organized a joint meeting to bring together the members of their groups in Brussels on 27 May 1989. During this meeting, André Goffeau, yet again, presented his plan to include a program to map and/or sequence *Bacillus*, yeast, *Arabidopsis* and other industrial micro-organisms under the forthcoming BRIDGE workprogram. He also suggested including a project on the development of new equipment and procedures for sequencing DNA. He pointed out the OTA's assertions that "understanding human genes will necessarily involve the study of other organisms"[92].

He also reported that a panel of renowned European scientists had gathered in Brussels on 2 May 1988 at the request of the Commissioner in charge of Research, M. Riesenhuber, and had concluded that the European answer to the Japanese Human Frontier Science Program would have to include sequencing activities, not only of the human genome but also those of yeast, *Bacillus subtilis*, *Drosophila melanogaster* and *Arabidopsis thaliana*.

Despite these arguments, and the results of a survey carried out by an industrial consultant that showed that 15 industrialists using yeast had expressed their intention in writing to a European yeast genome sequencing program, in some cases with promises of financial support, IRDAC remained opposed to the yeast project in particular, and more generally, to systematic sequencing proposals, because it considered them of no use to industry. It continued to hold this position throughout preparations for the BRIDGE program and during the IRDAC WP5 meeting of 16 December 1988. Despite this criticism, more and more support, positive assessments and indications of interest reached Goffeau from scientists in the member states of the community as well as from Switzerland and Austria (not then a member state) involved in yeast research and/or competent in molecular genetics and/or in the field of yeast molecular biology that were interested in participating in a major innovative European project. They asked Goffeau

---

[92] From page 9 of the report "Mapping our Genes; Genome Projects: How Big, How Fast?".

whether they could take part in the sequencing work. In contrast to IRDAC's opinion, 15 enterprises had expressed their enthusiasm for the project.

The network's organization, its management and the methodology necessary was becoming clearer, as were the roles to be taken by each participant. The project's future looked very bright, especially as the CGC members did not seem opposed to it and recognized its advantages and urgency with regard to European competition vis-a-vis the United States and Japan in this field.

The Pilot Sequencing Project for Chromosome III — A Model Project for Future Actions?

This idea was confirmed during the CGC meeting on 5 July 1988. At this meeting, André Goffeau presented the chromosome III sequencing project with its latest alterations, in particular in the network organization, its main elements and their functions, the laboratories that should participate in the sequencing (selected by the group of experts from the lists put forward by each delegation and the answers the laboratories had provided to the questionnaire) and the budgets recommended by the experts: 200,000 ECU to Mewes at MIPS and 200,000 ECU to Oliver at UMIST, plus another 50,000 ECU per unit of 10,000 bp sequenced.

The coordination for this work, with a common budget of 50,000 ECU, was recommended very favorably. The project was a well-defined large-scale task with measurable impact, criteria that looked good to many of the delegations. The degree of integration and the organizational level that had been reached impressed the CGC as an example of an initiative making a major contribution to a better understanding of the human genome. The Commission's staff thought that the Community could become even more involved in sequencing by using the various appropriate R&D programs:

- Yeast in the BAP revision, and then in BRIDGE;
- *Bacillus subtilis* in BRIDGE;
- *Arabidopsis thaliana* in BRIDGE;
- *Drosophila melanogaster* in the Stimulation Measures;
- *Homo sapiens* under the program for Predictive Medicine.

Although the yeast sequencing project's merits were widely recognized, starting the sequencing of other chromosomes in the course of the forthcoming BRIDGE program was being questioned. Furthermore, the fact that the entire BAP revision budget for feasibility studies would be spent on yeast blocked any other project on other questions, and this was duly noted. BRIDGE being a biotechnological program and sequencing work being considered strictly a question of fundamental biology, it was also asked whether it could not be carried out under another more fundamental framework.

As for industrial interest, opinions still differed. IRDAC's WP5 had voiced a strongly negative opinion, an echo of the 1974 situation when the new genetic technologies were not considered by industry to be of much use. Some delegations thought that this negative opinion should not receive too much attention since national authorities were confronted with the lack of a long-term view from their industrialists, particularly as some of those industrialists had paradoxically already signed expressions of interest in the project and had agreed to contribute US$ 10,000 per annum to be part of the YIP "Club". Other companies looked likely to adhere to the "Club" too. The role that could be played by the instrumentation companies and the effort needed to bring them together were also made clear, as automated procedures were to become increasingly important in sequencing.

The German delegation saw the project as being far more than just the sequencing work it would entail. It should lead, they said, through structural determination, to the elucidation of new biological functions, and contribute to the challenge of developing better automation. The potential of this enterprise would break down the artificial barriers between biological information and biotechnological information. This was indicated by Fasella and Carpentier's strategic decision to build an agreement between DGs XII and XIII on activities in this field receiving a lot of attention from the Commission, especially as European work in the bioinformatics field was not perceived as being very advanced by the United States[93].

---

[93] J. Franklin, *The role of information technology and services in the future competitiveness of Europe's bio-industry*, A report compiled for the Commission of the European Communities, January 1988.

Lastly, after evaluation by the Commission and its delegations, the pilot project for yeast genome sequencing was finally accepted. Goffeau was congratulated for his determination and enthusiasm for this highly successful initiative.

International Reactions to the Announcement of the European Yeast Chromosome III Sequencing Program

At the 14th International Conference on the Genetics and Molecular Biology of Yeast, held from 13 to 15 August 1988 in Helsinki and attended by 700 molecular geneticists, André Goffeau announced that a network of 35 laboratories would sequence yeast chromosome III under the aegis of the European Community BAP program. This news was received with enthusiasm, the work compared to that of Kepler, Linneus and Mendeleyev, who respectively, classified the planets, plants and chemical elements. The Canadian scientists immediately announced that they wanted to sequence chromosome I, which seemed realistic since they were able to spend CAN$ 5 million and planned to depend on the documentation gathered by Goffeau on the development of techniques, discussions and international enterprises in sequencing over the preceding two years. They looked as if they would use a similar sequencing approach to that taken by the BAP network for chromosome III. The conference was also a good time for most of the BAP contractors to meet, creating a sort of informal contractor's meeting at which the guidelines for sequencing chromosome III were written.

The Canadian reaction as well as the project that brought the Americans in St Louis and the Japanese of RIKEN together on chromosome VI, and in general, the international interest stirred up by the European effort during 1988 and 89, were fundamental. Apart from the interest expressed by some Swiss[94] and Australian[95] scientists, G. Simchen of the Hebrew University of Jerusalem[96] also asked about the project, and Bart Barrel (MRC,

---

[94] Letter from P. Linder to A. Goffeau of 20 May 1988.
[95] Letter from D. Clark-Walber to A. Goffeau of 5 April 1988.
[96] Letter from G. Simchen of A. Goffeau of 13 April 1988.

Cambridge) wrote to Goffeau on 12 January 1989 asking him to save him a small chromosome (number XI or V) to sequence when he had finished his work on the human cytomegalovirus in order not to duplicate work already in progress[97]. Later, there were further expressions of interest from R.W. Davis[98], D. Botstein, K. Hennessy, J.T. Mulligan and G. Hartzell of Stanford in the USA, to sequence chromosome V under a project being undertaken in a new center funded by the NCHGR to show the feasibility of yeast genome sequencing and to set up a new data base[99] in cooperation with Olson and Mortimer. This data base would integrate genetic and physical maps of yeast with new and existing sequences. All these projects, reactions and expressions of interest were fundamental for the goal of sequencing the entire yeast genome since the budget foreseen under BRIDGE would only pay for 20% of the genome to be completed, or 50% with the budgets of the Biotech I and II programs that would follow.

Research Contracts and the Significance of the Chromosome III Sequencing Programs

The contract for research project no. BAP 0346 (GDF) between the European Economic Community and the Université catholique de Louvain was signed on 5 December 1988, kicking off the beginning of sequencing work on yeast chromosome III officially for 24 months beginning the first of January 1989. The following conditions were to be observed:
- The Commission would pay 100% of the authorized marginal costs up to 2,235,000 ECU;
- There would be an advance payment of 1,341,000 ECU;
- There would be periodic reports to be submitted on 1 July 1989 and 1 July 1990 and a final report;

---

[97] Letter from B. G. Barrel to A. Goffeau of 12 January 1989.
[98] R.W. Davis, D. Botstein et al., *Stanford Yeast Genome Project,* announced at the Yeast Genetics and Molecular Biology Symposium, San Francisco, 23–28 May 1991.
[99] The proliferation of data bases (in addition to Botstein's project, there were also the MIPS, List A2, GGOEBL, the EMBL, SWISPROT and GENBANK databases) brought competition and completion into the issue. As A. Goffeau wrote to Dujon in his letter of 23 August 1992, "It would be entirely possibly for all of them to constitute incomplete data bases."

- A periodic cost statement was to be submitted after 12 months.

The contract was authorized to conclude contracts with the named subcontractors. The subcontracts foresaw that with regard to payment, there would be an advance of 50%, a periodic payment of 30% and a closing payment of 20%. Two very new management principles were adopted:

- A payment of five ECUs per final base pair fully sequenced and approved;
- A second contig to be received only when the first one was completed.

The Université catholique de Louvain, in accordance with annex one on the division of technical and financial aspects of the work, and annex two on the scientific, technical and administrative principles behind the project's organization, committed itself to

- Provide the Commission with a report on work in progress on 1 July 1989 and 1990;
- Make a second payment of 30% at the beginning of 1990 to each laboratory that had finished sequencing 40% of the original work assigned to it (the payment was to be made within 80 days of the Université receiving the second payment from the European Commission);
- Make a third payment of 20% to each contractor who had finished the work within 80 days of receipt of the final payment from the European Commission.

This contract, which was signed by a representative of the Commission and on behalf of the contractor by P. Macq, the Rector of the catholic University of Louvain, and Professor A. Goffeau, initiator and manager of the research program, was just a marker for the beginning of a major scientific enterprise, the success and evolution of which will be described in the next chapter. The full sequence of chromosome III was referred to in a *Nature* advertisement as one of the pieces of research which would change history, along with Darwin's article on natural selection in 1880 and Watson and Crick's on the structure of DNA in 1953. It was not only just a "world first" for fully sequencing a eucaryotic chromosome; it also indicated the Community's commitment to genome research, and more specifically to large-scale sequencing efforts on model organism genomes.

In fact, there were other projects foreseen under the European Community programs beside the sequencing of chromosome III that involved genomes of other organisms. These included:

- The physical mapping of the *Drosophila melanogaster* genome, under the SCIENCE program, involving three laboratories between 1988 and 1993 and 872,000 ECU from the EC;
- The complete physical map and a strategic approach to sequence the *Bacillus subtilis* genome, also under the SCIENCE program, involving five laboratories from 1989 to 1991 and 755,000 of the Community's ECU;
- The functional and structural analysis of the mouse genome under the SCIENCE program, involving three laboratories from 1989 to 1992 and 1,278,000 ECU from the Community;
- The yeast genome sequencing program that was to come under the BRIDGE program, involving 31 laboratories from 1991–1993 and 5,060,000 ECU from the Community;
- The development of the genetics of and a physical map of the pig genome under the BRIDGE program, involving 11 laboratories and 1,200,000 ECU from the Community;
- The molecular identification of new plant genes focusing on *Arabidopsis thaliana*, under the BRIDGE program, with 27 laboratories from 1991–1993, with a Community contribution of three MECU;
- And, of course, the human genome analysis program, at 15 MECU, which, despite many ethical difficulties at various levels, especially from the European Parliament, was adopted on 29 June 1990 for 1991–1992.

The recognition of gene mapping and genome sequencing as a fundamental priority in several Community programs and the adoption of the sequencing programs from 1988 onwards were essential from the point of view of European competition in these fields. In 1989, the USA assigned US$28 million to the human genome project with a promise of a hundred million for 1990, and Watson thought that a detailed genetic map of all human chromosomes could be available in fifteen years' time. But, instead of the human genome, the committee decided to start work on the study of simpler model genomes such as *E. coli*, yeast, plants and mice. According to Goffeau, "So, although the Americans had a large budget and two years' start on the Europeans, because we had been preparing various genome research programs during 1986–87 and managed to get the yeast genome sequencing program adopted in 1989 by the Commission, the Americans reached the same conclusions as the Europeans, but two years later".

The Europeans did not make human genome sequencing an absolute priority just to later see that the study of model organisms was a step that could not be bypassed. They moved directly to that step by sequencing small genomes. Before the Americans had even started sequencing the human genome, the Europeans had already sequenced large chunks of the yeast and *Bacillus subtilis* genomes. For Europe, this head start was very important. In fact, it would have led the Europeans to a leading place in the field of small genome sequencing and in the discovery of their structures and the functions of the newly identified genes if Craig Venter (and TIGR) had not suddenly converted from a human genome to a bacterial genome sequencer. This leading place could have given European scientists and industrialists priority access to some of the most interesting groups of genes. Even at a methods and infrastructure level, Europe could not have been as behind as had seemed in 1989 since the new databases were being created and sequencing equipment developed at EMBL seemed to be competitive. Furthermore, other researchers working under the BAP program were developing new genetic systems using other organisms of industrial interest.

Finally, the 1988 adoption of the yeast chromosome III sequencing project was important inside the Commission. It was both a model and a test for the scaled-up sequencing project to come under the BRIDGE program and other genome studies projects foreseen at the European level.

7
The Decryption of Life

7.1 The Structure and Organization of the European Yeast Genome Sequencing Network

By the time European research was getting organized to face the challenges of the globalized economy and international industrial and technological competition, the importance of research and development had been generally recognized at the European Commission.

In 1989, at the behest of the European Parliament, the Commission published the first report on the state of science and technology in Europe[1]. Their diagnosis of the situation can be summarized as follows:

- Great efforts had already been made to improve the situation in Europe by increasing the research and development budget and improving industrial performance through innovation. But these were still considered unbalanced and fragmented. Three member states (Germany, France and the United Kingdom) were responsible for three quarters of overall R&D spending in the community, and regional variations were very marked. Transnational concerted actions (Community programs such as COST, EUREKA, ESA, CERN) only constituted a very small part of the total research effort.
- European efforts were still very limited compared to those of the main competitors, the USA and Japan, who were pouring much

---
[1] First report on the state of science and technology in Europe, C.E., 1989.

larger budgets into R&D and taking initiatives to remedy their own weaknesses. The European position was also threatened by the new emerging scientific and technological powers, such as the CIS.
- Europe had come up against three main challenges:
    * to increase its ability to develop and follow, when necessary, its own technological and economic options;
    * to reinforce its international competitiveness especially in fields becoming increasingly important in the future;
    * to answer the social need for improvements in the quality of life.
- It was believed that Europe had the resources to rise to these challenges. It had the required scientific talent and organizational capabilities. In the economic climate as it stood, it could easily afford to invest greater funds in R&D. The question was how to obtain the best from these resources and on which areas they should be focused.
- There was a challenge to maximize the effective results of European work, and this meant that a series of problems had to be solved: that science and technology should be understood and accepted throughout European society, that the imbalances and fragmentations should be overcome through better coordination and that, through the provision of an adequate basis for technological progress in less-favored regions and countries, technological skills be distributed through society to bring more funding from the private sector; and, lastly, the diffusion of technology to industry, the encouragement of links between industry and universities and the exploitation of possibilities of international cooperation in domains where mutual benefit is clear.
- Regarding research, the report underlined five fields of research of major importance for the European economy:
    * technologies of information and telecommunications (particular efforts were needed to improve the situation of the European semi-conductor industry);
    * new materials and technologies with uses in the production industries (super-conductors looked particularly promising);

* aeronautics, an area in which Europe was facing a very large amount of competition;
* biology and biotechnology — which offer prospects of major transformation in industry and agriculture as well as in the medical field (specific efforts seemed needed in the field of fundamental plant biology, gene mapping, neurobiology and biotechnological applications);
* energy, a field in which Europe remained largely dependent on external sources (fusion seemed important in the long term, as well as precisely targeted research on new renewable energy resources and less energy-greedy technologies for the short term).

At the level of the different biotechnology sectors, Europe had begun to react by creating, at the behest of its scientific civil servants, the first BEP and BAP biotechnology programs. These were, of course, still very modest compared to the information technology programs, but they had the essential merit of starting a movement within the Community in a promising economic field in which other nations were already competing.

In the field of genome research, the adoption of the pilot program for yeast genome sequencing was a fundamental step forward for Europe. It allowed it to assume international rank in the genome sequencing field at a time when the USA was launching its Human Genome Program and other great nations were planning their own large sequencing programs. After two years of intensive preparation and the publication of a feasibility study, the BAP pilot project for sequencing chromosome III of *Saccharomyces cerevisiae* became active on 1 January 1989. A total of 2.6 million ECU covering 100% of the marginal costs was allocated to this task during the revision of the BAP program. The Université catholique de Louvain remained the main contractor, and the funds were redistributed amongst the 35 associated contractors (defined as subcontractors) to whom sequencing had been entrusted. The subcontractors, organized into a network, either communicated directly by e-mail (mainly through EARN) or by on-line access through the computer resources of the computer coordination center at MIPS. Other contracts had been negotiated to act as coordination centers with the following laboratories:

- Molecular resources: UMIST (Manchester UK) under Professor Steve Oliver;
- Computing resources: MIPS (Martinsreid DE) under Dr. H. Werner Mewes;
- Database: Professor Piotr Slonimski of the CNRS at Gif-sur-Yvette, France, was appointed coordinator of the search for functions for the 200 new genes it was hoped would be identified. As a preliminary task, Professor Slonimski (partially financed by a BAP contract) had compiled a database of all the known yeast genome sequences.

At the time of its launch, the yeast genome sequencing project was quite unique. However, a Canadian network working on chromosome I and two laboratories in the USA and Japan appeared interested in taking part. The goodwill of these nascent projects indicated that when the work continued under the BRIDGE program as a T-project, there would be an international approach to the program.

The European effort had several characteristics, as Bernard Dujon said: "First of all, the choice of the yeast genome to sequence for its intrinsic interest as a methodological stage to sequencing the human genome; then the decision to begin sequencing an entire chromosome rather than concentrate on the development or optimization of the technology; lastly, the decision to divide the sequencing task between laboratories known to be interested in and working on genetics and yeast molecular biology, rather than select and outfit a large genome research center"[2].

Mapping: A Prerequisite to Sequencing

The analysis of a genome is based on three approaches. The classical approach of genetic mapping began in the 1950s and is based on the frequency of segregation in the offspring of sexual crosses. It is still being used under the aegis of Bob Mortimer at Berkeley (USA). The second approach is that of

---

[2] B. Dujon, "Mapping and Sequencing the Nuclear Genome of the Yeast *Saccharomyces cerevisiae*: Strategies and Results of the European Enterprise", *Cold Spring Harbor Symposia on Quantitative Biology*, Volume LVIII, Cold Spring Harbor Laboratory Press, 1993, p. 357.

physical mapping based on the localization of restriction sites. It was begun in 1963 by M. Olson of St.-Louis (USA). This method allowed a general framework to be laid down for the systematic study of the yeast genome, with the availability of a map, using rare eight-base restriction sites (SfiI and NotI). The map allowed any gene cloned from *S. cerevisiae* by hybridization to be easily located on filters covering all of the genome, which were prepared by Olson's laboratory. However, since there were errors and non-contiguous regions for some chromosomes, the work would have to be continued; it was undertaken, from 1990, by the Unit for Yeast Molecular Genetics directed by B. Dujon at the Pasteur Institute. This work allowed a cosmid bank to be built by partial digestion by *Sau*3A of all the DNA of strain FY1679 (standard strain) of a cosmid bank (in fact, four complementary banks in two different cosmid vectors). This bank had a remarkable total of 168 genome equivalents in the form of a pool of primary clones, and secondary identified clones (genome equivalents).

As the European program progressed, the cosmid bank produced was distributed to several coordinators and provided single contigs for several yeast chromosomes (IV, VII, X, XI, XII, XIV and XV) which were used in the sequencing and covered 55% of the genome.

The MAT gene, to be found in the middle of the right arm of chromosome III, and the HML and HMR genes, respectively, sited at the left and right ends of the chromosome share sequence homologies. Consequently, intrachromosomal recombination may provide circular derivations. One of these, the small circular derivation formed by homologous recombination between HMC and MAT, was isolated and used by Newlon,[3,4] to create an organized clone bank in the YIP5 vector[5] for the whole of the left arm (except a small section) and half of the right arm of chromosome III.

---

[3] C.S. Newlon *et al.*, "Structure and Organisation of Yeast Chromosome III", *UCLA Symp. Mol. Cell. Biol. New Series*, vol. 33, 1986, pp. 211–223.

[4] C.S. Newlon *et al.*, "Analysis of a circular derivative of S. cerevisiae chromosome III: a physical map and identification and location of ARS elements", *Genetics*, vol. 129, 1991, pp. 343–357.

[5] K. Struhl, "High-frequency transformation of yeast: autonomous replication of hybrid DNA molecules", *Proc. Nat. Acad. Sci.*, USA 76, 1979, pp. 1035–1039.

Additional material from the right arm was obtained from a series of clones constructed in the lambda and cosmid vectors.[6]

These two gene banks, taken from yeast strains XJ24-24a and AB972, were ready to be used to generate most of the sequence of chromosome III (280 kb of the 315 kb). Other clones from strains A364A[7] and DC5[8] were used for the remaining 35 kb of the chromosome.

It was these banks that were used by the European sequencing network, which benefited from international contributions from the very beginning. Some characterization of clones as well as some preliminary sequences had already been undertaken at the laboratory of Professor Steve Oliver in Manchester, UK. The primary clones were organized using various techniques: direct subcloning, shotgun cloning and chromosome walking with synthetic oligonucleotides. Chromosome III was divided into pieces of 10 to 20 kb covering the chromosome from left to right telomere. Steve Oliver's laboratory distributed the clones between September 1988 and March 1989.

Each of the participating laboratories either received plasmids or bacteriophage DNA. When necessary, bacteriophage particles were also available. With the cloned material, Steve Oliver's laboratory provided a restriction map of the cloned section (a photo of an electrophoretic gel of a restriction digestion of the DNA confirmed the map, as well as a photo of a Southern Blot of a CHEF separation of the chromosomes, which confirmed that the cloned fragments did indeed come from chromosome III). There was no major problem with the plasmid clones, but there was a bit of difficulty and delay of the lambda clones from Olson's laboratory in St. Louis.

---

[6] M. Olson *et al.*, "Random-Clone, Strategy for genomic restriction mapping in yeast", *Proc. Nat. Acad. Sci.*, USA, vol. 83, 1986, pp. 7826–7830.

M. Olson *et al.*, "Genome Structure and Organisation in *S. cerevisiae*", in *The molecular and cellular biology of the yeast Saccharomyces*, ed. Broach, Pringle and Jone, vol. 1, Genome Dynamics, Protein Synthesis and Energetics, Cold Spring Harbor Laboratory Press, Cold Spring Harbor, N.Y., 1991, pp. 1–40.

[7] C.S. Newlon *et al.*, "Analysis of a circular derivative of *S. cerevisiae* chromosome III: a physical map and identification and location of ARS elements", *Genetics*, vol. 129, 1991, pp. 343–357.

[8] A. Yoshikawa and K. Isono, "Chromosome III of *S. cerevisiae*: an ordered clone bank, a detailed restriction map and analysis of transcripts suggest the presence of 160 genes", *Yeast*, vol. 6, pp. 383–401.

The bacteriophage strains were rather weak and tended to die off during their journey, which meant they had to be sent again. A few anomalies and teething problems were also encountered.

In addition to these activities, Steve Oliver's laboratory prepared, characterized and checked a complete set of clones for chromosome III supplied to one of the contractors (Dr. Alistair Brown, Aberdeen, UK). The latter used the clones to inventory the variations between the laboratory and industrial strains of yeast. Professor Oliver made two visits to MIPS, once accompanied by Dr. Gent and once by Dr. Greenhalf. During his visits, an analysis and evaluation of sequence data submitted by the contractors was carried out by John Sgouros from MIPS in order to identify possible problems and conflicts. A complete physical map of the chromosome was generated and this, aligned to the genetic map, was distributed to the participants to fill in remaining gaps.

The Communication and Coordination System

Before any research was undertaken, an efficient system of coordination and communication was set up. Each of the 35 laboratories was linked to MIPS. The latter, under Dr. H.W. Mewes, was in charge of the computer coordination, organization and analysis of sequence data received and exchange of information between laboratories.

MIPS began its activities in early 1988 as a member of the International Association of Protein Sequence Data Banks (PIR-International)[9] with the support of the German Ministry for Research and Technology (Bundesministerium fur Forschung und Technologie (BMFT)) and the European Commission. MIPS contributed to the collection of data (sequences from European sources) and distributed the database of protein sequences and other information and PIR software in Europe. Six scientists worked part-time to gather data and collect and distribute protein sequence data.

---

[9] The International Association of Protein Sequence Databanks (PIR-International) was founded by the National Biomedical Research Foundation (NBRF-PIR), Tokyo University and MIPS and MIRS in February 1988.

Three additional scientists were employed on the yeast genome project, the major one being John Sgouros. The yeast database had, in principle, three different levels of access:

- Confidential data accessible to project coordinators to allow them to check on progress (This data was the result of sequencing work submitted by the laboratories and would most likely remain at this level of confidentiality for a maximum of six months before publication);
- Group data accessible to every laboratory involved in the project, but unavailable to third parties (Data at this level could be kept for a maximum of 12 months, including the period involved under Level 1 or before publication);
- Data available to everyone, compiled from all the sources accessible to MIPS (This included sequences from the projects with their authorization and sequences which had been kept at Levels 1 and 2 for more than 12 months. This data was also forwarded to the EMBL's database — this third level was never implemented).

All the data were divided between protein and nucleic acid sequences. The formats were the following:

- VAX/VMS, formats accessible to the PIR and UWGCG programs;
- Standardized CODATA format;
- EMBL format (if required).

MIPS was equipped with DEC hardware and also used AT compatible PCs. Since the sequence databases traditionally used DEC hardware and the MPI für Biochemie had always had good luck with the DEC system, MIPS eventually used a local area VAX Cluster (Fig. 17).

In order to ensure that the data produced was correct, the following rules were laid down:
- The sequences had to be done on both DNA strands without gaps;
- Ambiguous nucleotides or regions had to be indicated;
- The sequence strategies had to be made clear to the computer center;

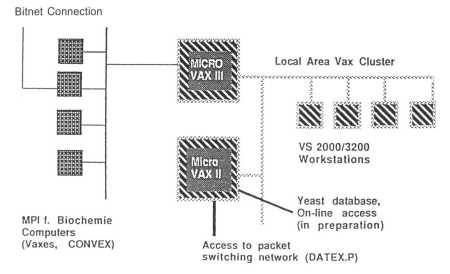

Fig. 17 MIPS hardware configuration

- The junctions between subclones had to be covered by overlaps;
- The subclones and autoradiographs (or any other raw data) were to be kept for use in checks should there be any conflict.

A quality statement had to be signed by all those involved in the yeast genome project.

While it assembled the new sequence data and analyzed the different elements of the sequences, MIPS scanned all the computer databases in existence as well as previous results and built the database with which the new results from the yeast genome project could be compared.

Publication of Data

Since all the data was to be made public within a maximum of one year from their deposition with MIPS, it was essential for the contractors to be able to publish their sequence data rapidly as short articles. Initially, these articles on the sequencing of chromosome III were published as annotated sequences in the journal *Yeast*. A single nomenclature was adopted for the designation of DNA fragments and Open Reading Frames (ORFs). In addition, a global article reporting the whole sequence for chromosome III was to be jointly signed by all the participants and submitted to an international journal.

Function Analysis

Yeast offers unique opportunities for rapid identification of genes since the tools required by inverse genetics to introduce deletions or other mutations in a particular gene are available. Professor Slonimski, coordinator of function analysis, organized a practical course funded by the European Commission. The participants brought their own disruption mutants to study during the course.

Links with Industry

There was further pressure within the IRDAC and CGC committees, and from certain representatives of the member states (links with industry were inventoried by Dr. Roy S. Tubb under contract BAP-0494N) between July 1989 and January 1990 to initiate and formulate an appropriate basis for the constitution of a European association of companies involved in yeast biotechnology (the Yeast Industry Platform, or YIP). The idea of YIP as an established organization (like the Green Industry Biotechnology Platform (GIBIP)), went back to 1987 [10], and meant that when a first inquiry was made, 18 companies were ready to support the European yeast genome sequencing initiative. The concept of a European club of "yeast industries" had also been proposed and was generally well-received. The need for the YIP was believed to be particularly apparent in research linked to yeast under the BAP and BRIDGE programs, with the need to provide concrete solutions to legal problems, and to answer consumer questions about these new technologies and products to be issued under the BAP and BRIDGE programs.

---

[10] R.S. Tubb, "The long-term benefits of genome sequence data to the yeast-dependent industries", *Sequencing the yeast genome: A detailed assessment*, ed. A. Goffeau, CEC XII-88/5J.

7 The Decryption of Life 431

During the course of July and August 1989, a letter was sent to the 32 European companies that had expressed an interest in participating in the Yeast Industry Platform. This letter presented the objectives and results of the GIBIP and outlined the needs of the future YIP, with particular reference to the yeast genome project.

The list included [11]:

- *The 17 following companies who responed positively or showed an interest in participating in a project to sequence the yeast genome:*

    Gist Brocades NV (NL), La Cruz del Campo SA (ES), Alko Ltd (FI), Arthur Guinness Son & Co (IE), Artois Piedboeuf Interbrew (BE), Burns Philp: Food and Fermentation Division (Australia), Unilever Research Laboratory (NL), Nestec SA (SE), Bass PLC (UK), ICI Pharmaceuticals (UK), Jastbolaget (SE), Moet et Chandon (FR), Transgene SA (FR), Idun A/S (NO), Norbio A/S, S.I. Lesaffre (FR).
    Later to join this list would be Arthur Guinness (IE) and British Biotechnology (UK).

- *Selected companies known for their major interest in biotechnology who did not respond or responed negatively to the preliminary 1987 survey:*

    Transgene (FR), Celltech (UK), Lesaffre (FR) contacted via Goffeau, Moet and Chandon (FR) contacted via M. Aigle, United Breweries (DE) (Carlsberg-Tuborg), Zymogenetics (USA), Unilever (NL).

- *Companies not approached during the initial survey (mainly from EFTA countries):*

---

[11] Although based in Switzerland, Nestlé was approached as early as 1987 because of its significant presence in EC countries, which is why Nestlé is included in Category 1. Distillers C° (Yeast) Ltd in Category 2 is now a Unilever company. Although the two firms were approached, they only count as one for statistical purposes.

Alko (FI), Cultor (FI), Pripps (SE), Jastbolaget (SE), Norbio (NO), Idun (NO), Norsk Hydro (NO), Apothekernes (NO), Ciba-Geigy (CH).

After a meeting of the *ad hoc* steering group[12] at Royal Gist Brocades NV (Delft, NL) on 12 December 1989, to which the first 17 companies were invited, along with 20 other major companies linked to yeast biotechnology, the YIP was founded in 1990, and by 1991, included 13 European firms (Table 19).

YIP's activities began on 1 January 1991, with a basic mandate:

- To identify and sustain possible applications of publicly funded research and development programs involving yeast;
- To improve communication between the regulatory agencies and industry in the field of yeast biotechnology;
- To provide solutions to the obstacles to the implementation of technology and/or products based on yeast, especially the problems of risk evaluation, environmental and consumer protection, intellectual property rights and the harmonization of European regulations;
- To enlighten the public as to the positive contribution and possibilities of biotechnology in procedures and products based or derived from yeasts, especially in the sectors linked to brewing and agro-foods and in the pharmaceutical and chemical sectors.
- To formulate the prospects for industry of the mechanisms of taxation and intellectual property rights that will emerge from sequencing the yeast genome and exploiting the results. Preliminary general information on sequences from the 35 laboratories was distributed regularly to each of the YIP members, who were able, after analysis, to contact the laboratory that had sequenced a gene that looked interesting. The companies guaranteed that data sent to them would

---

[12] This steering group consisted of Dr F. Pryklung from Kabo, Dr Hinchcliffe from Bass, Dr Iserentant from Artois, Dr Niederberger from Nestec, Dr Sander of Gist Brocades and Dr Vladescu from Pernod-Ricard. The presidency of the temporary committee was entrusted to Dr Matti Korhola of Alko.

Table 19 Members of YIP at the time of its foundation[13]

Company	Activity	Representative
Alko, FI	Alcoholic drinks, baker's yeast, industrial enzymes	Dr Matti Korhola director research lab and YIP director
Boehringer Mannheim GmbH, DE	Pharmaceuticals and biochemicals	Dr Werner Wolf, Vice President R&D
Champagne Moet & Chandon, FR	Beer, soft drinks, malt	Dr Bruno Dutertre
Interbrew NV Leuven, BE		Dr Dirk Iserentant
Kabivitrium AB, SD	Reagents, Procordia beer, baker's yeast	Dr Linda Frycklund
La Saffre et Cie, Marcq-en-Baroeul, FR	Baker's yeast, malt alcohol	Philippe Clement
Nestec Ltd, Vevey SW	R&D for Nestle food and drink	Dr Peter Niederberger
Pernod Ricard, Creteil, FR	Wine, distilled alcohol, soft drinks	Dr Barbu Vladescu
Rhone Poulenc Rorer SA, Anthony, FR	Pharmaceutical brews	Dr Reinhard Fleer
Royal Gist Brocades NV, Delft, NL	Yeast, industrial and pharmaceutical enzymes	Dr Klaus Osinga
Tepral BSN drinks research centre, Strasbourg, FR	BSN, beer, champagne, bread	Dr Frédéric Gendre
Transgene Strasbourg, FR	Research contracts	Dr R. Gloecker
Unilever NV, Vlaardingen, NL	Food, drink, chemical products, detergent, personal articles	Dr John M.A.Verbakel

remain confidential. Should possible industrial applications arise from sequence data, the laboratory was to give preference to European companies, in particular, YIP-affiliated companies. Each year a complete list of all the sequences sent to the EMBL database was to be sent to the YIP. Member companies were asked for an annual contribution of 2,000 ECU to finance the secretariat and other

---

[13] Later, other members joined the YIP, such as Spain's La Cruz del Campo S.A. represented by Dr Marco Delgado, Belgium's SmithKline Beecham Biologicals represented by Dr Nigel Hurford, France's C/O Orsan Eurolysine represented by Dr Daniel Pardo, and the Netherlands' Heineken Technical Services B.V. represented by Dr Jan Kempers.

operational costs, and from the BRIDGE yeast genome sequencing project, a further 1,000 ECU per year. Non-European member companies paid another 5,000 ECU membership fee (a total of 8,000 ECU) to make their contributions up to the amount already paid through taxation by the other member state companies to the European Community.

The total overall sum obtained (at least 28,000 ECU per year) was modest compared to the cost of sequencing itself, but it paid for the secretariat costs (Anne Marie Prieels) and contributed to the development of the computer network, the annotation and distribution of information, the coordination and promotion of activities. The YIP companies have an annual turnover of more than 180 billion ECU and a workforce of more than 600,000 employees. The initial industrial platform was set up for three years, the members meeting at least twice a year. Their opinions were channeled through the platform to the European Commission, the ECBG and the SAGB. The existence of the platform justified including the pilot project to sequence chromosome III in the Biotechnology Action Program (BAP). The interaction between contractors and communication between the consortium and the YIP were ensured through meetings and colloquia.

7.2 A World First — The Sequence for a Whole Eucaryote Chromosome: Chromosome III of the Yeast *Saccharomyces cerevisiae*

Between January 1989 and May 1991 the consortium of 35 European laboratories worked to determine the genetic sequence of only this chromosome. Each laboratory was entrusted with sequencing a prepared and accredited fragment consisting of one or two units, each of 8 kb of primary sequencing and 3 kb of overlapping sequencing. Thus, each laboratory had, on average, an obligation to sequence almost 6 kb per year, with an EC contribution of 5 ECU per base pair, 2.4 MECU in all, which was estimated to be half of the overall cost. The rest would be provided partially through national funding and partially by the laboratories themselves.

During these two years of work, the coordinators and contractors met to analyze the progress and problems in three plenary meetings[14] and several meetings of lesser importance. These allowed the coordinators to lean on the collective wisdom of all the scientists taking part in the program to maintain the motivation and flow of research.

During BAP, most of the sequencing was carried out using classical methods rather than automatic sequencers. Each of the functional components of the chromosome (centromere, telomere etc.) was isolated and analyzed.

After ten months of activity (at the Tutzing BAP meeting in October 1989 attended by 75 scientists from the 35 member laboratories of the sequencing network and scientists invited from Canada[15], Israel[16] and the USA[17]), some progress had been made. Sixteen industries had expressed an interest in participating in an industrial platform and had not ruled out taking part in the project.

As for international cooperation, the non-European scientists provided information on initiatives outside the European Community. A common project was being set up between the University of St. Louis (USA) and the RIKEN Institute (Japan) to sequence chromosome VI. However, there was, as yet, no information on its progress. In general, activities in the USA seemed behind the European initiative. The American and Canadian scientists tendered several possible reasons for this situation. Europe had taken the initiative in sequencing the yeast genome and, in consequence, the promotional credit. According to Nobel prizewinner James Watson (the Director of the Genome Office of the NIH), small genomes could only be sequenced when technological developments allowed them to be sequenced very quickly. Nor were there any well-known American yeast scientists

---

[14] "Yeast Sequencing Strategies; from BAP to BRIDGE", Louvain la Neuve, Belgium, 20–21 March 1989, the BAP meeting on "Sequencing Yeast Chromosome III", Tutzing, Germany, 31 October–2 November 1989, and the 2nd meeting of the European Network in charge of "Sequencing Yeast Chromosome III", Den Haag, the Netherlands, 26–27 July 1990.

[15] Dr J. Friesen of the Hospital for Sick Children, Toronto, Canada.

[16] Dr M. Goldway of the Hebrew University of Jerusalem, Israel, and Dr G. Simchem, of the Department of Genetics at the same institution.

[17] Professor C.S. Newlon, Dept. Microbiology and Molecular Genetics, New Jersey Medical School, Newark (USA).

ready to lead the setting-up of an initiative (in fact, at the time, some of them were flirting with the HGP). However, if a yeast genome project were to be launched, it seemed that most of the work could be carried out at the University of Stanford under Professors R. Davis and D. Botstein, in a centralized approach quite the opposite of the decentralized European approach. A Canadian project on the sequencing of chromosome I (240 kb) had been proposed to the relevant funding bodies by Professors J. Friesen, D. Thomas, H. Bussey and J. Von Borstel.

The Future Bridge Program

For the future European BRIDGE program, it was agreed that sequencing activities should continue in the decentralized style used for chromosome III. Possible exceptions (that is to say, one laboratory or a small number of laboratories sequencing an entire chromosome) could be considered if they could claim specific advantages. There were many advantages to the decentralized model:

- Rapid exploitation of results by the group of scientists who originally provided the data;
- A parallel gain of expertise and information shared throughout the European Community;
- Technological and methodological innovations resulting from the meeting of individual experience and ideas;
- An increased ease of access to European scientific competence (contractors were expected to develop new skills to share with other contractors confronted with specific problems.).

Individual laboratories involved were to sequence at least 25 kb/year, which was, at the time, the level of sequencing carried out per year by the world's most experienced scientists. During the establishment of the yeast genome project, a contribution of maximum 2 ECU/bp was originally expected. For the BRIDGE program the chromosomes to be worked on were chromosome II (840 kb), VII (1,140 kb), IX (455 kb) and XII (1,500 kb and some repetitive rDNA). Other chromosomes could be taken into consideration.

Detailed proposals for the construction of ordered chromosome libraries were to be made and include the coordination and distribution of clones. The financial support anticipated for these activities was of 0.1 ECU/bp. The informatics coordination, centralized for the entire program, was not to receive more than 400,000 ECU from the European Commission.

At the second BAP meeting of the European network sequencing yeast chromosome III[18] in July 1990, which brought together 67 participants, seven YIP representatives and five sequencing and mapping experts from Japan, Canada and the USA, almost 244 kb of the 320 kb ordered fragments of chromosome III DNA had been sequenced. A total of 196 ORFs longer than 80 amino acids had been identified. Eighteen genes (9%) previously sequenced had been confirmed and precisely mapped on chromosome III. Twenty-five (13%) unknown genes had some similarities to previously discovered and known genes. Six (3%) new genes were characterized by the phenotype of their disruption mutant. One hundred forty-seven (75%) of genes remained totally unknown. By July 1990, a total of 278,430 bp had been provided to MIPS, including 34,019 bp of overlap sequence. Several DNA fragments, corresponding to about 40 kb of original sequence, remained to be sequenced by 26 contractors, as well as 10 junctions.

It was during this meeting that Professor Slonimski introduced the "Integrated Gene Expression Network" (IGEN) as a new approach in the study of unknown ORF functions. This approach was based on 2-D gel electrophoresis systems presented by Helian Boucherie. More than a thousand soluble yeast proteins could be distinguished and digitalized. It was, therefore, possible to localize modifications caused by mutations, disruptions or amplifications of genes. This approach, which possibly could be extended to membrane proteins, was likely to help identify the different components of regulatory systems. In the meantime, the search for phenotypes after ORF disruption was undertaken spontaneously by most of the contractors, helped by Professor Slonimski's course. This was encouraged with the announcement that, for the final completion of contractual duties by laboratories that had

---

[18] Second meeting of the European Network in charge of "Sequencing Yeast Chromosome III", Den Haag, the Netherlands, 26–27 July 1990.

not received enough DNA, the disruption of one gene was going to be considered to be worth the sequencing of 3 kbp. The success of this approach was, however, limited at that time to the identification of a few phenotypes: gly (Grivell and Lucchini), letal (Glansdorf), drug (Thireos), cold (Jacquet), temperature sensitive (Grivell) and sterility (Hollenberg). The search for homologies by *in silico* analysis had had much more success.

A total of 196 ORF of a length superior to 80 amino acids were identified. Eighteen (9%) of the genes previously sequenced were confirmed and precisely mapped on chromosome III. Twenty-five (13%) of the new unknown genes had similarities to known genes that had already been discovered. Six (3%) new genes had been characterized by the phenotype of their disruptant. Finally, 147 (75%) of the genes remained totally unknown. The Japanese K. Isono, who had identified 156 transcripts of a length greater than 350 bp from chromosome III, showed that most of the identified ORFs were expressed[19]. The existence of totally unknown ORFs, not similar to each other nor to any other gene family known at the time, revealed that a whole set of new genes and gene products could not be elucidated by classical biochemical and genetic approaches. Elucidating the functions of the products of these genes looked like a good challenge for the future.

As for international cooperation, there had not really been any commitment from an American laboratory to take part in the European sequencing effort[20]. However, there was little doubt that M. Olson would

---

[19] K. Isono, A. Yoshikawa and S. Tanaka, "An approach to genome analysis of *Saccharomyces cerevisiae*", in 2nd meeting of the European Network in charge of "Sequencing Yeast Chromosome III", Den Haag, the Netherlands, 26–27 July 1990, Book of Abstracts, pp. 5–7.

A. Yoshikawa, K. Isono, "Chromosome III of *Saccharomyces cerevisiae*: An ordered clone bank, a detailed restriction map and analysis of transcripts suggests the presence of 160 genes", *Yeast*, 1991.

[20] Some time later, a letter addressed on 15 October 1990 by Mewes to Goffeau pointed out that D. Botstein had contacted Mewes to indicate that as far as the US genome project was concerned, they were hoping for cooperation rather than competition with the European project. Botstein had confided to Mewes that he hoped to start their side of the project with chromosome V, a chromosome for which Oliver had planned to build an ordered clone bank under the BRIDGE program. So the European consortium took over chromosome VIII while Botstein and Davis's group at Stanford sequenced chromosome V. Oliver was in charge of developing a bank of cosmid clones for chromosome VIII. A genomic bank for *S. cerevisiae* S288C had been prepared in the cosmid vector PWE15. A collaboration agreement was set up between Botstein's group and the consortium to make the European libraries, such as those of Oliver and Dujon, available to the Americans.

continue to provide access to his library for the European consortium (strain AB 972). Furthermore, K. Isono, who had worked for more than 10 years in Europe, was to make available his libraries of strain DC5, in particular, that of chromosome V, which was 90% complete, and VIII, which was 60% complete. He was also working on chromosomes IV and VIII. E. Soeda, of the RIKEN Institute in Japan, presented the sequencing strategy to be used by the Japanese venture on chromosome VI. In direct opposition to the European strategy, they had used the centralized model, basing their work on the new sequencing technologies and automatic sequencers. The chromosome VI contig map was prepared from lambda and cosmid clones by Olson's laboratory. The shotgun technique of sequencing was to be used on each clone. According to Dr Howard Bussey, work had started on sequencing chromosome I in Canada, although at that time, no decision had been made as to how this was going to be financed.

Finally, in May 1991, after 27 months of work, the sequence of yeast chromosome III was obtained, a world first for the completion of the entire sequence for a chromosome of a eucaryote organism.

Several well-known scientists had been somewhat skeptical of European ability to bring such a project to fruition. David Beach, of Cold Spring Harbor, felt that no-one took the European venture seriously because of the fragmentation of the work and because Stanford University was setting up a laboratory to focus exclusively on baker's yeast as part of a larger effort leading to the mapping and sequencing of the human genome[21]. M. Olson remarked that, "the idea of letting 1,000 flowers bloom is fine for the means. But this eclecticism has been allowed to spill over into the goal and that has the potential to spell disaster"[22]. J. Watson agreed that he did not think that coordination could be carried out by bureaucrats: "It must be done by scientists"[23]. Lennart Philipson[24], of the EMBL, bluntly announced that from his point of view: "One of the reasons for failure in the coordination

---

[21] D. Leff in *Biotechnology Newswatch*, 10, 6, 9 March, 1990.

[22] M. Olson, quoted by L. Roberts in *Science*, vol. 245, 1989, p. 1439.

[23] J.D. Watson, *Science*, vol. 237, 31 July 1987.

[24] L. Philipson, *Nature*, vol. 351, 9 May 1991.

of European research programs could be the existence of a group of scientific policy makers who influence national and international policies at the same time. Most of them have either failed in their own research or have a legal or economic training (...) their loyalty is towards the organization or the ministers for whom they work"[25]. In response to this pessimism, the sequencing pilot project had succeeded completely.

The sequence obtained was 315,356 nucleotides long with a total sequence of 385 kb. (This means that over a fifth of the chromosome had been verified by more than one laboratory.) This result, apart from being the first sequence of a whole chromosome, was also the longest piece of contiguous DNA described at that time.

Another record beaten on the same occasion was the number of authors of a scientific publication in molecular biology. The article, recounting the results of the sequencing, was published on 7 May 1992 in the famous scientific magazine *Nature*[26], and cosigned by 147 researchers from 37 laboratories, of which 35 were European, one Japanese and one American. *Nature* later was to list this article amongst the nine articles they had published in 120 years which shook the scientific world (namely, Darwin's theory of natural selection, the discovery of X-rays and monoclonal antibodies, the theory of tectonic plates, the structure of DNA...) On 15 May 1992, a press conference organized by the European Commission publicized the event[27]. Sixteen months later, hundreds of other articles had been published mentioning, commenting on and completing the analysis of the sequence, thereby highlighting the impact of these results on the international scientific community.

---

[25] It should be remembered that in 1987 the initiators of the project had asked the EMBL to join in its implementation. Making noises about a big yeast sequencing project in the United States and European difficulty in measuring up to American science, even in Europe-wide cooperation, L. Philipson declined the offer. So the EMBO and the EMBL were not involved in the project. Had they been the only input on the decision, an enormous opportunity would have been missed. It's another more general lesson to be taken from the success of the chromosome III sequencing project: no organization is immune to errors of judgement. What this example shows is the danger of entrusting one institution with all the decisions and funding for European biological research. As Philipson himself admits "some diversity may be beneficial".

[26] S. Oliver *et al.*, "The complete sequence of yeast chromosome III", *Nature*, vol. 357, 1992, pp. 38–46.

[27] A Consortium of European Laboratories sequences an entire chromosome of yeast, Brussels, 15 May 1992.

This success of the pilot sequencing program was unanimously praised by the world media, and thanks to Professor Goffeau's initiative, Europe had made some headway in the gene race in which it had been challenged by the USA. As André Goffeau pointed out, "We have shown that with modest funding, 2.67 MECU over two years, we can also do 'Big Science'."

Lessons Learnt from the Pilot Project

The success of the pilot phase of the yeast genome sequencing project taught several lessons. In the first place, this project demonstrated that the three-coordinator network, shown in Fig. 18 worked well.

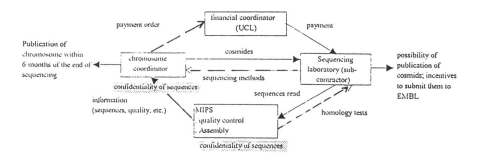

Fig. 18 Diagram of the network and its operation

(diagram from P.B. Joly and V. Mangematin in "Is the Yeast Model Exportable?", *manuscript provided to the author*)

This operation had the following main advantages (from a sociological point of view): it allowed the pleasure of small science to be maintained, i.e., the interactions, collaborations, excitement of confirming discoveries, avoidance of conflicts and problems involved in hierarchy and the maintenance of efficiency in work carried out at reasonable cost. Another advantage was that it allowed the promotion of molecular genetics in smaller European laboratories. More specifically, the distribution approach had the major advantage of automatically providing peer review of the results, which was the basis of guaranteeing the quality of sequencing and setting up methods to evaluate and/or increase it. The division of work allowed some laboratories to go to a deeper level of study immediately after the sequencing task was completed (genes discovered). It contributed to the development of researchers and improvement of laboratories, permitting progress and self-affirmation better than strong competition. Finally, despite what had been said beforehand, the distribution approach is not more expensive than the centralized approach, especially if one considers the real figures rather than the propaganda. In fact, when you look at the number of people working in the large sequencing centers and the cost of installation, many small European laboratories are more competitive, or at least as competitive as the centers, which is normal because the work was carried out by a consortium of extremely qualified personnel.

On a technical level, this success showed that completely manual sequencing was still competitive. One of the prerequisites was access to a complete and organized cosmid library. The limiting steps occurred, at that time, during subcloning, assembly, the resolution of problems such as gaps and in analysis. However, it was admitted that this approach was still too expensive and certainly too slow. Without automation, it would have taken over a century and US$150 million to fully decode the entire sequence of the yeast genome. This estimation showed that increased efficiency and network capacity was needed for the later stages of the yeast genome program.

From a scientific point of view, the *in silico* analysis of the sequence revealed some remarkable characteristics. There are 182 ORFs (which could be genes) longer than the more or less arbitrarily set length of 100 amino acid residues. At the time of publication, almost two thirds of the products of those genes had turned out to be completely new; for this reason, they were dubbed the "EEC genes" by Piotr Slonimski, EEC standing for Esoteric, Elusive and Conspicuous. Three ORFs were interrupted by introns; 37 ORFs, of which 34 were represented on the genetic map of *S. cerevisiae,* could be linked to known genes controlling essential functions, but 145 corresponded to new ORFs, potentially being 145 whole new genes; 15 of the 145 had similarities to genes of other organisms; 15 were homologous to other *S. cerevisiae* genes. The other 115 ORFs seemed totally new. Since the original report, further research keeps uncovering additional homologies.

The percentage of "EECs" dropped rapidly to 58%[28] and then down below the 50% mark[29] (Fig. 19). In 1993, of the 165 defined ORFs from chromosome III, 94 had been suggested by the presence of particular patterns or significant homologies with genes known in yeast or in other organisms[30].

The lower limit for the length of ORFs had been set at 100 codons, but it is certain that genes coding for shorter proteins do exist, for example, the gene PMP1, which codes for a proteolipid of only 40 amino acids[31]. Without clear information on their functions, the shorter genes could not easily be identified. Some short genes were, however, identified from their sequence because of introns, which are usually flanked by codons, or by the existence of corresponding transcripts. In all, there were only a few of them, and all indicated that the established 100 codon limit was close to the optimum to minimize either the number of genes omitted or the number of incorrect ORFs.

Even if the percentage of "EECs" has been greatly reduced since the sequence of chromosome III was published, one of the major lessons to be learnt was how ignorant we in fact were, rather than the discovery of lots of new genes. As B. Dujon said, "The idea that a large portion of the genes was escaping the methods of the yeast geneticists was not a new one. In the mid-1980s, a paradox was brought to light with regard to the number of genes when the results of synthetic mutageneses were compared to the number of transcripts in chromosome I[32]. It was also a shock to see that only a very small portion of yeast genes led to a lethal haploid phenotype when they were disrupted"[33].

---

[28] P. Bock et al., "Comprehensive sequence analysis of the 182 predicted open reading frames of yeast chromosome III", *Protein Sci.*, vol. 1, 1992, pp. 1677–1690.

[29] E. Koonin et al., "Yeast Chromosome III, new gene functions", *EMBO J.*, vol. 13, 1994, pp.493–503.

[30] P. Slonimski and S. Brouillet, "A data base of chromosome III of *Saccharomyces cerevisiae*", *Yeast*, vol. 9, 1993, pp. 941–1029.

[31] C. Navarre, M. Ghislain, S. Leterne, C. Ferroud, J.P. Dufour and A. Goffeau, *J. Biol. Chem.*, vol. 267, 1992, pp. 6425–6428.

[32] D.B. Kaback, P.W. Oeller, H.Y. Steensma, J. Hirshman, D. Ruezinsky, K.G. Coleman and J.R. Pringle, *Genetics*, vol. 108, 1984, pp. 67–90.

[33] M.G. Goebl and T.M. Petes, *Cell*, vol. 46, 1986, pp. 983–992.

444 *From Biotechnology to Genomes: A Meaning for the Double Helix*

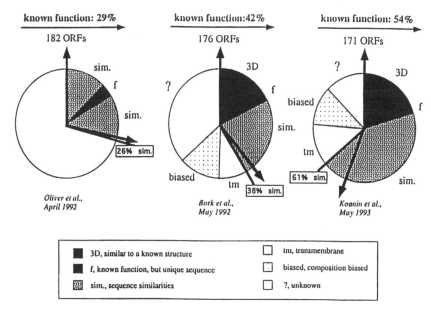

Fig. 19 Evolution of sequenced ORF of chromosome III of *S. cerevisiae* (between April 1992 and May 1993)
Source: A. Goffeau

Before sequencing, we knew that the yeast genome contained a lot of "FUN" (Function UNknown) genes which, for reasons which remain obscure, could only be brought to light by molecular methods. But the fundamental discovery, made during the reading of chromosome III, was that a substantial proportion of "FUN" yeast genes seemed to have escaped the attention of the geneticists. In comparison, a much smaller proportion (about a quarter) of yeast genes previously identified with traditional genetic methods did not, when analyzed, evoke homologies from the databases. This result was news, both on the content of the yeast genome and the database[34]. A large group

---

[34] B. Dujon, "The yeast genome project; what did we learn?", *Trends in Genetics*, vol. 12, no. 7, July 1996, p. 266.

of orphan genes which for the most part are common to lots of organisms, seemed to make up most of the yeast genome not being studied by geneticists. It must, however, be noted that even when there is no fully significant homology, *in silico* analyses are able to provide some clues to the nature of some of the orphans. So, the prediction of transmembrane segments led to the astonishing conclusion that 35 to 40% of the proteins predicted from chromosome III had transmembrane helixes[35]. Lastly, the predicted function of each ORF seemed to be demonstrable only by experimentation, a challenge to be taken up by the functional analysis program. Although they were carried out haphazardly on chromosome III, the gene disruptions led to another fundamental discovery: the majority of disrupted genes conferred no phenotype on the cell (only 6% of disrupted genes turned out to be essential to the cell's survival and only 14 of the 42 viable disruptants had altered phenotypes). In any case, the experimental study carried out by Isono on chromosome III[36] showed that the total number of transcripts was close to that of the ORFs sequenced, eliminating the possibility that the orphans might be less often expressed than the well-known genes. Some unexpected homologies were also discovered such as a homology to the *white* gene in *Drosophila* and a homologue of the *nifs* gene of nitrogen-fixing bacteria. Even more astonishing was the discovery that *nifs'* function was essential for the cell's existence even though yeast does not fix nitrogen. The number of genes with homologies, with proteins from other organisms or other *S. cerevisiae* proteins, were, at the time of the sequence's publication, already impressive. Ten ARNt genes and several transposable elements were also identified. A series of *Ty* insertions were confirmed, mainly in two regions present in all the strains (dubbed "left arm and right arm transposon hot spots"), although in other strains, another distal transposon was discovered on the right arm.

---

[35] A. Goffeau, P. Slonimski, K. Nakai and J.L. Risler, *Yeast*, vol. 9, 1993, pp. 691–702.

[36] A. Yoshikawa and K. Isono, "Chromosome III of *Saccharomyces cerevisiae*: An ordered clone bank, a detailed restriction map and analysis of transcripts suggests the presence of 160 genes", *Yeast*, 1991, vol. 6, 1990, pp. 383–401.

The results of the alignment of the sequences, carried out by the famous University of Wisconsin Genetics Computer Group Software Package, revealed differences between the physical and genetic maps of chromosome III[37]. Except for one difference for a particular gene site, the variations mainly concerned the distance between genes. The comparison of distance on genetic and physical maps gave an average slip of 0.51 centimorgans per kb, but this varied according to the siting on the chromosome, occurring the farthest from the centromere and mostly in the middle of each arm. At chromosome level, the recombination frequency varied enormously and "hot spots" and "cold spots" of recombination were identified[38]. The sequence for chromosome III suggested new molecular bases for local variations in recombination frequency.

The polymorphism of sequences not identified by traditional methods but by the comparison of sequences obtained from various strains of *S. cerevisiae,* were also interesting from an evolutionary point of view. Research on this phenomenon continued on the strains used in the BAP project and in other laboratories and industrial facilities. These studies pinpointed two sites, simple restriction polymorphisms mostly due to single-base mutation and, far less common, polymorphisms all along the chromosome, caused in chromosome III by rearrangements involving the distal transposon to be found in some strains. Subsequent analyses of chromosome III confirmed a very high gene density. The chromosome is made up of a succession of segments, rich and poor in G and C, with a peak G and C content greater than 50% in each arm[39].

Following the publication of the sequence of chromosome III, calculations were made based on the results obtained led Goffeau, Slonimski, Nikai and

---

[37] R.K. Mortimer, R. Contopoulou and J.S. King, "Genetic and physical maps of *Saccharomyces cerevisae*, Edition II", *Yeast,* vol. 8, 1992, pp. 817–902.

[38] T.D. Petes, E.R. Malone and L.S. Symington in *The Molecular and Cellular Biology of the Yeast* Saccharomyces: *Genome Dynamics, Protein Synthesis and Energetics*, ed. J.R. Broach, J.R. Pringle and E.W. Jones, Cold Spring Harbor Laboratory, New York, 1991, pp. 407–521.

[39] P. Sharp and A. Lloyd, "Regional base composition variation along yeast chromosome III. Evolution of chromosome primary structure", *Nucleic Acid Res.*, vol. 21, 1993, pp. 179–183.

Risler[40], estimating that the entire yeast genome could be about 12,540 kbp long, including 16 centromic regions of 300 base pairs each, some 300 ARS (autonomously replicating sequences) and about 6,800 ORFs.

Although the success of the European project was lauded worldwide, there was some criticism. Doubts were voiced as to the quality of the sequence. It is, of course, impossible that any enterprise beyond a certain complexity and size would be error-free. In the case of the sequence obtained, it was almost a given that there would be inaccuracies in the original data due to experimental error, and because of this, from the start of the program, the analysis was accompanied by a constant check on, and validation of, the data. Some scientists were worried about the percentage of error. The results obtained during the BAP project were later submitted to retrospective analysis, and a low rate of error was found. Wicksteed et al.[41] analyzed the clones used during the sequencing project and compared the published sequence with the sequence found in each of the six strains to determine whether the composite chromosome used for the original sequencing enterprise was a sufficiently exact representation of the chromosome of yeast. They also searched for any significant cloning artifact that might have been mixed up with the original sequence. They pointed out that 0.69% of the restriction sites predicted from the DNA sequence were not detected, suggesting an error rate in the published sequence of about 1.3%. They failed to discover a cloning artifact. The quality standards were, in all cases, at the level of those imposed by first-rate scientific journals, including 100% sequencing on both strands, the checking of junctions and detailed description of the sequencing strategy. As an additional precision, all the authors of the final article signed a quality statement on the sequence results.

In addition to these concerns, there were further, essentially American, criticisms, accusing the network of having kept the sequences confidential for too long. *Science* reported:

---

[40] A. Goffeau, P. Slonimski, K. Nakai, and J.L. Risler, "How many yeast genes code for membrane-spanning proteins?", *Yeast*, vol. 9, 1993, pp. 691–702.

[41] B.L. Wicksteed, I. Collins, A. Dershowitz, L.I. Stateva, R.P. Green, S. Oliver, A.J.P. Brown and C.S. Newlon, "A physical comparison of chromosome III in 6 strains of *Saccharomyces cerevisae*", *Yeast*, vol. 10, pp. 39–57.

"... With this incentive, the sequence came in rapidly and the Europeans began to talk about their work at a conference back in 1989. But at that point, "some of the shine began to go from a magnificent achievement", as Russel Doolittle of the University of California at San Diego puts it. The reason? The sequences weren't being published, and more and more researchers began to ask, "Where's the data?" Some scientists were privately furious: "This sequence information could make or break some people's careers", said one. It's scandalous. If they've been paid good money to do it, they should get out and make the information available"[42].

Against this accusation, those running the project raised the following points:

- A total of 135 kb of chromosome III sequence data was published by the European Consortium in articles between March 1990 and March 1992 in *Yeast* journal;
- All the requests of other scientists on specific regions of the chromosome were fulfilled;
- The completion of the sequence was announced to the scientific community during a meeting in the United States in Summer 1991 and in the article submitted to *Nature* at the end of that same year;
- The period between the announcement and submission of results was due to rigorous checks of data through later analysis of the sequences and restriction analysis carried out by Professor Caroll Newlon (considerable efforts were made to ensure that the data provided to the research community was of the highest quality);
- Finally, the complete sequence was archived in the EMBL database on 16 March 1992 under the access code X59720 (the article describing the fundamental discoveries appeared on 7 May 1992 in *Nature*)[43].

---

[42] From *Science*, 24 April 1992, p. 462.

[43] H.W. Mewes' answer, entitled "European Science", to the criticisms put forward, in particular, those that appeared in the above-mentioned *Science* "Science in Europe" article, and which appeared in *Science*, vol. 256, 1992, p. 1378.

These points clarified the negative impressions that might have been engendered by the *Science* article and showed that the consortium had respected international conventions for the dissemination of genome data, as none had remained confidential for more than a year.

Despite this criticism, the European Community's success marked a new stage in molecular level life sciences. The sequence of chromosome III was, at the time, the only successful genome project in the world; and it proved that the sequencing programs on other organisms, being started by several nations and considered at the European level, could really be brought to term.

Genome Activity in 1992

By June 1992, the UK-USA bilateral cooperation to sequence *Caenorhabditis elegans* had already resulted in 100 kb of contiguous sequence, and Wellcome had decided to continue funding most of this work at the MRC; it did not look as if EC funding would be needed. Thanks to the BRIDGE program, *Arabidopsis* scientists were able to start sequencing under the BIOTECH I program (at the same contribution levels and on the same principles as the yeast project). In addition, there were plans to sequence some cDNA. In comparable organisms such as *C. elegans*, and with the help of powerful analysis software, it was believed to be no longer necessary to sequence cDNA to obtain the ORF sequences of a genome. As for *Bacillus*, the physical map and strategic approach having been established (under the SCIENCE (89–91) program financed to 609,000 ECU by the EC, although that contract could not pay for further work), the participants had already begun systematic sequencing at the Institut Pasteur (100 kb) and the University of Pavia (40 kb). Work on *Bacillus* would be continued under the BIOTECH I and II programs, from 1993–96, by 16 laboratories, with a total EC contribution of 2,870,000 ECU.

In the USA, no large-scale program had been set up, but in Japan, H. Yoshikawa, of the University of Osaka, had already sequenced 100 kb of the *Bacillus subtilis* genome. Yoshikawa, who had recently met some of the more prominent characters of DG XII, made sure that enough MITI funding was set aside to continue sequencing both *Bacillus* and yeast simultaneously.

At a strategic level, it looked as if the most efficient and advisable European scientific policy would be to keep the bipolar configuration of genome research activity between man and associated model genomes on the one hand and on the other, the smaller genomes with a certain biotechnological potential. This approach took into account the specific characteristics of the work (due to the size and state of progress of each genome program — physical and genetic mapping and systematic sequencing) and to the diverse scientific and industrial communities involved. Although the various entities involved had to remain separate, a continuous flow of information between projects was encouraged, especially regarding the new technologies, as well as computer management and analysis of data to prevent any reinvention of the wheel. For example, there were at least four independent databases for yeast genome sequences. This tendency amongst the scientific community to do just that, reinvent the wheel, was due in part to the lack of communication and/or trust in initiatives that were going on and also, in part, to the positions and influences that affected decisions in this sort of science.

As for yeast, the publication of the sequence for chromosome III attracted the interest of the international scientific community. This encouraging success contributed to the continuation of work under the BRIDGE program launched in 1991 by the EC. Officially, seven T-projects (large coordinated projects with the aim of eliminating bottlenecks braking or preventing the exploitation of data, material or methods of modern biotechnology by agriculture or industry) had been foreseen. Yeast genome sequencing was one of them[44].

---

[44] A. Vassarotti, A. Goffeau, E. Magnien, B. Loder, P. Fasella, "Genome research activities in the EC", *Biofutur*, vol. 85, 1990, pp. 1–4.

D. de Nettancourt, "The T-projects of BRIDGE, a new tool for technology transfer in the Community", *Agro Industry High-Tech*, vol. 3, 1991, pp. 3–9.

D. de Nettancourt, *BRIDGE Report: Sequencing the Yeast Genome*, 1992.

A. Vassarotti and A. Goffeau, "Sequencing the yeast genome: The European effort", *Trends Biotechnol.* vol. 10, 1992, pp. 15–18.

7.3 The Complete Sequence of the Genome and the Intensification of European Efforts

With 5,060,000 ECU of European Commission money, the yeast sequencing project intended to constitute cosmid libraries for, and sequence, chromosomes II (820 kb) and XI (670 kb). Lessons learned from the pilot phase dictated that a significant increase in efficiency and a reduction in sequencing costs were necessary. During this second phase, each laboratory would have to sequence at least 25 kb per year with an EC contribution of no more than 2 ECU per base pair analyzed and received by the MIPS analysis center. A much larger quantity of the sequencing was carried out automatically.

The consortium, which attained this objective in three years of work, involved 11 European nations; 17 laboratories worked on chromosome XI and 19 on chromosome II. Most of these research groups had been involved in the BAP program. A certain number of new laboratories which were not in fact yeast centers but brought significant expertise in macrosequencing and the use of automated equipment, were welcomed.

The approach used was similar to that in the BAP program. One coordinator per chromosome (H. Feldmann for II and B. Dujon for chromosome XI[45]) controlled a group of laboratories and sequencers. The management principles were as follows: one team per chromosome, 2 ECU per base pair obtained, quality control (less than one error per 10,000 base pairs) and publication of data within six months.

A new principle intended to increase the efficiency of the teams was the "first come first served" principle, which was applied to the laboratories involved. When a given cosmid had been completely sequenced, a second cosmid clone could be provided and a second payment made, and so on. The speedy laboratories compensated for the slow ones, and how much each laboratory sequenced determined its status in the consortium.

---

[45] The coordinator's task consisted of building, characterizing and distinguishing the clones needed for sequencing, to map the chromosomes, and coordinate the sequencing work.

Construction and Use of Yeast Cosmid Banks

Contrary to what many had thought at the outset, the pilot project had shown that the success of a genome program essentially depends on the existence of a complete and correctly mapped library of appropriate DNA fragments. So it was to this task that the Unit of Yeast Molecular Genetics at the *Institut Pasteur*, under B. Dujon, turned from 1990 onwards. A cosmid bank (in fact four complementary banks in two different cosmidic vectors) was built by partial digestion of EY1679 strain (standard strain) DNA by *Sau*3A, providing a total of 169 genome equivalents, which is exceptionally high. It was established as a pool of primary clones and as identified secondary clones (nine genome equivalents). It was sent to several coordinators in the European program and later allowed the construction of unique contigs of several yeast chromosomes (IV, VII, X, XI, XII and XV — 52% of the genome), which were used in sequencing. The general properties of the bank were studied using chromosomes VII, XI and XV, mapped by B. Dujon's unit.

A restriction enzyme that snipped long DNA, but not too often, would make the mapping and sequencing of complex genomes easier, and the new endonuclease *I-Sce*I discovered and studied in Dujon's[46] laboratory and put onto the market by Boehringer Mannheim, was just that enzyme. The meganuclease *I-Sce*I is not just your average restriction enzyme. For a start, the recognition site is asymmetric and very long (18 nucleotides) and the enzyme cuts within this sequence. Secondly, from the statistical point of view, such a sequence of 18 base pairs only occurs once in 68,7x10^9 base pairs (50% composed of G and C), i.e., once in every 20 human genomes strung end to end. *I-Sce*I was, therefore, the first endonuclease capable of cutting most genomes only in one place — at sites specially inserted (the probability of a wild site in the genome or chromosome being studied being very slight). The extremely rare frequency of the cutting site contributed to make *I-Sce*I the ideal enzyme to study genomes, map large fragments of

---

[46] L. Colleaux *et al.*, *Cell*, vol. 44, 1986, pp. 521–533, L. Colleaux *et al.*, *Proc. Natl. Acad. Sci.*, vol. 85, 1988, pp. 6022–6026, C. Monteilhel *et al.*, *Nuc. Acids. Res.*, vol. 16, 1990, pp. 1407–1413.

DNA or dissect chromosomes after inserting the cutting site[47].

When Dujon and his team began their work, the genome program was just starting. They developed the use of the *I-SceI* enzyme at the same time and applied the new technique to chromosome XI[48]. When cosmid clones had been obtained, they were mapped by hybridization, using chromosome fragments as a probe. When the cosmid clones had been correctly ordered against the restriction sites, it was possible to build a physical map, the resolution of which depended on the number and distribution of the *I-SceI* sites (Fig. 20).

The cosmid library of chromosome II was prepared by H. Feldmann, in Munich. The program objective was to sequence the two chromosomes completely by the end of December 1993. In all, the consortium included 36 laboratories, 19 and 17 sequencing YCII[49] and YCXI, as mentioned above. The laboratories chose their own sequencing technique. In 1994, chromosome XI was the first to be completely sequenced (Fig. 21) and published[50], the article cosigned by 104 authors. This sequence was

---

[47] A. Thierry and B. Dujon, "Nested chromosomal fragmentation in yeast using the meganuclease *I-SceI*: a new method for physical mapping of eucaryotic genomes", *Nucleic Acids Research*, vol. 20, no. 21, 1992, pp. 5625–5631.

A. Perrin, A. Thierry and B. Dujon, "La méganucléase *I-SceI*, premier exemple d'une nouvelle classe d'endonucléases utiles pour les grands génomes", 2 Décembre 1991, (texte communiqué par B. Dujon).

[48] The methodology used was the following. The first step was to show that there were no natural I-SceI sites in the entire yeast genome. Then the first vectors allowed a site to be inserted in order to optimise the *in vitro* digestion conditions. Lastly, a systematic *I-SceI* site integration strategy in yeast chromosomes was developed, and on the basis of this and the transgenic clones obtained, a new physical mapping strategy called "Nested chromosomal fragmentation" evolved. In short, it consisted of building transgenic yeasts with homologous recombination once the site of the *I-SceI* enzyme was integrated at several points along the chromosome. This allowed the nested chromosome fragments to be purified and then used as probed to order the physical maps of chromosomes XI, VII and XV, a quarter of the yeast genome. The process was extended, thereafter, to mapping the inserts of YACs with large fragments of mammalian DNA in cooperation with the Unit of Marine Molecular Genetics (P. Avner). L. Colleaux, C. Rouguelle, P. Avner and B. Dujon, "Rapid physical mapping of YAC inserts by random integration of *I-SceI* sites", *Nucleic Acids Research,* vol. 20, no. 21, 1993, pp. 265–271.

[49] Yeast chromosomes are sometimes designated by the acronym YC and the number.

[50] B. Dujon *et al.*, "Complete DNA sequence of yeast chromosome XI", *Nature*, vol. 369, 2 June 1994, pp. 371–378.

454 From Biotechnology to Genomes: A Meaning for the Double Helix

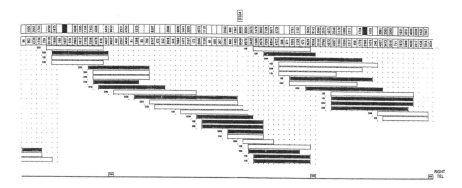

Fig. 20 Map of chromosome XI

(Diagram taken from A. Thierry *et al.*, "Construction of a complete genomic library of *S. cerevisiae* and physical mapping of chromosome XI at 3.7 kb resolution", *Yeast*, vol. 11, 1995, p. 129.)

From top to bottom the figure shows 1) the positions of the *I-Sce*I sites inserted into chromosome XI (uppermost line) and an *Eco*R1 map deduced from this work (second line down) or the complete sequence (third line down). All the *Eco*R1 fragments are represented with their size in base pairs. On the second line, the size of fragments has been estimated on the basis of calibrated electrophoresis gels. Note that during the initial assembly of this map, fragments less than a kilobase in length were not considered, leading to the empty rectangles. Some *Eco*R1 fragments were not aligned with each other (rectangles). They were placed in the appropriate positions on the map deduced from the complete sequence. This diagram is taken from the work of B. Dujon.

determined using a group of 29 clones from the library built by Dujon. The sequence was 666,448 nucleotides long. A total of 331 ORFs were identifie[51], 93 (28%) of which corresponded to genes with known functions. Another 93 had homologies with other genes, and 37 (11%) with homologous genes of unknown function. After a check intended to show up the homologies, it was

---

[51] This number was later to be slightly reduced. Of the 331 ORFs determined, it was estimated that only 316 were real genes; in addition, a certain number of ORFs had less than 100 codons. In the chromosome III sequence, 9 ORFs were excluded because they showed similarities to non-coding sequences. Furthermore, in chromosome IX 3 sets of 2 ORFs and a set of 3 ORFs were fused together because they contained portions of the same genes. At least three reading errors were identified and a pseudogene was found in the YC III sequence. On the other hand, additional ORFs were found, including some shorter ones.

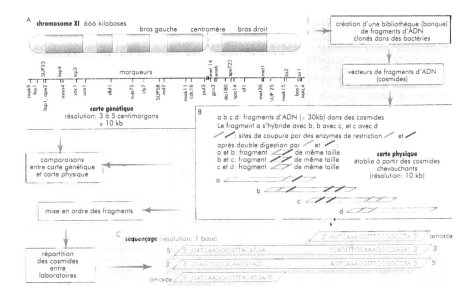

Fig. 21 The sequencing of chromosome XI
(taken from Europe, supplement from no. 276 of *La Recherche*, May 1995, p. 24)

estimated that of the 331 ORFs, 108 (33%) were potential genes with no homologue in the data bases. No complete *Ty* element was discovered. The percentage of chromosome XI genes with no clear function was similar to that for chromosome III. Furthermore, the sequence showed a dense distribution of the genes. Seventy two percent of the chromosome's sequence were ORFs, confirming that a large majority of the genome was genetically functional. The waist of the right arm of the chromosome was confirmed to be the same as the waist of the left. The size of the intergenic regions discovered were seven nucleotides to 3 kb with an average inter-ORF distance of 804 base pairs for the promoter and 381 base pairs for the terminator.

Analysis of this chromosome showed a regular distribution of ORFs presenting a succession of segments rich and poor in G and C, each of 50 kb. The ORF with the highest average content of C and G had 40.2%, followed by divergent promoters (36.3%) and convergent terminators

(29.8%)[52]. This periodic variation in the composition of the ORFs and intergenic regions did not seem to be linked to local gene density but seemed rather to represent a typical pattern in the genome. This was confirmed by the analysis of chromosome III's sequence. For Dujon *et al.*, this was an indication that there are rules for siting that the genes obey, rules that determine expression, mutation bias and recombination phenomena. This periodicity could also underlie certain organizational phenomena such as the spatial form of the chromosome or the attachment to the nuclear matrix. Another lesson taught by the sequences was subtelomere intrachromosomal duplication. The comparison of the last *S. cerevisiae* genetic map (which assigned 1,046 genes to specific sites[53]) with the physical map obtained, provided complete agreement with the list of genes sited on chromosome XI by Mortimer. However, there was some disparity with regard to site, especially on the left arm and at the centromere, disparities showing sequence inversions at certain sites on the genetic map. These differences between the genetic and physical maps (Fig. 22) can, in certain cases, show up real inversions and translocations in strains used by individual laboratories. The number of disparities, most of all, underlined the need for each genome projects to have independent physical and genetic mapping data.

As for the accuracy of the sequence obtained, after careful checking, this was estimated at 99.7%, a far higher accuracy than that estimated for the sequences from chromosome XI in the databases before systematic sequencing began. The sequence for chromosome II was published during 1994[54]. The consortium of laboratories worked on a series of overlapping cosmid clones from strain S288C prepared and distributed by R. Stucka and H. Feldmann of the University of Munich. The sequence was 799,312 base

---

[52] B. Dujon, "Mapping and sequencing the nuclear genome of the yeast *Saccharomyces cerevisiae*: Strategies and results of the European enterprise", *Cold Spring Harbor Symposium on Quantitative Biology*, vol. LVIII, Cold Spring Harbor Laboratory Press, 1993, pp. 357–366.

[53] R.K. Mortimer, C.R. Contopoulou and J.S. King, "Genetic and physical maps of *Saccharomyces cerevisiae*, Edition II", *Yeast*, vol. 8, 1992, pp. 817–902.

[54] H. Feldmann *et al.*, "Complete sequence of chromosome II of *Saccharomyces cerevisiae*", *EMBO J.* vol. 13, 1994, pp. 5795–5809.

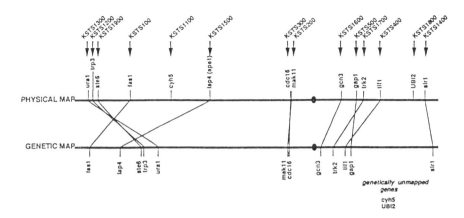

Fig. 22 Comparison of physical and genetic maps of chromosome XI (diagram from the work of A. Thierry and B. Dujon)

The physical and genetic maps have been drawn to the scale of the completed sequence. On the genetic map level, only genes that have been used as probes against the entire set of cosmid clones have been indicated.

pairs long, 845,723 base pairs if including the 39,039 base pairs of overlapping sequence and the checking sequencing. Three hundred and ninety ORFs were identified (11 of those interrupted with introns) and 14 tRNA genes and 3 *Ty* elements were identified, as well as an autonomously replicating sequence (ARS). The compilation, analysis and representation of the data was facilitated by the use of "X chromo", an interactive graphics program developed by Dr. S. Liebl of MIPS. Eighty nine of the ORFs (23%) were known genes, 65 (17%) had similarities with sequences in the database and 36 (60%) had no correspondence with known genes or proteins. Lastly, 75% of the overall chromosome II DNA sequence consisted of coding regions. These proportions were similar to those found for chromosomes III and XI.

The Perfect Gentleman Sequencer: The Development of Deontological Rules

The progress of the project between 1990 and 1992 provoked the constitution of a group of simple recommendations and rules to ensure high accuracy in sequences it obtained. At the same time, this pronouncement ensured that the rules for sequencing in the international scientific community would be followed. This group of rules and recommendations, entitled "The perfect gentleman sequencer; a set of deontological rules submitted to the European yeast sequencing network (BRIDGE and BIOTECH programs)", was presented during the Cosmid Group meeting in Brussels on 29 June 1992. It was elaborated, by B. Dujon from the initial technical annex, into the chromosome III contracts written by André Goffeau in 1989.

It stipulated that each participant had rights and duties laid out under the terms of the contract. The sequencers were to be remunerated in proportion to the amount of sequencing. A first advance payment of 45,000 ECU was made with the delivery of the first cosmids, with another payment of 45,000 ECU to follow only after 25 kb of sequencing had been submitted to MIPS. The sequences were to be made public, as mentioned previously, within 12 months. The contractors were provided with work on a first come first served basis. When a contractor had completely sequenced the segment it had been assigned, and it had been submitted to MIPS and the sequencing strategy approved by the coordinator, the contractor could have another segment to work on. The segments were reserved for the contractor who received them during the entire sequencing project for that chromosome, but this could be canceled if the contribution were too small not of high enough quality or not fast enough. It would be the coordinator's responsibility to cancel, after consultation with the informatics coordinator, Mewes, at MIPS, and the general coordinator, Goffeau, at the UCL.

The contractors were encouraged to publish, patent or distribute their sequences as they saw fit. However, they had to be submitted to MIPS before any of this could be done. It was the DNA coordinator's task to build a physical map of the entire chromosome of sufficiently high resolution. This map would be distributed to all the contractors but remain confidential until the coordinator published it. For optimal results, the DNA supplied to contractors had to overlap, by about 1–3 kbp per cosmid. The overlappings of assigned segments were to be distinguished from clone overlaps. Both contractors were to be notified of the existence of the overlap and specific limits assigned by the DNA coordinator. The contractor submitting the region first in the form linked to phase 2 (see further on) had priority on intellectual property rights until the contractors involved had reached agreement, which they would be notified of at once, in writing, by MIPS and the DNA coordinator. Such agreement normally leading to publication, and commercial exploitation was encouraged. The overlaps were checked by MIPS. Should there be disparities, the two contractors were informed simultaneously. As for the problem of confidentiality, the sequence results were to be submitted to MIPS and examined by the DNA coordinator, both responsible for the strictest confidentiality in their handling of the situation[55]. Inversely, any information provided by MIPS or the DNA coordinator was to be considered confidential. In reality, no such conflicts arose.

The sequence analysis was systematically carried out by MIPS, both for coding and non-coding regions. The analysis results were quickly provided to the contractors, who were encouraged to make their own comparisons and alignments. A summary of the analysis was available to the contractors and to YIP, but it was to remain confidential. Each contractor was to ensure that his close colleagues respected that confidentiality. Each contractor was also encouraged to carry out functional analysis with reverse genetics.

For the following program, BIOTECH, the quality benchmark adopted for the sequences was 99.99%. The authors were responsible for the quality of the data and were asked to sign a quality statement on the fragments they had sequenced. There were three levels to check:

Phase 1: The original sequences of at least 5 kb in length were to be entirely sequenced over the two strands with sufficient overlaps. All uncertainties were to be resolved experimentally as much as possible. The latter sequences could be submitted to MIPS.

Phase 2: The sequences would be analyzed by MIPS and by the DNA coordinator. They were classified as "phase 2" after checks on the sequencing strategy, the correspondence of estimated restriction sites with the DNA coordinator's map, alignments with overlaps or

---

[55] MIPS proposals for limited and controlled access to unpublished sequence data had been considered.

460   *From Biotechnology to Genomes: A Meaning for the Double Helix*

sequences published either by other contractors or pre-existing literature and after checking the ORFs in the six registers and analyzing potential reading bias.

The quality evaluation provided 95% to 99% accuracy. After a 25-kb segment was classified as phase 2, a second payment was made.

Phase 3: The data reached its last accreditation (phase 3 or final) only after resequencing and checks of about 15% of the total sequence in regions chosen by the DNA coordinator and distributed anonymously on the basis of his evaluation of the quality of the original work, his understanding of disparities with sequences carried out prior to the project or other criteria such as the specific biological interest of certain regions. The checking of sequences was funded on the same basis as the original sequences; however, they could not be the property of the laboratory which carried out the work. They could not be published, nor patented, nor provided to a third party. They could only be sent back to MIPS, which after comparing the sequences with the DNA coordinator, notified the two contractors of the result.

Once the data was classified as Phase III[56], the publication was to occur within six months under the leadership of the DNA coordinator. As soon as it was accepted, the final sequence of the entire chromosome was to be brought into the public domain by MIPS[57]. This stage could only be reached once all the contractors had had the time to react and sign the final quality statement. Under no circumstances was it to occur more than 12 months after the end of the project.

---

[56] Sequences submitted to MIPS were kept confidential throughout the sequencing of the entire chromosome, except when a contractor decided to publish, patent or disseminate the sequences of his assigned segment. In such cases, the sequences were disseminated under the sequencer's responsibility and did not affect the evaluation of the sequence within the network.

[57] Only the complete sequence implied the collective responsibility of the network in terms of quality standards. After the complete sequence was published, contractors automatically lost their rights over the sequence (unless they had patented them previously) and became free, like the DNA coordinator, to send the DNA and any other material to a third party for any request linked to academic research in accordance with the rules in force in the international scientific community.

This group of deontological rules was not applicable to BRIDGE, already under way, except for those rules linked to the distribution of tasks and intellectual property. A more gradual application of the rules on quality control could only be truly implemented in the next Biotechnology program. As chromosomes II and XI were being sequenced, B. Dujon (Paris, France) and P. Philippsen (Giessen, Germany) were required to prepare additional chromosome libraries for the forthcoming BIOTECH I (1993-1995) and BIOTECH II (1994-1996) programs.

The BIOTECH Projects

These projects provided funding totaling 12,113,000 ECU. The targets of these projects, of which the administrative coordinator was officially Dr. F. Foury of the UCL in Belgium (but André Goffeau kept the control), were the following.

For the BIOTECH I contract (BIOT CT 920063):

- Subcontracts to the DNA coordinators (chromosomes VII, X, XIV and XV) and the 56 sequencing laboratories;
- Initial advance payment to all the subcontractors;
- Detailed account of the base pairs produced by each laboratory;
- Final payment on a basis of 2 ECU per base pair sequenced and approved;
- Subsequent payments to the DNA coordinator according to progress;
- Continued interaction with each DNA coordinator, informatics coordinator and sequencing laboratory;
- Preparation of organized library of chromosome IV (Claude Jacq, France);
- Complete sequencing of chromosomes X (720 kb) and XIV (810 kb) and partial sequencing of chromosomes VII (1,150 kb) and XV (1,150 kb);
- Construction of two overlapping clone libraries to cover chromosomes VII and XV;

- Organization of clones and construction of high resolution physical maps;
- Distribution of clones to participants;
- Implementation of quality controls for completed sequences;
- Assembly of completed sequences for chromosomes X and IV from the groups of overlapping clones and interpretation of that assembly;
- Organization of contractors' meetings (Manchester, UK, February 1994, Lisbon, 1995 and Trieste, 1996).

For the BIOTECH II contract (BIO2 CT 942071):

- Subcontracts to the DNA coordinators (chromosomes IV, VII, XII, XV and XVI, telomeres) and the 74 sequencing laboratories;
- Advance payments to each subcontractor;
- Detailed account of the base pairs produced by each sequencing laboratory;
- Subsequent payment 2 ECU per final sequenced base pair approved by the DNA and informatics coordinators;
- Subsequent payment to contractors in accordance with the projects;
- Continued interactions with each DNA and informatics coordinator and sequencing laboratory;
- Preparation of organized libraries for chromosomes IV and XII (joerg hoheisel, DKFZ, heidelberg);
- Construction of telomeric clones for all the yeast chromosomes;
- An end to sequencing chromosomes VII (1,091 kb) and XV (1,091 kb) and partial sequencing for chromosomes IV (598 kb), XII (472 kb) and XVI (283 kb);
- Construction of overlapping clone libraries to cover the parts of the yeast chromosomes to be sequenced as a single contig;
- Organization of clones and construction of high resolution physical maps;
- Distribution of clones to participants for the sequencing of entire chromosomes;
- Implementation of quality controls for sequences determined by all participants;

- Assembly of complete sequences for chromosomes VII and XV and partial sequencing of chromosomes IV, XII and XVI from groups of overlapping clones and the interpretation of sequences
- Organization of contractors' meetings (Lisbon, 1995 and Trieste, 1996);
- Informatics coordination (assembly and analysis of chromosomes IV, XII and XVI).

A new characteristic of these rule-governed project was a minimum level of overlap in sequencing leading to a low proportion of whole chromosomes sequenced by more than one laboratory. Experienced researchers from the past BRIDGE and BIOTECH programs were sufficiently confident in the accuracy of sequencing techniques and validation processes that they no longer felt the need to resequence everything. The implementation of the new rules also meant lower costs and greatly increased speed and efficiency. The BIOTECH sequencing network had greatly evolved from the BAP and BRIDGE networks. The number of laboratories was far larger (too large according to André Goffeau, the initiator of the European venture), 56 laboratories in BIOTECH I and 74 laboratories in BIOTECH II. The chromosomes were assigned as follows.

BIOTECH I:

- Chromosome VII (1,150 kb), coordinators André Goffeau and Hervé Tettelin (Louvain la Neuve, Belgium);
- Chromosome X (720 kb), coordinator Francis Galibert (Rennes, France);
- Chromosome XV (820 kb), coordinator Peter Philippsen (Basel, Switzerland), cosmid library provided by B. Dujon;
- Chromosome XV (1,130 kb), coordinators Bernard Dujon and Agnes Thierry (Paris, France).

BIOTECH II:

- Sequencing of part of chromosome IV, coordinator C. Jacq, (ENS, France) Sequencing of part of chromosome XII, coordinator J. Hoheisel, (DKFZ, Heidelberg, Germany)

– Sequencing of part of chromosome XVI, coordinator A. Goffeau (Louvain-la-Neuve, Belgium)

The efficiency of the network had risen to 1,500 kb per year. The choice of sequencing procedures had been left to individual contractors although the sequencing strategies still had to be approved by the DNA coordinators before MIPS would accept them. MIPS continued to receive sequence data, assemble it according to the restriction map and analyze it. Each chromosome was managed by a MIPS informatics coordinator (Fig. 23). The telomeres of each chromosome were separately isolated, cloned and sequenced, the enterprise coordinated by E.J. Louis of ICRF[58], who had already pointed out[59] that the unusual structure found on the right telomeric region of chromosome III was incorrect and had resulted from an event during cloning. The comparison of sequences from several telomeres showed the repetition of several elements that might be involved in the stability and conservation of the chromosome.

By the end of the BIOTECH program, with the sequences obtained during BAP and BRIDGE, the involvement of over 100 laboratories (Table 21) and with a total cost of 19,110 MECU (BAP 1989–1991, 2,235 MECU; BRIDGE 1991–1993, 4,760 MECU, and BIOTECH 1993–1996, 12,115 MECU) for the entire program, the EC consortium had sequenced 55% of the yeast genome. Fig 10a shows the pattern of submission of sequences in kb per year to the EMBL, Genbank and MIPS. The curve is a testimony to the progression of the network's efficiency, a progression also clear in Fig. 24b.

Meanwhile ... Beyond the Borders of Europe

Despite the effort and money invested (Table 20), Europe could not hope to sequence more than a fair portion of the yeast genome in such a limited

---

[58] E.J. Louis, "A nearly complete set of marked telomeres in S288C for mapping and cloning", in *Report of the Manchester Conference of Yeast Genome Sequencing Network*, CEC DG XII Biotechnology, February 1994, p. 122.

[59] E.J. Louis, "Corrected sequence for the right telomere of *Saccharomyces cerevisiae* chromosome III", *Yeast*, vol. 10, 1994, pp. 271–274.

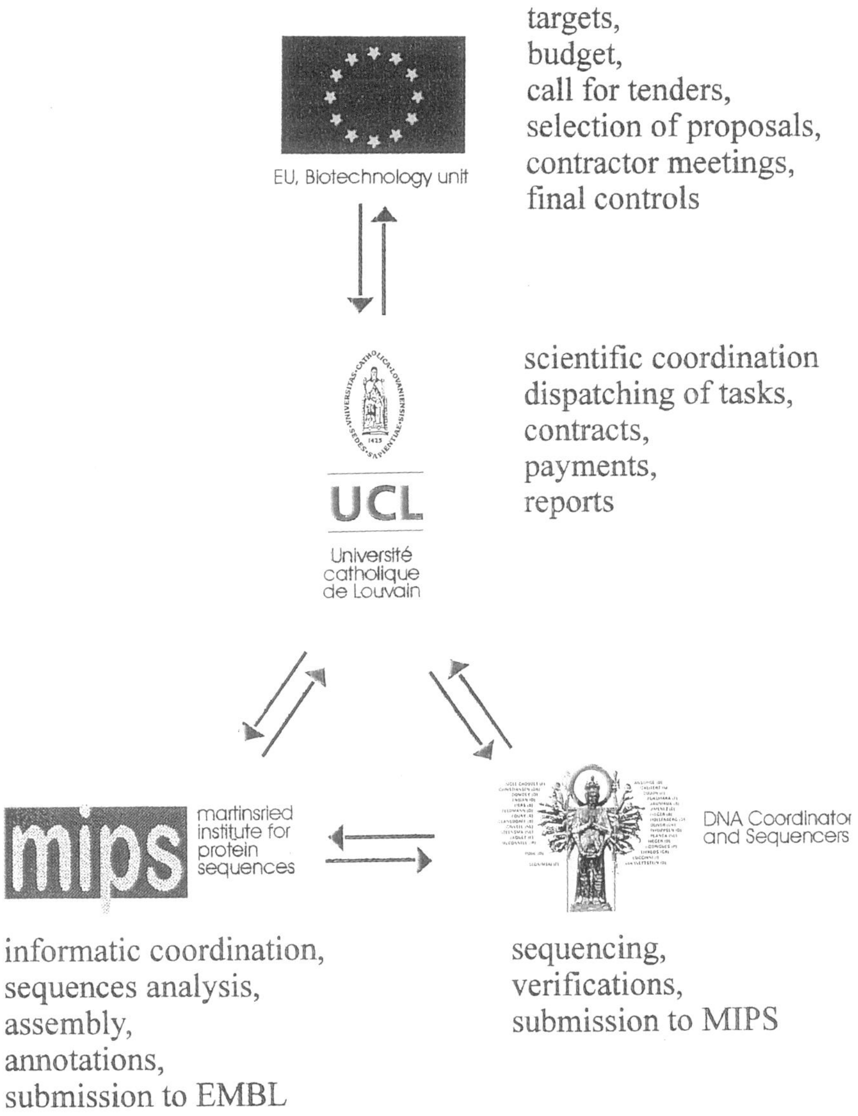

7 The Decryption of Life 465

EU, Biotechnology unit

targets,
budget,
call for tenders,
selection of proposals,
contractor meetings,
final controls

Université
catholique
de Louvain

scientific coordination
dispatching of tasks,
contracts,
payments,
reports

martinsried institute for protein sequences

DNA Coordinator and Sequencers

informatic coordination,
sequences analysis,
assembly,
annotations,
submission to EMBL

sequencing,
verifications,
submission to MIPS

Fig. 23 The management structure
(source: EU and A. Goffeau)

Table 20 Sequencing costs

(diagram from P.B. Joly and V. Mangematin in "Is the yeast model exportable?", manuscript provided to the author)

Ch	Coordin.	Size in Kb	Coord. In Kecu	MIPS in Kecu	sub-contractor in Kecu	overlap. and verific.	Add. Funds by national authority	Total EU and nat. funds
2	Feldmann	807,0	80,7	30,0	1 614,0	227,1	920,6	
3	Oliverr	314,0	47,6	156,0	1 570,0	178,5	874,3	
4 L2	Jacq	600,0	30,0	15,0	1 200,0	168,9	684,4	
7	Tettelin	1 150,0	114,0	15,0	2 300,0	323,7	1 311,8	
10	Galibert	720,0	72,0	15,0	1 440,0	202,6	821,3	
11	Dujon	666,0	66,6		1 332,0	187,4	759,7	
12L	Hoheisel	450,0	22,5	15,0	900,0	126,6	513,3	
14	Philippsen	810,0	81,0	15,0	1 620,0	228,0	924,0	
15	Dujon	1 150,0	114,0	15,0	2 300,0	323,7	1 311,8	
16 L2	Goffeau	300,0	15,0	15,0	600,0	84,4	342,2	
	fixed cost			21,0				
	Total EU	6 967,0	643,4	312,0	14 876,0	2 050,9	8 463,5	26 345,8
	USA	2 500,0						
	UK	2 190,0						
	Canada	535,0						
	Japan	270,0						
	Total	12 462,0						

Estimation of costs :
- sub-contractor : 2 Ecus/base except BAP with 5 Ecus
- verification and overlaps : 658 Kb overlaps inta EEC and 671 Kb diverse verifications at 1 ECU
- estimation of additional funds by national authorities : 50% of costs financed by EEC on the basis of 2 Ecus per pair of bases (rough estimation considering the different types of funding - e.g. purchase of material, but not researchers' salaries).

time. The sequencing of the whole genome required some international assistance. For this reason, during BRIDGE and at the onset of BIOTECH, it was agreed that the sequencing of the sixteen yeast chromosomes would be shared between the European Union, the USA, Canada, Japan and

Cambridge (UK)[60]. Table 22 and Fig. 25 show the distribution of DNA between these parties.

A certain number of longer (or more complex) chromosomes were the subject of cooperation between some of the parties. Chromosome XII was shared between the EU and the USA, IV between the EU, the USA and the Cambridge group and XVI between the USA, the EU, Canada and the Cambridge group.

The non-European programs generally had different types of organizations than the distribution system being used by the European consortium. Often they were highly centralized and automated, which is believed to be more efficient at sequencing. Davis and Botstein's team, for example, in charge of sequencing chromosome V and the right arm of chromosome VI (600 kb), claimed to sequence one cosmid every two weeks[61]. In reality, their final output was far lower.

The chromosome I sequencing project (220 kb)[62], managed by H. Bussey of McGill University in Canada, worked on the smallest chromosome of the yeast genome, which had been studied intensively by David B. Kaback. The genetic, physical and transcription maps were already available[63]. Sequencing showed a similar gene density as for chromosome III. This group also sequenced the left arm of chromosome XVI using clones from Dujon's genome libraries as well as carrying out a functional analysis of chromosome I.

---

[60] A. Goffeau, *Nature*, vol. 369, 1994, pp. 101–102.

[61] F. Dietrich *et al.*, "Status of the sequencing of chromosomes V and VI", in *Report of the Manchester Conference of Yeast Genome Sequencing Network*, CEC DG XII Biotechnology, February 1994, p. 23.

[62] The references on the articles about the various chromosomes can be found in Fig. 15.

[63] B.E. Diehl and J.R. Pringle, "Molecular analysis of *Saccharomyces cerevisiae* chromosome I: Identification of additional transcribed regions and demonstration that some encode essential functions", *Genetics*, vol. 127, 1991, pp. 279–285.

S.D. Harris and J.R. Pringle, "Genetic analysis of *Saccharomyces cerevisiae* chromosome I: On the role of mutagen specificity in delimiting the set of genes identifiable using temperature-sensitive lethal mutation", *Genetics*, vol. 127, 1991, pp. 279–285.

R.K. Mortimer, R. Contopoulou and J.S. King, "Genetic and physical maps of *Saccharomyces cerevisiae*, Edition II", *Yeast*, vol. 8, 1992, pp. 817–902.

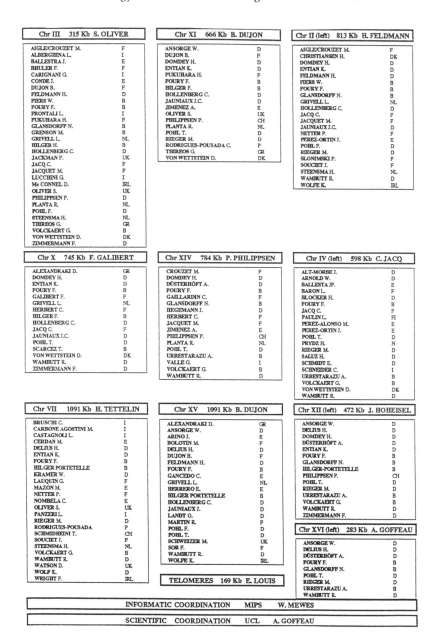

Table 21 European sequencing teams per chromosome
(from A. Goffeau's archives)

7 The Decryption of Life 469

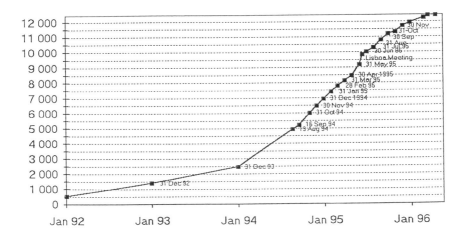

Fig. 24A Rate of EU sequence submission to the EMBL, Genbank and the MIPS database (from A. Goffeau's archives)

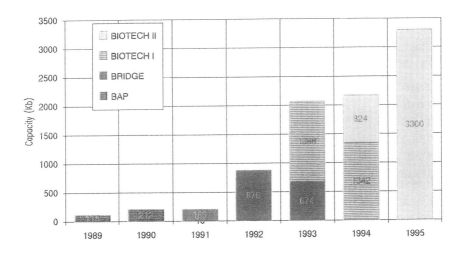

Fig. 24B The growth in capacity of the European sequencing network (from A. Goffeau's archives)

Table 22 World agreement on the distribution of sequencing tasks by country and coordinator (from A. Goffeau's archives)

CHR	KB		COORDINATOR	
I	228	CAN	H. BUSSEY	McGill University, Montreal
II	813	EU	H. FELDMANN	Universität Munchen
III	315	EU	S. OLIVER	UMIST, Manchester
IV right 1	240	USA	M. JOHNSTON	Washington University, St Louis
IV right 2	290	USA	R. DAVIS	Stanford University
IV middle	541	UK	B. BARREL	Sanger Center, Cambridge
IV left	598	EU	C. JACQ	ENS, Paris
V	569	USA	R. DAVIS	Stanford University
VI	270	JAP	Y. MURAKAMI	RIKEN, Tsukuba
VII	1091	EU	H. TETTELIN	UCL, Louvain-la-Neuve
VIII	563	USA	M. JOHNSTON	Washington University, St Louis
IX	440	UK	B. BARRELL	Sanger Center, Cambridge
X	745	EU	F. GALIBERT	CNRS, Rennes
XI	666	EU	B. DUJON	Institut Pasteur, Paris
XII right	782	USA	M. JOHNSTON	Washington University, St Louis
XII left	472	EU	J. HOHEISEL	DKFZ, Heidelberg
XIII	924	UK	B. BARRELL	Sanger Center, Cambridge
XIV	784	CH	P. PHILIPPSEN	Universität Basel
XV	1091	EU	B. DUJON	Institut Pasteur, Paris
XVI right 1	166	UK	B. BARREL	Sanger Center, Cambridge
XVI right 2	261	USA	M. JOHNSTON	Washington University, St Louis
XVI left 1	289	CAN	H. BUSSEY	McGill University, Montreal
XVI left 1b	34	USA	R. DAVIS	Stanford University
XVI left	283	EU	A. GOFFEAU	UCL, Louvain-la-Neuve
30 TELOMERS	153	EU	E. LOUIS	John Radcliffe Hospital, Oxford

The Stanford (USA) Yeast Genome project, coordinated by R. Davis and D. Botstein, was entrusted with the sequencing of chromosome V's 600 kb. The sequences were determined from a group of overlapping lambda clones and cosmid clones provided by Olson and Riles.[64]

---

[64] L. Riles et al., Genetics, vol. 134, 1993, pp. 81–150.

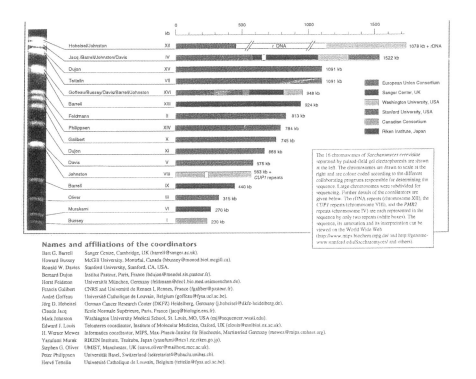

Fig. 25 Distribution of DNA between international coordinators and sequencing teams (from A. Goffeau's archives)

The RIKEN (Japan) Yeast Genome project to sequence chromosome VI was carried out in the RIKEN Tsukuba Life Science Center (Institute of Physical and Chemical Research). Initially, it was run by Dr. Eiichi Soeda, in cooperation with Drs. Olson and Schlessinger of the University of Washington at St Louis (USA). Research funds mainly came from the latter institution until September 1990. In the first phase, Dr. Soeda led three technicians himself. In October 1990, Dr. Murakami took over the project, bringing funding from RIKEN as part of the Human Genome Project. Four technicians were taken on, and the sequencing was continued by these new team members under Dr. Y. Murakami. Two of these technicians also worked as support scientists in the Human Genome Analysis project, which used entirely automated sequencing. The clone contigs were provided by the Washington University group; the clones covering the gaps in the contigs were provided by Dr. Hiroshi Yoshikawa's lab. They used the shotgun sequencing strategy, and most of the clones were produced with ABI 373A DNA sequencers. The data was gathered either with Shotgun software, originally developed by Mitsui Knowledge Co., or with ATSQ, developed by SDC Co., Japan.

The yeast genome project at the University of Washington, St. Louis (USA), was coordinated by M. Johnston. Most of the sequencing was carried out by the St. Louis Genome Sequencing Center[65], which brought a total of 1,845,101 nucleotides of DNA sequence to the world yeast genome sequencing effort. This figure includes the complete sequence of chromosome VIII (562,638 bp) a large portion of chromosome XII (781,866 bp) and a fair chunk of chromosome IV (239,913 bp). They used the shotgun sequencing technique on cosmids mapped by Riles and Olson. About 800 clones were sequenced using ABI automated sequencers, giving a rough coverage of about six to eight times. The error rate on this sequencing was estimated at being slightly less than one in 10,000 nucleotides. The sequence analysis allowed 840 ORFs to be predicted and 43 tRNA genes. The analysis of chromosome VIII showed 269 ORFs[66], of which 210 were new and 59 corresponded to genes with known or predicted functions; 124 seemed to code for proteins not significantly similar to sequences in public databases and 21 coded for proteins similar to proteins with unknown functions; 11 tRNAs were identified. Three were interrupted by introns and there were delta and sigma elements and tau sequences. Sixteen genes seemed to have been duplicated quite recently. The high density of coding seemed similar

---

[65] The GSC was involved in other large-scale sequencing projects, including the final sequencing of the genome of *C. elegans* in collaboration with the Sanger Center foreseen for the end of 1998. By mid April 1996, the GSC has sequenced 20,099,119 nucleotides of the *C. elegans* genome. For 1996, it was expected the St Louis group would have reached 12 Mb of completed sequence at a cost of less than $0.40 per base. Furthermore, the GSC had begun sequencing the human genome (hoping to have 25 Mb done by August 1997) and continued to identify human and mouse genes (285,000 human ESTs were submitted to the dbEST database towards mid-April 1996).

[66] M. Johnston *et al.*, "Complete nucleotide sequence of *Saccharomyces cerevisae* chromosome VII", *Science*, vol. 265, 1994, pp. 2077–2082.

to that found in the other chromosomes (III and XI, in particular) although the alternate pattern of segments, rich and poor in C and G, was not repeated.

Chromosomes IX, XIII and a part of chromosomes IV and XVI were sequenced at the Sanger Center in Cambridge. These projects, and the center itself, were financed by the Wellcome Trust and the Medical Research Council. The Sanger Center sequenced 2,340 kb with the shotgun technique on overlapping cosmids using ABI 373A machines, which had a rate of up to 72 templates per gene per machine, all nine of their machines being dedicated to the yeast genome work. In addition, the Sanger Center began work on sequencing *Schizosaccharomyces pombe*. Researchers at the center are also working on the nematode and human genomes.

The largest chromosome, IV, was sequenced in a coordinated USA, European Union and Cambridge initiative.

The World Sequencing Enterprise was informally coordinated by A. Goffeau and Mark Johnston. Its efforts provided the complete sequence of the yeast genome (Table 24). The "world first" of completely sequencing the genome of a complex organism, in which Europe had carried 55% of the weight (Fig. 26) was announced on 24 April 1996 at two separate press conferences, one held in Brussels by the European Commission and the other in Bethesda, USA by the NIH.

During these press conferences, there was also a presentation of the anticipated benefits for fundamental research, industry and health. The press conferences had enormous impact, with almost 200 articles appearing in the European press. Of the 188 articles clipped into the press book, 70% cited the European Commission as director of the project, although the press corps of the various member states with important research activities such as Britain, Germany and France, tended to highlight the importance of the teams from their country within the overall project. Curiously, only 40% of articles mentioned the European Commission[67].

The success of this world effort showed, above all, what could be achieved when international cooperation overcomes the barriers of national interest.

---

[67] Percentages and analyses from the Press Book. Press Conference 24 April 1996. Le génome de la levure, CGI Europe, 1996, Analyse quantitative, pp. 3–4.

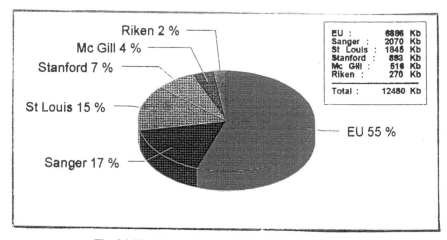

Fig. 26 World contributions to the sequencing project
(EU documentation)

What the Sequence Has Taught Us

The availability of the entire sequence of a eucaryote is without a doubt a fantastic source of information for the entire biology community, in particular, of course, for the yeast scientists, but also for molecular geneticists working on other organisms. Yeast, an extensively studied organism widely used in industry and a great friend of man, has once more been pressed into service as a model organism. We now know every one of the 1,206 million base pairs, 6,275 ORFs, of this complex organism. The data obtained has already provided a generous harvest of information (Table 23), leading to major progress in our understanding of the fundamental mechanisms of life in complex cells.

The sequence allows us to make a first analysis of the anatomy of any eucaryote genome. The yeast genome is moderately biased in its composition, with an average content of G and C to 39%, and has a characteristic pattern of dinucleotides that show up in a lot of eucaryotes with a significant deficiency of 5'CG3' and 5'TA3'. Another characteristic of the dinucleotides of the yeast genome, compared to other organisms, is its high density of coding sequences. On average, the ORFs take up 72% of the yeast genome

Table 23 Distribution of genes and other elements of sequences
(from A. Goffeau's "Life with 6,000 genes", *Science*, vol. 274, 25 October 1996, p. 564.)

	\multicolumn{17}{c}{CHROMOSOME NUMBER}
	I	II	III	IV	V	VI	VII	VIII	IX	X	XI	XII	XIII	XIV	XV	XVI	TOTAL
Elements																	
Sequenced lenght (Kb)	230	813	315	1,532	577	270	1,091	563	440	745	667	1,078	924	784	1,091	948	12,068
nonsequenced identical repeats																	
name of unit				ENA2 and Y'		tel		CUP1				RDNA and Y'					
lenght of unit (Kb)				4 and 7		<1		2				9 and 7					
number of unitis				2 and 2		1		13				±140 and 2					
lenght of repeats (kb)				8 and 14		<1		26				1,260 and 14					1,321
total length (Kb)	230	813	315	1,554	577	271	1,091	589	440	745	667	2,352	924	784	1,091	948	13,389
ORF (n)	110	422	172	812	291	135	572	288	231	387	334	547	487	421	569	497	6,275
Questionable proteins (n)	3	30	12	65	13	5	57	12	11	29	20	41	30	23	3	36	390
Hypothetical proteins (n)	107	392	160	747	278	130	515	276	220	358	314	506	457	398	566	461	5,885
Introns in ORF (n)	4	18	4	30	13	5	15	15	8	13	11	17	19	15	15	18	220
Introns in UTR (n)	0	2	0	1	2	0	5	0	0	0	0	3	0	2	0	0	15
Intact Ty1 (n)	1	2	0	6	1	0	4	1	0	2	0	4	4	2	2	4	33
Intact Ty2 (n)	0	1	1	3	1	1	1	0	0	0	0	2	0	1	2	0	13
Intact Ty3 (n)	0	0	0	0	0	0	0	0	1	0	0	0	0	0	0	0	2
Intact (Ty4 (n)	0	0	0	0	0	0	0	1	0	1	0	0	0	0	0	1	3
Intact Ty5 (n)	0	0	1	0	0	0	0	0	0	0	0	0	0	0	0	0	1
tRNA genes (n)	2	13	10	27	20	10	36	11	10	24	16	22	21	16	20	17	275
snRNA genes (n)	1	1	2	1	2	0	3	1	1	4	1	3	8	3	7	2	40

(without counting ribosomal DNA), leaving little room for the non-coding DNA or other structural and functional elements. The average length of the ORFs is 1,450 bp. The longest known ORF is from chromosome XII and has 4,910 codons of unknown function. The second longest is a gene coding for dyneine (4,092 codons) from chromosome XI. Very few ORFs are longer than 1,500 codons. Small ORFs cannot easily be differentiated from the haphazard appearance of short pseudo-ORFs. The lower limit in the yeast program was arbitrarily fixed at 100 codons. Genes coding for smaller proteins do exist, for example, PMP1, which codes for a 40-amino acid proteolipid. On average, divergent ORFs are separated by 618 nucleotides, whereas convergent ORFs are only separated by 326 nucleotides. The average length of intergenic regions is halfway between these two figures. A typical yeast gene is, therefore, an ORF of 1,450 base pairs preceded by a region of 309 base pairs and succeeded by 163.

Six thousand two hundred and seventy-five ORFs were identified (including overlapping ones). MIPS tried to classify yeast proteins according to function[68]. Analysis has shown that a yeast cell dedicates 17% of its proteome (the complete set of proteins a cell is capable of making) to metabolism, 3% to energy, 14% to cell division and the synthesis of DNA, 10% to transcription, 5% to protein synthesis, 7% to protein destination, 5% to transport, 5% to intracellular transport, 28% to cell organization and biogenesis, 2% to signal transduction and 4% to cell repair.

About 6.7% of those ORFs are not real genes[69], leaving about 5,800 genes coding for proteins. When they are present, the introns are short (the longest known intron is only one kilobase long) and almost always at the 5' end of the genes just after the ATG initiator codon, and sometimes even within it or in the untranslated 5' region. The known introns are often in

---

[68] For a detailed analysis of genome composition, please read "The yeast genome directory", *Nature*, supplement vol. 387, 29 May 1997.

[69] These 6,275 ORFs theoretically code for proteins longer than 99 amino acids. However, 390 ORFs appear not to translate into proteins. So only 5,885 genes coding for proteins seem to exist. In addition, the yeast genome contains about 140 ribosomal RNAs in large assemblies in tandem with chromosome XII and 40 genes coding for small nucleic RNA dispersed across the 16 chromosomes. 275 tRNAs belong to 43 families and are also dispersed across all the chromosomes.

Table 24 List of articles for the complete sequence and the chromosomes of the yeast *Saccharomyces cerevisiae*

Chromosome I: H. Bussey *et al.*, "The nucleotide sequence of chromosome I from *Saccharomyces cerevisiae*", *Proc. Acad. Sci.*, USA, vol. 92, 1995, pp. 3809–3813.

Chromosome II: H. Feldmann *et al.*, "Complete DNA sequence of yeast chromosome II", *EMBO J.*, vol. 13, 1994, pp. 5795–5809.

Chromosome III: S. Oliver *et al.*, "The complete DNA sequence of yeast chromosome III", *Nature*, vol. 357, 1992, pp. 38–46.

Chromosome IV: C. Jacq *et al.*, "the nucleotide sequence of chromosome IV from *Saccharomyces cerevisiae*", *Yeast Genome Directory, Nature*, vol. 387, 29 May 1997, pp. 75–78.

Chromosome V: F.S. Dietrich *et al.*, "The nucleotide sequence of *Saccharomyces cerevisiae* chromosome V", *Yeast Genome Directory, Nature*, vol. 387, 29 May 1997, pp. 78–81.

Chromosome VI: Y. Murakami *et al.*, "Analysis of the nuclotide sequence of chromosome VI from *Saccharomyces cerevisiae*", *Nature Genet.*, vol. 10, 1995, pp. 261–268.

Chromosome VII: H. Tettelin *et al.*, "The nucleotide sequence of *Saccharomyces cerevisiae* chromosome VII", *Yeast Genome Directory, Nature*, vol. 387, 29 May 1997, pp. 81–84.

Chromosome VIII: M. Johnston *et al.*, "Complete nucleotide sequence of *Saccharomyces cerevisiae* chromosome VIII", *Science*, vol. 265, 1994, pp. 2077–2082.

Chromosome IX: B.G. Barrell *et al.*, "the nucleotide sequence of *Saccharomyces cerevisiae* chromosome IX", *Yeast Genome Directory, Nature*, 1997, pp. 84–87.

Chromosome X: F. Galibert *et al.*, "Complete nucleotide sequence of *Saccharomyces cerevisiae* chromosome X", *EMBO J.*, vol. 15, 1996, pp. 2031–2049.

Chromosome XI: B. Dujon *et al.*, "Complete DNA sequence of yeast chromosome XI", *Nature*, vol. 389, 1994, pp .371–378.

Chromosome XII: M. Johnston *et al.*, "The complete sequence of *Saccharomyces cerevisiae* chromosome XII", *Yeast Genome Directory, Nature*, vol. 387, 29 May 1997, pp. 87–90.

Chromosome XIII: B.G. Barrel *et al.*, "The nucleotide sequence of *Saccharomyces cerevisiae* chromosome XIII", *Yeast Genome Directory, Nature*, vol. 387, 29 May 1997, pp. 90.93.

Chromosome XIV: P. Philippsen *et al.*, "The nucleotide sequence of *Saccharomyces cerevisiae* chromosome XIV and its evolutionary implications", *Yeast Genome Directory, Nature*, vol. 387, 29 May 1997, pp. 93–98.

Chromosome XV: B. Dujon *et al.*, "The nucleotide sequence of *Saccharomyces cerevisiae* chromosome XV", *Yeast Genome Directory, Nature*, vol. 387, 29 May 1997, pp. 98–102.

Chromosome XVI: H. Bussey *et al.*, "The nucleotide sequence of *Saccharomyces cerevisiae* chromosome XVI", *Yeast Genome Directory, Nature*, vol. 387, 29 May 1997, 1997, pp. 103–105.

short ORFs containing frequently expressed genes such as those coding for the many ribosomic proteins. The number of other genetic elements of the yeast genome can also be worked out from information given from the sequencing efforts. The 52 Ty elements, of various sorts, are relatively rare. They show a significant preference for integrating prior to the tRNA genes, probably because of their interaction with the machinery of the RNA polymerase III. This preference is even clearer for the long terminal repeats (LTRs) which are much more abundant. Other recognizable structures which constitute a meaningful proportion of the yeast genome include the subtelomeric elements X and Y', the latter being a member of the line type family of elements found in mammal genomes. Pseudogenes are very rare in the yeast genome. Aside from a few overlapping ORFs, there are some containing one or two stop codons (the sequence of which has been checked). Most of these are to be found in subtelomeric regions[70].

Gene density is not uniform across the chromosomes. In most chromosomes, there are segments of several dozen kilobases, in which the gene density is significantly greater than average, reaching more than 85%, in some cases. Such chromosome segments are separated by other regions (generally shorter ones) with a far lesser gene density (50 to 55%). In certain chromosomes, the spacing between these regions is more or less regular. The pericentromeric and subtelomeric regions contain fewer genes. For most chromosomes, except chromosome II, the overall ratio of coding on the Crick and Watson strands is very similar; and in general, the orientation of ORFs is completely haphazard.

There are variations in composition from one yeast chromosome to another as found for the first time in the analysis of chromosome III, where it was discovered that there were rich zones of G and C in the middle of each arm. The analyses carried out on chromosome XI confirmed this, but because it is much larger, they also revealed that several G and C rich regions might be found on one of the chromosome arms. Other chromosomes have also shown similar variation in pericentromeric and subtelomeric regions

---

[70] Bussey *et al.*, *Proc. Nat. Acad. Sci.*, USA, vol. 92, 1995, pp. 3809–3813.

rich in A and T, although the spacing between regions rich in G and C is not always regular. In most cases, however, there is a strong correlation between the regions rich in G and C and high in gene density, a correlation also found in more complex genomes in which compositional GC rich isochores are much more frequent.

Overall, the yeast genome is remarkably poor in long identical repeats. Other than the Ty elements and their long terminal repeats (LTRs), the ribosomic DNA provides most of the repetition. Ribosomic DNA repetitions line up in a long row occupying half of chromosome XII; there is also a free circular form due to an ARS in each repetition. In chromosome VIII, a cluster of about 15 repetitions of 1,998 base pairs can be found. It is strongly polymorphically variant from strain to strain. There are repeated chunks of short oligonucleotides such as poly-A or poly-T or poly-ATs. The yeast genome has many of these repetitions of trinucleotides in the ORFs and intergene regions. Such repetition is particularly interesting because some human genetic illnesses are caused by an expansion of these trinucleotides. As B. Dujon points out, "It was quite a surprise to see the extent of genetic redundancy. Its quantity, and the different forms in which it is to be found, make it a remarkable characteristic of the yeast genome. The redundancy most easily diagnosed is due to the duplication of genes that have more or less diverged and which are often found in repetition in tandem or inverted within the chromosome, but mainly in subtelomeric positions. There are also blocks of duplication between chromosomes, probably from events that occurred long ago and involved several neighboring genes, between which new genes inserted themselves. In all, close to a third of the yeast genes are repeated, the gene families mainly only having two members, but in some cases several dozens. The functional upshot of this structural redundancy could be part of the explanation as to why only a very small proportion of essential genes have been found. Less than 15% of the yeast genes seem essential for the cell to continue living. Amongst the others, less than half have phenotype consequences once they are deleted. More than half of the yeast genes, therefore, are not responsible for a recognizable phenotype. Some of this may be due to a lack of precision in phenotype testing currently available, and some from redundancy, but it is not out of the question that it might be an essential property of eucaryote genes, without which evolution would be impossible"[71].

But the most typical result, from the first completion of a chromosome, was the recognition of large numbers of genes, which, when subtracted from the sequence and studied with the usual comparative methods (*in silico* analysis), still looked like nothing else yet seen. The quantitative importance of the phenomenon only became more obvious as other chromosomes were sequenced. A little less than half of the 5,800 proteins in yeast are coded for by known genes, "known" meaning those that have been genetically and biochemically characterized. For 20% of the remaining proteins, experimental data is mixed. It only has a little light to shine on *in vivo* function. Thirty eight percent of the others either have similarities to other uncharacterized proteins or have no similarities at all. These latter cases are "orphans" of unknown function. Therefore, with the complete sequence, about 2,200 genes will remain to be characterized[72]. As they are defined by the absence of known function and structural similarities, their number will, of course, lessen as time goes on. Indeed, the functions of several genes initially classified as orphans were clarified during sequencing itself, as were similarities for some orphans during the sequencing of other organisms, such as *C. elegans*, *Haemophilus influenzae* and *Bacillus subtilis*. A few homologies with yeast orphan genes were also found amongst the many human ESTs. But the way the number is dropping still suggests that there will be a group of orphans specific to the yeast genome for which no home will be found. As pointed out by B. Dujon, "If this proves to be the case, the fact that the orphans constitute the very group of genes that have escaped the attention of the yeast geneticists will not be the only ironic lesson to emerge from the yeast genome sequencing program"[73].

---

[71] B. Dujon, "The yeast genome project", *TIBs*, vol. 12, October 1996.

[72] Some orphans have structural similarities, and can therefore be grouped into small families of genes with unknown functions.

[73] B. Dujon, "The yeast genome project: What did we learn?", *TIBs*, vol. 12, no. 7, 1996, p. 267.

The Yeast Genome Sequence — A Potential Goldmine?

The sequence results are not only a source of information for academic research, providing a clearer understanding of the mechanisms at work at the heart of life, but they also have potential benefits for industry. It seems obvious that some genes will be a lot more useful from an industrial point of view than others. For the agro-alimentary industry, the completed yeast sequence may accelerate the study of the production or assimilation of metabolites likely to improve the quality, quantity and stability of food products. In other cases (for example brewing), different aspects of the physiology of yeast could be important, such as flocculation or survival under extreme conditions. Overall, the yeast genome sequence could contribute to the following improvements in industry:

* cost — it could provide an easy way to get around the need for certain growth factors, allowing the duplication of specific parts of the genome, thus increasing productivity and providing improvements in the robustness of procedures (stable DNA, understood and applied changes in physiological regulation);
* availability — of new combinations of procedures leading to new components;
* purity of products/natural foodstuffs — it could provide a way to inactivate possible derivative products and synthesize products using genes that were previously silent or brought in from other organisms;
* new concepts — new tastes, for example yeast extracts;
* recognition specificity — the easy identification of strains (in taxonomy and for industrial properties).[74]

* Pharmaceutical industrials interested in the use of yeast to make medicines or vaccines may pay more attention to:

---

[74] Dr J.P.M. Sanders, Gist-Brocades. "Will industry benefit from the yeast DNA sequence analysis?", *1st BRIDGE meeting on* "Sequencing the yeast chromosome II and XI", Bruges, Belgium, 22–24 September 1991, p. 106.

1. Factors important in the expression and stability of genes.
   Expression of genes
   – transcription factors
   – translation
   – transport signals
   – targeting signals
   – retention signals

   Post-translation modification
   – proteases
   – glycosylation signals
   – chaperones (heat shock proteins)
   – recognition/degradation (N-end rule...)

   Stability
   – between extrachromosomic segments
   – segregation systems

   Moving elements
   – Ty elements

2. Factors linked to the cell cycle and transduction signals
   Cell cycle
   – oncogenes

   Transduction signals
   – G proteins
   – receptors
   – modulator/effector signals

   cAMP systems
   – kinases

3. Physiological and growth factors
   Morphology
   – genes responsible for hyphal growth
   – branching genes

Cell metabolism
- inducers

Permeation, Transport and Membrane Traffic
- permease
- components of cell and biogenic membranes

These industries also find in yeast a simple system in which to evaluate new chemical or binding substances and study multiple drug resistance phenomena, thanks to the presence of certain simple mechanisms that exist in both yeast and man. It is just that similarity that makes some yeast genes important for human genetics too.

The yeast genome is 200 times smaller than the human genome, but if the respective density of gene distribution is taken into account, it is only nine to ten times less complex in terms of its protein coding ability. It is hard to guess how many human genes will be similar to yeast genes, but there are already a number of significant cases (Table 25).

Yeast can also be a tool in the search for new genes in other organisms since some of its mutants can be completed with heterologous cDNA. Several human genes have been isolated in this way. Even more interesting, many mutant yeasts can now be created through reverse genetics precisely for those genes which had, until now, escaped geneticists' attention. As André Goffeau points out, "Yeast, the faithful servant of man for whom it has produced bread and wine for thousands of years, is now not only contributing to a better understanding of life at a molecular scale, but also clearer vision of some pathological mechanisms to be found everywhere in the living world[75]... This is how the function of one of the first carcinogenic genes, the *ras* gene, was elucidated in 1987 because of its similarity to the CDC25 gene in yeast, which is involved in cell reproduction cycle control. Another, more recent, example also reveals the importance of sequencing. While chromosome XI was being systematically sequenced, a Portuguese researcher, Claudina Rodriguez-Pousada, discovered a gene in yeast similar to the one

---

[75] A. Goffeau, P. Mordant, A. Vassarotti, "Le génome de la levure bientôt déchiffré", supplément to *La Recherche*, vol. 276, May 1995, p. 23.

Table 25 Positional correspondences of sequences between cloned human genes and *S. cerevisiae* proteins

Enzyme deficiency or disease name a)	MIM b)	Yeast gene or ORF c)	P-value d)	Brief disease description e)
Positional cloning				
Achondroplasia (FGFR3)	100800	IPLI 1)	e-16	Membrane Ser/Thr protein kinase
Adrenoleukodystrophy (ADL)	300100	PXA1	e-108	ABC transporter; neurodegenerative disease
		PAT1	e-92	
Amyotrophic lateral sclerosis (SOD1)	105400	SOD1	e-58	Superoxide dismutase
Ataxia telangiectasia (ATM)	208900	TEL1	e-85	Phosphatidylinositol kinase-related protein
		ESR1	e-57	
Barth syndrome (G4.5.)	302060	YPR140w*	e-16	Unknown function; cardioskeletal myopathy
Bloom syndrome (BLM)	210900	SGS1	e-105	RecQ DNA helicase-related protein; growth defect; predisposition to all types of cancer
		YABCD	e-143	
Chediak-Higashi syndrome (CHS)	214500	BPH1	e-83	Unknown function; "Beige" protein; decreased pigmentation; immunodeficiency
Choroideremia (CHM)	303100	CD11 k)	e-42	Component A of RAB geranylgeranyltransferase
Cystic fibrosis (CFTR)	219700	YCF1	e-167	ABC transporter; impaired clearance in a variety of organs
Deafness, DFN-1 (DDP)	304700	YJR135w-a *1)		
		SUL1	e-44	Sulfate transporter; undersulfation of proteoglycans
Fanconi syndrome (CLCN5)	300009	GEF1	e-118	Kidney chloride channel; nephrolithiasis
Fragile histidine triad protein (FHIT)	601153	HNT2	e-26	Dideadenosine tetraphosphate hydrolase; cancer
		aph1m)	e-43	
Friedreich ataxia (FRD)	229300	YFH1*n)	e-16	Unknown function; neurodegenerative disease
Glycerol kinase (GK)	307030	GUT1	e-124	Hyperglycerolemia; poor growth; mental retardation
HNPCC (MSH2)	120436	MSH2	e-254	Mismatch-repair; hereditary nonpolyposis colon cancer
HNPCC (MLH1)	120436	MLH1	e-190	Mismatch repair; hereditary nonpolyposis colon cancer
Lissencephaly (LIS1)	247200	MET30	e-44	Subunit of platelet-activating factor acetylhydrolase
Lowe syndrome (OCRL)	309000	SJH1	e-48	Inositol polyphosphate 5 phosphatase-related protein; cataracts and glaucoma
		PIE3	e-47	

Continue Table 27

Enzyme deficiency or disease name a)	MIM b)	Yeast gene or ORF c)	P-value d)	Brief disease description e)
Menkes disease (MNK)	309400	CCC2	e-186	Copper-transporting ATPase; neurodegenerative disease and death
Migraine (CACNL1-A4)	601011	CCH1	e-44	Calcium channel; familial hemiplegic migraine and episodic ataxia
Monocytic leukemia (MOZ)	601408	TAS1 SAS3	e-79 e-72	Acetyltransferase; erythrophagocytosis
Multiple endocrine neoplasia (RET)	171400	PH085	e-14	Related to transmembrane receptors with a cytoplasmic tyrosine kinase domain
Myotonic dystrophy (DM)	160900	YNL161w* p)	e-79	Ser/thr protein kinase; neurodegenerative disease
Myotubular myopathy (MTM1)	310400	YJR110w*	e-78	Probable tyrosine phosphatase; muscle specific disease
NBCC syndrome (PTC)	601309	NCR1*	e-26	Homologue of Drosophila patched; nevoid basal cell carcinoma syndrome
Neurofibromatosis (NF1)	162200	IRA2	e-42	GTPase-activating protein
Niemann-Pick disease (NPC1)	257220	YPL006w*	e-135	Fatal neurovisceral disorder
Pallister- Hall syndrome (GLI3)	165240	ZAP1*	e-17	Defect in development of multiple organ systems
Retinitis pigmentosa (RPGR)	312610	SRM1 q)	e-10	RCC1-related protein; progressive retinal degeneration
Thomsen disease (CLCN1)	160800	GEF1	e-29	Muscle chloride channel; myotonic disorders
Werner syndrome (WRN)	277700	SGS1	e-64	DNA helicase Q-related protein; premature aging and strong predisposition to cancer
Wilms tumor (WT1)	194070	FZF1	e-20	Zinc finger protein; nephroblastoma
Wilson disease (WND)	277900	CCC2	e-152	Copper transporting ATPase; toxic accumulation of copper in liver and brain
Wiskott-Aldrich syndrome (WASP)	301000	LAS17	e-24	Effector for CDC42H GTPase; immunodeficiency

(table from the MIPS database web page compiled by Françoise Faury (UCL) http://www.mips.biochem.mpg.de/mips/yeast/)

*a) The names of the enzymes involved in diseases (cloning without reference to map) or the names of the diseases (positional cloning) are given in alphabetical order.

b) Disease accession number in OMIM database. The OMIM database was scanned using the following key words: disease; cDNA; gene; complementation; blood; bone; brain; heart; kidney; liver; muscle; pancreas; skin. The XREF database was also used. The accession number of the sequence of disease-accociated genes can be found using the OMIM disease accession number.

c) The names of the yeast ORFs discovered by genome sequencing were retrieved from the MIPS database (http://www.mips.biochem.mpg.de/) and annotated with an asterisk. The sequence and accession number of the yeast genes can be found in the MIPS database.
d) BLASTp and tBLASTn searches were carried out on the Stanford server against the *Saccharomyces* genome database The SEG algorithm was used to filter sequence bias. Reciprocal BLAST using the yeast gene as query and additional analyses were carried out using the NCBI server
e) Protein deficiency and clinical synopsis of the disease.
f) P-values are given for the rat sequence (accession L40624); the human sequence is incomplete.
g) Accession number of the human succinate dehydrogenase is P31040.
h) Date of identification of the genes is given in parentheses.
i) Accession number of the closest human partner of IPL1, X85545 (P=e-53).
j) *S. pombe* ORF SPA2G11.12 (accession Q09811 or Z54254).
k) Accession number of the closest human partner of GDI1, X79354 (P=e-175)
l) New ORF located on Watson strand between YJR135C AND YJR136c at positions 676660-676920 of the nucleotide sequence of chromosome X. The accession number of DDP is U66035
m) *S. pombe* FHIT homologue (accession U32615).
n) Accession number of the *C.elegans* homologue is U53332 (cosmid FS9G1, cDNA CEESF63F) and the accession number of the *E.coli* cyay is P27838.
o) Accession number of the human closest partner of PH085, X66364 (P=e-121).
p) Accession number of the human closest partner of YNL161w, Z35102 (P=e-164).
q) Accession number of the human closest partner of SRM1, P18754 (P=e-40).

responsible for adrenoleucodystrophy (an illness that kills young boys through the degeneration of their nervous systems). Biochemical studies showed that when this gene was deactivated in yeast, there was a change in the transportation of fatty acids across the membranes of the cellular organisms, the *peroxysomes*. This confirmed and backed up some studies already carried out in man and led to a clearer understanding of the illness. It may be hoped that in the long term, this knowledge will allow early diagnosis of the illness and better orientation of future therapies"[76].

---

[76] *Ditto*, p. 24.

7.4 After The Sequence — The Challenge of Functional Analysis

The complete sequence of an organism brings a flood of new information that can, and should, be exhaustively analyzed with *in silico* computer comparisons. In the case of the yeast genome, these comparisons occurred spontaneously amongst the existing structures without the need for financial stimulation or planning within the EEC programs.[77] But they weren't sufficient. Once all the possible conclusions as to the function of new genes have been subtracted from computer analysis and the similarities detected, continued research depends on experimentation. As about a third of the genes did not have clear homologues and were candidates for new and unknown functions, it was natural that someone should want to study them. Since the beginning of the program, P. Slonimski had pointed out to André Goffeau that systematic functional studies would be needed. This concern was reinforced by the industrialists of the Yeast Industrial Platform supporting the sequencing program, for whom the sequencing information was interesting, but difficult to exploit, and for whom potential functional analysis results were much more promising.

Given the high number of orphan genes, the problem, initially, was to define the strategy, objectives and methods for such an analysis. Thanks to efficient homologous recombination and the ease with which it mutates, yeast can be used in reverse genetics on a scale without measure, compared to the flexibility of other organisms. It's also very easy to suppress all or a part of a chosen gene (by disrupting it)[78] to study the effects of the lack of its function (Fig. 27, left-hand diagram). A gene (or genes) can also be replaced by a mutant version *in vivo*, or the version present in the chromosome can be picked out by the shuttle vector (Fig. 27, right-hand diagram).

---

[77] A. Goffeau, Mission report, Spetsai, September 1992, European Commission, DG XII, Brussels.

[78] R. Rothstein, "One step disruption in yeast", *Methods. Enzymol.*, vol. 101, 1983, pp. 202–211.

S.R. Sikorski and P. Hieter, "A system of shuttle vectors and yeast host strains designed for efficient manipulation of DNA in *Saccharomyces cerevisiae*", *Genetics*, vol. 122, 1989, pp. 19–27.

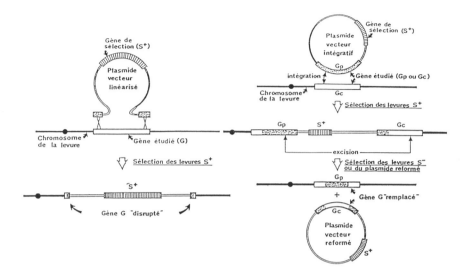

Fig. 27 Techniques of microsurgery in yeast genes
(Diagrams from B. Dujon)

Therefore, all the orphans can be mutated, either singly or as a group[79]. The basic strategy is the inactivation of targeted DNA fragments by simple or multiple deletion or disruption[80]. The resulting organisms can then be studied to see whether there are phenotypic differences, either structural or functional, that could be associated with the altered gene. For example, simple phenotype tests act on the composition of environment, temperature and the inhibitions to reveal the fundamental nature of some genes. Several techniques have been developed to improve this method of investigation. One of the most significant advances has come from the Snyder group at Yale and involves

---

[79] A. Wach, A. Brachat and P. Philippsen, "Plasmids for simultaneous testing of yeast essential alleles and strength of transcription (pSt-Yeast) a strategic proposal", *in Report of the Manchester Conference of the Yeast Genome Sequencing Network*, European Commission DG XII Biotechnology, February 1994, p. 127.

[80] R. Rothstein, "One step disruption in yeast", *Methods. Enzymol.*, vol. 101, 1983, pp. 202–211.

S.R. Sikorski and P. Hieter, "A system of shuttle vectors and yeast host strains designed for efficient manipulation of DNA, in *Saccharomyces cerevisiae*", *Genetics*, vol. 122, 1989, pp. 19–27.

gene fusion, during which yeast genes are systematically replaced by marker genes to create a series of disruptants. Using this method, thousands of genes can be examined in parallel[81]. Other disruption techniques have also been invented[82]. Also the probing of all 6,000 genes at once on a glass slide or a chip with total cell mRNA (or cDNA: see DeRisi and Brown, *Science*, October 1997).

However, despite its obvious interest, the analysis of disruption phenotypes was too dispersed. In the mid-1980s, it was noted that the disruption of only a few genes led to fatal modifications[83].

Functional Analysis of the Genes of Chromosome III

The rapid advance of chromosome sequencing and the final attainment of the entire sequence for chromosome III allowed the functional analysis of a certain number of ORFs to be begun. The work of disrupting ORFs and looking at the phenotypes was undertaken spontaneously by most of the contractors and facilitated by the workshop organized by Slonimski. It was also promoted by the announcement that laboratories that had not received enough DNA could consider the disruption of a gene the equivalent of sequencing 3 kbp in order to finalize their side of the contract with the EEC.

---

[81] N. Burns *et al.*, *Genes and Development*, vol. 8, 1994, pp. 1087–1105.

[82] Such as the new disruption method based on transposition events developed by M. Bolotin-Fukuhara (M. Bolotin-Fukuhara, *Functional studies of yeast genes, proposal for mass construction and screening of descriptant strains*, provided by A. Goffeau) or the new knock-out allele construction strategy. There is also the systematic direct cloning of wild or mutated genes based on new *E. coli*-yeast shuttle vectors which contain partially overlapping fragments of the selection market and which can receive fragments of the yeast genome inserted in one easy step. These vectors also include sites for the *I-Sce*I endonuclease around a selected gene to permit its direct elimination (C. Fairhead, B. Llorente, F. Denis, M. Soler and B. Dujon, "New vectors for combinatorial deletions in yeast chromosomes and for gap-repair cloning using "split-marker" recombination", *Yeast*, 1996).

[83] D.B. Kaback *et al.*, "Temperature-sensitive lethal mutations on yeast chromosome I appear to define only a small number of genes", *Genetics*, vol. 108, 1984, pp. 67–90.

M.G. Goeble and T.D. Petes, "Most of the yeast genomic sequences are not essential for cell growth and division", *Cell*, vol. 46, 1986, pp. 983–992.

This approach had limited success[84]. Of the 55 ORFs already disrupted[85] or deleted, only three had proved to be vital; and later analysis of 42 genes showed that only 21 of them had a different phenotype. In total, as of 1993, two-thirds of the disrupted genes on chromosome III led to no obviously different phenotype[86] and only 6% of the genes were essential for the life of the cell.

It is obvious that the conditions under which disruptant phenotypes were tested were far too limited. Nevertheless, the production of such mutants is the first step to any strategic attempt to associate a function to the deleted gene through successive waves of increasingly sophisticated analysis. So, in 1992, P. Slonimski, coordinator of the functional analysis work in the European sequencing network, developed a manual in which he recommended, and provided details on, standard initial procedures that had demonstrated their efficiency and trustworthiness. This manual included descriptions of strains to be used for disruptions, the appropriate techniques and several relatively simple analyses (growth in various media and at different temperatures and stresses), which provided a few clues to direct later research into the functions of the genes in question. It was very useful to standardize this domain so that laboratories could compare their results and strategies. However, from all the disruptions of chromosome III, it would seem that a particular phenotype modification could be associated with a disruptant for only a minority of potential genes. Most often, the new genes brought to light by sequencing coded for traits whose absence could not be detected with phenotype testing. How could one show the function of these mysterious genes?

Specialists were brought together by the European Commission at a small meeting in Brussels and they said that the answer to the problem was the technique of analyzing proteins with 2D gels. In 1990, Stephen J. Fey and Peter Mose Larsen had already sent a letter to the Director General of

---

[84] S. Olivier *et al.*, "The complete DNA sequence of yeast chromosome III", *Nature*, vol. 35, 7 May 1992, p. 43.

[85] R. Rothstein, "One step disruption in yeast", *Methods. Enzymol.*, vol. 101, 1983, pp. 202–211.

[86] On 14 June 1991.

DG XII, then Paolo Fasella, underlining the importance of 2-D analysis for the development of biotechnology and the need for European commitment in that field if Europe did not want to be dependent on other markets. They voiced their regret that the Commission had not financed three proposals they had presented, although they had been well received and rated highly by the evaluators. 2D analysis was also strongly advocated by P. Slonimski at the BAP meeting on the sequencing of yeast chromosome III, at which he presented "The Integrated Gene Expression Network (IGEN)" as a new approach to studying unknown ORF functions. IGEN, based on a 2D electrophoresis gel system, presented at the same meeting by H. Boucherie, increased the number of proteins that could be distinguished and digitalized. When later reports are considered, it can be seen that their proposals were correct and why 2D analysis took on greater importance.

The analysis of proteins by 2D electrophoresis gel allows 2,000 to 3,000 proteins (in this case, yeast proteins) to be visualized at the same time and separated on the basis of their isoelectric spots (first dimension) and their molecular weight (second dimension). The typical pattern of individual points on the gel can be reproduced[87] if the same procedures and conditions are used; the differences between the strains can also be seen at this protein level. It is possible to detect the absence of a spot (for example, if the corresponding gene has been deleted) or a change in migratory capacity (for example, due to a different-sized isoelectrical point from a mutation in the gene responsible or another gene in charge of an event in the post-translation process). Furthermore, differences of intensity in the spots can be pinpointed and quantified. It's even possible to see parallel variations in the expression of enzymes of the same route controlled by a common transcriptional regulator if the gene coding for the regulator has been deleted.

P. Slonimski thought that 2D analysis of all the disruptants without a recognizable phenotype could provide useful information, which could be optimized if the data were completed with an analysis of mutants where the target gene is overexpressed. In deletion, not only the gene product disappears

---

[87] A. Blomberg *et al.*, "Interlaboratory reproducibility of yeast protein patterns analysed by immobilised pH gradient two dimensional gel electrophoresis", *Electrophoresis,* 1995.

from the pattern of the spot on the gel, but often there is a cascade of effects linked to the deletion that can be recorded. In other terms, several, and sometimes several dozen, products are quantitatively modified (diminished or increased by factors between 3 and 20) not only by deletions of several genes coding for transcriptional regulators, but also by genes coding for structural or enzymatic proteins. Subgroups of interconnected gene products can then be identified.

The biochemical significance of these new types of network remains to be understood, but probably corresponds to a certain number of physiological needs. The opportunity to collect information on functionally or structurally linked proteins and characterize them as being part of the same network, is another interesting contribution of 2D analysis. It will allow biologists to better clarify the number of networks there are, and how they interconnect and contribute to the way the cell works.

The availability of sequences for genes makes finding the gene product on gels easier (you determine the isoelectrical point and its molecular weight by the unaltered form of the protein). By comparing the theoretical and observed mobility of a protein, you can tell whether and how a protein has been cleaved or modified (phosphorylation, acetylation, methylation or glycosylation). Several examples were presented at the meeting in Brussels on 14 June 1991 (Boucherie, Mose Larsen and S.J. Fey), examples that showed that this approach could be used to confirm a protein's identity. A series of tests carried out in several laboratories and collected by André Goffeau had found candidates for five of the 12 genes disrupted in a pilot study being run by Fey and Mose Larsen[88]. These studies showed the performance of 2D analysis, a technique appropriate to fundamental systematic research as long as procedure and nomenclature were standardized and a common frame of reference was available to which the 2D analysts of yeast could refer[89].

---

[88] P. Mose Larsen and S.J. Fey, "Two dimensional gel electrophoresis and yeast genetics", *Comments to the meeting held at the European Commission*, vol. 21, 14 June 1991.

[89] A. Vassarotti and A. Goffeau, "A second dimension for the European Yeast Genome Sequencing Project", *Biopractice*, vol. 1, 1992.

In accordance with these evaluations and the need for a systematic functional study of unknown ORFs, P. Slonimski put together and submitted a pilot project entitled "Functional analysis, newly discovered and putative protein-coding-genes of chromosomes III of *S. cerevisiae*"[90]. It was accepted and financed to 800,000 ECU by the European Union over three years, beginning on 1 November 1993. At the beginning, six laboratories were involved[91]: the Center for the Molecular Genetics of the CNRS at Gif sur Yvette in France, under P. Slonimski; the Institute of Genetics and Microbiology of the CNRS at Orsay in France, under M. Bolotin-Fukuhara; the Laboratory of Genetics at Talence in France, under H. Boucherie; the University of Dusseldorf in Germany at the Laboratory of G. Michaelis; the Institute for Medical Microbiology of Aarhus University in Denmark, under P. Mose Larsen and S. Fey; and finally, the Section of Biochemistry, Cellular Biology and Genetics at the University of Athens, Greece, under T. Pataryas.

The main objective of this pilot study was finding the most appropriate strategies to discover the functions of the new genes brought to light during systematic sequencing, in particular, that of chromosome III. The research strategy developed was based on the study of perhaps 80 mutant yeast strains in which the target gene was either inactive or amplified by P. Slonimski and M. Bolotin-Fukuhara[92]. In each case, a battery of 100 phenotype tests — growth conditions from the substrates (sources of carbon and nitrogen, metabolites, stresses such as heat shock, superoxides and osmotic sensitivity) to predominant inhibitors and substances of various cell processes — was applied by P. Slonimski, the RNA transcripts analyzed by Michaelis, the 2D maps inventoried by H. Boucherie, S. Fey and P. Mose Larsen, and the partially sequenced proteins linked to the mutations analyzed (by A. Haritos). The co-segregation in diploid heterozygotes of targeted

---

[90] The final official title is "Experimental pilot study for a European co-operation on gene function search in *Saccharomyces cerevisiae*".

[91] The 6th laboratory eventually pulled out.

[92] Coppée *et al.*, "Construction of a library of 73 individual gene deletions in view of the functional analysis of the new genes discovered by systematic sequencing of *S. cerevisiae* chromosome III", *Yeast*, 1995.

deletion and the resulting phenotypes were checked[93]. This systematic distribution approach looked appropriate for application on a large scale on both solid and liquid media. The technique in question allowed hundreds of strains to be treated at the same time. The large number of pleiotropic and monotropic phenotypes obtained as well as the large number of additional proteins identified on the reference protein map (by overexpression of genes, composition of amino acids or mass spectrometry) and the experience gathered, proved that this method constituted a powerful tool for hunting down biochemical and physiological functions of unknown genes[94]. Since there were so many genes to analyze, it was foreseen that the cooperation of many laboratories, in the sort of network that had worked so well for the European enterprise sequencing chromosome III, would be a good way to go[95]. From the experience of the pilot project and that acquired during sequencing work, a large functional analysis project based on a distribution network was set up.

From 1995 on, over 100 laboratories submitted applications to the European Commission to support this functional analysis network[96] and take part in it.

The EUROFAN I Project

To discover the function of the new ORFs (new genes brought to light by the sequencing effort), and given the depth of technical and biological competence needed to complete the task, in the second part of the genome

---

[93] Also to be included is the creation of a gene-protein index. H. Boucherie *et al.*, "Two dimensional protein map of *S. cerevisiae*: Construction of a gene-protein index", *Yeast*, vol. 11, 1995, pp. 601–613.

[94] K-J. Rieger, "Basic physiological analysis key-step between genomic sequencing and protein function", in *Final European Conference of the Yeast Genome Sequencing Network*, Trieste, Italy, 25–28 September 1996, p. 68, European Commission, DG XII, Brussels.

[95] S.J. Fey, A. Nawrocki and P. Mose Larsen, "Function analysis by 2D gel electrophoresis: From gene to function, yeast genome sequencing network", 8–10 June 1995, DG XII, European Commission, *BIOTECH programme book of abstracts*, p. 214.

[96] A. Vassarotti *et al.*, "Structure and organization of the European Yeast Genome Sequencing Network", *Journal of Biotechnology*, vol. 41, 1995, p. 134.

program, the network approach invented by G⸱ffeau was used to set up the new EUROFAN I and II projects (EUROpean Functional Analysis Network). EUROFAN was launched in January 1996 for 24 months to clarify the biological function of 1,000 new *S. cerevisiae* genes. EUROFAN has a hierarchical structure that allows approaching the function of a particular gene with a specificity that increases with its progression down the analytical paths. Such a hierarchy has intrinsic efficiency since it means that not all the analyses must be carried out on each gene. It is important to emphasize that such a systematic approach to the problem of functional analysis can't replace normal biological research. Its goal is to lend refinement to the analysis of each new gene so that a specialized research laboratory can confidently incorporate it into its research program. In this way, EUROFAN plans to accelerate normal research by creating unparalleled information and material resources of fundamental use to European biomedical science and bioindustry. Those resources are:

– A targeted database on yeast gene function;
– A center for genetic stocking and archiving, including a depot and distribution site in Frankfurt with the 1,000 deleted strains, the clones, the disruption cassettes etc. The material stocked there is to be at the disposal of all EUROFAN researchers. After a time, to allow the members to publish their results and for YIP to evaluate the commercial potential for European industry, this material will be publicly available. It forms the basis of a permanent European collection.

The EUROFAN project involves 144 laboratories in 14 European nations and is organized into a two-consortium format, one service consortium and one resource consortium, with nodes.

A: Resource consortium
A1: Central coordination
– scientific coordination (UMIST - S. Oliver, UK)
– financial coordination (F. Foury and P. Mordant at the UCL, Belgium)

A2: Informatics coordination center
– MIPS, Germany (H.W. Mewes)
A3: YIP Industrial Liaison (AM Prieels, Brussels, Belgium)
A4: Genetic archives and stock center (K.D. Entian, Frankfurt, Germany)

B: Research consortium
B0: Constitution of deletants
P. Philippsen, Basel, Switzerland and the coordinators of the sequencing projects.
B1: Quantitative phenotype analysis
P. Slonimski, Gif sur Yvette - 3 laboratories
B2: Analysis of RNA-level expression
R.J. Planta, Amsterdam, the Netherlands - 4 laboratories
B3: this consortium has been concealed
B4: Level II protein expression analysis
M. Bolotin-Fukuhara, France - 3 laboratories
B5: Gene interaction; analysis of 2 hybrids
D. Alexandraki, Heraklion, Greece - 2 laboratories
B6: Metabolic control analyses
S.G. Oliver, UMIST, UK
H.V. Westerhoff, Amsterdam, the Netherlands - 4 laboratories
B7: Subcellular structure and organelles
L.A. Grivell, Amsterdam, the Netherlands
M. Veenhuis, Groningen, the Netherlands
B8: Relations with other genomes
A. Hinnen, Iena, Germany - 5 laboratories
B9: Development and evaluation of new methodologies for genome analysis
B. Dujon, Paris, France - 19 laboratories

Nodes for functional analysis:

N1: DNA synthesis and the cell cycle
Giovanna Lucchini, Milan, Italy - 5 laboratories

N2: RNA synthesis and processes
Bernard Dujon, Paris, France - 3 laboratories
N3: Translation
Rudi J. Planta, Amsterdam, the Netherlands - 1 laboratory
N4: Stress responses
Horst Feldmann, Munich, Germany - 6 laboratories
N5: Cell wall synthesis/morphogenesis
Cesar Nombela, Madrid, Spain - 5 laboratories
N6: Transport
Bruno Andre, Brussels, Belgium - 2 laboratories
N7: Energy and the carbohydrate metabolism
Karl-Dieter Entian, Frankfurt, Germany - 1 laboratory
N8: Lipid metabolism
G. Daum, Graz, Austria - 4 laboratories
N9: Special metabolism
Carlo Bruschi, Trieste, Italy - 1 laboratory
N10: Development
M. Breitenbach, Salzburg, Austria - 4 laboratories
N11: Mutagenesis (repair/recombination/meiosis)
Alain Nicolas, Paris, France - 2 laboratories
N12: Chromosomal structure
Edward J. Louis, Oxford, UK and P. Philippsen, Basel, Switzerland - 3 laboratories
N13: Cell architecture
E. Scheibel, Martinsried, Germany - 2 laboratories

Structure of the EUROFAN network

The general management structure is to be found in Fig. 28. The overall scientific coordination was carried out by S. Oliver of UMIST, GB, who was responsible for: 1) the definition of objectives; 2) fixing prices for tasks completed in the various nodes and consortia; 3) evaluating payments; and, 4) drawing up and amending the budget (total EU contribution — 7,320,000 ECU). The other two coordination roles were those of informatics and finance.

The Université Catholique de Louvain (UCL) in Belgium set up the EUROFAN management office. It was run by Philippe Mordant under the supervision of Françoise Foury, the scientist in charge. The UCL was the main link to the European Communities and was responsible for all communication with the European Commission. The UCL made the payments to the participating laboratories on the basis of reports submitted to UMIST by the coordinators of consortia and nodes and at the submission of data approved and validated in the EUROFAN data base at MIPS. MIPS had two roles in EUROFAN: scientific and administrative.

Its administrative contribution, linked to the acquisition of data, is, on the basis of prearranged criteria, to allow payments to be made for results. The evaluation of data was the joint responsibility of MIPS, the scientific, consortia, node and chromosome coordinators. The consortia and node coordinators organized the scientific program for their charges, distributed the tasks to participating laboratories, assigned (in coordination with the scientific coordinator) prices to those tasks, evaluated and accepted data, developed validation procedures and drew up scientific and financial reports to be sent to the EUROFAN scientific and financial coordinators.

There was believed to be a great potential for commercial exploitation in the data generated by EUROFAN, but there is also great potential in the various «tools» it generated. In accordance with the major objective of the 4th Framework Program (1994–1998), which is to increase the competitiveness of European industry, it was important that all be done to make such exploitation easier. This is why close cooperation was coordinated with the YIP. Through a private data base, members of the YIP have access to the preliminary data and information from projects still under way before they are published or made available in data bases open to the general public. Each YIP member, having beforehand signed a confidentiality agreement with the MIPS center and the EUROFAN Trust[97] to guarantee that the exchange of information would take place safely, could contact the laboratory that had obtained the results for a given function found during the EUROFAN work. Then the firm and the trust could negotiate conditions for

---

[97] *See below.*

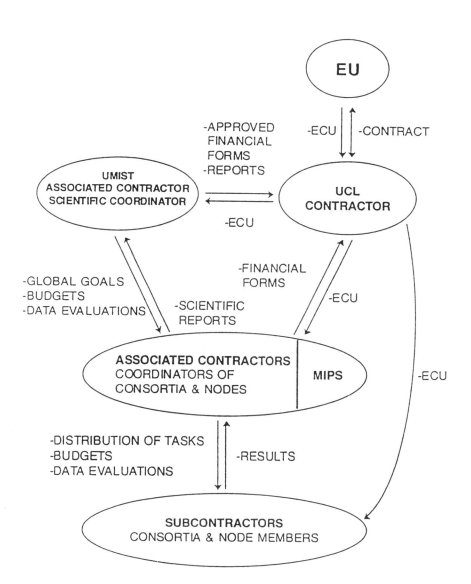

Fig. 28 General management structure for EUROFAN 1
(Diagram from the European Commission, provided by A. Goffeau)

further research at that laboratory or an additional delay before the data was published to allow for a patent or license request.

If no agreement was reached, the firm could not exploit the results. The first company to show interest and come to an agreement had priority over the other companies. The question of intellectual property rights over EUROFAN data has always been very complex. The clarification of the function of a given gene necessarily involves a large number of laboratories connected to various consortia and nodes. Furthermore, the assignment of a given gene to a laboratory at the B0 consortium level was more or less a matter of chance; therefore, the fact that a laboratory might discover something of commercial interest was merely the luck of the draw. It was, therefore, difficult or even impossible to assign rights to the individual members of EUROFAN and their home institutions. For this reason, a legal entity was created, the EUROFAN Trust, upon which the intellectual property rights of EUROFAN research results were to be bestowed. The objective of the trust is to promote yeast research in the European Union and Switzerland by providing grants and fellowships to young researchers. The EUROFAN Trust acts as a "one-stop shop" for any firm hoping to acquire the rights to exploit research data. The member firms of YIP have no right to exploit the data, just the opportunity. Any money resulting from EUROFAN rights, royalties or other income is paid to the trust, which assigns the money to the institutions involved in proportion to their contribution to the research program (for example, in direct proportion to the sums received from the EC for work carried out under EUROFAN). The institutions must spend the money in ways that fulfill the principal objectives of the trust.

The Research Structure

EUROFAN involves 144 laboratories in 14 European and 13 non-European nations and is probably the largest scientific network ever assembled. Its size and structural complexity are a direct consequence of the problem of the functional analysis of new genes discovered during the systematic sequencing of the yeast genome. The network is so complicated that it is difficult to provide a clear systematic representation of its nature and extent without being very long-winded indeed.

EUROFAN II

EUROFAN I was only a first phase in the functional analysis program. In order to continue systematic analysis for new yeast gene functions, as initiated under EUROFAN I, a large second-phase project was being drawn up. With a planned budget of nearly 8,000,0000 ECU, EUROFAN II has involved 93 laboratories in 13 different nations[98]. Its aim was to make a detailed analysis of the biological functions of 1,000 new genes of the yeast *S. cerevisiae*, an analysis accomplished through the use of analytical tools (deleted strains, clones and disruption cassettes) and primary phenotype data obtained during EUROFAN I. At the same time, EUROFAN II has participated in a world effort to produce deletion mutants for all the yeast genes that code for proteins. This wave of scientific effort made a major contribution, fulfilling a number of the objectives of the 4th Framework program (2.2 function search, 2.2.1. resource consortia, 2.2.2. functional analysis network 2.2.3. heterologous expression network, 2.2.4. new methods of genome analysis, 1.1 cell factories, 1.1.1. microbial industry, 8.2 genetic archives and stock centers). EUROFAN has also tried to clarify the apparent genetic redundancy, which is central to understanding gene function.

Like EUROFAN I, EUROFAN II has a hierarchical structure. This structure also allows the function of a given gene to be approached with increased efficiency as it passes through the layers of analysis. EUROFAN II has accelerated the normal research process by creating unparalleled information and material resources which were of fundamental importance for biomedical science and the bio-industries. These resources were:

- A targeted data base on yeast gene function (This data base, at MIPS in Germany, has included modeling tools to allow the information to be used in simulations and models of bioprocesses in academic and industrial projects. It has been set up to integrate with other existing

---

[98] A reduction in the number of laboratories involved in relation to EUROFAN I is due to certain rationalisations (fusions of consortia or nodes, elimination of research directions that have not proved satisfactory and the departure of some laboratories that proved unable, or did not want, to work within the regulatory framework of EUROFAN or had not managed to reach the norms and objectives required).

data bases of DNA and protein sequences as well as structural data libraries on genetics and proteins);
- A genetic stock and archive center (This has include a center for depositing and distributing information (EUROSCARF, Frankfurt, Germany) which will hold the deleted strains and plasmid tools. This material has been available to all EUROFAN I and II researchers. After a short length of time to allow EUROFAN members to publish their results and YIP to evaluate the potential utility for industry, the material has been available to all as a permanent European collection. The creation of this resource was a radical new approach to biology, and can be carried out by a large scale research network.

The EUROFAN II structure (running and organization) was very complex (Fig 19). It includes:

Service consortia providing service or materials to the entire EUROFAN II program

A1.1 Central coordination under S. Oliver, (UMIST), UK, scientific and financial administration including the implementation of paying for results

A1.2 Financial coordination under F. Foury and office management under Philippe Mordant, UCL, Belgium

A1.3 Informatics coordination under H.W. Mewes, MIPS, Germany

A2: Informatics resource center under H.W. Mewes, MIPS
1. data acquisition with a quality control procedure to pay according to results (international cooperation and exchange of information)
2. creation and management of an object-oriented data base
3. development of new computer tools to allow intelligent use of the data

A3: YIP, A.M. Prieels (Belgium). YIP will undertake the commercial evaluation of EUROFAN's work and ease technology transfer. It will ensure the liaison between DG XII (now Research) of the European

Commission and EUROFAN II and edit a "patent newsletter" and an annual report on EUROFAN II with an industrial viewpoint.

A4: Genetic archives and stock center
EUROSCARF, K.D. Entian and P. Kotter (Frankfurt, DE)

B: Fundamental basic analysis
The resource consortia are only the first level of the series of functional analyses. They carry out fundamental genome analysis including generic screening of all the deletion mutants and studies of the action and interaction of all *S. cerevisiae* genes. Such generic screening with *in silico* analysis allow functions to be more or less assigned, leaving individual genes to progress up the series of investigations and reach the level of specific analysis, that of the functional analysis clusters.

B0. Generation of deletion mutants and corresponding plasmid tools
A group of 16 laboratories will take part in the creation of this basic resource, working on the constitution of internationally accepted protocols and verified norms of cooperation with the US and Canadian consortia (P. Philippsen, Basel, CH) 9 laboratories.

B1. Qualitative and quantitative phenotypic analysis, coordinator P. Slonimski, Gif-sur-Yvette, France, 4 managers and 2 laboratories
B1a: qualitative phenotypic analysis, coordinator P. Slonimski
B1b: quantitative phenotypic analysis and metabolic testing, coordinator S.G Oliver (Manchester, GB) 5 laboratories.

B2: Analysis of transcription, coordinator R.J. Planta (Amsterdam, NL) 6 laboratories

B3: Essential genes, coordinator J.H. Hegemann (Giessen, DE) 9 laboratories

B4: Integrated approach to function, coordinator B. Dujon (Paris FR) 13 laboratories

F: Clusters of functional analysis

Each cluster gets a group of 250 genes of the 1,000 genes analyzed by EUROFAN I. It has carried out a basic analysis of these genes. Depending on the results, certain genes has been passed on to nodes of the cluster for detailed second-level testing. Genes with no positive results from the first tests were the first tested in the later functional analysis clusters.

F1: Nuclear Dynamics, Convenor A. Nicolas, Paris, France
    N1: Genome plasticity/nuclear structure/meiosis, coordinator A. Nicolas (Paris, FR) 7 laboratories
    N2: Chromosome structure, architecture and biology, coordinator E.J. Louis (GB) and P. Philippsen (Basel, CH)

F2: Metabolic actions and interactions, Convenor L.A. Grivell (Amsterdam, NL)
    N3: Transport, coordinator Andre (Brussels, BE) 3 laboratories
    N4: Organelles, coordinator L.A. Grivell, 4 laboratories

F3: Cell organization and development, Convenor Howard Riezmann (Basel, CH)
    N5: Secretion and protein transport, coordinator Howard Riezmann, 4 laboratories
    N6: Cell architecture, coordinator Barbara Winsor/Robert Martin (Strasbourg FR) 5 laboratories
    N7: Cell walls and morphogenesis, coordinator F.M. Klis (Amsterdam NL) 2 laboratories
    N8: Development, coordinator Michael Breitenbach (Salzburg, AT) 3 laboratories.

The estimated budget for EUROFAN I and II totals 14,803,000 ECU.[99] A European charity trust was in charge of distributing the funds, rights, budget incomings, investments and loans. A council to advise the trustees has also been created to: 1) inform the trustees in all of the aspects of the research programs set up by the charity and controlled by the trustees, and

---

[99] EU contribution foreseen: 13,490,000 ECU, with another 950,000 ECU from the Swiss.

of all facts relevant to the publication of results; 2) advise the trustees on the works of the charity trust; 3) pay the members of the council all the costs of their participation in committee meetings and the business of the Charity Trust.

The overall management structure of EUROFAN II can be seen in Fig. 29. The coordinators of the consortia and nodes are indubitably the key to EUROFAN's success. They were in charge of the organization of the scientific work of their sector, the distribution of tasks to the laboratories under their supervision, the attribution of budgets for the tasks in consultation with the scientific coordinator, the basic evaluation and acceptance of the data, the development of validation procedures and the production of scientific and financial reports for the scientific and financial coordinators of EUROFAN II.

EUROFAN II, like EUROFAN I, has been organized as a hierarchy both scientifically and structurally. Only with such a structure could a collaboration the size of EUROFAN II could succeed. The innovation since EUROFAN I has been the introduction of the F-level clusters, which has increased the efficiency of analysis by cutting out redundancy in connected nodes, improving communication and the convergence of data and creating a smaller and more efficient group capable of advising the scientific coordinator.

The Scientific Benefits of EUROFAN

The scientific benefits of EUROFAN are many. EUROFAN I set up a new genetic approach. Traditionally genetic research proceeds in the following pattern: 1. mutant phenotype; 2. definition of phenotype; 3. isolation and sequencing of the gene whose mutant is responsible for the phenotype. But systematic sequencing has provided genes not associated with a phenotype, and this means new approaches must be developed to discover the functions of these genes. This problem is one of the great challenges of modern biology because the rate at which new genes are discovered during the systematic sequencing of genomes is much higher than that of the classical approach. It is even more urgent now that there are complete sequences for pathogenic bacteria and archae, organisms which do not have as easily manipulated systems and are not as useful for studying human, plant and

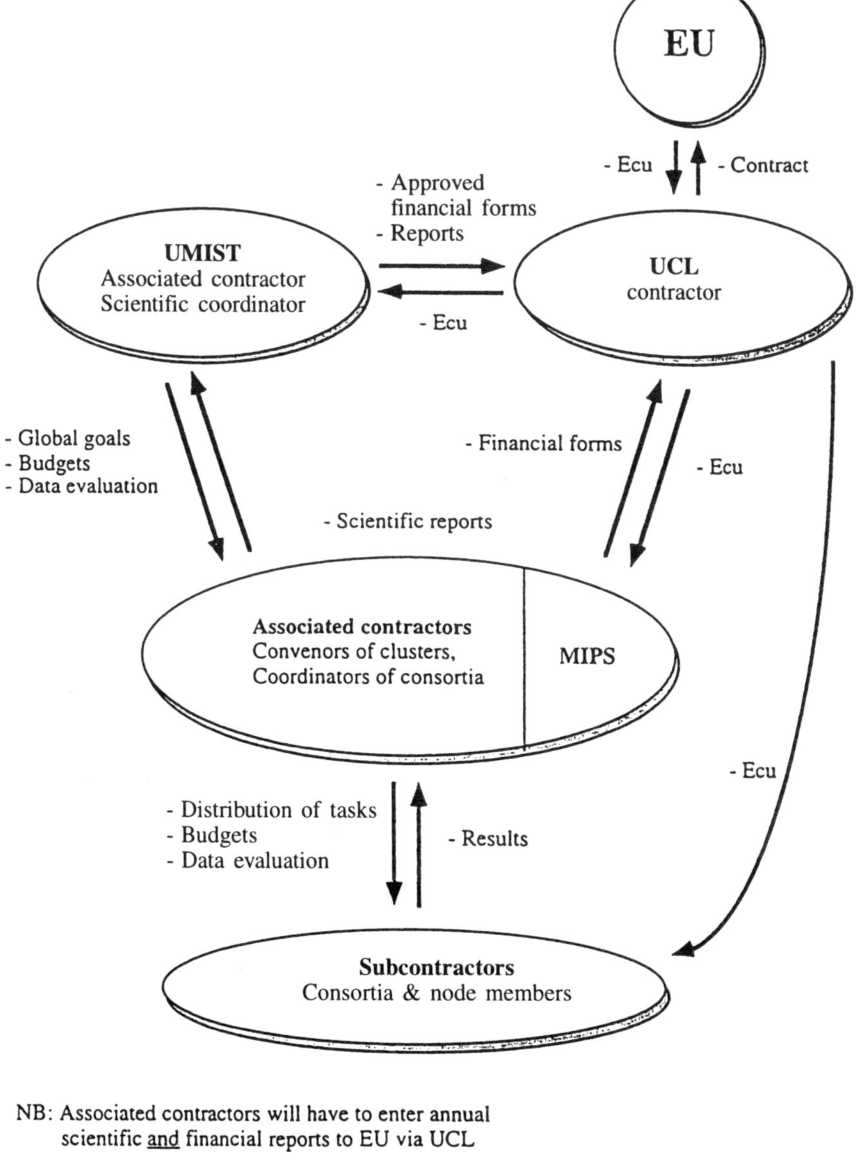

506 *From Biotechnology to Genomes: A Meaning for the Double Helix*

EU

- Ecu
- Contract

UMIST
Associated contractor
Scientific coordinator

- Approved financial forms
- Reports

UCL
contractor

- Ecu

- Global goals
- Budgets
- Data evaluation

- Financial forms
- Ecu

- Scientific reports

Associated contractors
Convenors of clusters,
Coordinators of consortia | **MIPS**

- Distribution of tasks
- Budgets
- Data evaluation

- Results

- Ecu

Subcontractors
Consortia & node members

NB: Associated contractors will have to enter annual
scientific <u>and</u> financial reports to EU via UCL

Fig. 29 General management structure for EUROFAN II
(Diagram from the European Commission, provided by A. Goffeau)

stock animal genomes as is yeast. It is the ease with which the yeast genome can be manipulated, the understanding of it that has been reached and its use as an experimental organism with no major difficulties that make yeast an excellent choice for systematic functional analysis. EUROFAN II's contribution to the study of larger genomes goes way beyond being a test case for new approaches and research methods.

The entire collection of 6,000 deletion mutants to be generated by EUROFAN II through international collaborations with other functional study ventures in the US and Canada will be an important resource for the study of other genomes since these lesions can be complemented by cDNA from other organisms, thus mapping functions on genomes larger than that of yeast. Furthermore, the functional analysis of 1,000 genes will provide a unique set of functional assignments allowing the interpretation of similarities between the yeast genome and that of higher organisms. The combination of sequence and function data will make *S. cerevisiae* a basic tool for research on other larger genomes, in particular in the Human Genome Project.

A large number of genes involved in hereditary illnesses have similarities with yeast genes. The study of their physiological roles and their interactions with other yeast genes or gene families will improve understanding of human genetic disease syndromes and might suggest treatment either through the classical approaches or gene therapy. The recognition of sequence homologies or structures between the product of a single human gene involved in the control of a particular illness and its equivalent in yeast could provide an important key to significant improvement of its diagnosis or therapy.

In silico analysis of the yeast genome sequence and *in vivo* studies can quickly reveal the interactions between genes and their products. For example, the identification of extragenic suppression or synthetic phenotypes and the study of expression control (consortium B2, B4, cluster F2) as well as the use of 2-hybrid systems (consortium B5) can show up all the genes whose structural or functional homologies could clarify the molecular and biochemical basis of human illness and provide a set of tools (DNA probes and antibodies) to rapidly identify the corresponding important genes.

An Opportunity for Industry

EUROFAN I and II are an opportunity for industry to have access to information that can be exploited (in any case, more easily than raw sequence data) either through genetic manipulation or by the development of more conventional strains to improve traits such as rate and efficiency of fermentation (consortium B1, cluster F2) and flavor and alcohol tolerance (consortium B1). They also provide the possibility of understanding the metabolism (consortium B1, cluster F2) to improve the specificity, rate and efficiency of processes such as specific biotransformations. With this new knowledge, yeast will become the preferred organism for genetic engineering to improve the production of existing or novel substances of commercial value.

The European pharmaceutical industry will also be a major beneficiary of the EUROFAN programs. For them, the benefits will initially be more general; a greater understanding of eucaryotic molecular genetics, the deletion mutant library for function mapping and greater ease in identifying genes involved in illnesses. EUROFAN II will help identify genes which, from their similarities quantity or type of functions discovered, may represent realistic targets for antifungal chemotherapy (consortia B3, clusters F1 and F3). The similarities of these genes can be isolated from fungal pathogens with PCR technology. And as yeast, after EUROFAN, will see an affirmation of its role as favorite cell factory, EUROFAN II will contribute to the rise in its use as a choice organism for drug screening trials.

Yeast is also the favorite organism for large scale production of many recombinant proteins of human and viral origin. The regulated expression of these proteins is often required, and the expression analyses carried out by EUROFAN II's consortium B2 and cluster F2 should allow the identification of new promoters for such applications, which could provide regulated expression of heterologous coding sequences under physiological conditions suitable for large scale industrial processes. Such promoters are also useful in that they allow European industry to circumvent certain restrictive patents.

Dynamic Research

Besides the scientific industrial and technical benefits[100], one of the great results of the sequencing program and network approach was the atmosphere of friendly collaboration that developed between the laboratories involved, generating a large number of cooperative ventures throughout yeast research. EUROFAN I, of course, extended this aspect, increasing the efficiency of European yeast research. One of the fundamental benefits of EUROFAN II will be the improvement of technical competence in participating laboratories by the creation of common protocols and norms and the exchange of researchers and technicians between EUROFAN laboratories, an exchange encouraged by funding from the European Commission, the Yeast Industry Platform and other sources. Finally, EUROFAN II is expected to generate a major worldwide collaboration allowing the establishment of even more material resources (deletion mutant strains and disruption cassettes) available for European science and industry. These world resources will be immediately accessible because EUROFAN is considered a major player in the international scientific field.

Without a doubt, the European genome program started by André Goffeau in 1989 has not only allowed Europe to defend its respected place in international genome research, but has also contributed to a revitalization of yeast research. Several coordinated research efforts are currently underway worldwide. The USA has started a functional analysis research program, but it is a lot more loosely organized than the European programs. Parallel programs are also underway in Germany[101], Canada and Japan[102]. Broad

---

[100] Please read N. Williams' article "Yeast genome sequence ferments new research", in *Science*, vol. 272, no. 5261, 26 April 1996, p. 481.

[101] The German Science Ministry (BMBF) had funded a functional analysis network working on yeast since July 1994 (this program ended in June 1997) to DM 8 million. The target of the program was to disrupt and study 440 genes.

[102] In Japan, the Ministry of Education, Science, Sports and Culture (MESSC) supports several large functional research ventures: a *B. subtilis* project under the Human Genome Project (1996–2000), a *C. elegans* project (Kobara, US$3 Million) in co-operation with the Europeans at the Sanger Center and the USA, and a systematic piece of research by Murakami on the ORFs of chromosome V of *S. cerevisiae* over five years in close collaboration with the European programs.

complex functional analyses are ongoing or have already been completed. There are, for example, analyses based on transposons, including a method described by D. Botstein to use the Ty1 transposon. A random insertion allows researchers to create a PCR genomic footprint fragment. Changes in the structures of the PCR bands can show overall gene expression changes in varied growth conditions. By using another transposon, Tn3, M. Snyder of Yale[103] has created *S. cerevisiae* strains with a lacZ insertion at a random genome location. To increase the transposon's ability to immunolocalize the products of mutant genes, it has been modified. The new version *mTmlacZ* is present in the transposon. However, during expression of pre-recombinase in a yeast strain with such a transposon genome insertion, all the sequence elements between *lox* sites are removed, leaving only one *lox* site and the adjacent transposon sequences. The latter have been constructed to contain three tandem copies of the HA epitope of the influenza hemaglutinine virus, generating an insert of 93 AAs in the product of the mutant gene. The use of HZ antibodies allows the localization of proteins with most of the insertions and is being incorporated into the large-scale genome analysis program.

A more systematic approach to labelling each gene was described by D. Shoemaker and R. Davis of Stanford[104]. A single label sequence of "20-meres" is added during the generation of each deletion strain allowing PCR analysis of a mixture of deletion strains, possibly up to 6,000 at any one time. This technique, applied initially to chromosome V, is now being used over the whole yeast genome[105]. The amplification of the labels, hybrid to a segment of DNA containing an ordered matrix of all the labels, will rapidly and precisely identify strains that can grow in a variety of conditions. Additional genomic analysis is carried out, including the fluorescence study of the hybridization of individually amplified ORFs, producing an evaluation

---

[103] P. Ross-Macdonald, S. Azarwal, S. Erdman, N. Burns, A. Sheehan, M. Malczynski, G.S. Roeder and M. Snyder, "Large functional analysis of the *S. cerevisiae* genome", *EUROFAN meeting*, ref. above.

[104] Summary of a description of the "bar code" technique sent to A. Goffeau by R. Davis.

[105] V. Smith, K. Chou, D. Lashkari, D. Botstein and P. Brown, "Large-sclae investigation of gene function in yeast using genetic footprinting", *EUROFAN, Louvain la Neuve*, 28–30 March 1996, *Book of abstracts*, European Commission DG XII Biotechnology.

of changes in expression under different growth conditions and in different strains (D. Lashkari, Stanford) or allowing the mapping of recombination events throughout the genome (F. McKusker, Duke University) and protein analysis such as 2D analysis of each protein (J. Garrels, proteome). The development of a database of protein interactions using the 2-hybrid system (S. Field, Washington University, Seattle) for yeast proteins (M. Vidal, Harvard University) and mammal proteins is also being created.

It is clear that the yeast scientific community works in a climate of cooperation rather than competition in their attempts to reveal how the cell of a eucaryote works. In any case, they have taken up the challenge, and the information they obtain will greatly accelerate the attribution of functions to new genes discovered in larger genomes. Yeast will become the navigation instrument for the voyages being undertaken across the world into large genomes of higher complexity.

The fruit of world collaboration and a unique network of laboratories, conceived and established by André Goffeau and sustained by the European Community through its various programs, has ensured the success of the yeast program and allowed the modest creature used for centuries by bakers and brewers to enter the pantheon of sequenced organisms. It is the first eucaryote to be sequenced, although it will probably not remain so for long. The time of systematic genome studies has truly begun, and sequencing an entire genome will soon not be considered much of a scientific feat. Sequencing is, in fact, becoming more and more of an industrialized activity. Typically, large-scale sequencing is becoming the domain of big sequencing centers, smaller laboratories only being called in to resolve specific problems where their specialized understanding might be useful.

7.5 Sequences, Sequences and More Sequences

With all these international programs, yeast was not, of course, the only organism being sequenced. Before the release and distribution of the numerical data of the complete sequence of the *S. cerevisiae* genome on 14 April 1996, two other entire bacterial genome sequences had been released to the public domain by TIGR: the 1.8 Mb sequence of the *Haemophilus influenzae*[106] genome and 0.6 Mb sequence of *Mycoplasma genitalium*[107] (Table 26). The same year, another prokaryote genome, that of *Methanococcus jannaschi* (1.7 Mb) was published[108] and the Kazusa DNA Research Institute published the sequence of *Synechocystis* sp. (3.6 Mb)[109]. Also, in 1996, Himmebreich *et al.* of the University of Heidelberg published the sequence of *Mycoplasma pneumoniae*[110] (0.81 Mb). Blattner's group at the University of Wisconsin finally finished their sequencing of the genome of *Escherichia coli* K-12 at 4.60 Mb after more than six years of laborious sequencing by several talented people. Although there were many errors made during this project, the sequencing of *E. coli* is very important for the whole field of microbiology. Several other genomes are now sequenced, for examples: *Methanococcus jannaschii* (3.57 Mb, TIGR[111]), *Helicobacter pylori* (1.66 Mb, TIGR[112]), *Methanobacterium thermoautotrophicum* (1.75 Mb, Genome Therapeutics and Ohio State University[113]), *Bacillus subtilis* (International Consortium, funding: EC[114]), *Archaeoglobus fulgidus* (2.20 Mb, TIGR[115]), *Borrelia burgdorferi* (1.44 Mb, TIGR[116]), *Aquifex aeolicus* (1.50 Mb[117]), *Pyrococcus korikoshii* (1.80 Mb, NITE[118]), *Mycobacterium tuberculosis*

---

[106] R.D. Fleishmann *et al.*, *Science*, vol. 269, 1995, pp. 496–512.

[107] C.M. Fraser *et al.*, "The minimal gene complement of *Mycoplasma genitalium*", *Science*, vol. 270, 20 October 1995, pp. 397–403.

[108] C. J. Bult *et al.*, "Complete genome sequence of the methanogenic archaeon, *Methanococcus jannaschii*", *Science*, vol. 273, no. 5278, 23 August 1996, pp. 1058–1074.

[109] Kaneko *et al.*, *DNA Res.*, vol. 3, 1996, pp. 109–136.

[110] Himmebreich *et al.*, *Nuc. Acid Res.*, vol. 24, 1996, pp. 4420–4449.

[111] Bult *et al.*, *Science*, 273, 1996, pp. 1058–1073.

[112] Tomb *et al.*, *Nature*, 388, 1997, pp. 539–547.

[113] Smith *et al.*, *J. Bacteriology*, 179, 1997, pp. 7135–7155.

[114] Kunst *et al.*, *Nature*, 390, 1997, pp. 249–256.

[115] Llenk *et al.*, *Nature*, 390, 1997, pp. 364–370.

[116] Fraser *et al.*, *Nature*, 390, 1997, pp. 580–586.

[117] Deckert *et al.*, *Nature*, 392, 1998, p. 353.

[118] Kawarabayasi *et al.*, *DNA Research*, 5, 1998, p. 55.

(4.40 Mb, Sanger Center[119]) and *Treponema pallidum* (1.44 Mb, TIGR / Univ. Of Texas[120]).

Table 26 Data on *Mycoplasma pneumoniae, Mycoplasma genitalium* and *Haemophilus influenzae*

Genome size: 816,394 bp
G+C content: 40.1%
Number of hypothetical ORFs: 677 (716,176 bp)
Average MP–ORF size: 39,500 Da
Number of other coding regions (tRNA ; rDNA): 39 (7,998bp)
Length of all coding regions: 88.7%
Gene density: one gene per 1.14 kb
Number of ORFs with a *M. genitalium* homolog: 541 (74.9%)
Number of *M. pneumoniae* specific ORFs : 136 (20.1%)
Average percentage of homology: 64.7%
Lenght of MG homologue ORFs: 611,388 bp (74.9ù)
Lenght of *M. pneumoniae* specific ORFs: 104,788 bp (12,8%)
Mycoplasma pneumoniae

Genome size: 580070 bp
G+C content: 32%
Number of hypothetical ORFs: 468
Lenght of all coding regions: 520,602 bp
Percentage of all coding regions: 89.7%
Mycoplasma genitalium

	M. pneumoniae	M. genitalium	H. influenzae
Genome size	816,394 bp	580,070 bp	1,830,137 bp
number of sequence reactions	>6,385	8,472	24,304
Number of nucleotides	2,415,202	3,806,280	11,631,485
Average reading length	378nte	375 nte	
Number of oligonucleotides	5,095		
Average redundancy	2.95	6.56	6.35

---

[119] Cole *et al.*, *Nature*, 393, 1998, p. 537.

[120] Fraser *et al.*, *Science*, 281, 1998, p. 375.

The systematic sequencing of some sixty genomes is under way (Table 27). Most of them are between 1 and 3 Mb. TIGR's successes show that today shotgun sequencing of small bacterial genomes can be carried out in under six months at less than US$0.50 per base pair [121]. Larger genomes, however, are much more expensive and complex, the production of a contiguous DNA clone library becoming a problematic factor rather than the sequencing.

> "In the absence of such a library, long-range PCR amplification, direct PCR sequencing or both must be employed. The sequencing of genomes larger than 6 Mb typically requires the time-consuming construction of clone libraries in cosmids or other high-capacity vectors and the tedious filling-in of the unavoidable, sometimes numerous and occasionally intractable, gaps in clone coverage. Plans for the determination of medium-sized genome sequences (10–100 Mb) almost always underestimate the costs of these essential steps. The existence of two complementary, well-organized and almost gapless libraries of yeast DNA in cosmid vectors [122] was a major factor in the unexpected speed at which the full genome sequence was obtained" [123].

With the experience that has been gained, the sequences obtained and the progress made in new sequencing techniques, new tool-vector-sequencers, new processors, new *in silico* analysis programs, a sequencing rate of 10Mb contiguous per year costing less than US$5 million should quickly become the norm.

---

[121] R.D. Fleischmann *et al.*, "Whole-genome random sequencing and assembly of *Haemophilus influenzae* Rd", *Science*, vol. 273, no. 5223, 1995, pp. 496–512.
  C. Bult *et al.*, *Science*, vol. 273, p. 1058.
  As A. Goffeau points out, it is difficult to determine whether such estimations are for total or marginal costs (A. Goffeau, in "Life with 6000 genes", *Science*, vol. 274, October 1996, p. 566).

[122] L. Riles *et al.*, *Genetics*, 134, 1993, p. 81.
  A. Thierry, L. Gaillon, F. Galibert, B. Dujon, *Yeast*, 11, 1995, p. 121.

[123] A. Goffeau *et al.*, "Life with 6000 genes", *Science*, vol. 274, October 1996, p. 566.

Table 27 Microbial genomes currently being sequenced
(table from the TIGR database)

▣ TIGR Microbial Database:
a listing of microbial genomes and chromosomes completed and in progress
Published microbial genomes and chromosomes (scroll down for genomes in progress)

	Link	Genome	Strain	Domain	Size (Mb)	Institution	Funding	Publication
1	▣	*Haemophilus influenzae* Rd	KW20	B	1.83	TIGR	TIGR	Fleischmann *et. al.*, *Science* **269**:496–512 (1995)
2	▣	*Mycoplasma genitalium*	G-37	B	0.58	TIGR	DOE	Fraser *et. al.*, *Science* **270**:397–403 (1995)
3	▣	*Methanococcus jannaschii*	DSM 2661	A	1.66	TIGR	DOE	Bult *et. al.*, *Science* **273**:1058–1073 (1996)
4	▣	*Synechocystis sp.*	PCC 6803	B	3.57	Kazusa DNA Research Inst.		Kaneko *et. al.*, *DNA Res.* **3**: 109–136 (1996)
5	▣	*Mycoplasma pneumoniae*	M129	B	0.81	Univ. of Heidelberg	DFG	Himmelreich *et. al.*, *Nuc. Acid Res.* **24**: 4420–4449 (1996)
6	▣	*Saccharomyces cerevisiae*	S288C	E	13	International Consortium	EC, NHGRI, Welcome Trust, McGill U., RIKEN	Goffeau *et. al.*, *Nature* **387** (Suppl.) 5–105 (1997)
7	▣	*Helicobacter pylori*	26695	B	1.66	TIGR	TIGR	Tomb *et. al.*, *Nature* **388**:539–547 (1997)
8	▣	*Escherichia coli*	K-12	B	4.60	University of Wisconsin	NHGRI	Blattner *et. al.*, *Science* **277**:1453–1474 (1997)
9	▣	*Methanobacterium thermoautotrophicum*	delta H	A	1.75	Genome Therapeutics & Ohio State Univ.	DOE	Smith *et. al.*, *J. Bacteriology*, **179**: 7135–7155 (1997)
10	▣	*Bacillus subtilis*	168	B	4.20	International Consortium	EC	Kunst *et.al.*, *Nature* **390**: 249–256(1997)
11	▣	*Archaeoglobus fulgidus*	DSM4304	A	2.18	TIGR	DOE	Klenk *et al.*, *Nature* **390**:364–370 (1997)
12	▣	*Borrelia burgdorferi*	B31	B	1.44	TIGR	Mathers Foundation	Fraser *et al.*, *Nature*, **390**: 580–586 (1997)
13	▣	*Aquifex aeolicus*	VF5	B	1.50	Diversa	DOE, Diversa	Deckert *et al.*, *Nature* **392**:353 (1998)
14	▣	*Pyrococcus horikoshii*	OT3	A	1.80	NITE		Kawarabayasi *et al.*, *DNA Research* **5**: 55–76 (1998)
15	▣	*Mycobacterium tuberculosis*	37Rv (lab strain)	B	4.40	Sanger Centre	Wellcome Trust	Cole *et al.*, *Nature* **393**:537 (1998)
16	▣	*Treponema pallidum*	Nichols	B	1.14	TIGR / Univ. Texas	NIAID	Fraser *et al.*, *Science* **281**: 375–388 (1998)
17	▣	*Chlamydia trachomatis*	serovar D (D/UW-3/Cx)	B	1.05	UC Berkeley & Stanford	NIAID	Stephens *et al.*, *Science* **282**: 754–759 (1998)
18	▣	*Plasmodium falciparum* **Chr2** (isolate 3D7)		E	1.00	TIGR / NMRI	NIAID	Gardner *et. al.*, *Science* **282**: 1126–1132 (1998)
19	▣	*Rickettsia prowazekii*	Madrid E	B	1.10	University of Uppsala	SSF / NFR	Andersson *et al.*, *Nature* **396**: 133–140 (1998)
20	▣	*Helicobacter pylori*	J99	B	1.64	Astra Research Center Boston / Genome Therapeutics	Astra Research Center Boston / Genome Therapeutics	Alm *et.al.*, *Nature* **397**:176–180 (1999)
21	▣	*Leishmania major* **Chr1**	Friedlin	E	0.27	SBRI		Myler *et al.*, *Proc Natl Acad Sci USA* **96**: 2902–2906 (1999)
22	▣	*Thermotoga maritima*	MSB8	B	1.80	TIGR	DOE	Nelson *et al.*, *Nature* **399**: 323–329 (1999)

516   From Biotechnology to Genomes: A Meaning for the Double Helix

Microbial genomes and chromosomes in progress (Searches available for some TIGR genomes)

Genome	Strain	Domain	Size (Mb)	Institution	Funding	Anticipated Completion
Aeropyrum pernix	K1	A	1.67	Biotechnology Center	NITE	Complete
Actinobacillus actinomycetemcomitans	HK1651	B	2.2	University of Oklahoma	NIDR	
Aspergillus nidulans		E	29	Cereon Genomics		
Bacillus anthracis	Ames	B	4.5	TIGR	ONR	
Bacillus halodurans	C-125	B	4.25	Japan Marine Science and Technology Center		1999
Bartonella henselae	Houston 1	B	2.00	University of Uppsala	SSF	1999
Bordetella bronchiseptica	RB50	B	4.9	Sanger Centre	Beowulf Genomics	
Bordetella parapertussis		B	3.9	Sanger Centre	Beowulf Genomics	
Bordetella pertussis	Tohama I	B	3.88	Sanger Centre	Beowulf Genomics	
Campylobacter jejuni	NCTC 11168	B	1.70	Sanger Centre	Beowulf Genomics	Complete
Candida albicans	1161	E	15	Sanger Centre	Beowulf Genomics	
Candida albicans (1.5X)	SC5314	E		Stanford	NIDR / NIH / Burroughs Wellcome Fund	
Caulobacter crescentus BLAST Search		B	3.80	TIGR	DOE	
Chlamydia pneumoniae	CWL029	B	1.23	UC Berkeley	Incyte	
Chlamydia pneumoniae		B	1.00	TIGR / University of Manitoba	NIAID	
Chlamydia trachomatis BLAST Search	MoPn	B	1.00	TIGR / University of Manitoba	NIAID	
Chlorobium tepidum BLAST Search	TLS	B	2.10	TIGR	DOE	
Clostridium acetobutylicum	ATCC 824	B	4.1	Genome Therapeutics	DOE	
Clostridium difficile	630	B	4.4	Sanger Centre	Beowulf Genomics	
Deinococcus radiodurans BLAST Search	R1	B	3.2	TIGR	DOE	Complete
Dehalococcoides ethenogenes		B		TIGR	DOE	
Desulfovibrio vulgaris		B	1.70	TIGR	DOE	
Dictyostelium discoideum Chr 2	AX4	E	7.0	University of Cologne/ University of Jena	DFG	
Dictyostelium discoideum Chr 6	AX4	E	4.0	Baylor College of Medicine/ Sanger Centre	NIH/ EU	
Encephalitozoon cuniculi		E	2.9	GENOSCOPE		
Enterococcus faecalis BLAST Search	V583	B	3.00	TIGR	NIAID	1999
Giardia lamblia	WB	E	12	Marine Biological Laboratory	NIAID	
Francisella tularensis	schu 4	B	2.00	European & North American consortium		
Halobacterium sp.	NRC-1	A	2.50	University of Massachusetts / University of Washington		
Halobacterium salinarium		A	4.0	Max-Planck-Institute for Biochemistry		
Klebsiella pneumoniae	M6H 78578	B		Washington University Consortium		
Lactobacillus acidophilus	ATCC 700396	B	1.9	Environmental Biotechnology Institute	Dairy Management, Inc. / California Research Foundation / Environmental Biotechnology Institute	
Lactococcus lactis		B		GENOSCOPE		
Legionella pneumophila		B	4.10	TIGR		

7 The Decryption of Life 517

Genome	Strain	Domain	Size (Mb)	Institution	Funding	Anticipated Completion
Leishmania major Chr3	Friedlin	E		SBRI		
Leishmania major Chr4	Friedlin	E	0.5	Sanger Centre	Beowulf Genomics	
Leishmania major Chr5,13,14,19,21,23	Friedlin	E		Sanger Centre/European Consortium	European Commission	
Leishmania major Chr27	Friedlin	E		SBRI		
Leishmania major Chr35	Friedlin	E		SBRI		
Listeria monocytogenes	EGD-e	B	3.2	EC Consortium	EC	2000
Methanosarcina mazei	Gö1	A	2.8	Goettingen Genomics Laboratory	Ministry of Lower Saxony for Science and Culture	1999
Mycobacterium avium BLAST Search	104	B	4.70	TIGR	NIAID	2000
Mycobacterium leprae		B	2.80	Sanger Centre	The New York Community Trust	
Mycobacterium tuberculosis BLAST Search	CSU#93 (clinical isolate)	B	4.40	TIGR	NIAID	1999
Mycoplasma mycoides subsp. mycoides SC	PG1	B	1.28	The Royal Institute of Technology, Stockholm & The National Veterinary Institute, Uppsala	SSF	1999
Mycoplasma pulmonis		B	0.95	GENOSCOPE		
Neisseria gonorrhoeae		B	2.20	University of Oklahoma	NIAID	
Neisseria meningitidis BLAST Search	MC58	B	2.30	TIGR	TIGR	
Neisseria meningitidis	serogroup A strain Z2491	B	2.30	Sanger Centre	Wellcome Trust	
Photorhabdus luminescens	TT01	B	5.0	GMP		2000
Plasmodium falciparum Chr 1,3,4,5,6,7,8,9,13 (isolate 3D7)		E	0.8	Sanger Centre	Wellcome Trust	
Plasmodium falciparum Chr 10,11 (isolate 3D7)		E	2.10	TIGR / NMRI	NIAID/ NIH/DOD	
Plasmodium falciparum Chr 12 (isolate 3D7)		E	2.4	Stanford University	Burroughs Wellcome Fund	
Plasmodium falciparum Chr 14 (isolate 3D7)		E	3.4	TIGR/ NMRI	Burroughs Wellcome Fund/ DOD	
Pneumocystis carinii	f. sp. carinii	E	7.7	Univ. of Cincinnati / National and International Consortium	NIAID	2004
Pneumocystis carinii	f. sp. hominis	E	7.5	Univ. of Cincinnati / National and International Consortium	NIAID	2004
Porphyromonas gingivalis BLAST Search	W83	B	2.20	TIGR/ Forsyth Dental Center	NIDR	1999
Pseudomonas aeruginosa	PAO1	B	5.90	University of Washington PathoGenesis	Cystic Fibrosis Foundation PathoGenesis	
Pseudomonas putida BLAST Search		B	5.00	TIGR	DOE	
Pyrobaculum aerophilum		A	2.22	Caltech / UCLA	ONR / DOE	
Pyrococcus abyssi	GE5	A	1.8	GENOSCOPE		Complete
Pyrococcus furiosus		A	2.10	Center of Marine Biotechnology / Univ. Utah	DOE	
Ralstonia solanacearum		B		GENOSCOPE		
Rhodobacter capsulatus	SB1003	B	3.70	University of Chicago / Institute of Molecular Genetics		1999
Rhodobacter sphaeroides	2.4.1	B	4.34	Univ. of Texas - Houston Health Science Center		

518 From Biotechnology to Genomes: A Meaning for the Double Helix

Genome	Strain	Domain	Size (Mb)	Institution	Funding	Anticipated Completion
Salmonella typhimurium	SGSC1412	B	4.80	Washington University Consortium		
Schizosaccharomyces pombe		E	14	Sanger Centre	Wellcome Trust	
Shewanella putrefaciens BLAST Search	MR-1	B	4.50	TIGR	DOE	
Shigella flexneri 2a	301	B	4.7	Microbial Genome Center	Chinese Ministry of Public Health	2000
Staphylococcus aureus BLAST Search	COL	B	2.80	TIGR	NIAID / MGRI	
Staphylococcus aureus	8325	B	2.80	University of Oklahoma	NIAID / MGRI	
Streptococcus mutans	UAB159	B	2.20	University of Oklahoma	NIDR	
Streptococcus pneumoniae BLAST Search	type 4	B	2.20	TIGR	TIGR / NIAID /MGRI	1999
Streptococcus pyogenes		B	1.98	University of Oklahoma	NIAID	
Streptomyces coelicolor	A3(2)	B	8.0	Sanger Centre / John Innes Centre	BBSRC/ Beowulf Genomics	2000
Sulfolobus solfataricus		A	3.05	Canadian & European Consortium		
Thermoplasma acidophilum		A	1.7	Max-Planck-Institute for Biochemistry		
Thermus thermophilus	HB27	B	1.82	Goettingen Genomics Laboratory	Ministry of Lower Saxony for Science and Culture	1999
Thiobacillus ferrooxidans	ATCC 23270	B	2.90	TIGR	DOE	
Treponema denticola		B	3.00	TIGR/ Univ. Texas		
Trypanosoma brucei Chr1		E		Sanger Centre	Wellcome Trust	
Trypanosoma b. rhodesiense BLAST Search	TREU 927/4	E	35	TIGR	NIH	
Ureaplasma urealyticum	serovar 3	B	0.75	U. Alabama / PE-ABI	PE-ABI / NIH / UAB	1999
Vibrio cholerae BLAST Search	serotype O1, Biotype El Tor, strain N16961	B	2.50	TIGR	NIAID	1999
Xanthomonas citri		B	5.00	Brazilian Consortium	FAPESP	2001
Xylella fastidiosa	8.1.b clone 9.a.5.c	B	2.00	Brazilian Consortium	FAPESP	2000
Yersinia pestis	CO-92 Biovar Orientalis	B	4.38	Sanger Centre	Beowulf Genomics	
Domain	A: *Archaea*			B: *Eubacteria*		E: *Eucaryote*

Nota bene: Other work under way should be added to this table, notably the work on Biovariant, L2/ 434/BU, Stanford, *Chlamydia* group, Ron Davis, *Candida albicans*, NIH and NIDR funding, Burroughs Welcome Fund, *Mycobacterium leprae*, Sanger Center, funding from the Heiser Program for Research in Leprost and Tuberculosis of the New York Community Trust, and for France, the *Association Française* Raoul Follereau and, until its disbanding, the GIP-GREG, *Rhodobacter capsulatus* (3.7 Mb and 134 kbs of plasmid) at the University of Chicago's Department of Molecular Genetics and Cell Biology, with funding from the DOE and the Tchec Republic Academy of Sciences , *Trypanosome brucei* (TIGR with funding from NIH, Sanger. Funding has been secured from the Wellcome Trust (Beowulf Genomics) to sequence the 1.05 Mb chromosome 1 of TREU 927/4 (Barell, Melville and Gull). The determination of the complete sequences of chromosome 2-6 + 9 is planned).

A Second Eukaryote Genome Sequenced

If the money lasts, the next yeast genome to be completed should be *Schizosaccharomyces pombe*. The genome of the yeast *S. pombe* has some 15 million base pairs in three chromosomes. Several projects have been set up aiming at its eventual complete sequencing (Table 28). At first, there was only the Cold Spring Harbor Laboratories *Schizosaccharomyces pombe* Sequencing Project, which had, for some time prior to this been working on the characterization of this genome. The sequencing program began in the laboratories of David Beach and Tom Marr who developed a high resolution physical map of the *S. pombe* genome[124]. It has minimal cover and a greater redundant number of cosmids. Another map was independently developed at the same time by Hoheisel, Mitzutami *et al.*[125]. Their map was the basis of a pilot sequencing project that began in the Lita Annenberg Hazen Genome Center at Cold Spring Harbor, starting on a site close to the telomere of chromosome II.

Another *S. pombe* sequencing project was initiated in 1995 on chromosome I by Bart Barrell and his colleagues at the Sanger Center. During 1995 and early 1996, over 80% of that chromosome was sequenced with funds from the Wellcome Trust. The European Commission now funds the Sanger Center as well as twelve other European laboratories to continue sequencing the genome of *S. pombe*. In 1997, work started on chromosome 2. The aim is to finish the *S. pombe* genome sequence within the next two years. It looks like a realistic one, especially now that a new large scale *S. pombe* sequencing program in Japan at the Medical Radiation Institute started in 1996 in collaboration with US groups, funded by the MHW and supervised by M. Yanagida.

Now that *S. cerevisiae* is completely sequenced and *S. pombe* almost finished (Table 29), complete sequences for the genomes of other yeasts of industrial or medical importance should follow. As A. Goffeau and others point out, "Such knowledge should considerably accelerate the development

---

[124] T. Mizukami *et al.*, *Cell*, vol. 73, no. 109, 1993, p. 121.

[125] J.D. Hoheisel *et al.*, *Cell*, vol. 73, no. 109, 1993.

Table 28 Laboratories involved in the European *Schizosaccharomyces pombe* genome sequencing project

Sanger Center, Wellcome Trust, Cambridge, B. Barrell (coordination of distribution of clones) analysis sequencing	Imperial Cancer Research Fund, London, Paul Nurse (Scientific coordinator)	Katholieke Universiteit Leuven, G. Volckaert (control of sequencing quality
Unit of Physiological Biochemistry, UCL, Louvain-la-Neuve, A. Goffeau, sequencing	Laboratory of Genetic Recombination, Faculty of Medicine, Rennes, F. Gallibert, sequencing	INRA-CBAI, Lab. of Cellular and Molecular Genetics, Thivernahal Grignon, C. Gaillardin, sequencing
GATC GmbH, Konstanz, Germany, T. Pohl, sequencing	Biotechnologishe und Molekular biologishe, Wilhelmsfeld, DE, M; Rieger, sequencing	Quiagen GmbF, Hilden, DE, A. Düsterhoft, sequencing
Max Plank Inst. for Molecular Genetics, Berlin, DE, R. Reinhart, sequencing	Dept. of cellular and genetic biology, Malaga, ES, J. Jimenez Martinez, sequencing	University of Salamance, Inst. of Biochemical Microbiology, ES, S. Moreno
Exeter University, Dept. of Biological Sciences, UK, S. Aves		

Table 29 *Schizosaccharomyces pombe*
(data generated 25 January 1999)

	Size Mb	Redund %(estm)	Finished %(adj'd)	Finished (total bp)	Unfinished (total bp)	1999	1998	1997	1996	1994
total	14	9.4	80.6	12,451,455	0	416,570	4,207,912	2,742,808	1,055,822	1,211,978

of more productive strains and the search for badly needed antifungal drugs" [126].

It is for this reason that the current sequencing of *Candida albicans* has considerable importance. Initially supported by limited industrial

---

[126] A. Goffeau *et al.*, "Life with 6000 genes", *Science*, vol. 274, 25 October 1996, p. 566.

funding[127], a new project backed by the NIDR, the NIH and the Burroughs Wellcome Fund was set up at Stanford.

Candida albicans is one of the most common human pathogens. It causes a large number of infections in people of generally good health from mucosal infections and dangerous systemic infections in the immunodeficient. Oral and esophageal infections are very common in people with AIDS. Few medicines are really of use against fungal infection, and that use is limited by side-effects. The objective of the project is to sequence one and a half times the haploid genome from a complete genomic DNA library from strain SC5314, made available without restrictions by the firm of Bristol Myers Squibb. Sequencing to such an extent should provide at least partial sequences for most of the genes of *C. albicans* and allow them to be identified by their similarities to genes of other species, particularly *S. cerevisiae*. The identification of most of the *C. albicans* genes should greatly increase our understanding of the organism and facilitate the development of new therapies for fungal infection.

We must add that the Sanger Center is carrying out a pilot sequencing project on the *Candida albicans* strain 1,161 genome in collaboration with Prof. Duncan Shaw at the University of Aberdeen. A set of 10 cosmids is being sequenced for this pilot project. The project is funded by Beowulf Genomics.[128] Sequenced cosmids from the Sanger Center will be submitted to the EMBL database as soon as they have been analyzed and annotated. All unfinished sequence contigs over 1,000 bp are available from FTP as individual files or as database in FASTA format. Sequencing is also in progress for chromosome III.

"Complete genome sequencing may be unnecessary when a yeast or fungal genome displays considerable synteny (conservation of gene order) like that of *S. cerevisiae*. For instance, studies on *Ashbya gossypii* (a filamentous fungus that is a pathogen of cotton plants) have revealed that

---

[127] W.S. Chu, B.B. Magee, P.T. Magee, *J. Bacteriol.*, vol. 175, no. 6637, 1993. Http://alces.med.umn.edu/Candida.html

[128] For more details on the mapping project see: "A *Candida albicans* Genome Project: Cosmid Contigs, Physical Mapping and Gene Isolation.", *Fungal Genetics and Biology,* 21, 1997 Article no. FG970983, pp. 308–314.

most of its ORFs show homology to those of *S. cerevisiae* and that at least a quarter of the clones in an *A. gossypii* genomic bank contain pairs or groups of genes in the same order or relative orientation as their *S. cerevisiae* counterparts[129]. This gives considerable hope for the rapid analysis of the genomes of a large number of medically and economically important fungi through the use of the *S. cerevisiae* genome sequence as a paradigm. However, this optimism is tempered by the lack of apparent synteny between the *S. cerevisiae* and *S. pombe* genomes. This is perhaps not surprising, as the two species probably diverged from a common ancestor some 100 million years ago"[130, 131].

Towards the Entire Genome Sequence of the First Multicellular Eucaryote: The C. elegans Genome

Despite the difficulties of technical as well as political and budgetary problems that affect large sequencing programs, several systematic sequencing programs, which are proceeding well, have been launched to sequence complex (multicellular) eucaryote genomes.

The first such program was on *Caenorhabditis elegans*. A collaboration between the teams of John Sulston and Alan Coulson at the MRC Laboratory of Molecular Biology in Cambridge (UK) and Robert Waterston at the Washington University School of Medicine in St Louis (USA), it is very far advanced and has been completed during 1999[132]. As of January 1999, a total of 84,160,365 base pairs have been sequenced (finished) by the two

---

[129] R. Altman-Jöhl and P. Philippsen, *Molec. Gen. Genet.*, 250, 1996, p. 69.

[130] P. Russell and P. Nurse, *Cell*, 45, 1986, p. 781.

[131] A. Goffeau *et al.*, "Life with 6000 genes", *Science*, vol. 274, 25 October 1996, p. 566.

[132] **10 December 1998**: The first sequence of an animal genome is *essentially* complete. The article titled "*C. elegans*: Sequence to biology" was published in *Science*, 11 December 1998: 2011 Funded by the Medical Research Council and America's National Institutes of Health, the Sanger Center and the Genome Sequencing Center at St Louis have completed a fifteen year project to sequence the complete genome of the nematode worm *Caenorhabditis elegans*.

This completed gene sequence gives scientists and health practitioners world-wide valuable information to aid the study of the human body in health, as well as in illness and may, for example, lead to new treatments for disease.

groups, with another 38,380,793 to go, so more than 80% of the 100 Mb genome has been sequenced (Tables 30 and 31). The progress they have made has allowed some conjecture as to the contents of the sequence; 16,610 genes are expected to be found, only a little more than double the number of genes estimated for yeast, despite the fact that the *C. elegans* genome is eight times longer than that of *S. cerevisiae*. The lower density of coding sequences compared to that found in unicellular organisms reflects an increase in the size and number of introns (about six per gene) and the length of the regions between genes, some of which can stretch for dozens of kilobases.

Table 30 *Caenorhabditis elegans*
(Data generated January 25 1999)

	Size Mb	Redund %(estm)	Finished %(adj'd)	Finished (total bp)	Unfinished (total bp)	1999	1998	1997	1996	1995
total	100	0.9	83.4	84,160,365	38,380,793	58,864	13,956,839	14,987,301	30,172,025	17,574,277

There are some systematic differences between the genome regions. For example, the genes on one region of the X chromosome seem to be less dense and expressed to a lesser extent than those in autosomic (non-sex-gene) clusters. There is also a very unequal distribution of certain repetition families between the chromosomes. Forty six percent of the predicted proteins are similar to non-nematode proteins with known function and many more seem to be similar to each other, defining new protein families. It is expected that 185 RNAs will be found.

Introducing the press conference at the Royal Society, Professor George Radda, MRC Chief Executive said: "This is an exciting day for British science. The first complete genomic sequence of a complex organism — an animal, with which the human body can be compared, promises to open a new chapter in the understanding of human health and disease."

Lord Sainsbury, Minister for Science said: "The completion of this project is a terrific scientific achievement. Not only is it an example of international partnership and co-operation with strong British involvement, but a world scientific first — the first multicellular animal to be completely sequenced. This research will ultimately contribute towards interpretation of other genomes, including the human, and help to ensure that we revolutionise healthcare."

Table 31 Status of the *Caenorhabditis elegans* genome sequencing
(data generated 25 January 1999)

	Size (Mb)	Finished %	Finished (total bp)	Unfinished (total bp)	1999	1998	1997
Chr. I	13.4	145.7	19530397	0	7216369	3586741	3461786
Chr. II	16.1	125.5	20209376	0	7213723	1713844	2640806
Chr. III	11.8	124.2	14653901	0	4996366	1322007	856668
Chr. IV	16.0	122.8	19643208	0	7148159	2356282	2244679
Chr. V	21.3	124.5	26524803	0	9016740	3012723	4550882
Chr. X	18.5	100.7	18634847	0	2209441	982266	1146893
Subtotal:	97.1	122.8	119196532	0	37800798	12973863	14901714
Chr. ??	n.a.	n.a.	3179956	0	579995	982976	85587
Total:	97.1	126.0	122376488	0	38380793	13956839	14987301

As pointed out by Jonathan Hodgkin, Ronald Plasterk and Robert Waterston in their article "The nematode *Caenorhabditis elegans* and its genome", "It is reasonable to expect that the complete genome sequence *of C. elegans* will provide, in some sense, the basic formula for constructing a multicellular animal, in much the same way as the complete sequence of *S. cerevisiae* will reveal the basic ingredients for making and maintaining a eucaryotic cell. The phylogenetic position of *C. elegans* is convenient in this regard because it appears that nematodes diverged at an early point from the rest of the metazoan radiation. Consequently, they provide a universal out-group for the rest of the animal kingdom[133]. What this means is that if a gene can be identified both in *C. elegans* and in any other kind of animal, whether it be vertebrate, insect or mollusk, then it must also have been present in the common ancestor of all animals.

With time, it should, therefore, become clear how much of the *C. elegans* genome is devoted to this basic animal construction kit and

---

[133] A. Sidow and W.K. Thomas, *Curr. Biol.*, 4, 1994, p. 596.

how much is associated with specializations that are unique to the phylum *Nematoda* or to *C. elegans* itself. For example, the compactness of this genome may be correlated with idiosyncratic features such as operons, which seem to be absent from larger genomes. Other molecular or biological properties may turn out to be unique to nematodes in general, but absent from other animal groups. Such properties will also be valuable to discern because nematodes are a large and important animal group in their own right. Many nematode species have considerable medical or economic significance as agents of disease and as major agricultural pests. However, it is already clear that a large part of the genome is doing universal things, so much of what is learned from *C. elegans* will also apply to all multicellular organisms" [134].

The extent of current understanding of the biology, ontogenesis and physiology of *C. elegans* provides an overall explanation of what the biology community expects of the forthcoming completion of the sequence for the genome in question. This makes *C. elegans* the first completely sequenced pluricellular eucaryote and confirms its place as the first pluricellular eucaryote model organism. In order to illustrate the potential usefulness of *C. elegans* as a model system, BLASTX has been used to show that 32 of the 44 genes involved in human hereditary diseases identified by positional cloning have significant correspondences with *C. elegans* genes. Table 32 gives a few examples. In certain cases, such as the genes with a possible link to Alzheimer's, the *C. elegans* gene is the only gene that correlates in the databases.

The Genomes of Mycobacterium tuberculosis and Plasmodium falciparum

As work ended on sequencing the *S. cerevisiae* genome, the Wellcome Trust financed a project to sequence two of the most serious pathogenic organisms

---

[134] J. Hodgkin, R.H.A. Plasterk, R.H. Waterston, "The nematode *Caenorhabditis elegans* and its genome", *Science*, vol. 270, 20 October 1995, p. 414.

Table 32 Correspondence between some human genes and *Caenorhabditis elegans* genes

Human genes	C. elegans genes
+ *Alzheimer (S182)*	Spe-4
+ *Achondroplastic dwarfism (FGFR3)*	ZK938.5
+ *Amyotrophic lateral sclerosis (SOD1)*	F55H2.1
+ *Cystic fibrosis (CFTR)*	DH11.3
* *Predisposition to breast and ovarian cancer (BRCA1)*	TO2C1.1
+ *Duchenne's muscular dystrophy (DMD)*	YK26d3.5
− *Lowe's occulocerebrorenal syndrome (OCRL)*	C50C3.7
+ *Congenital dominant myotonia (CLC1)*	EO4F6.11
+ *Polycystic Kidney Disease 1 (PKD1)*	ZK945.9
* *Tuberous sclerosis (TSC2)*	T14B4.7
+ *Wilson's disease (ATP7B)*	Yh29a9.5
+ *Klein-Waardenburg syndrome (PAX3)*	F26C11.2
+ *WilM Tumour (WT1)*	F54H5.4
+ *Type 1 neurofibromatosis (NF1)*	CO7B5.1
+ = strong correspondence − = moderate correspondence * = weak correspondence	

— those responsible for tuberculosis and malaria. The Wellcome Trust has dedicated over US$2 million[135] to *sequencing Mycobacterium tuberculosis*. In 1997, an estimated 7.25 million people developed tuberculosis (TB)[136]. Around one in every three people in the world is infected with *Mycobacterium tuberculosis*, the bacterium that causes TB, and each of these has a 10% lifetime risk of progressing from infection to clinical disease. TB occurs in every country of the world, but the highest incidence is found in Asia and

---

[135] The fact that the NIH had planned, in 1996, to set up an *M. tuberculosis* sequencing project, knowing that a similar programme was under way, highlights the problems of the international coordination of genome research, especially when direct interest is at stake.

[136] *WHO The World Health Report*, 1998 (WHO, Geneva, 1998). 2. WHO Global tuberculosis control; *WHO report* 1998 (WHO, Geneva, 1998). 3. WHO *Treatment of tuberculosis: guidelines for national programmes* (WHO, Geneva, 1997). 4. Murray, C. & Lopez, A. Global burden of disease (WHO, Geneva. 1996). Pablos-Mendez, A. *et al. New Engl. J. Med.* 338, 1641–1649 (1998).

Africa, while in recent years it has re-emerged as a major public health issue in Russia and other Eastern European countries. This conspires to make TB one of today's most pressing global health problems. Resistance of *M. tuberculosis* to antibiotics is a result of man-made amplification of spontaneous mutations in the genes of the bacterium. A high prevalence of multidrug-resistant TB (MDR-TB) is the result of poor therapeutic management of TB cases. Patients with MDR-TB can be extremely difficult to cure and require antibiotics that are much more toxic and expensive than are needed for drug-sensitive strains. Dramatic outbreaks of MDR-TB in HIV infected patients in the USA and in Europe have focused international attention on the emergence of multidrug-resistant strains of *M. tuberculosis* and their threat to clinical management and control programs.

In 1994, the WHO/IUATLD Global Project on Anti-tuberculosis Drug Resistance Surveillance commenced with the aim of measuring the prevalence of *M. tuberculosis* drug resistance in several countries worldwide. Representative surveys from 1994–97 in 35 countries identified several "hot spots" where MDR-TB prevalence is high and threatens control programs. For example, in Latvia, 22% of all TB patients were carrying strains resistant to at least isoniazid and rifampicin, the two most potent antituberculosis drugs.

Researchers hope to develop new approaches to the treatment of the illness, but are held back by the slow growth rate of the organism. In fact, there is little information on the proteins and metabolism of this bacteria and researchers are searching for new treatments "in the dark". The sequence will allow them to identify all the proteins of the bacteria and develop rational targets. The *M. tuberculosis* genome was estimated at 4.5 million base pairs, which is small compared to the 14 million base pairs of yeast (counting the ribosomal DNA), but still represents a challenge to sequencers. There is an imbalance in the composition of the bases. Two-thirds of the bacteria's genome is made of cytosine and guanine bases, and this makes sequencing very difficult since long strands of DNA can take on awkward shapes. The Wellcome Trust, which funds the Sanger Center, has assigned US$2 million to Barrel's group and another 100,000 dollars to Cole's group at the Institut Pasteur, which collaborated on this project. In a report on p537 [137], Stewart T. Cole of the Institut Pasteur (Paris, France), and colleagues present the complete DNA sequence of H37Rv, the best-characterized strain of the organism. The sequence and annotation has been deposited in the public databases (ID MTBH37RV / accession number AL123456). The sequence is 4,411,529 bp long with an overall GC content of 65.6% and 3,924 predicted protein coding genes.

This slow-growing pathogen is amongst the most recalcitrant in terms of clinical treatment; moreover, the recent emergence of strains highly resistant to antibiotics currently available has caused great concern. There is, therefore, particular interest in what the DNA sequence may reveal about the biology of the organism.

A remarkably large proportion of the organism's coding capacity is devoted to the production of enzymes involved in the synthesis and breakdown of fats. The genome also reveals two new families of proteins (PE and PPE) with a repetitive structure that may represent a source of antigenic variation. An intensive search for clues to aid the design of new therapeutic agents may now begin.

M. tuberculosis did indeed present a challenge, but it is the Wellcome Trust's other target, *Plasmodium falciparum*, which is the real adventure. The problem is similar to that with *M. tuberculosis* in that over 80% of the genome consists of the other two bases, adenine and thymine. When the parasite's genes are cloned, the DNA strands tend to recombine, clouding the information. In addition, the genome is much larger at 27 million base pairs, twice the size of the yeast genome. For this reason, the Wellcome Trust is starting slowly with a pilot project to sequence chromosome III of *P. falciparum* at about 1.2 million bp. Barell's team, in collaboration with Chris Newbold and Alastair Graig of the Molecular Parasitology Group (Institute of Molecular Medicine, Oxford, GB) has overcome the problem of such high levels of A + T.

The Malaria Genome Project now in progress has been undertaken by an international consortium of scientists and funding agencies. The goal of

---

[137] S.T. Cole, R. Brosch, J. Parkill *et al.*, "Deciphering the biology of *Mycobacterium tuberculosis* from the complete genome sequence", *Nature*, vol. 393, 11 June 1998, pp. 537–544.

the project is to produce the sequence of the genome of the human malaria parasite *Plasmodium falciparum* (clone 3D7). The project work in the USA is funded by the National Institute of Allergy and Infectious Disease, the Burroughs Wellcome Fund and the US Department of Defense. The project work in the UK is funded by the Wellcome Trust. High-throughput DNA sequencing and genome analysis is being carried out at three centers: The Institute for Genomic Research/Naval Medical Research Institute (USA), The Sanger Center (UK) and Stanford University (USA). Additional investigators at other sites are contributing to the goal of the project by developing alternative methods of library construction, producing genetic and physical maps and developing databases and bioinformatics resources. In parallel, informatics personnel at the National Center for Biotechnology Information are developing a general malaria website to allow common access to the sequence and other relevant information on malaria.

The scientists participating in the Malaria Genome Project have agreed to release *P. falciparum* sequence information (raw shotgun reads and assembled contigs in progress) as rapidly as possible as a resource for the entire malaria research community. These data releases do not constitute scientific publication. In fact, these data are preliminary and will contain errors and possible contamination from other species; e.g., yeast and *E. coli*. It is anticipated that the *P. falciparum* sequence data will assist scientists in their research, particularly in the search for genes and the studies of the genes' biological functions. They are encouraged to share their results with the members of the Stanford DNA Sequencing and Technology Center in order to achieve optimal annotation of the *P. falciparum* genome.

The Stanford DNA sequence and Technology Center also contribute to the *Chlamydia* genome project. The goal of the *Chlamydia* genome project is to determine the DNA sequence of the chromosome of *Chlamydia trachomatis*, serovar D (D/UW-3/CX [138]), *Trachoma biovar* and L2/434/BU, LGV biovar. This project is a collaborative effort involving scientists at the

---

[138] The genome consists of a 1.05 Mb chromosome and a 7 Kb plasmid and has G+C content of approximately 45%.

University of California at Berkeley (Dr. Richard Stephens' group, Program in Infectious Diseases and the School of Public Health) and Stanford University (Ron Davis, Departments of Biochemisry and Genetics, Sue Kalman, Project Leader, Jun Fan, Research Associate and Rekha Marathe, Research Associate). It is supported by USPHS/NIH/NIAID grant number AI39258.

Europe and Its Programs

The success enjoyed by the European consortium in sequencing chromosome III of yeast (especially as a pilot project for keeping Europe in the world gene race and gold rush) was very encouraging. By showing that the distribution network approach could work, the yeast program stimulated and contributed to the birth of several sizable new systematic sequencing programs (Table 33). The yeast program alerted scientific policymakers to the importance of this new field for biotechnology and European industries and economies. In particular, the biotechnology programme (BIOTECH 2) gave a great place to genome sequencing. It implemented research, technological development and demonstration activities in line with the priorities established by the Council Decision of 15 December 1995 (Official Journal No L 361, 31 December 1994). Covering the period from December 1994 to December 1998, it was one of the three specific programs implementing the life sciences and technologies domain of the fourth RTD framework program. The total EU contribution for research activities in the biotechnology sector under the fourth RTD framework program is ECU 595.5 million.

The BIOTECH Program

BIOTECH 2 aimed to improve our basic biological knowledge of living systems and increase productivity in a sustainable fashion, with respect to applications in agriculture, industry, health, nutrition and the environment. It considered the ethical and socio-economic implications of biotechnology.

The areas of research covered by the Biotechnology program were:

> 1. Cell factories 2. Genome analysis 3. Plant and animal biotechnology 4. Cell communication in neurosciences 5. Immunology, transdisease vaccinology 6. Structural biology 7. Prenormative research, biodiversity and social acceptance 8. Infrastructures

Concerning horizontal activities:

- Demonstration activities in biotechnology • Ethical, legal and social aspects (ELSA) • Public perception • Socio-economic impacts

Four calls for proposals were made under the program with different areas targeted at given times so as to address key issues and enable scientists to plan their participation. For instance, in the first call, only specific areas and sub-areas were open for funding applications. Likewise, the annual budgetary allocation varied depending on the areas open. Looking at the outcome of the calls, 60 projects were funded in the first call with a total EC contribution of ECU 73 million, while the second call was larger with 150 projects funded for a total of ECU 206.6 million. In 1997, 96 projects received a total of ECU 113.6 million following the third call for proposals. In 1998, 154 projects will get ECU 138 million in total after the fourth and final call. Below are the key figures from the Biotechnology Program (1994–98).

Since 1995, a total of 456 projects have been funded, with a total contribution by the EU of ECU 533.23 million. This represents an average EU contribution of ECU 1.17 million per project. Industrial participation has risen to a satisfactory level with two thirds of the projects involving at least one company. In parallel, BIOTECH 2 has funded training activities by awarding almost 100 training grants per year and funding a number of workshops.

Table 33 EU genome research activities
(source DGXII, European Commission)

Title	Program	N° of lab	Period of execution	EC financial contribution Ecu
Sequencing of chromosome III from Yeast	BAP	35	89-90	2,635,000
Sequencing of the Yeast genome (chromosome II and XI)	BRIDGE	31	91-93	5,060,000
Molecular identification of new plant genes (focused on the *Arabidopsis* genome)	BRIDGE	27	91-93	4,654,000
Establishment of a complete physical map and strategic approach to the sequencing of *the Bacillus subtilis* genome	SCIENCE	5	89-91	609,000
A complete physical map of *the Drosophila melanogaster* genome	SCIENCE	3	88-93	718,000
Functional and structural analysis of the mouse genome	SCIENCE	3	89-92	1,278,000
Development of a genetic and physical map of the porcine genome (PigMap)	BRIDGE	11	91-93	1,200,000
Eucaryote genome organization: repeated DNA elements and evolution in the genome of *Caenorahbditis*	SCIENCE	2	91-93	250,000
Human Genome Analysis Program	HGAP	> 90	91-92	11,700,000
Genetic/linkage mapping	BIOMED I	25	93-96	3,500,000
Technology development/applications of human genome analysis	BIOMED I	76	93-96	9,052,875
Physical mapping	BIOMED I	7	93-96	1,149,000
DNA sequencing	BIOMED I	19	93-96	4,165,000
Data handling and databases	BIOMED I	47	93-96	5,432,910
HUGO: Single Chromosome Workshop	BIOMED I	1	93-96	700,000
Gene mapping and genome analysis	BIOMED 2	80	96-2000	10,599,923
Gene function and interaction	BIOMED 2	103	96-2000	6,500,081
Role of gene and gene products in disease Etiology/pathogenesis	BIOMED 2	148	96-2000	10,129,359
Somatic gene therapy	BIOMED 2	36	96-2000	6,442,166
Information management	BIOMED 2	13	96-2000	2,350,000
HUGO/workshop organization	BIOMED 2	5	97-2000	300,000
Sequencing of the yeast genome (chromosomes VII, X, XIV, and XV + 1/3 rd of IV and XII	BIOTECH I	80	93-95	12,100,000

Table 33 continued

Title	Program	N° of lab	Period of execution	EC financial contribution Ecu
Sequencing of the *Bacillus* genome	BIOTECH I	116	93-95	2,870,000
European Scientists Sequencing *Arabidopsis* (ESSA)	BIOTECH I	22	93-95	5,750,000
Development of genetic and physical marker maps of the bovine genome (BovMap project)	BIOTECH I	31	93-96	1,200,000
Genome mapping informatics infrastructure (Gemini) (pig and bovine mapping)	BIOTECH I	2	93-95	300,000
Mapping of Apple	AIR 1	10	93-96	1,900,000
Interesting genes of *Prunus*	AIR 2	~10	under neg.	
Experimental pilot study for a European cooperation on gene function search in Yeast	BRIDGE	6	93-97	800,000
Arabidopsis genome sequencing project	BIOTECH II	20	96-98	6,038,000
Sequencing of the *Bacillus subtilis* genome	BIOTECH II	15	96-98	2,013,000
Finishing the yeast genome sequencing (MIT-DNA)	BIOTECH II	4	96-99	410,000
EUROFAN I: European Network for the Functional Analysis of yeast genes discovered by systematic DNA sequencing - EUROFAN	BIOTECH II	144	96-98	7,320,000
Systematic function analysis of *Bacillus subtilis* genes	BIOTECHII	16	95-98	3,199,000
The pig gene mapping project (PigMap II) - identifying trait genes	BIOTECH I	17	94-96	1,665,400
A comprehensive molecular resource for quantitative Trait Loci mapping in poultry (ChickMap)	BIOTECH II	7	95-98	1,614,800
A search for the function of MYB transcription factors in plants	BIOTECH II	6	95-98	1,122,998
Arabidopsis insertional mutagenesis for functional analysis of sequences identified by ESSA	BIOTECH II	6	95-98	1,320,000
Construction of a resource protein database for the *Arabidopsis* plasma membrane	BIOTECH II	7	95-98	1,625,000
Construction of protein linkage maps between yeast *Saccharomyces cerevisiae* nuclear proteins involved in RNA processing	BIOTECH II	6	95-98	1,003,000

Table 33 continued

Title	Program	N° of lab	Period of execution	EC financial contribution Ecu
Methods and software for evolutionary analysis of genome sequence data	BIOTECH II	6	96-99	1,225,000
Characterization of regulatory genomic regions. Development of database and sequence analysis tools	BIOTECH II	6	96-99	600,000
MITBASE: a comprehensive and integrated mitochondrial database	BIOTECH II	7	96-99	650,000
Sequence of divisions 1-3 of the *Drosophila* genome	BIOTECH II (IInd call)	7	96-99	3,500,000
European *Schizosaccharomyces pombe* genome sequencing project	BIOTECH II (IInd call)	13	96-99	6,901,000
Completion of the genome of *Sulfolobus solfataricus*	BIOTECH II (IInd call)	4	96-99	1,050,000
EUROFAN II: European Network for the functional analysis of yeast genes discovered by systematic DNA sequencing – phase 2	BIOTECH II (IIIrd call)	79	97-99	7,485,000
Comparative analysis for the understanding of molecular evolution. Emphasis on gene families, gene duplications and functional restrictions	BIOTECH II (IIIrd call)	5	97-2000	1,290,000
Using grids of PNA oligomers for quantitative analysis of complex variations in the expressions pattern of all genes of the yeast *Saccharomyces cerevisiae*	BIOTECH II (IIIrd call)	2	97-2000	428,000
Transposon mutagenesis in rice for the identification of agronomically important genes in cereals	BIOTECH II (IIIrd call)	6	97-2000	1,360,000
Proteonme analysis group in Europe: 2D gel electrophoresis of the yeast *Saccharomyces cerevisiae* proteins	BIOTECH II (IIIrd call)	8	97-2000	2,642,000
Sequencing of 3.8 MB of *Leishmania* major genome	BIOTECH II (IVth call)	12	98-2000	2,305,000
Sequencing of the *Rhizobium meliloti* genome (3.7 Mb)	BIOTECH II (IVth call)	6	98-2000	2,442,000
Sequencing of 1 Mb of the *Dictyostelium* genome	BIOTECH II (IVth call)	13 ?	98-2000	750,000
Sequencing of the *Listeria monocytogenes* genome (3.15 Mb)	BIOTECH II (IVth call)	10	98-2000	2,130,000
	TOTAL			175,434,512

Concerning the genome activities the objectives were stated as below:

Area 2 - Genome Analysis

2.1 Sequencing Objectives:

To secure the Community's strategic position through continuing EU sequencing initiatives on the Bacillus subtilis, Yeast *and* Arabidopsis *genomes, assuring high quality standards. These activities will be planned and implemented in close coordination with other international efforts and are expected to lead, by the end of the program, to the full description of the yeast and* B. subtilis *genomes while almost 20% of the* Arabidopsis *genome will be completed.*

Research tasks:

Regions of the *Arabidopsis, B. subtilis* and yeast genomes will be sequenced as will other small (less than 500 Mb) genomes of biotechnological interest. The work will be carried out by highly integrated networks with laboratories or consortia capable of producing at least 100 Kb/year reimbursed to a maximum of 1.6 ECU/bp. Appropriate quality controls will be implemented to aim at an error rate below 1/10.000 bp. The networks will be fully self-sufficient both in DNA distribution and coordination and a unique centre for data collection, assembly and analysis should be created for each genome. The following goals have been identified as appropriate:

- 2.1.1 Sequencing of contiguous regions spanning 10,000 Kb of the *Arabidopsis* genome. DNA coordinator(s) should be responsible for contigs of at least 2,000 Kb. The network is welcome to include appropriate cDNA sequencing activities.

- 2.1.2 Sequencing of 2,000 Kb of the *B. subtilis* genome. Although in view of the international agreements on the subdivision of the genome, the full sequence output does not need to be contiguous, individual participating laboratories are bound to sequence contigs.

- 2.1.3 Sequencing of 2,000 Kb of the yeast genome. The network will concentrate on the finalisation of the DNA sequence of the entire genome by sequencing contiguous regions of chromosomes whenever needed to complete the sequence (i.e., to compensate for other unachieved international initiatives).

- 2.1.4 Physical mapping and sequencing of other small (less than 500 Mb) genomes of biotechnological interest to be carried out on a limited basis: Construction of appropriate organised libraries preparing possible future sequencing initiatives based on particularly innovative technologies and/or particularly promising organisms for biotechnological applications. Sequencing work will be carried out on a pilot scale.

2.2 Function search

Objectives

Unknown genes sequenced by the European networks in the preceding programmes will be analysed, aiming at the identification of their role and looking, in particular, for genes encoding function with an applied potential in biotechnology. Development of new tools and methodologies for function search.

Research tasks

2.2.1 Resource consortia

2.2.2 Function-driven analytical networks

The function-driven analytical networks will be constituted by laboratories providing a critical mass of complementary research approaches geared towards the identification of gene families of interest. Although they will have full access to the toolkits produced in the resource consortia, picking up, for instance mutants, with appropriate preliminary phenotypes, they might be compelled to construct additional resources specifically adapted to their aims. Analytical networks are expected to be active, for example, in the function search of:

- genes directly relevant to industrial processes: flocculation, secretion, ethanol resistance, fermentation properties, etc;
- genes involved in transcriptional regulation;
- genes involved in the faithful replication, segregation, recombination, mutation and repair of DNA.

2.2.3 Heterologous expression networks

The function search toolkits set up should also be exploited to analyse genes from higher organisms by complementation studies using genomic or cDNA libraries as appropriate. Although heterologous expression networks are expected to carry on function searches, systematically taking advantage of all mutants available for complementation from the resource consortia, individual laboratories interested only in specific genes (or families) should introduce their proposals jointly with the corresponding thematic analytical network.

2.2.4 New methods for genome analysis

Development of new tools for function search studies (in particular in higher model organisms); for instance: transposon tagging, antisense inhibition, molecular scissors, plantibodies.

Synergies with other specific programs

Although research on the human genome (i.e., mapping, sequencing, analysis of gene function and regulation and determination of genetic/ multifactorial disease) and related model animal genomes (i.e., mouse) will be carried out under the specific programme on biomedicine and Health, a few opportunities are open here. Laboratories wishing to express human cDNAs in microbial models can apply under 2.2.3. Also, comparative analysis of human genes is expected to be carried out under 2.3.

2.3 In *silico* (computer based) comparative analysis

Objectives

New tools for comparative analysis will be provided assuring that the existing DNA sequence information is properly exploited to ensure the most productive cross fertilization among various model genomes under study and to understand molecular evolution at the level of genomes and individual genes.

Research tasks

Development of innovative approaches and algorithms for comparative analysis to determine alignments, conserved structural or regulatory regions/patterns, phylogenic trees, etc. Interspecies comparison of functional/structural domains, entire genes and genomic regions which constitute a key to understanding molecular evolution patterns and how they have influenced the biological structure of living organisms. Databases built up in this "genome analysis" area as well as relevant databases from area 8 will constitute the core resource upon which projects under this sub-area 2.3 will be established.

Area 3 - Plant and Animal Biotechnology

3.1 Plant molecular and cellular biology

Objectives

Progress with plant characteristics is dependent on the discovery of new genes and on the further study of genomic organisation, gene functions and signalling or regulatory circuits. This recognized need has recently brought plant geneticists and molecular biologists to the heart of an international endeavour leading to rapidly expanding lists of available genes, gene constructs, transgenic plants and elucidated pathways. Before this new knowledge can be put to the service of plant breeders, substantial gaps remain to be filled.

Topics of particular importance are as follows: developmental biology, extrapolation of knowledge acquired with model genomes, quality characteristics, source-link relationships, biosynthetic pathways, functional analyses, responses to stress, transduction cascades from environmental signals to biochemical reactions, resistance traits and corresponding genes, microorganisms and lower organisms interacting with plants, general and molecular microbiology, etc.

In addition to the above research priorities, an important effort of integration is required to combine complementary disciplines and techniques and eventually bridge basic science and applications. This is where this program is expected to produce specific advantages by bringing into projects a number of key players capable of associating independently produced results which may bear on a common application. As well as shedding new light on some of the above topics, the projects will have to be particularly well designed to provide cross-fertilization of academic and engineering approaches over the background of anticipated social expectations.

The following examples of combined approaches served as guidelines:

- Modern eukaryotic biology (genome analysis, protein engineering, biocomputing, transgenic models of gene expression, etc.) linked with the most rigorous investigations at genetic, biochemical or physiological levels.
- Models (particularly *Arabidopsis thaliana*) producing data and technology for use in, for example, rice or maize genome research and, more generally, in the genetics of other agricultural species important for Europe through optimal communication between the different genome networks.
- Combination of product-linked research (gene regulation and product targeting, cellular transport and partitioning, interaction of sugar and nitrogen metabolisms, sink-source relationships, polymerization and storage) and technology-driven approaches favoured in industry which rest on structural work and chemical/biochemical analysis.
- Advances in gene cloning and mapping, genetic screening technology, reproductive biology, seed physiology, transgene expression, disease resistance strategies, all put to the service of plant breeding or genetic diversity surveys.
- Regulation of morphogenesis through molecular signals and genes; e.g., in light of new models of intra- and inter-cellular matrix formation, and their changing composition and structure leading to pattern formation (cell wall activities, short-distance signal transduction, differentiation and cell cycle, etc.).

- Linkage of mechanistic studies of plant-microbe interactions with dynamic studies of microbial populations, colonization, genetic acquisition of resistance, microbial identification and taxonomy and nutrient up-take cycles.

Proposals originally incorporating large-scale coordination of numerous participants might be given a pivotal role in implementing fully integrated action. Large-scale coordination of this type is considered advantageous since it facilitates the management, and thus the implementation, of such projects.

Research tasks

3.1.1 Molecular genetic maps
Molecular genetic maps are required in order to locate and select genes of agricultural importance as well as QTLs (quantitative trait loci) supporting useful phenotypes. The work will consist in using maps for gaining rapid access to novel genes.
- Molecular genetic maps of plant genomes: positional cloning of pest and disease resistance genes
- Syntenic maps of plant genomes to be used for:
- Precise localization of a mutant in a species and isolation of the corresponding gene in an appropriate model plant genome;
- Understanding the function of major genes based on availability of mutants in other species and on synteny.

3.1.2 Development and morphogenesis
- Studies of plant development by analysis of:
- The perception and transduction of chemical and physical signals in plant development
- Somatic and zygotic embryo and seed development
- The genetic and molecular basis of seedling development
- The genetic and molecular control of the transition to flowering
- Study of the mechanisms controlling plant architecture (size, internode length, branching, leaf morphology, plant canopy, tuber formation, etc.).

- Molecular studies of cell wall synthesis and degradation.

3.1.3 Resistance to stress and pathogens
- Studies to reveal the nature of transduction pathways leading to desiccation, cold and salt tolerance, response to U.V., ozone, in connection with the conditions of diverse habitats including both the cool and humid north and the dry and hot south, etc.
- Studies to reveal the nature of transduction pathways leading to disease resistance.
- Isolation and characterization of plant genes controlling resistance against nematodes, insects and microbes.
- Isolation of bacterial and fungal genes coding for pathogenesis factors and elicitors synthesizing enzymes.
- Studies on the cell biology of virus synthesis and transmission in plant cells.

3.1.4 Metabolisms
- Understanding of link-source relationships in order to optimize crop yield.
- The regulation of metabolic pathways in relation to plant storage organs.
- Elucidation in both *Rhizobium* and its plant host of the regulatory and metabolic pathways which are intimately connected with the process of nitrogen fixation during bacteroid life inside the plant.

3.1.5 Gene expression
- Mechanisms affecting the tissue-specifity and stability of transgene expression in plants.
- Identification of plant genes by transposon and transcriptional activator mutagenesis.
- Use of sense and antisense approaches to study plant physiology relevant to agricultural productivity.
- Mechanisms of antisense and ribozyme actions: gene silencing.
- Study of homologous recombination in plants.

3.2 Synergies with other specific programs

Concerning the topics which might impinge on future medical and veterinary applications, the emphasis of this program will remain on the design and development of new experimental tools prior to evaluation of their general applicability in specific disease-related research. Development of models for specific human disorders will be considered under the biomedical and health research program.

3.2.1 Genome mapping and improvement of farm animal selection
Objectives
Genetic and physical mapping will be very useful in selecting animals for the traits under multiple gene control (quantitative trait loci or QTL). The use of new techniques derived from studies on the basic reproduction mechanism, as well as the exploitation of new information arising from the genome mapping projects, should improve the reproduction and selection of farm animals. No research modifying the genetic constitution of farm animals will be considered under this programme.

Research tasks

European networks will be established (or extended) to map the genomes of farm animals chosen for their importance in agriculture and fisheries. The use of already established farm animal maps, and particulary for QTL analysis, will be invited; e.g., in relation to disease resistance, improvement of the quality of production, etc. Basic reproductive mechanisms from spermatogenesis, oogenesis and oocyte maturation to implantation in farm animals will also be considered, provided animal welfare and animal genetic diversity principles are respected.

3.2.2 Animal models
Objectives
Studies will be conducted to allow the development of new techniques to raise animal models with precise and predictable genetic characteristics designed to provide information of high quality and specificity in relation to pathological disorders. Research will be encouraged where it produces

evidence on the physiological roles of regulated/deregulated pathways or genetically-encoded factors during the evolution of any particular disease.

Research tasks

Projects considering the development of new techniques of genome modification will be considered first, especially those derived from classical laboratory animal species. Preference will be given to models having large application and usage. Rat genome mapping will be encouraged as a tool for future research in this sector.

3.2.3 Somatic gene therapy
Objectives
The identification and isolation of genes associated with specific human and animal diseases has enabled better diagnosis and paved the way to somatic gene therapy.

Research tasks

The development of generic techniques for the complementation or restoration of weakened or missing gene functions of potential medical importance — particular emphasis will be given to the construction of vectors and their controlled expression in predetermined recipient cells. Models which could be used for the evaluation of the method from the point of view of efficacy and safety will also be considered.

As the sequences are read, it is worth looking at the other ongoing European sequencing programs and their state of play, in particular, *B. subtilis*.

The Bacillus subtilis *Genome Sequencing Program*

All sequencing programs have a checkered history and none more so than *B. subtilis*. The story of this program goes back to 1985, which was when discussions began in the USA on sequencing the human genome. In view of the success of virus genome sequencing, it was thought that it might be possible to understand how genetic expression was collectively regulated at the highest hierarchical level from an analysis of the complete genome sequence. Two technical limitations were foreseen: the basic sequence determination itself and the computer analysis of the sequence. At the time, there was much talk about artificial intelligence (AI), but analysis programs in the early 1980s were still only conceptual and or experimental. Were the ideas of computer analysis and artificial intelligence just illusory or realistic? Only time would tell.

A discussion between Antoine Danchin and Olivier Gascuel, then busy building expert systems for medical use, led to an outline for a project to evaluate the interest in a computing approach normally classified as "artificial Intelligence work." At the time, it was known that human proteins could be expressed and secreted in bacteria. Since this is so, the general description model for secretion was the same in bacteria and animal cells. A signal peptide indicated where the secretion took place and was sufficient to allow it. For the industrials, however this model was not fully satisfactory because having a human protein secreted by a *coli* bacillus from its own peptide signal is often inefficient. Was it then possible to provide a pertinent description of the peptide signal of the *coli* bacillus and distinguish it from the human peptide signal? To answer this question, Danchin and Gascuel produced a document containing 17 specific descriptions of the bacterial peptide signal, when experts had only counted three. This study led Danchin to believe that this *in silico* approach, complementary to *in vitro* and *in vivo* experimentation, was the major advantage of considerable development in computing and would aid in making discoveries in the genome text.

For these reasons, during a spring conference organized by the French Microbiological Society (at the Pasteur Institute), Danchin suggested that France launch a systematic genome sequencing program. At the time, it was recognized that *E. coli* genome sequencing had already advanced quite well on its own. Another organism could, therefore, be chosen, using the implicit know-how gleaned from *E. coli* studies. The next bacteria on the list was *B. subtilis*, but contacts made with French laboratories specializing in it had proved negative. So Danchin, with the help of the bio-informatics workshop of the Pasteur Institute and Alain Henaut of the CNRS Center of Molecular Genetics at Gif sur Yvette (also the head of computer studies at the DESS of computing and biology of Paris University VI), decided to start a research group to train students for the computing part of the sequencing work to come. This effort, necessary but at the time greatly underestimated, was launched in 1986 in close collaboration with André Goffeau, who was then setting up the European *S. cerevisiae* program. Goffeau asked Danchin to draw up the report that appeared in 1989 to help justify the sequencing programs to EEC experts[139]. As *La Lettre du Greg*, "Itinéraire d'un projet financé par le GREG: le sequencage du genome de *Bacillus subtilis*"[140] (from which much of this early history has been taken) reports, the proposal was greeted with wide skepticism. Simon Wain-Hobson, who was responsible for the sequencing work on the AIDS virus genome, and who was convinced of the virtues of a systematic approach, proposed that a bacteria with a shorter genome be sequenced first, such as that of the most widespread sexually transmitted disease, *Chlamydia trachomatis*. With this support and the proposal, Danchin championed the project to the various possible funding agencies, particularly to the French Ministry of Research, but without any luck.

At the end of June 1987, at the bi-annual meeting for *Bacillus subtilis* specialists in San Diego, Jim Hoch proposed that an American-European consortium begin sequencing its genome. R. Dedonder, then Director of the Institut Pasteur, was convinced that the program could work. Upon his return to France, he remembered Danchin's proposal and gave Hoch a positive answer. A quick survey of the scientists likely to be involved in such a project provided a consortium of 10 laboratories, five from Europe and five from the United States, that were ready for the project. All that was needed was the money.

During a meeting of the Science Council of the above-mentioned Center for Molecular Genetics in November 1987, Danchin met Goffeau, who was then busy setting up the yeast genome sequencing program. The two scientists were convinced that *B. subtilis* was also important, but if EEC funding were requested, the two sequencing programs would be in competition for funding. Furthermore, Dedonder had mentioned it to the appropriate authorities and

---

[139] A. Danchin, "Complete genome sequencing: future and prospects", in A. Goffeau's *Sequencing the Yeast genome, a detailed assessment*, CEC, 1989.

[140] *La Lettre du GREG*, no. 8, December 1996, p. 12.

had managed to obtain funding for a pilot study for the *B. subtilis* sequencing program. The European project was being supported by the EEC SCIENCE[141] program under the administrative aegis of Dedonder, and in France, through the construction of a sequencing laboratory within the ambit of the Regulation of Gene Expression Unit at the Pasteur Institute. Five groups were involved, and they had agreed on the following strategy: 1) the strain chosen for the work was *B. subtilis* 168, a favorite in most laboratories; 2) regions of the chromosome were assigned to each of the participating groups — each region (between 200 and 400 kb) began and ended at a gene that had already been cloned, and if possible, sequenced. The objectives foreseen (and attained) were the construction of genome banks for strain 168, two separate banks in lambda bacteriophages (F. Kunst and K. Devine); alternative vectors such as YAC vectors should be tested, then the ordering of the DNA insertions in the lambda bank. Each group should identify the overlapping lambda phages in the assigned DNA region by screening with DNA probes for genes that had already been sequenced. Lastly, sequencing methods and strategies and *in silico* analyses were to be optimized. The project coordinator was R. Dedonder, to be succeeded later by F. Kunst.

In the USA, the situation was not going as well. For various reasons, Hoch's request for funding was refused. This, of course, had serious consequences for the European project. The project received favorable opinions from Paolo Fasella and André Goffeau and very positive conclusions from the national experts who gathered at Martinsried 2–3 November 1989[142] (except for the opposition of one representative of a large European nation)

---

[141] Science Program, "Establishment of a physical map and strategic approach to the sequencing of the *Bacillus subtilis* genome", between 1989–1991 with 5 laboratories, and a European Community contribution of 609,000 ECUs. The support from the SCIENCE program allowed the project to avoid going into competition for funding with the yeast project, which was then being financed for its pilot study under the BAP program.

[142] Report of the meeting held at Martinsreid 2–3 November 1989 to consider the feasibility of including the sequencing of the *Bacillus subtilis* genome as a T-project in the BRIDGE programme.

and industrial recognition of the value of such a program[143]. However, under these circumstances, the competition for limited funds and the single but firm veto from a national representative, the project could not be accepted as a BRIDGE T-project[144]. Although the pilot project was well on its way, funding from the SCIENCE program was to terminate at the end of the year. Luckily, it was prolonged for another year. In the meantime, the American groups gathered supportively around Hoch and tried several times to have the program funded by the NIH or the DOE, but with no success.

There was a very different situation in Japan. During a meeting at the Institut Pasteur in 1990 to summarize the initial results of these collaborative efforts, H. Yoshikawa protested that he didn't understand why Japan had been left out of the project. This settled the project's future. Yoshikawa was determined to get hold of the money needed. At the European group's request, a Japanese project was organized by N. Ogasawara and Yoshikawa, who had already sequenced 10,000 bases of the replication origin region of *B. subtilis* in 1985, at the time, the longest piece of bacterial chromosome to be sequenced. The Japanese *B. subtilis* systematic sequencing program began in 1991, funded by the Monbusho as part of the Japanese Human Genome Program. This program consisted of three parts: physical and genetic mapping of the human genome, systematic analysis of human cDNA and the development of sequencing technologies. The *B. subtilis* program was assigned to the third part as a model system for large-scale sequencing efforts involving several laboratories. International cooperation was a *sine qua non* for the allocation of funding. Five contiguous sections of 1Mb in total and another separate section of 0.2 Mb were originally assigned to the six Japanese groups under the supervision of Fukuyama, Ogasawara, Takahashi, Yamane, Sadaie and Kobayashi. In addition, another section of

---

[143] Several large European firms are interested in the biology of the Bacilli: they include Novo Nordisk, Gist Brocades, Solvay, Rhone-Poulenc, Orsan, ENI, Lepetit and Celltech who have all expressed marked interest in the sequencing of *B. subtilis*. An industrial club has been considered.

[144] E. Magnien, M. Bevan and K. Planqué, "A European BRIDGE project to tackle a model plant genome", *Trends in Biotechnology*, vol. 10, 1992, pp. 12–15.

0.1 Mb was assigned to another Japanese group[145]. In total, 1.3 Mb of the *B. subtilis* genome was to be sequenced by the Japanese, and by March 1996, 90% of that had been done.

Back in Europe, between the end of the SCIENCE program funding and the beginning of the BIOTECH program, *B. subtilis* work happily continued, especially in France (thanks to the modest, but regular, support provided by the Pasteur Institute and some assistance obtained through a call for proposals from the committee prefiguring the GIP genome, assistance that allowed the Pasteur Institute to complete 100 kb of sequencing before the end of 1992). After this year without EEC support, the project was again granted European funding in August 1993 for another three years under BIOTECH area 2 program: gene structure, to the tune of 1,960,000 ECU. In this way, the original project developed from a group of five laboratories to a major international initiative involving cooperation between the European and Japanese sequencing networks. The European labs involved are: four French groups, including A Danchin and G Rapoport at the Insitut Pasteur, S.D. Ehrlich at INRA Jouy en Josas, J. Haiech and F. Denizot, LCB, CNRS Marseille, and S. Seror at University of Paris 11, Orsay; two Italian groups under A. Galizzi at the University of Pavia and G. Grandi at Eniricherche, Milan; three British groups including J. Errington at Oxford, C. Harwood at Newcastle and I. Connerton at the BBSRC in Reading; one Irish group under K. Devine at Trinity College Dublin; two Dutch groups under Sierd Bron at Groningen University and B. Oudega at the University of Amsterdam; one Swiss group under D. Karumata and C. Mavel, Lausanne; two German groups under R. Borris of the Humboldt Universitat Berlin and K. Entian at the Insitut for Mikrobiologie of the J.W. Goethe Universitat Frankfurt; one Belgian group under B. Joris of the University of Liege; one Spanish group under R.P. Mellado of the Centro Nacional de Biotecnologia (CSIC) at the Madrid Cantoblanco Campus.

In addition, two biotechnology firms also participated in this project with funding of their own, Genencor International of the USA and Novo

---

[145] N. Ogasawara *et al.*, "Systematic sequencing of the *B. subtilis* genome: Progress report of the Japanese group", *Microbiology*, vol. 141, 1995, pp. 257–259.

Nordisk of Denmark as well as a laboratory of the Korean Research Institute of Bioscience and Biotechnology, which contributed with funding from the Korean government.

The objective of the BIOTECH *B. subtilis* project was to sequence 33% of the genome between 1 August 1993 and 1 August 1997[146]. In order to facilitate the bilateral contacts between the European sequencing network and European biotechnology industrials, a *Bacillus* Industrial Platform (BACIP)[147] was set up based on the YIP model. A European database nicknamed "Subtilist" was begun at the Institut Pasteur by A. Danchin and I. Moszer[148]. By March 1996, it contained some 2,497,000 kb of non-redundant sequences. Finally a new two-year contract under BIOTECH II has allowed the sequencing to be completed[149] to solve the remaining questions and validate the quality of the results, a continuation of the exhaustive annotation of the sequence and a start on functional analysis. The Japanese group, now that it has completed the sequencing it was assigned, has been reorganized to undertake the systematic analysis of the functions of genes located on the chunk of sequence they have been working on. This work comes under the second phase of the Japanese Human Genome Project, as a five-year project begun in 1996 and further supported by a grant for "Special Project Research" from the Ministry of Education, Science and Culture. From this program, which has gathered together 25 laboratories in Europe, Switzerland, Japan and Korea, over 40 publications have emerged.

---

[146] The European *B. subtilis* programme has unfurled in 4 phases. Initially the SCIENCE program ran from Sept 1989 to Dec 1991, with 5 laboratories. Then the BIOTECH program ran from 1 August 1993 to 1 August 1996, with 9 laboratories. Then a second period under BIOTECH I funded 7 laboratories. The 4th period under BIOTECH II would be for 2 years.

[147] The following firms have accepted the principle of this form of cooperation: Eniricerche (Italy) Gist-Brocades (the Netherlands) Genencor (Finland and the USA), Hoffman-La Roche, (Switzerland), Novo Nordisk (Denmark), Puratos (Belgium) and Solvay (Belgium).

[148] I. Moszer, P. Glaser and A. Danchin, "Subtilist: A relational database for the *Bacillus subtilis* genome", *Microbiology*, vol. 141, 1995, pp. 261–268. The Japanese data is organised according to the Japanese consortium's own approach, on the mirror site http://www.pasteur.fr:8008/, which, in interaction with the Swiss protein sequence database Swiss Prot, provides a reference point for the construction of other specialist bases.

[149] Kunst *et al.*, *Nature*, 390, 1997, pp. 249–256.

The project still continues today along the same lines — an approach through a specific network for most of the European genome research organizations with some additional participants in Europe (coordinated and administrated by S.D. Ehrlich) and Japan (coordinated and administrated by N. Ogasawara) for functional research.

Completing the sequencing of *B. subtilis* is of great interest. Of all the gram positive bacteria, *B. subtilis* is certainly the most understood. When the sequencing began, its genetic map was better drawn than any other except that of *E. coli*. Unlike the enteric bacteria, *B. subtilis* provides a chance to inventory some unique processes at the molecular level such as spore-forming and germination, the regulation of which seems to be linked to the differentiation systems of simple bacteria. While *E. coli* remains the bacterial paragon, there will always be the difficulty of using inverse genetics with it, which means another bacteria has to be considered in which inverse genetics is easier, and *B. subtilis* fits that bill. Furthermore, *B. subtilis* will, without a doubt, make great contributions to the solution of fundamental problems in the study of processes mentioned above (spore-forming, protein secretion, etc.). It will be intrinsically interesting for the study of chromosome structure (rearrangement, composition, constraints) and evolution (since the sequences for other genomes have already been completed). Lastly, the availability of a sequence for *E. coli* and its database (Colibri), and *B. subtilis* with its database (Subtilist), will permit comparisons of the coding strategies of the two organisms[150].

It seems from the results that *B. subtilis* has a very different coding strategy from that of *E. coli*. Comparison analysis of the frequency of dinucleotides have shown that regions with unusual frequencies can be found along the chromosome at intervals of some 80 kb. As in the other genome sequences that have been obtained, many genes have been found with no

---

[150] This comparison is even more important as *B. subtilis* and *E. coli* diverged in their evolution at least 1,500 million years ago.

functional similarities[151]. This shows that, as in the case of other sequences, classical genetics has left aside a large section of our understanding of these organisms that we would never know we had missed without systematically sequencing large genomes.

After the yeast success, Europe began planning other genome sequencing programs concerning small genomes, in particular, the completion of the genome of *Sulfolobus solfataricus*.

The Sulfolobus solfataricus *Genome Program*

The primary objective of the general project was to contribute to the completion of the genome sequence of the archaeal hyperthermophile *Sulfolobus solfataricus*. This project was started in 1993 by three Canadian laboratories sponsored by the Canadian National Research Council. Twenty five percent of the 3 Mbp genome has been sequenced and about one third will be completed when their three year grant terminates in Summer 1996. They have been promised a three-year renewal at the same financial rate as before, which means that they will have sequenced two-third of the genome by the end of 1999. Three European laboratories and one industrial laboratory that have negotiated with the Canadian group in charge of sequencing the remaining one third of the genome are involved. The Canadians have a linear cosmid library covering almost the whole genome, and they will provide the European laboratories with cosmids covering about one-third of the genome. The three European university laboratories will perform subcloning and sequencing, and Novo-Nordisk A/S will help both with processing of the sequencing data and developing software for accessing regulatory signals, motifs and coding regions in the genome sequences which

---

[151] The analysis of 1,037 kb of the Japanese project has shown, on the total of 1,007 ORFs determined, 512 ORFs (51%) had no known function. 262 ORFs were already genetically or biochemically characterised. The function of 233 additional ORFs was suggested by similarities in product to sequences of proteins known to the databases (H. Yoshikawa, N. Ogasawara, F. Kunst and A. Danchin, "The systematic sequencing of *B. subtilis* genome", in *Final European Conference of the Yeast Genome Sequencing Network*, Trieste, Italy, 25–28 Sept 1996, p. 61 of the Book of Abstracts, European Commission, DG XII, Brussels).

are of potential industrial interest. They will make these sequences available to the industrial platform. The new sequences will be exchanged with the Canadians and made available, as quickly as possible to university and industrial laboratories working on archaeal hyperthermophiles and thermostable gene products. In particular, they will be made available to other laboratories working on EU cell factory projects concerned with thermophilic organisms. The objective is to complete the *S. solfataricus* genome, which is of considerable interest to industry because of the thermostable nature of all its enzymes, as a joint European-Canadian effort by the end of the three-year project in December 1999.

In the latest round of funding, between 1996/1997 and 2000, sequencing of several other small genomes has been initiated. These include: *Leishmania, Rhizobium meliloti, Dictyostelium, Listeria* and extremophiles.

After its victory on the yeast genome, Europe has not limited itself to sequencing smaller genomes but has taken steps towards sequencing far longer genomes, namely those of *Arabidopsis thaliana* and the fruit fly *Drosophila melanogaster.*

The Arabidopsis thaliana *Program*

Worldwide, research on the genetics of *Arabidopsis thaliana* (Fig. 30) has always been very active. *Arabidopsis thaliana* is a small plant in the mustard family with the smallest genome and the highest gene density so far identified in a flowering plant. It is an ideal model organism because of its many advantages for experimentation and genetic studies. The entire genome of *Arabidopsis* is about 100,000,000 base pairs (100 Mb) in length, with genes occupying approximately 50–60% of this DNA in contrast to humans, in whom genes comprise only 5% of the DNA. Humble *Arabidopsis*, with its minimal number of genes to perform all necessary functions and the highest ratio of genes to non-coding DNA, is one efficient little organism. It has a short life cycle of about two months, is autogamous and very prolific. Its pod contains about 20 seeds and each plant bears several hundred pods. It's also not a big plant, which allows researchers to do away with the need for greenhouses or experimental gardens. In addition, it's easy to alter *Arabidopsis thaliana* with agro-bacteria, which provides a way of proving the identity of any previously-identified gene. According to Mary Clutter, NSF assistant director for biological sciences, what scientists learn from the study of *Arabidopsis* genes will be immediately applicable to economically important plant species and will lead to the creation of new and improved plants and plant-based products. "Because plants are vital to our existence, increased understanding of the biology of plants will impact every facet of our lives, from agriculture, to energy, to the environment, to health," says Clutter.

For Catherine Woteki, USDA acting Undersecretary of Agriculture for research, education and economics, "Mapping the *Arabidopsis* genome will enable us to use biotechnology to develop a host of new plant varieties for agriculture and other purposes. This research is like exploring a continent for the first time; each step leads on to several others, with tremendous possibilities. We're going to see productive results for years to come."

There are approximately 2,000 so-called "essential genes," which are vital to cellular function. Rob Martienssen, of Cold Spring Harbor Laboratory,

Fig. 30 *Arabidopsis thaliana*

has found that these essential genes in *Arabidopsis* bear uncanny similarities to the analogous essential genes in yeast. Using a technique called "gene traps" which Rob and colleague Venkatesan "Sundar" Sundaresan developed, Rob and postdoctoral fellow Patricia Springer demonstrated, with the sequencing help of W. Richard McCombie, that *Prolifera*, an *Arabidopsis* gene, was closely related to a yeast gene of known function. In yeast, *Prolifera* encodes a DNA replication factor that is involved in cell proliferation.

The essential genes are complemented by vital "survival" genes. Important among these are genes which control growth and development in response to the plant's environment, an aspect which is crucial to the organism's survival. Plants cannot remove themselves from an immediate environmental change as animals can by running away, thus many plants have adapted to endure dry seasons (as a cactus would, for example), extremes of temperature (grasses which grow in the Alaskan tundra or the Gobi dessert), and salinity, acidity, fire, flooding, wind and nearly any other environmental alteration, depending upon where the plant evolved.

Although *Arabidopsis* is a wild plant, it is part of the cress family, which has several agriculturally interesting species, including colza. From the beginning, the transfer of results to closely-related crop species has been clear. *Arabidopsis* is easily subject to genetic analysis.[152] There are five chromosomes in its genome, which is one of the smallest known plant genomes, estimated at 100 to 145 Mbp per haploid genome, compared to rice, thought to have has 440 Mbp, cabbage 600 Mbp, sweet corn 2,500 Mbp and wheat 1,600 Mbp. The difference is normally because of a lot of DNA sequence repetition. The *Arabidopsis* genome has only a few repeated sequences, which is an important advantage in the construction of a map of the genome. There are about the same number of genes as in most plants, within a factor of two or three, so it is easier to isolate a gene from *Arabidopsis* than from another species.

---

[152] M. Boutry, "*Arabidopsis*: une mauvaise graine sort de l'ombre", *Biofutur*, vol. 94, 1990, pp. 55–57.
J. Giraudat, "*Arabidopsis*: état des lieux", *Biofutur*, vol. 95, 1990, p. 54.

While *Arabidopsis* may not have all of the genes which allow for such developmental and environmental responses, it does have representatives of each of the gene classes found in other plants. In fact, in the tiny, compressed genome of *Arabidopsis*, one can find genes and gene classes homologous to nearly all of the genes found in other flowering plants, sort of a basic tool box of plant genes.

Initial studies rapidly convinced researchers of the need for common tools, the constitution and interest of which go way beyond the confines of any one project, such as the production of a mutant collection, a recombinant clone bank, gene catalogues, genetic and physical maps, and these have always, and will always, be the object of intense international collaboration, which only continues to grow. It is for this reason that, in 1990, a multinational, coordinated, *Arabidopsis thaliana* genome research program began. Its basic target is understanding the plant's physiology, biochemistry, growth and development at the molecular level. It will include obtaining new data on the basic biology of the plant and developing new resources for studying plants and communal resources to further scientific progress. The project was set up by an international group of scientists who saw the need to coordinate the different national programs on *A. thaliana*. Each year, since 1990, the group, known as "the multinational steering committee", produces a detailed report on progress made the previous year and lists targets for the coming year. The scientists and administrators involved come from five continents.

Because of the relatively small size of *Arabidopsis thaliana* and its lack of repetition, it was soon possible to clone a gene through the chromosome walking method normally used for micro-organisms. This chromosome stroll allows otherwise inaccessible genes to be defined and can be greatly enhanced if there is an ordered gene library, that is to say a group of genome clones that overlap to provide a physical map of all five *Arabidopsis* chromosomes. Several American and European laboratories have been working on this library, namely Goodman in Boston and Somerville at Stanford (in the USA) and Flavell in Norwich, UK. This research aims to provide a complete physical map. While other sequencing projects were being setting up, it was asked whether systematic sequencing would be appropriate. In 1990, many were still reticent about this since there were still many genes to identify with the methods available. Despite this, the systematic sequencing of yeast provided encouragement, and as long as the techniques could be automated even further, as pointed out by M. Boutry, "It seemed obvious that *Arabidopsis* was the best candidate to become the first vegetable species to have its intimacy unveiled"[153]. As for other organisms then being mapped and sequenced (*E. coli*, *S. cerevisiae*, unicellular eucaryotes, fruit flies, the mouse and man), there was a flurry of concentration and coordination of research. In the international climate at the end of the 1980s and the emergence of worldwide coordination of *Arabidopsis* work, the European Commission had to provide a contribution.

Under the aegis of Alessio Vassarotti, a DGXII scientific officer and a former student of André Goffeau's, a large European project to study *A. thaliana* was set up. This project, entitled "The molecular identification of new plant genes", was integrated into the BRIDGE program. It began in 1991 with a two-year period of funding and involved 27 European laboratories financed by the European Commission to 4,654,000 ECU. It represented a coordinated attempt to explore and exploit the *Arabidopsis* genome alongside other international initiatives, such as those of the AFRC, NSF, DOE[154] and so forth. There was a world movement going on at the time in plant molecular biology. The project had been constructed in such a way that it could also contribute to other projects. It was to provide a specific long-term competitive advantage to the European biotechnological industry. It kept the approach developed for the yeast program and included at the outset six biotechnology companies that contributed directly, others having expressed interest. The BRIDGE project was supported by a good base. Its main objectives were the development of methods to identify and isolate genes of agricultural

---

[153] M. Boutry, "*Arabidopsis*: une mauvaise graine sort de l'ombre", *Biofutur*, vol. 94, 1990, pp. 55–57.

[154] The Department of Energy is supporting the plant sequencing effort because the applications of the genetic information learned could be used to meet a number of agency mission needs. Potential applications include improved quality and quantity of biomass products such as alternative fuels and chemical feedstocks (which can conserve petroleum resources) and using plants to clean up contaminated soil (phytoremediation) at DOE's former nuclear weapons production sites.

importance by using the results from the model plant *A. thaliana* and to apply these methods to the study and manipulation of plant development physiology, an important field in plant productivity. In order to reach these targets, the T-project was also expected to set up an infrastructure that could support future pan-European progress in *Arabidopsis* research and maintain it after the project was over. A future goal would be to set up an industrial platform which would get the research results and from which researchers could get ideas and support for advanced studies to benefit European agriculture and biotechnology.

This project provides a framework linking the scientists working on identification techniques to the scientists whose research answers the basic questions in plant biology, both through genetics and physiology. It was hoped this would accelerate the dissemination of gene research techniques and their application to the characterization of genes involved in development, the flow of seeds and embryogenesis. Unlike the "sequencing the yeast genome" T-project, in which the goal was uniquely and perfectly defined (sequencing chromosomes II and XI and preparing libraries of other chromosomes for the BIOTECH program), the goals of the T-project "molecular identification of new plant genes" were far wider, which is why the following subdivisions were necessary:

A. Physical Mapping (C. Dean, Norwich, GB)
1. Physical mapping of the *Arabidopsis* genome (C. Dean)
2. Construction of an *Arabidopsis* genome library in the P1 phage (T. Kavanagh, Dublin)
3. Construction of physical and genetic maps (P. Vos and M. Zabeau, Keygene, NL)

B. Resource center (B. Mulligan, Nottingham University, GB)
1. *Arabidopsis* resource center (B. Mulligan, Nottingham University, UK)
2. *Arabidopsis* DNA resource center, (J. Dangl at the MPI Koln)

C. Gene replacement (P. van Elzen, Leiden, NL)
1. Directed mutagenesis of embryons (P. van den Elzen, Mogen International, Leiden, NL)

2. Directed mutagenesis using *Agrobacterium* (P.J.J. Hooykaas, Leiden, NL)
3. Targeting genes with YACs (P. Meyer, MPI Koln)
4. Gene replacement (L. Willmitzer, IGF, Berlin)
5. Stimulation of recombination with rec A (B. Reiss, MPI Koln)

BRIDGE T-Project
Molecular Identification of New Plant Genes
Overall Coordinator, M. Bevan of Norwich, GB

By the end of the project, most of the assigned goals had been reached, especially those of setting up resource centers and developing labeling and mapping methods based on gene cloning. Significant progress had been made in the incorporation of molecular genetic techniques in established plant physiology fields (seed and flower development), and a promising start had been made on the isolation and characterization of genes involved in these important processes. The mapping work, supported by the European Commission, was a central contribution to the world effort of mapping the genome. The success of this work also led several BRIDGE program participants to put together a BIOTECH project as the first real attempt to sequence the whole of the *Arabidopsis* genome.

The European project on *Arabidopsis* acquired a more clearly international aspect which reflected the success of the integration of European national programs. The consortia set up under BRIDGE had become productive as the project matured, and extended its results to various aspects of European agricultural and biotechnological research for beyond the field of *Arabidopsis* research.

Despite this progress, the international links developed between the European Stock Center at Nottingham and the Ohio Stock Center in the United States, the establishment of a Plant Industrial Platform, the extent of the project and the many targets it envisaged, the project was criticized, especially when the constitution of a European systematic sequencing network was being considered. For some, the impression was that the project had been of most use to the promotion of molecular biology, and it was suspected that in 1991, the scientists were really not quite able to present a coherent plan for the systematic sequencing of *Arabidopsis*. A fringe effort had been made by Goodman, in the USA, to build the library, which did not have to be complete for sequencing to begin. Furthermore, the AMICA project had started receiving its funding from the European Union. Its aim was to stimulate plant molecular biology research in Europe through grants and multinational contracts under the supervision of a scientific committee, and it was thought possible that the sequencing could be integrated into it.

Despite these criticisms, a systematic sequencing project for *Arabidopsis* was funded under the BIOTECH program. Starting on 1 September 1993, for a three-year period, "European Scientists Sequencing *Arabidopsis* (ESSA)" (Fig. 31) was granted 5,751,000 ECU by the European Union.

ESSA's initial goal was the creation of contiguous YACs for chromosomes IV and V as its contribution to EU-US cooperative drawing of a physical map of the five chromosomes. Secondly, it intended to create contigs ready for sequencing, covering two regions of chromosome IV. Several other areas of biological interest were pursued and methods developed to build libraries for all of chromosome IV. Thirdly, at MIPS, a specialized center for the gathering of genome resources was set up and entrusted with analyzing the ESTs and the genome sequence, thus providing a quality control service and setting up links for public access to the sequences. The fourth target, which was attained, was the constitution of a network of EST sequencers, which should add 3,000 EST sequences to public data bases. Fifth, a network of laboratories was to be set up to sequence genomic DNA both in regions of individual interest and systematically, on parts of chromosome IV, with the aim of sequencing 2.5 Mbp of the genome and providing expertise for future sequencing projects. Lastly, international agreements were made to sequence the entire genome as soon as possible.

Cosmids were sequenced, although at first it was slow going due to the learning curve on new methods and the shake-down of new research teams, but production is rising and will soon reach a constant rate of output. Most sequenced cosmids have come off of current Goodman contigs and the YAC subclone libraries. BAC libraries, created by digesting the DNA with Hind3 or EcoR1 with inserts of 100 kb, are now available. They have significant advantages in terms of DNA stability and for shotgun strategies, and their

560 *From Biotechnology to Genomes: A Meaning for the Double Helix*

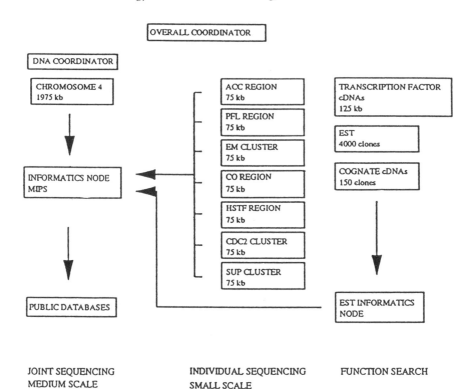

Fig. 31 Organization and flow of information in the ESSA project
(from the archives of A. Goffeau at the UCL)

use is an important factor in future plans to finish sequencing the 100 Mb of genome left (in a given time frame at acceptable costs). BAC contigs covering 600 Kb have been assembled from four BAC groups into an FCA region. This success shows that obtaining complete coverage of the genome using x20 libraries with a few gaps will be possible, and even those gaps can be covered through cosmid clones. By the end of 1996, the FCA region (1,800 Kb) had been completed, as had 500 Kb of other regions and 450 Kb of the AP2 region on chromosome IV.[155] The sequence of the central region

---

[155] In ESSA II 56 BAC (~ 5Mb non redundant sequence) adjacent to these contigs are currently sequenced and annotated.

of the lesser arm of chromosome IV is very interesting, with a potential or known gene on average every 5 Kb, the regions between the genes being between 0 and 5 Kb. Over 200 genes have been sequenced in a 1,000 Kb sector of chromosome IV. Clusters of functionally-related genes have also been discovered. About 60% of the possible genes have homologues in other species, and about half of these homologies have never before been observed in plants. The high density of the genome, the almost complete absence of repetition and the complex structure of the genes themselves show that a precise approach to sequencing allows interpreting the sequence for the maximum information.

The sequencing work has extended to the United States, where a project to sequence large regions has been funded through an inter-agency project involving the DOE, NSF and USDA.

Sequencing Arabidopsis *in the USA*

The *Arabidopsis* Sequencing Project brought together two types of researchers — plant geneticists like Rob Martienssen and experienced sequencers including W. Richard "Dick" McCombie, also of Cold Spring Harbor laboratory. Prior to his arrival at the lab in 1992, Dick worked for the NIH, developing new techniques for large-scale automated DNA sequencing. His group was the first to sequence more than 50 kilobases (50,000 base pairs) of human DNA using the new automated techniques for the Human Genome Project. He also played a role in the first Expressed Sequence Tags (ESTs) project. He and his research group used this approach, now widely used in sequencing efforts, to jet propel the identification of genes in another organism, the *C. elegans* worm. During his time at Cold Spring Harbor Laboratory, Dick has worked on the genomic sequencing of *S. pombe*, a fission yeast which served as both a model system and a method of technological development for the human genome project, and the sequencing of segments of DNA obtained from human tumors, frequently in collaboration with CSH scientists interested in particular genes.

In 1993, Dick was approached with a new and highly appealing proposal: "Rob and Sundar approached me at Rich Roberts' Nobel Prize party in

Blackford Bar... they wanted to sequence their transposon insertion sites." Dick remembers it as an ideal collaboration: "The combined use of sequencing and transposons gives you the ability to functionally analyze any gene in the genome". After substantial initial support from CSHL institutional funds, the project took off when Rob, Dick and Sundar were awarded a grant by the National Science Foundation (NSF).

However, it soon became apparent that while the sequencing of the transposon insertion sites would be fruitful, the sequencing of the *Arabidopsis* genome should be the ultimate goal. The idea of systematically sequencing the *Arabidopsis* genome seemed both plausible and necessary, and the shifting winds within the scientific community favored the formation of a collaboration. Rob Martienssen, with novel and elegant gene traps, and Dick McCombie, with powerful methods of sequencing, joined forces and set out to unravel the genetic puzzle within *Arabidopsis*.

Late in 1995, laboratory trustee and Chairman of the Board, David L. Luke III, and Westvaco Corporation provided seed money of $290,000 to begin purchasing the equipment that would get the *Arabidopsis* project off the ground. In 1996, he and his wife Fanny gave an additional $362,250. These very generous donations allowed the lab to purchase more of the state-of-the-art sequencing equipment necessary for such an endeavor, setting the project into full swing and putting the lab in a position to qualify for much needed government funds. The project hit pay dirt when, in 1996, an interagency panel comprised of the NSF, the US Department of Agriculture (USDA) and the US Department of Energy (DOE) awarded $12.7 million dollars to a consortium of institutions worldwide for *Arabidopsis* research. Cold Spring Harbor Laboratory heads one group within the consortium (other members of the group include the Genome Sequencing Center at Washington University and the ACGT Group at Applied Biosystems); other consortium components include The Institute for Genomic Research (TIGR) in Maryland, a group headed by Stanford University, and groups in Japan and Europe. The Cold Spring Harbor group, however, is the only one using gene traps to isolate and characterize genes (although this is being done under separate funding from NSF and the USDA).

Formation of the consortium in 1996 culminated years of speculation and a handful of previous attempts at such a large-scale organization of resources.

Laboratory President James Watson, as founding director of the US Human Genome Project (HGP), had proposed sequencing model organisms including *Arabidopsis* as part of the HGP endeavor, but the NIH opted not to fund plant genomics through this project. In 1994, Dr. Watson held a meeting at the lab's Banbury Center; with organizers Joe Ecker (University of Pennsylvania), Mike Bevan (John Innes Center, UK) and Rob Martienssen. The meeting brought together *Arabidopsis* researchers and sequencing people. About two months later, there was a policy meeting at the National Science Foundation, attended by many of the Banbury participants. The NSF decided to launch the *Arabidopsis* sequencing project. The three US consortiums were established, and a global, broad-scale effort was finally put into motion. The coalition is in many ways the first of its kind, not only in its focus on plant genetics, but also in its scope, its well-organized broad international base and its use of the Internet.

The Arabidopsis *Genome Initiative*

In order to coordinate the sequencing work properly, the *Arabidopsis* Genome Initiative (AGI) (Table 32) has been set up to obtain the complete sequence for the genome by 2004. The *Arabidopsis* Genome Initiative (AGI) is the name of the multinational effort to sequence the *Arabidopsis* genome.

On 20–21 August 1996, representatives of six research groups committed to sequencing the *Arabidopsis* genome met in Washington DC to discuss strategies for facilitating international cooperation in completing the genome project. All six groups have secured major funding to pursue large-scale genomic sequencing of *Arabidopsis*, and the EU and Japanese groups have been engaged in large-scale sequencing for some time. The primary objectives of this meeting were to establish *Arabidopsis* as a model for international coordination of sequencing efforts and to develop guidelines for rapid and efficient completion of the sequencing project by the year 2004. Representatives from Japan, France, the EU, and the USA were present at the meeting.

A remarkable degree of consensus was reached by the end of the meeting on the general strategy for the *Arabidopsis* sequencing project. All parties agreed to follow several practices that were seen as facilitating international cooperation. This document was drafted to serve as a *modus operandi* for the participating groups until such time as it is modified by mutual agreement of representatives of the participating groups. All signatories to this document have agreed to the following:

1. The Arabidopsis *Genome Initiative (AGI) is intended to be an inclusive international collaboration. Any group that intends to engage in the sequencing of hundreds of kilobases of contiguous* Arabidopsis *genomic DNA will be invited to participate as a coequal collaborator in the AGI and will be expected to follow the guidelines outlined in this document.*

2. A coordinating committee with representation from each of the participating groups was formed. This committee will be responsible for making all decisions that affect the overall goals and operations of the AGI. In particular, it is anticipated that the AGI coordinating committee will be a planning and brokering system for establishing efficient ways of completing the genome. The committee will endeavor to apportion regions of the genome to the various groups in such a way as to minimize needless duplication of effort while maximizing progress toward complete sequencing of the genome. The committee will also be responsible for keeping the Arabidopsis *community informed of continuing advances in the sequencing project.*

Members of the committee for 1996-97 are Mike Bevan (Chair; EU consortium), Satoshi Tabata (Kazusa DNA Research Institute), Joe Ecker (Stanford University of Pennsylvania-Plant Gene Expression Consortium [the SPP consortium]), Dick McCombie (Cold Spring Harbor-Washington University, Applied Biosystems Consortium {CSH-WU-ABI}), Steve Rounsley (The Institute for Genomic Research {TIGR}), Francis Quertier (French Genome Center) and David Meinke (Multinational Arabidopsis Steering Committee). Each member of the committee will be responsible for arranging a temporary or permanent replacement from the represented group when appropriate. New members will be invited to join the committee based on a nomination from one member of the committee and an affirmative vote by a majority. It is anticipated that the committee will maintain regular communication and will meet annually. Mike Cherry (Curator of ATDB) will develop an email server to facilitate correspondence between members of the committee.

3. The six research groups are expected to complete different amounts of finished sequence because they have different capabilities and levels of funding devoted to this project. In order to prevent duplication of effort, it was considered useful to have the various groups initiate sequencing in different well-defined regions of the genome. It was agreed that each group should begin by nucleating sites over a contiguous region of a size that could be completed with the funding available. It was recognized that it may not be possible to define such a region with high accuracy because of variation in the ratio of genetic distance to physical distance. The goal in this respect should be to avoid situations where one group obtains scattered regions of sequence that must eventually be finished (i.e., linked up) by other groups. Exceptions to this strategy are noted elsewhere in this document.

The SPP group will begin nucleating on chromosome 1. The EU group will nucleate the bottom arm of chromosome 4. The CSH-WU- ABI group will nucleate a 4 Mb region on the top arm of chromosome 4 and a 2 Mb region on the top arm of chromosome 5 (the latter in collaboration with the EU group and the Kazusa group). The TIGR group will nucleate chromosome 2. The Kazusa group will nucleate the lower part of chromosome 5. The region at the top of chromosome 5, of mutual interest to the EU, CSH-WU-ABI and Kazusa group will be sequenced collaboratively. The Kazusa group, which anticipates a monthly sequencing rate of approximately 500 Kb, expects to begin nucleating a region of chromosome 3 in 1997. The EU, TIGR, SPP and CSH-WU-ABI groups anticipate an average monthly rate of approximately 200, 220, 150 and 150 Kb per month, respectively. Thus, when all the groups are operating at full capacity, the average monthly rate for the entire AGI collaboration is expected to exceed 1.2 Mb per month. The philosophy of the AGI collaboration is that as the assigned regions near completion, the coordinating committee will designate new regions of unfinished sequence to the groups in proportion to their sequencing capabilities. For example, the French Genome Center is tentatively interested in sequencing BAC ends during the first year or two of operation, but after that time, it is anticipated that they will engage in sequencing a contiguous region of genomic DNA that will be decided at a later date.

Several of the participants had differing views about the relative merits of sequencing unique sequences versus regions of repetitive sequence such as centromeres and telomeres. On the one hand, it may be expected that the maximum number of coding sequences will be found by sequencing the regions of low copy number. On the other hand, it will be interesting to know the structure of the centromeric and telomeric regions. The majority view appeared to be that it was not necessary at this time to resolve this issue. However, the majority view was that renewals of existing grants should take into account the fact that some regions of sequence will be more difficult to complete than others and large stretches of contiguous sequence are more difficult to achieve than small scattered regions.

Sequencing efficiency should be the sole criterion for choosing which clone to sequence. It was agreed by all parties that none of the groups should perform service sequencing for outside groups interested in particular clones. The reason for this is that the sequencing groups should not be seen to be favoring certain colleagues.

4. The most efficient strategy for sequencing the Arabidopsis genome is to shotgun sequence large clones such as BACs, YACs or inserts from P1 clones. Most of the groups have had preliminary experience with BACs and YACs and preferred BACs. The fact that most of the groups are currently satisfied with the available public BAC libraries will facilitate coordination and exchange of information. In particular, in order to minimize the requirement for additional physical mapping, it is desirable to obtain several hundred base pairs from the ends of a large number of BAC clones so that the minimum tiling path from a region of sequence to an overlapping clone can be determined by database analysis. The groups led by Craig Venter (TIGR) and Francis Quertier (French Genome Center) agreed to sequence the ends of approximately 14,000 BACs from public BAC libraries during the next two years. and to make the information freely available to the community.

All of the groups will use public BAC, YAC or P1 libraries constructed from the Columbia ecotype that will be freely available to the world community. A suitable BAC library to begin with is the TAMU BAC library constructed by Choi et al. that is currently available at the Ohio Stock Center. The other BAC library was constructed by Thomas Altmann and collaborators and is also publicly available. A P1 library (the 'M library') developed by Bob Whittier and colleagues at Mitsui is also available at the Ohio Stock Center and a second library (the 'K library') is being tested at the Kazusa Institute.

5. The objective of the AGI is to obtain high accuracy sequence of the entire genome. There was general agreement that it was not possible to set a standard for exactly what high accuracy means or for mechanisms to enforce high accuracy. However, it was generally agreed that a minimal standard would be that >97% of all sequence would be obtained on both strands or by two chemistries. It was the opinion of the group that these criteria were of similar importance and that with most clones, about sevenfold redundancy of sequencing would be required for shotgun sequencing.

An unknown factor affecting the accuracy of the sequence concerns the fidelity of the BAC clones. Preliminary experience suggests that the BACs are generally faithful clones of the genome. However, it will be essential to verify the integrity of each BAC. A minimum criterion is that both ends of the BAC should map to the same region of the genome, typically to the same YAC. When 14,000 BAC ends are sequenced, it is expected that, on average, we will have 500 bp of sequence every 5 kb on average throughout the genome. The resulting library of end-sequenced BACs will represent a check on BAC integrity that will assist in revealing any major rearrangements, deletions or additions. No standard was agreed upon for BAC (or P1) integrity checking. However, most groups indicated that comparing fingerprints of tiled BACs would be the most appropriate criterion for integrity.

After some discussion, it was agreed that a single-pass shotgun sequence of the entire genome would not be worthwhile because the combination of available ESTs and the high output rate of the AGI collaboration would obviate much of the value of single-pass shotgun sequencing for gene

568 From Biotechnology to Genomes: A Meaning for the Double Helix

discovery. However, because chromosome 3 will be sequenced later than the other regions, the group endorsed a proposal by the SPP consortium to do a feasibility study involving shotgun sequencing of clones from chromosome 3. After the meeting was concluded, the SPP consortium decided that this was not a good idea and reverted to their original plan in which "limited testing of some of the new instrumentation being developed at Stanford will utilize clones from a whole-genome shotgun library".

6. All of the participating laboratories are committed to early data release via the Internet. One approach discussed at the meeting involved daily release of preliminary sequence information (ie., sequences that have been edited to remove vector and regions of high ambiguity and condensed into 1 kb contigs). The C. elegans *sequencing groups follow this approach and the community has found it very useful. Two of the US groups, the SPP consortium and the CSH-WU-ABI consortium intend to release data in this way. Both groups anticipate release of finished, annotated sequence within six months of beginning to sequence a clone. The EU group does not consider it feasible, at the moment, to do daily releases because the consortium is composed of seventeen relatively small sequencing groups with varying levels of technical capabilities. The EU anticipates release of finished annotated sequence within one month of completion. The TIGR and Kazusa groups do not wish to release unfinished sequence because they believe that carefully edited sequence will be most useful to the community. Both groups promised release of information on a given clone to public databases within three to six months after sequencing began. The TIGR group will release finished, annotated sequence within three months of beginning to sequence a BAC. The Kazusa group estimates that they will release finished, annotated sequence within four to six months of beginning to sequence a clone. In all cases, the start date for sequencing a specific clone will be announced on linked WWW sites so that members of the community will know when to expect the finished sequence. In summary, all of the groups agreed to establish linked WWW pages for posting complete lists of all clones that have been sequenced to date, along with the start dates of those clones that are still in progress, and the anticipated start dates for the next set of clones to be sequenced in the future. Each clone will, therefore, have a start date that will be widely advertised to the community. All of the groups anticipate that it will take less than six months to completely sequence and annotate a BAC, YAC or P1 clone, and that they will deposit the complete annotated sequences in a public database (eg. GenBank, EMBL, JDB). No sequence information will be withheld from the community for the sole purpose of benefitting selected individuals, groups, or private companies.*

7. There was consensus that the value of the sequence obtained is proportional to the quality of annotation. Thus, each group will attempt to achieve a common standard of annotation. Each group will perform BLAST (or FASTA) searches to align ESTs and known genes and gene products to the genomic sequence. In addition, each group will use programs such as GRAIL and GeneFinder to identify ORFs. Annotation should be presented to the community in a format that can be readily accessed and understood by plant biologists worldwide.

It was agreed that all unassigned ORFs would be named according to the C. elegans system. A provisional agreement was reached that the following rules of nomenclature will apply: The first letter is the library name. T = TAMU BAC, F = IGF BAC, M = Mitsui P1 clone, K = Kazusa P1 clone, C = cosmid clone from Goodman library. The first letter is followed by the microtiter plate number, then the row and column numbers followed by a dot and the number of the ORF (numbered sequentially from one side of the clone to the other). Thus, a typical ORF might be called t23a11.12 (i.e., a TAMU BAC from plate 23, well a11, the 12th ORF from one end). It was agreed that zeros will not be included (i.e., t23a11.12 but not t23a11.012). It was also suggested that the names be all lower case for consistency. Sometimes it will happen that after all the ORFs have been named, a new one will be found by some functional test or other criteria. In this case, the two ORFs will be named with an extension to the name (eg., t23a11.12.1 and t23a11.12.2). When two ORFs are found to belong to the same gene or an ORF is found not to be expressed, the name will be deleted. When one ORF spans two or more clones, the entire ORF will be given the name of the 5' region of the ORF.

It was recognized that annotation of a clone at the time of deposit in public databases will rapidly be rendered obsolete because of information about genes being discovered by the community at large. Thus, there will be an ongoing need for annotation of previously sequenced clones. Because most of the groups are funded to produce new sequence, it will be difficult for the groups producing sequence to also take responsibility for revising the annotation of previously completed sequence. There was broad agreement that the task of annotation revision should be institutionalized by assigning responsibility for revision to the curators of the Arabidopsis *database (ATDB). The group expressed its strong enthusiasm and support for the continued funding of ATDB to make certain that essential informatics components of the* Arabidopsis *genome project are not overlooked. Mike Cherry agreed that it was a suitable responsibility for ATDB and agreed to accept the task to the extent that resources permit.*

8. Because the US groups associated with the Arabidopsis *Genome Initiative will need to reapply for funding within 2.5 years, there was concern about the criteria that will be used to evaluate success. It was agreed that each of the groups will be evaluated based on their overall contribution to the AGI collaboration and that the criteria will not simply be dollars per kb.*

9. It is considered essential to keep the entire community well informed of technical advances and practical applications of the genome project. Each group will mount a WWW page that will report the contribution of the group to the multinational sequencing effort. Each group will also work through Mike Cherry (ATDB) and the coordinating committee to make certain that community members receive the training required to make efficient use of the extensive sequence data that will be generated over the next several years. In addition, the coordinating committee will evaluate the feasibility of appointing a part-time public relations specialist to produce user-friendly documentation about the progress of the Arabidopsis *Genome Initiative. These efforts should help to advertise the dramatic impact that sequencing the* Arabidopsis *genome will have on basic and applied research in plant biology.*

Signed: Mike Bevan, Ian Bancroft (EU consortium) Satoshi Tabata, Kiyotaka Okada (Kazusa DNA Research Institute) Joe Ecker, Sakis Theologis, Nancy Federspiel (SPP consortium) Dick McCombie, Rob Martienssen, Rick Wilson, Ellson Chen (CSH-WU-ABI) Craig Venter, Steve Rounsley, Owen White, Chris Somerville (TIGR) Francis Quertier (French Genome Center) David Meinke (Multinational Arabidopsis Steering Committee). 14 September 1996

As of August 1996, the AGI has six groups (Table 34). Each group will initiate sequencing in a defined region of the genome using BAC, YAC, cosmid or P1 clones. Certain groups will also sequence the ends of ~14,000 BACs for mapping purposes. One critical objective is to obtain very accurate sequence. The minimum standard is that 97% of the sequence will be obtained on both strands, or by two chemistries. The AGI will produce 45 Mb of finished sequence in the next three years, and the projected year of completion of the entire genome is 2004 (see Table 35 for the 1999 status).

Table 34 The *Arabidopsis* Genome Initiative (AGI): Who, Where, How

Group	Partners	Region	Scale	Strategies/Features
COLD SPRING HARBOR SEQUENCING CONSORTIUM (CSHSC)	Dick McCombie Rob Martienssen Rick Wilson Bob Waterston Ellson Chen	chr 4 & 5	6.5-7.0 Mbp/3yrs	BAC fingerprinting OSS strategy for YAC sequencing
EU *ARABIDOPSIS* GENOME PROJECT	Mike Bevan (Coordinator)	chr 4	5 Mbp/2yrs	Network of 18 labs
KAZUSA INSTITUTE	Satoshi Tabata	chr 5 & 3	15 Mb/3yrs	P1 libraries
SPP CONSORTIUM (Fig. 32)	Ron Davis Nancy Federspiel Joe Ecker Sakis Theologis	chr 1	5 Mbp/3yrs	Large-scale-automation networked functions
TIGR	Steve Rounsley Chris Somerville Craig Venter	chr 2	6-8 Mbp/3yrs	BAC end-sequencing
GENOSCOPE -- CENTRE NATIONAL DE SEQUENAGE	Marcel Salanoubat Francis Quetier	chr 3	3 Mb/yr	BAC end-sequencing
GENETICS DEPT STANFORD UNIV	J. Michael Cherry			Development of a general Arabidopsis database, ATDB

Beyond the systematic genome sequencing program, the plan to sequence *Arabidopsis* cDNA is also progressing. The objective of the project is to identify the 20,000 estimated expressed *Arabidopsis* genes. Two major groups have shared in this work — one at Michigan State University under Thomas Newman (funded by the NSF and DOE), and the other a consortium of laboratories, funded initially by the French government and later integrated into the ESSA consortium run by the EU. These two groups have identified over 14,000 non-redundant ESTs and sent over 23,000 ESTs to the dbEST database managed by the National Center for Biotechnology Information (NCBI). These figures date from 1995 and by now most (70%) of the genes expressed in *Arabidopsis* have been identified. The homology analyses carried out show probable functions or identities for about 40% of unique sequences.

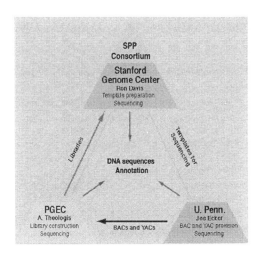

Fig. 32 The SPP consortium

Table 35 *Arabidopsis thaliana*
(data generated 25 January 1999)

	Size Mb	Redund %(estm)	Finished %(adj'd)	Finished (total bp)	Unfinished (total bp)	1999	1998	1997	1996	1995
Total	100	2.3	34.1	34,923,219	20,187,155	629,789	17,800,153	14,437,368	796,467	340,928

In assessing the implications for agricultural applications, we can offer the example of the relationship between the corn plant, maize, and teosinte, a grass common to Mexico and Latin America. At a glance, the two plants appear quite distinct: teosinte is a small bladed grass, maize is a large, thickly stalked, cob bearing plant. Yet plant geneticists have determined that the genetic difference between the two species is confined to alterations of only five fundamental genes and a number of minor modifiers. This would suggest that using a genetic map and today's genetic tools, scientists could manipulate specific genes to create desired changes, ultimately creating "designer plants" with various desirable qualities.

So, despite its unassuming appearance, *Arabidopsis*, with its neat little toolbox of genes, may hold the key to countless future advances in science. Perhaps, in only a matter of years, we will be living in a world where deserts are colonized by fruit-bearing plants. It's perhaps the norm of genetics, as has been seen with fruit flies, yeast and nematodes, that it is through the apparently simple that great discoveries are made. *Arabidopsis* is no exception, and it is only a matter of time before the far-reaching effects of these fundamental studies will begin to be seen.

Aside from *Arabidopsis*, there is another multicellular eucaryote that is being systematically sequenced. It's the geneticist's favorite model organism, the fly *Drosophila melanogaster*.

The Drosophila melanogaster Program

The genome of *D. melanogaster* is, like *Arabidopsis*, an average size. It is estimated to be 165 Mb of which about 125 Mb are euchromatic.[156] It is to be found in four chromosome pairs, and according to G. Rubin and G. Miklos, has about 10,000 to 12,000 genes. The *Drosophila* research community, working individually or in genome projects, has produced a large amount of information on its genetic and molecular organization as well as on the structure, expression and function of individual genes. The major groups involved in the project are listed in the Table 36.

---

[156] Unlike euchromatic segments, the heterochromatic segments contain very few genes.

Table 36 The major groups involved in the *Drosophila* Genome project

Groups	Senior Members	Funding sources
Berkeley *Drosophila* Genome Project (BDGP)	*Drosophila* Genome Project (DGC)	NCHGR, DOE (consortium of the DGC, HHMI and LBNL)
DGC *Drosophila* Genome Center	G. Rubin, S. Lewis, Berkeley; A. Spradling, Carnegie Inst.; Washington; M. Palazzo and C. Martin, LBNL D. Hartl, Harvard; before 7/95, I. Kiss, Szeged, Hungary, collaborating on the gene description project	NCHGR
Lawrence Berkeley National Laboratory LBNL Center	M. Narla, M. Palazzo, C. Martin, J. Jaklevic, F. Feckman	DOE
Howard Hughes Medical Institute (HHMI)	G. Rubin and A. Spradling	HHMI
European *Drosophila* Mapping Consortium	M. Ashburner, Cambridge; D. Glover, R. Saunders, Dundee J. Molodell, CSIC, Madrid; F. Kafatos, EMBL Heidelberg; K. Louis, B. Savakis and I. Siden-Kiamos, IMBB (Heraklion)	Fondation Schlumberger, Paris, HRC, UK, EU, Fundacion Ramon
Karpen	G. Karpen, Salk Institute, La Jolla, USA	NCHGR
Univ. Alberta, Canada	J. Locke, A. Ahmed, J. Bell H. McDermid, D. Nash, D. Pilgrim, K. Roy and R. Hodgetts	Canadian Genome Analysis and Technology Program
McGill Univ., Canada	P. Lasko and B. Suter	Canadian Genome Analysis and Technology Program
Duncan	I. Duncan (Washington Univ.)	NCHGR
Flybase	W. Gelbart (Harvard), M. Ashburner (Cambridge, UK), T. Kaufman and K. Mathews (Indiana University)	NCHGR MRC (UK)

Drosophila has now been a model organism for genome research for over 80 years. The fact that the recombination frequency could be used to rank the genes on a linear map was shown, and the first genetic maps were drawn, using *Drosophila* back in 1913[157]. Since then *Drosophila* has remained, until the recent genome projects, the metazoan with the fullest and most accurate genetic map. The first physical maps were polytene chromosome maps of *Drosophila* made 60 years ago by C. Bridges[158]. They had a resolution of about 100 Kb but allowed the siting of hundreds of genes at short physical intervals with classical cytogenic methods. In the early 1970s, when recombinant DNA methods became available, *Drosophila* was the only organism with a physically mapped genome. This map, and the precise mapping that could be carried out by hybridizing the polytene chromosomes *in situ*[159], led to a number of innovative studies in David Hogness's laboratory in the mid-1970s. Among these, he made the first chromosome map of repetitive, unique or dispersed and cloned DNA segments[160] and developed procedures for sieving clones to build large chromosomal contigs and for positional cloning[161]. Dozens of *Drosophila* genes identified by mutants with physiologically interesting or developmental phenotypes were positionally cloned by the early 1980s.

Physical maps based on the polytene clones are, of course, valuable as accurate reference maps, but they are merely cytogenetic maps, not the tools clone-based maps are. The first systematic attempts to draw a physical map

---

[157] A.H. Sturtevant, "The linear arrangement of six sex-linked factors in *Drosophila*, as shown by their mode of association", *J. Exp. Zool.*, vol. 14, 1913, pp. 43–59.

[158] C. Bridges, "Correspondence between linkage maps and salivary chromosome structure, as illustrated in the tip of chromosome 2R of *Drosophila melanogaster*", *Cytologia Fugii Jubilee Volume*, 1937, pp. 745–755.

[159] M.L. Pardue *et al.*, "Cytological localization of DNA complimentary to ribosomal RNA in polythene chromosome of diptera", *Chromosoma*, vol. 29, 1970, pp. 268–290.

[160] P. Wensink *et al.*, "A system for mapping DNA sequences in the chromosomes of *Drosophila melanogaster*", *Cell*, vol. 3, 1974, pp. 315–325.

G.M. Rubin *et al.*, "The chromosomal arrangement of coding sequences in a family of repeated genes", *Proc. Nucleic. Acid Res. Mol. Biol.*, vol. 19, 1976, pp. 221–226.

[161] M. Grunstein and D.S. Hogness, "Colony hybridization: A method for the isolation cloned DNAs that contain a specific gene" *Proc. Nat. Acad. Sci*, USA, vol. 72, 1975, pp. 2961–2965.

based on *Drosophila* genome clones were those of the Hartl and Duncan groups and the cosmid maps of the European *Drosophila* Mapping Consortium (EDMC). In 1992, the Berkeley *Drosophila* Genome Project (BDGP) began work on a P1 STS map. All these maps are cross-referenced, the EDML and BDGP maps with additional cross reference STSs.

YAC Maps

YAC maps were built using *in situ* hybridizations of polytene chromosomes with individual YACs. The Hartl group[162] mapped 1,193 YAC clones with an average insert size of 207 Kb, and Duncan's group[163] mapped 855 euchromatic YACs with an average insert size of 211 Kb. Together, these maps cover about 90% of the euchromatic genome of the autosomes and 80% of the X chromosome. However, the overlaps between the YACs have not been confirmed by molecular methods. The distribution of available YAC clones seems more or less at random over most of the euchromatic genome, and some regions have not or have barely been covered by the clones.

Cosmid Maps

The approach used by the European consortium[164] (EDMC) to make the cosmid maps was to produce individual and contiguous maps, each covering a chromosome division of about 1 Mb, with added STS markers generated from the ends of the inserts of the mapped cosmids[165]. The X chromosome

---

[162] D.L. Hartl *et al.*, "Genome structure and evolution in *Drosophila*: Applications of the framework P1 map", *Proc. Natl. Acad. Sci.*, USA, vol. 91, 1994, pp. 6824–6829.

J.W. Ajioka *et al.*, "*Drosophila* genome project: One hit coverage in yeast artifical chromosome", *Chromosoma*, vol. 100, 1991, pp. 495–509.

[163] H. Cai, J. Kiefel, J. Yee and I. Duncan, "A yeast artificial chromosome clone map of the *Drosophila* genome", *Genetics*, vol. 136, 1994, pp. 1385–1401.

[164] The European venture began in 1988 and ended in 1993 under the SCIENCE program, with a European Community funding of 718,000 ECU.

[165] I. Siden, R.D.C. Saunders *et al.*, "Towards a physical map of the *Drosophila melanogaster* genomic divisions", *Nucleic Acids Res.*, vol. 18, 1990, pp. 6261–6270.

is the best mapped at present, with 62% of the euchromatic part of the chromosome covered and about 560 STS markers[166]. The autosomal maps are at about the same stage. Over 1,300 STS markers with an average length of 400 bp have been discovered by sequencing the ends of the cosmids. Eight percent of these STSs are either known *Drosophila* genes or P1 clones and BDGP STS markers providing links for future maps, whereas 3% of the STS markers are very similar to genes in other organisms.

The P1 Map

The BDGP is building a physical map with clones using a P1 *bacteriophage* genome library. The first stage was to build a work base map with *in situ* hybridizations on the *polythene* chromosome of the 267 P1 clones which have an average size of 80 Kb[167]. This map covers about 70% of the euchromatic genome. The second stage used STS markers made from the ends of the genome inserts of the P1 clones of the working map. Nearly 300 STSs obtained from the ends of P1 clones have been mapped to date. When the marker pair is adapted to the library, the resulting contigs extend in both directions from the clone and on average cover some 200 Kb of the genome. This average contig size is larger than that available from non-random mapping methods or by unique extremities. Furthermore, simulations indicate that the number of markers needed to assign all the clones in the library to contigs will be fewer using this approach, which is called the "double-ended" approach[168]. These phases are over now. 746 P1 clones have been assigned to 682 contigs covering about 90% of the euchromatic genome

---

[166] E.G. Madueno *et al.*, "A physical map of the chromosome X of *Drosophila melanogaster*: Cosmid contigs and sequence tagged sites", *Genetics*, vol. 139, 1995, pp. 1631–1647.

[167] D.A. Smoller *et al.*, "Characterization of bacteriophage P1 ubiary containing inserts of *Drosophila* DNA of 75–100 kilobase pairs", *Chromosoma*, vol. 100, 1991, pp. 487–494.

D.L. Hartl *et al.*, "Genome structure and evolution in *Drosophila*: Applications of the framework P1 map", *Proc. Natl. Acad. Sci.*, USA, vol. 91, 1994, pp. 6824–6829.

[168] (Double end clone-limited approach) M.L. Palazzolo *et al.*, "Optimized strategies for STS selection in genome mapping", *Proc. Natl. Acad. Sci*, vol. 88, 1991, pp. 8034–8038.

with an STS about every 50 Kb[169]. The positions and approximate sizes of the gaps are known since all the contigs have been traced to the polytene map. Two thousand and two hundred of the 9,216 clones have not yet been assigned to contigs and are used as STS markers, which will help fill the remaining 10% of the euchromatic genome. It is true that most of the gaps that remain are due to overlaps that have not been found yet rather than regions that have not been cloned, and work is ongoing to fill those gaps.

In order to provide a more direct link between the physical and genetic maps, the ranked group of P1 clones in the physical is being used as a substrate for other STS maps in which the STS have been obtained from markers that are also genetically mapped, including individual genes which have been cloned and sequenced by researchers (354 STS) and the sites of P element insertions which disrupt essential genes (232 STS).

The Problem of Heterochromatic Segments

One of the most striking and enigmatic aspects of the way pluricellular eucaryote genomes are organized is the chromosome division in the euchromatic and heterochromatic regions. Heterochromatic segments can be differentiated from euchromatic ones by the rarity of genes, the tightly packed chromatin structure throughout the cell cycle, unusual coloration, late replication in the S phase and a high content of repeated sequence. In the fruit fly, about a quarter of the genome is heterochromatic including a quarter of the X chromosomes, the 2nd and 3rd chromosomes, most of the Y chromosome and the 4th chromosome. The essential functional components are in the heterochromatic regions, such as the centromeres, the telomeres, the ribosomal RNA genes and 30 to 50 protein coding genes.

Although progress in the comprehension of the structural and molecular composition of the heterochromatins has been slow because most of this DNA, especially the satellite DNA, can't be cloned into current cosmids, YACs or P1 vectors with any stability, some progress has been made. The

---

[169] BDGP www homepage.

data obtained from the construction of the P1 working map suggests that the P1 library has a large quantity of non-satellite heterochomatin sequences[170].

Karpen and his colleagues are working on the analysis of the structure and function of the heterochromatic regions of the minichromosome Dp1187 (1.3 Mb), which is a derivative of the X chromosome[171]. Recently Dp1187 derivatives generated through irradiation mutagenesis have had their structure determined. Mini-chromosome derivatives with a breach in the euchromatin and the heterochromatin provide access points for pulsed field restriction mapping of central heterochromatin regions previously inaccessible. The map showed three large central islands containing some repeat and some unique sequences separated by oceans of satellite sequence. DNA field pulse blot analysis showed that in general the fruit fly's heterochromatin is made up of an alternation of complex DNA blocks and blocks of simpler satellite DNA, each hundreds of Kb long. The complex DNA blocks themselves have a complex substructure and many transposable insertion elements. The conclusion of these studies is that a surprisingly significant amount of substructure can be found in the central heterochromatin of the fruit fly. DNA repeats have made heterochromatic genetic analysis very difficult. Several molecular and cytological studies have linked satellite DNA to the centromeres in mammals, but the function of this satellite DNA is not clear, mainly because the behavior of components molecularly defined during transmission has not been analyzed directly yet, except for some regions of the central heterochromatin[172].

The Drosophila "EST" Project

The objective of the project is to provide at least 40,000 ESTs from sequencing the 5' ends and some 3' of high quality cDNA. The target is to be able to

---

[170] D.L. Hartl *et al.*, "Genome structure and evolution in *Drosophila*: Applications of the framework P1 map", *Proc. Natl. Acad. Sci., USA*, vol. 91, 1994, pp. 6824–6829.

[171] G.H. Karpen and A.C. Spradling, "Reduced DNA polytenisation of a minichromosome region undergoing position-effect variegation in *Drosophila*", *Cell*, vol. 63, 1990, pp. 97–107.

[172] T. Murphy and G.H. Karpen, "Localization of centromere function in a *Drosophila* minichromosome", *Cell*, vol. 82, 1995, pp. 599–609.

identify cDNA clones from regions as they are sequenced by the DGC. It is thought that obtaining an EST for each of the 40,000 cDNAs will allow the cDNA of up to 70% of the fruit fly genes to be identified. The long term goal of the BDGP EST project is to provide a map of transcripts containing information on the exon-intron structure, the beginning and end sites of transcription, etc., by sequencing the cDNA of regions where the genome sequence is available. These cDNA chunks will be completely sequenced, but only on one of the strands, and with sufficient precision to allow the genomic sequence to be aligned without any ambiguity. By 21 April 1997, 7,504 ESTs had been submitted to the dbEST; 1,350 ESTs were being produced per month and this should rise at the BDGP to 2,000 a month.

The Genome Sequence

The BDGP were the first group to use a directed sequencing strategy for *Drosophila* genome sequencing (Fig. 34). A two-year pilot project involving a team of seven researchers has just ended. During its two years, about 2 Mb of genome sequence was achieved and registered in the data bases. In December 1995, the BDGP obtained funds for a further three years from the NHCGR. Technological progress and economic support should allow output to be increased relatively fast. It may be possible to finish sequencing the 120 Mb euchromatic genome within five to seven years, depending on future funding allocations and international collaboration. As a partner of the European *Drosophila* Genome Project, the Sanger Center started sequencing chromosome X; the cosmids were provided by Dr. Inga Siden-Kiamos of the Institute of Molecular Biology and Biotechnology, Crete[173].

Table 37 *Drosophila melanogaster*
(Data generated 25 January 1999)

	Size Mb	Redund %(estm)	Finished %(adj'd)	Finished (total bp)	Unfinished (total bp)	1999	1998	1997	1996	1995
Total	165	18.9	13.5	27376790	9869973	115931	18176086	4255849	519013	1272815

---

[173] For the strategy, see Fig. 32.

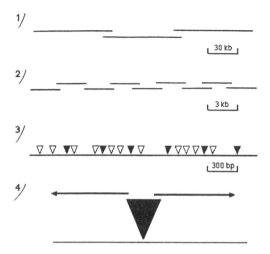

Fig. 33 The managed sequencing strategy
(taken from Gerald M. Rubin's "The *Drosophila* Genome Project",
Genome Research, 6, 1996)

Unlike the sequencing projects that use a shotgun sequencing strategy, the BDGP uses a directed sequencing approach which has been developed and implemented by the Palazzolo and Martin group at the LBNL. This strategy, which is represented in the above diagram, has four stages: A) P1 physical map is produced, providing a minimal set of overlapping clones through an STS mapping strategy; B) the DNA of the individual P1 clones is broken up into sections of 3 kb and subcloned in a plasmid vector, and a set of subclones is identified (The initial approach for generating this subclone set involves the use of a sifting technique based on PCR to identify the overlapping clones from three pools of 960 different clones. Recently, a strategy developed by Bruce Kimmel has been used in which 192 subclones are selected and the two ends are then sequenced. The information to be found in the end sequences is then used to build 3 kb clone contigs, used as transposon targets and completely sequenced. Subsequent contig construction sessions and subclone sequencing is carried on until the 80 kb insert of the P1 is contiguous); C) gamma delta transposons are mobilized in target sequences of 3 kb by appropriate bacterial conjugation (each clone has an independent insertion and the insertions across the clone set are mapped with PCR by using the YS element and primer-vectors; D) a minimal set of clones with transposons at every 400 bp (indicated by the black triangles) is selected and sequenced to provide the complete double-strand sequence of each 3 kb insert. The link sites of the sequencing primers are provided by the sequences close to the ends of the transposons. This approach has several potential advantages, it requires less sequencing and the assembly can be carried out by machine. It is based on redundant information which increases precision; and the robust nature of it allows it to be easily considered for automation.

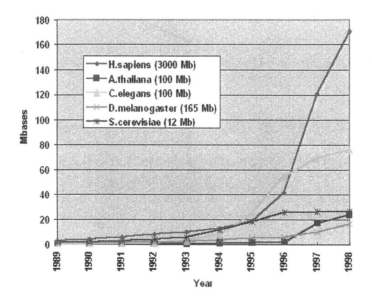

Fig. 34 Progress of major genome sequencing projects

This figure shows a cumulative sequencing progress plot. The sequencing of large genomes clearly took off during the last couple of years. The plot also confirms that the sequencing of the *Saccharomyces cerevisiae* genome was completed in 1996. Please note that the contribution from the current year is from 1 January until the date the graph was created (using datasets from 29 June 1998).

(graph from the "genome Monitoring table": http://www.ebi.ac.uk/~sterk/genome-MOT)

Homo sapiens : 2006
Arabidopsis thaliana : 2002
Caernorhabditis elegans : 1999
Drosophila melanogaster : 2005

Last completion dates

(The finishing dates for the ongoing projects shown in the plot above were predicted using the exponential regression trendline option in Microsoft Excel. This resulted in the above predicted completion dates (based on data drom 17 January 1998.)

The End of "Handmade" Genome Sequencing

Although researchers are only just beginning to systematically sequence the human genome[174], great successes in sequencing micro-organisms and small eucaryote genomes are an important step to the development and optimization of the technologies and strategies needed to sequence the three billion base pairs of the human genome (Fig. 34).

In view of these successes, we can be reasonably optimistic for a reference sequence of the human genome within ten years (see Fig. 35 and Table 38), even without any revolutionary advances in technique. Below are the goals of the US HGP for the period 1998–2003.

New Goals for the US Human Genome Project: 1998–2003

Francis S. Collins, Ari Patrinos, Elke Jordan, Aravinda Chakravarti, Raymond Gesteland, LeRoy Walters, the members of the DOE and NIH planning groups

The Human Genome Project has successfully completed all the major goals in its current five-year plan, covering the period 1993–1998. A new plan, for 1998–2003, is presented, in which human DNA sequencing will be

---

[174] Only some human chromosomes (16, 19, 21, 22, and X) have enough clones ready for sequencing to start immediate sequencing work and only partial maps are available for some of them. Some feasibility study pilot projects on large-scale sequencing have been set up, including the pilot project at the NIH National Center for Human Genome Research.

the major emphasis. An ambitious schedule has been set to complete the full sequence by the end of 2003, two years ahead of previous projections. In the course of completing the sequence, a "working draft" of the human sequence will be produced by the end of 2001. The plan also includes goals for sequencing technology development; for studying human genome sequence variation; for developing technology for functional genomics; for completing the sequence of Caenorhabditis elegans *and* Drosophila melanogaster *and* starting the mouse genome; for studying the ethical, legal, and social implications of genome research; for bioinformatics and computational studies; and for training of genome scientists.

The Human Genome Project (HGP) is fulfilling its promise as the single most important project in biology and the biomedical sciences — one that will permanently change biology and medicine. With the recent completion of the genome sequences of several microorganisms, including Escherichia coli *and* Saccharomyces cerevisiae, *and the imminent completion of the sequence of the metazoan* Caenorhabditis elegans, *the door has opened wide on the era of whole genome science. The ability to analyze entire genomes is accelerating gene discovery and revolutionizing the breadth and depth of biological questions that can be addressed in model organisms. These exciting successes confirm the view that acquisition of a comprehensive, high-quality human genome sequence will have unprecedented impact and long-lasting value for basic biology, biomedical research, biotechnology, and health care. The transition to sequence-based biology will spur continued progress in understanding gene-environment interactions and in development of highly accurate DNA-based medical diagnostics and therapeutics.*

Human DNA sequencing, the flagship endeavor of the HGP, is entering its decisive phase. It will be the project's central focus during the next five years. While partial subsets of the DNA sequence, such as expressed sequence tags (ESTs), have proven enormously valuable, experience with simpler organisms confirms that there can be no substitute for the complete genome sequence. In order to move vigorously toward this goal, the crucial task ahead is building sustainable capacity for producing publicly available DNA sequence. The full and incisive use of the human sequence, including comparisons to other vertebrate genomes, will require further increases in sustainable capacity at high accuracy and lower costs. Thus, a high-priority commitment to develop and deploy new and improved sequencing technologies must also be made.

The Planning Process

The last five-year plan for the HGP, published jointly by NIH and DOE in 1993, covered fiscal years 1994 through 1998. The current plan is again a joint effort and will guide the project for fiscal years 1999 through 2003.

The goals described below have resulted from a comprehensive planning and assessment process that has taken place over the past year in both agencies. Each agency identified a group of advisors to oversee its process, and eight workshops were held to address specific areas of the plan. A large number of scientists and scholars as well as public representatives participated in these events, including many who had no historical ties to the HGP. Comments were also sought from an extensive list of biotechnology and pharmaceutical companies. A draft of the goals was presented for evaluation at a public meeting in May 1998. Suggestions and comments from that meeting were incorporated into the plan. Finally, the new goals were reviewed and approved by the National Advisory Council for Human Genome Research at NIH and the Biological and Environmental Research Advisory Committee at DOE. Summaries of the workshops that contributed to this plan are available at www.nhgri.nih.gov/98plan and www.ornl.gov/hg5yp.

Specific Goals for 1998–2003

The following sections outline eight major goals for the HGP over the next five years.

Goal 1 — The Human DNA Sequence

Providing a complete, high-quality sequence of human genomic DNA to the research community as a publicly available resource continues to be the HGP's highest priority goal. The enormous value of the human genome sequence to scientists and the considerable savings in research costs its widespread availability will allow are compelling arguments for advancing the timetable for completion. Recent technological developments and experience with large-scale sequencing provide increasing confidence that it will be possible to complete an accurate, high-quality sequence of the human genome by the end of 2003, two years sooner than previously predicted. NIH and DOE expect to contribute 60 to 70% of this sequence, with the remainder coming from the effort at the Sanger Center, funded by the Wellcome Trust and other international partners.

This is a highly ambitious, even audacious goal, given that only about 6% of the human genome sequence has been completed thus far. Sequence completion by the end of 2003 is a major challenge, but within reach and well worth the risks and effort. Realizing the goal will require an intense and dedicated effort and a continuation and expansion of the collaborative spirit of the international sequencing community. Only sequence of high accuracy and long-range contiguity will allow a full interpretation of all the information encoded in the human genome. However, in the course of finishing the first human genome sequence by the end of 2003, a "working draft" covering the vast majority of the genome can be produced even sooner, within the next three years. Though that sequence will be of lower accuracy and contiguity, it will nevertheless be very useful, especially for finding genes, exons, and other features through sequence searches. These uses will assist many current and future scientific projects and bring them to fruition much sooner, resulting in significant time and cost savings. However, because this sequence will have gaps, it will not be as useful as finished sequence for studying DNA features that span large regions or require high sequence accuracy over long stretches.

Availability of the human sequence will not end the need for large-scale sequencing. Full interpretation of that sequence will require much more sequence information from many other organisms, as well as information about sequence variation in humans (see also Goals 3, 4, and 5). Thus, the development of sustainable, long-term sequencing capacity is a critical objective of the HGP. Achieving the goals below will require a capacity of at least 500 megabases (Mb) of finished sequence per year by the end of 2003.

a) *Finish the complete human genome sequence by the end of 2003.* The year 2003 is the 50th anniversary of the discovery of the double helix structure of DNA by James Watson and Francis Crick (2). There could hardly be a more fitting tribute to this momentous event in biology than the completion of the first human genome sequence in this anniversary year. The technology to do so is at hand, although further improvements in efficiency and cost effectiveness will be needed, and more research is needed on approaches to sequencing structurally difficult regions (3). Current sequencing capacity will have to be expanded two- to threefold, but this should be within the capability of the sequencing community.

Reaching this goal will significantly stress the capabilities of the publicly funded project and will require continued enthusiastic support from the Administration and the US Congress. But the value of the complete, highly accurate, fully assembled sequence of the human genome is so great that it merits this kind of investment.

b) *Finish one-third of the human DNA sequence by the end of 2001.* With the anticipated scale-up of sequencing capacity, it should be possible to expand finished sequence production to achieve completion of 1 Gb of human sequence by the worldwide HGP by the end of 2001. As more than half of the genes are predicted to lie in the gene-rich third of the genome, the finishing effort during the next three years should focus on such regions if this can be done without incurring significant additional costs. A convenient, but not the only, strategy would be to finish bacterial artificial chromosome (BAC) clones detected by complementary DNA (cDNA) or EST sequences.

In addition, a rapid peer-review process should be established immediately for prioritizing specific regions to be finished based on the needs of the international scientific community. This process must be impartial and must minimize disruptions to the large-scale sequencing laboratories.

To best meet the needs of the scientific community, the finished human DNA sequence must be a faithful representation of the genome, with high base-pair accuracy and long-range contiguity. Specific quality standards that balance cost and utility have already been established. One of the most *important uses for the human sequence will be comparison with other human and nonhuman sequences. The sequence differences identified in such comparisons should, in nearly all cases, reflect real biological differences rather than errors or incomplete sequence. Consequently, the current standard for accuracy — an error rate of no more than one base in 10,000 — remains appropriate. Although production of contiguous sequence without gaps is the goal, any irreducible gaps must be annotated as to size and position. In order to assure that long-range contiguity of the sequence will be achievable, several contigs of 20 Mb or more should be generated by the end of 2001. These quality standards should be reexamined periodically; as experience in using sequence data is gained, the appropriate standards for sequence quality may change.*

c) Achieve coverage of at least 90% of the genome in a working draft based on mapped clones by the end of 2001. The current public sequencing strategy is based on mapped clones and occurs in two phases. The first, or "shotgun" phase, involves random determination of most of the sequence from a mapped clone of interest. Methods for doing this are now highly automated and efficient. Mapped shotgun data are assembled into a product ("working draft" sequence) that covers most of the region of interest but may still contain gaps and ambiguities. In the second, finishing phase, the gaps are filled and discrepancies resolved. At present, the finishing phase is more labor intensive than the shotgun phase. Already, partially finished, working-draft sequence is accumulating in public databases at about twice the rate of finished sequence.

Based on recent experience, the rate of production of working draft sequence can be further increased. By continuing to scale up the production of finished sequence at a realistic rate and further scaling up the production of working draft sequence, the combined total of working draft plus finished sequence will cover at least 90% of the genome at an accuracy of at least 99% by the end of 2001. Some areas of the genome are likely to be difficult to clone or not amenable to automated assembly because of highly repetitive sequence; thus, coverage is expected to fall short of 100% at this stage. If increased resources are available or technology improves, or both, greater than 90% coverage may be possible.

The individual sequence reads used to generate the working draft will be held to the same high-quality standards as those used for the finished genome sequence. Assembly of the working draft should not create loss of efficiency or increases in overall cost.

Recently, two private ventures[175] *announced initiatives to sequence a major fraction of the human genome, using strategies that differ fundamentally from the publicly funded approach. One of these ventures is based upon a whole genome shotgun strategy, which may present significant assembly problem . The stated intention of this venture to release data on a quarterly basis creates the possibility of synergy with the public effort. If this privately funded data set and the public one can be merged, the combined depth of coverage of the working draft sequence will be greater, and the mapping information provided by the public data set will provide critically needed anchoring to the private data. The NIH and DOE welcome such initiatives and look forward to cooperating with all parties that can contribute to more rapid public availability of the human genome sequence.*

d) Make the sequence totally and freely accessible. The HGP was initiated because its proponents believed the human sequence is such a precious

---

[175] In May 1998, J. Craig Venter [The Institute for Genomic Research (TIGR)] announced plans to form a new company with Perkin-Elmer's Applied Biosystems Division (PE-ABD) to sequence a large portion of the human genome in three years for $300 million. The company plans to use several ressources generated in the government-sponsored Human Genome Project. A report by Venter in *Science [280 (5369),1540–42]* called the plan a *"mutually rewarding partnership between public and private institutions."* Although DOE and NIH managers welcomed the promise of substantial private-sector investment, they noted that the planned venture will be more like a rough draft rather than the publicly available, detailed "A-to-Z" recipe book promised by the genome project. Positioning of sequencing subclones into the chromosomal puzzle is expected to be less certain than in more conservative strategies. Data release will be quarterly rather than immediate, as required of government-funded sequencing centers. Because commercial interests will guide the path of the new venture, the focus will be on such potentially lucrative genomic regions as susceptibility and disease-associated sequences that can guide the development of new diagnostic and pharmaceutical products. The company expects to seek intellectual-property protection for 100 to 300 of these sequenced regions. Generating data on biologically important genomic locations can be of great value to researchers and consumers alike. However, genome-maps at the highest level of resolution (those promised by the genome project) still will be needed as the ultimate tools for scientists to embark on a thorough investigation into human biological function in all its complexity.

scientific resource that it must be made totally and publicly available to all who want to use it. Only the wide availability of this unique resource will maximally stimulate the research that will eventually improve human health. Public funding of the HGP is predicated on the belief that public availability of the human sequence at the earliest possible time will lead to the greatest public good. Therefore, NIH and DOE continue to strongly endorse the policy for human sequence data release adopted by the international sequencing community in February 1996 (5), and confirmed and expanded to include genomic sequence of all organisms in 1998 (6). This policy states that sequence assemblies 1 to 2 kb in size should be released into public databases within 24 hours of generation and that finished sequence should be released on a similarly rapid time scale.

Goal 2 — Sequencing Technology

DNA sequencing technology has improved dramatically since the genome project began. The amount of sequence produced each year is increasing steadily; individual centers are now producing tens of millions of base pairs of sequence annually. In the future, de novo sequencing of additional genomes, comparative sequencing of closely related genomes, and sequencing to assess variation within genomes will become increasingly indispensable tools for biological and medical research. Much more efficient sequencing technology will be needed than is currently available. The incremental improvements made to date have not yet resulted in any fundamental paradigm shifts. Nevertheless, the current state-of-the-art technology can still be significantly improved, and resources should be invested to accomplish this. Beyond that, research must be supported on new technologies that will make even higher throughput DNA sequencing efficient, accurate, and cost-effective, thus providing the foundation for other advanced genomic analysis tools. Progress must be achieved in three areas:

a) Continue to increase the throughput and reduce the cost of current sequencing technology. Increased automation, miniaturization, and integration of the approaches currently in use, together with incremental, evolutionary improvements in all steps of the sequencing process, are needed to yield *further increases in throughput (to at least 500 Mb of finished sequence per year by 2003) and reductions in cost. At least a twofold cost reduction from current levels (which average $0.50 per base for finished sequence in large-scale centers) should be achieved in the next 5 years. Production of the working draft of the human sequence will cost considerably less per base pair.*

b) Support research on novel technologies that can lead to significant improvements in sequencing technology. New conceptual approaches to DNA sequencing must be supported to attain substantial improvements over the current sequencing paradigm. For example, microelectromechanical systems (MEMS) may allow significant reduction of reagent use, increase in assay speed, and true integration of sequencing functions. Rapid mass spectrometric analysis methods are achieving impressive results in DNA fragment identification and offer the potential for very rapid DNA sequencing. Other more revolutionary approaches, such as single-molecule sequencing methods, must be explored as well. Significant investment in interdisciplinary research in instrumentation, combining chemistry, physics, biology, computer science, and engineering, will be required to meet this goal. Funding of far-sighted projects that may require five to 10 years to reach fruition will be essential. Ultimately, technologies that could, for example, sequence one vertebrate genome per year at affordable cost are highly desirable.

c) Develop effective methods for the advanced development and introduction of new sequencing technologies into the sequencing process. As the scale of sequencing increases, the introduction of improvements into the production stream becomes more challenging and costly. New technology must, therefore, be robust and be carefully evaluated and validated in a high-throughput environment before its implementation in a production setting. A strong commitment from both the technology developers and the technology users is essential in this process. It must be recognized that the advanced development process will often require significantly more funds than proof-of-principle studies. Targeted funding allocations and dedicated review mechanisms are needed for advanced technology development.

Goal 3 — Human Genome Sequence Variation

Natural sequence variation is a fundamental property of all genomes. Any two haploid human genomes show multiple sites and types of polymorphism. Some of these have functional implications, whereas many probably do not. The most common polymorphisms in the human genome are single base-pair differences, also called single-nucleotide polymorphisms (SNPs). When two haploid genomes are compared, SNPs occur every kilobase, on average. Other kinds of sequence variation, such as copy number changes, insertions, deletions, duplications, and rearrangements also exist, but at low frequency, and their distribution is poorly understood. Basic information about the types, frequencies, and distribution of polymorphisms in the human genome and in human populations is critical for progress in human genetics. Better high-throughput methods for using such information in the study of human disease is also needed.

SNPs are abundant, stable, widely distributed across the genome, and lend themselves to automated analysis on a very large scale, for example, with DNA array technologies. Because of these properties, SNPs will be a boon for mapping complex traits such as cancer, diabetes, and mental illness. Dense maps of SNPs will make possible genome-wide association studies, which are a powerful method for identifying genes that make a small contribution to disease risk. In some instances, such maps will also permit prediction of individual differences in drug response. Publicly available maps of large numbers of SNPs distributed across the whole genome, together with technology for rapid, large-scale identification and scoring of SNPs, must be developed to facilitate this research. The early availability of a working draft of the human genome should greatly facilitate the creation of dense SNP maps (see Goal 1).

a) Develop technologies for rapid, large-scale identification or scoring, or both, of SNPs and other DNA sequence variants. The study of sequence variation requires efficient technologies that can be used on a large scale and that can accomplish one or more of the following tasks: rapid identification of many thousands of new SNPs in large numbers of samples and rapid and efficient scoring of large numbers of samples for the presence or absence of already known SNPs. Although the immediate emphasis is on SNPs, ultimately technologies that can be applied to polymorphisms of any type must be developed. Technologies are also needed that can rapidly compare, by large-scale identification of similarities and differences, the DNA of a species that is closely related to one whose DNA has already been sequenced. The technologies that are developed should be cost-effective and broadly accessible.

b) *Identify common variants in the coding regions of the majority of identified genes during this five-year period.* Initially, association studies involving complex diseases will likely test a large series of candidate genes; eventually, sequences in all genes may be systematically tested. SNPs in coding sequences (also known as cSNPs) and the associated regulatory regions will be immediately useful as specific markers for disease. An effort should be made to identify such SNPs as soon as possible. Ultimately, a catalog of all common variants in all genes will be desirable. This should be cross-referenced with cDNA sequence data (see Goal 4).

c) *Create an SNP map of at least 100,000 markers.* A publicly available SNP map of sufficient density and informativeness to allow effective mapping in any population is the ultimate goal. A map of 100,000 SNPs (one SNP per 30,000 nucleotides) is likely to be sufficient for studies in some relatively homogeneous populations, while denser maps may be required for studies in large, heterogeneous populations. Thus, during this five-year period, a map of at least 100,000 SNPs should be created. If technological advances permit, a map of greater density is desirable. Research should be initiated to estimate the number of SNPs needed in different populations.

d) *Develop the intellectual foundations for studies of sequence variation.* The methods and concepts developed for the study of single-gene disorders are not sufficient for the study of complex, multigene traits. The study of the relationship between human DNA sequence variation, phenotypic variation, and complex diseases depends critically on better methods. Effective research design and analysis of linkage, linkage disequilibrium, and association data are areas that need new insights. Questions such as which study designs are *appropriate to which specific populations, and with which population genetics characteristics, must be answered. Appropriate statistical and computational tools and rigorous criteria for establishing and confirming associations must also be developed.*

e) Create public resources of DNA samples and cell lines. To facilitate SNP discovery, it is critical that common public resources of DNA samples and cell lines be made available as rapidly as possible. To maximize discovery of common variants in all human populations, a resource is needed that includes individuals whose ancestors derive from diverse geographic areas. It should encompass as much of the diversity found in the US population as possible. Samples in this initial public repository should be totally anonymous to avoid concerns that arise with linked or identifiable samples.

DNA samples linked to phenotypic data and identified as to their geographic and other origins will be needed to allow studies of the frequency and distribution of DNA polymorphisms in specific populations and their relevance to disease. However, such collections raise many ethical, legal, and social concerns that must be addressed. Credible scientific strategies must be developed before creating these resources (see Goal 6).

Goal 4 — Technology for Functional Genomics

The HGP is revolutionizing the way biology and medicine will be explored in the next century and beyond. The availability of entire genome sequences is enabling a new approach to biology often called functional genomics — the interpretation of the function of DNA sequence on a genomic scale. Already, the availability of the sequence of entire organisms has demonstrated that many genes and other functional elements of the genome are discovered only when the full DNA sequence is known. Such discoveries will accelerate as sequence data accumulate. However, knowing the structure of a gene or other element is only part of the answer. The next step is to elucidate function, which results from the interaction of genomes with their environment. Current methods for studying DNA function on a genomic scale include comparison and analysis of sequence patterns directly to infer function, large-scale analysis of the messenger RNA and protein products of genes, and various approaches to gene disruption. In the future, a host of novel strategies will be needed for elucidating genomic function. This will be a challenge for all of biology. The HGP should contribute to this area by emphasizing the development of technology that can be used on a large scale, is efficient, and is capable of generating complete data for the genome as a whole. To the extent that available resources allow, expansion of current approaches as well as innovative technology ideas should be supported in the areas described below. Large-scale characterization of the gene transcripts and their protein products underpins functional analysis. Therefore, identifying and sequencing a set of full-length cDNAs that represent all human genes must be a high priority.

a) Develop cDNA resources. Complete sets of full-length cDNA clones and sequences for both humans and model organisms would be enormously useful for biologists and are urgently needed. Such resources would help in both gene discovery and functional analysis. Unfortunately, neither cloning full-length cDNAs nor identifying rare transcripts is yet a routine task. High priority should, therefore, be placed on developing technology for obtaining full-length cDNAs and for finding rare transcripts. Complete and validated inventories of full-length cDNA clones and corresponding sequences should be generated and made available to the community once such technology is at hand.

b) Support research on methods for studying functions of non-protein-coding sequences. In addition to the DNA sequences specifying protein structure, there are numerous sequences responsible for other functions, such as control of gene expression, RNA splicing, formation of chromatin domains, maintenance of chromosome structure, recombination, and replication. Other sequences specify the numerous functional untranslated RNAs. Improved technologies are needed for global approaches to the study of non-protein-coding sequences, including production of relevant libraries, comparative sequencing, and computational analysis.

c) *Develop technology for comprehensive analysis of gene expression. Information about the spatial and temporal patterns of gene expression in both humans and model organisms offers one key to understanding gene expression. Efficient and cost-effective technology needs to be developed to measure various parameters of gene expression reliably and reproducibly. Complementary DNA sequences and validated sets of clones with unique identifiers will be needed for array technologies, large-scale* in situ *hybridization, and other strategies for measuring gene expression. Improved methods for quantifying, representing, analyzing, and archiving expression data should also be developed.*

d) *Improve methods for genome-wide mutagenesis. Creating mutations that cause loss or alteration of function is another prime approach to studying gene function. Technologies, both gene- and phenotype-based, which can be used on a large scale* in vivo *or* in vitro, *are needed for generating or finding such mutations in all genes. Such technologies should be piloted in appropriate model systems, including both cell culture and whole organisms.*

e) *Develop technology for global protein analysis. A full understanding of genome function requires an understanding of protein function on a genome-wide basis. Development of experimental and computational methods to study global spatial and temporal patterns of protein expression, protein-ligand interactions, and protein modification needs to be supported.*

Goal 5 — Comparative Genomics

Because all organisms are related through a common evolutionary tree, the study of one organism can provide valuable information about others. Much of the power of molecular genetics arises from the ability to isolate and understand genes from one species based on knowledge about related genes in another species. Comparisons between genomes that are distantly related provide insight into the universality of biologic mechanisms and identify experimental models for studying complex processes. Comparisons between genomes that are closely related provide unique insights into the details of gene structure and function. In order to understand the human genome fully, genomic analysis on a variety of model organisms closely and distantly related to each other must be supported.

Genome sequencing of E. coli *and* S. cerevisiae, *two of the five model organisms targeted in the first five-year plan, has been completed. Availability of these sequences has led to the discovery of many new genes and other functional elements of the genome. It has allowed biologists to move from identifying genes to systematic studies to understand their function. Completion of the DNA sequence of the remaining model organisms,* C. elegans, D. melanogaster, *and mouse, continues to be a high priority and should proceed as rapidly as available resources allow. Additional model organisms will need to be analyzed to allow the full benefits of comparative genomics to be realized. This ongoing need is a major rationale for building sustainable sequencing capacity (see Goals 1 and 2).*

a) Complete the sequence of the C. elegans *genome in 1998. The DNA sequence of the* C. elegans *genome is well on the way to completion, with a target date of December 1998. Some difficult-to-close regions may remain at the end of this year and should become the subject of research projects aimed at closing them. The lessons learned from this project will be crucial in devising strategies for larger genomes.*

b) Complete the sequence of the Drosophila *genome by 2002. The wealth of information accumulated about* Drosophila *over many decades makes it a critically important genetic model. Its DNA sequence is eagerly awaited by all biologists. A significant increase in investment in* Drosophila *sequencing capacity will be needed to achieve this goal, and the benefits of early completion to comparative biology will be tremendous. Anticipated contributions from the private sector may enable the completion of this goal even earlier than 2002.*

c) The mouse genome. The mouse is currently the best mammalian model for studies of a broad array of biomedical research questions. The complete mouse genome sequence will be a crucial tool for interpreting the human genome sequence because it will highlight functional features that are conserved, including noncoding regulatory sequences as well as coding

598   *From Biotechnology to Genomes: A Meaning for the Double Helix*

sequences. Comparisons between mouse and human genomes will also identify functionally important differences that distinguish mouse from human. Therefore, this is the time to invest in a variety of mouse genomic resources, culminating eventually in full-genome sequencing, to allow development of whole-genome approaches in a mammalian system.

1) Develop physical and genetic mapping resources. The integrated mouse yeast artificial chromosome (YAC)/STS map that has been developed provides a useful framework for the more detailed mapping resources now needed for positional cloning and sequencing projects. These resources should include mapped STSs, polymorphic markers, cDNA sequences, and BACs. The usefulness of SNPs as polymorphic markers in the mouse should also be explored in the near term.

2) Develop additional cDNA resources. More cDNA libraries and cDNA sequences are needed. These should derive from a variety of tissues and developmental stages and have good representation of rare transcripts. The mouse offers an opportunity to capture cDNA sequences from developmental stages, anatomical sites, and physiological states that are under-represented in human cDNA collections, and these should receive particular attention. Full-length cDNAs should be developed and sequenced once the technology for doing this efficiently becomes available (see also Goal 4).

3) Complete the sequence of the mouse genome by 2005. Mouse genomic sequence is an essential resource for interpreting human DNA sequence. For this reason, the centers sequencing human DNA are encouraged to devote up to 10% of their capacity to sequencing mouse DNA. Additional capacity for mouse DNA sequencing should be built up over the next few years with a goal of finishing the mouse sequence by 2005. Initially, a working draft of the mouse genome should be produced even sooner (see Goal 1 for a discussion of a working draft of the human genome sequence).

d) Identify other model organisms that can make major contributions to the understanding of the human genome and support appropriate genomic studies. As DNA sequencing capacity becomes available, new model organisms that can contribute to understanding human biology should be identified for genomic sequencing. Even if such capacity is not available during this five-year period, development of other useful genomic resources *should be considered. The scientific community will need to establish criteria for choosing those models that can make the greatest contribution. Characteristics such as phylogenetic distance from other models, genome size, transfection capability, ability to mutagenize, and availability of experimental material should all be considered. Because different characteristics will be useful for different purposes, organisms that are phylogenetically distant from each other and those that are close should be studied.*

Goal 6 — Ethical, Legal, and Social Implications (ELSI)

While recognizing that genetics is not the only factor affecting human well-being, the NIH and DOE are acutely aware that advances in the understanding of human genetics and genomics will have important implications for individuals and society. Examination of the ethical, legal, and social implications of genome research is, therefore, an integral and essential component of the HGP. In a unique partnership, biological and social scientists, health care professionals, historians, legal scholars, and others are committed to exploration of these issues as the project proceeds. The ELSI program has generated a substantial body of scholarship in the areas of privacy and fair use of genetic information, safe and effective integration of genetic information into clinical settings, ethical issues surrounding genetics research, and professional and public education. The results of this research are already being used to guide the conduct of genetic research and the development of related health professional and public policies. The ELSI program has also stimulated the examination of similar issues in other areas of the biological and medical sciences.

Continued success of the ELSI program will require attention to the new challenges presented by the rapid advances in genetics and its applications. As the genome project draws closer to completing the first human genome sequence and begins to explore human sequence variation on a large scale, it will be critical for biomedical scientists, ELSI researchers, and educators to focus attention on the ethical, legal, and social implications of these developments for individuals, families, and communities. The new goals for

ELSI research and education can be visualized as a pyramid of interrelated issues and activities. Given the complexity of the issues encompassed by the ELSI goals, only a summary of the major areas is presented here. To illustrate more fully the breadth and range of the issues that will be addressed, a Web site has been created that provides examples of the types of research questions and education activities envisioned within each goal .

The major ELSI goals for the next five years are:

a) Examine the issues surrounding the completion of the human DNA sequence and the study of human genetic variation.

b) Examine issues raised by the integration of genetic technologies and information into health care and public health activities.

c) Examine issues raised by the integration of knowledge about genomics and gene-environment interactions into nonclinical settings.

d) Explore ways in which new genetic knowledge may interact with a variety of philosophical, theological, and ethical perspectives.

e) Explore how socioeconomic factors and concepts of race and ethnicity influence the use, understanding, and interpretation of genetic information, the utilization of genetic services, and the development of policy.

<u>Goal 7</u> — Bioinformatics and Computational Biology

Bioinformatics support is essential to the implementation of genome projects and for public access to their output. Bioinformatics needs for the genome project fall into two broad areas: (i) databases and (ii) development of analytical tools. Collection, analysis, annotation, and storage of the ever-increasing amounts of mapping, sequencing, and expression data in publicly accessible, user-friendly databases is critical to the project's success. In addition, the community needs computational methods that will allow scientists to extract, view, annotate, and analyze genomic information efficiently. Thus, the genome project must continue to invest substantially in these areas. Conservation of resources through development of portable software should be encouraged.

a) *Improve content and utility of databases.* Databases are the ultimate repository of HGP data. As new kinds of data are generated and new biological relationships discovered, databases must provide for continuous and rapid expansion and adaptation to the evolving needs of the scientific community. To encourage broad use, databases should be responsive to a diverse range of users with respect to data display, data deposition, data access, and data analysis. Databases should be structured to allow the queries of greatest interest to the community to be answered in a seamless way. Communication among databases must be improved. Achieving this will require standardization of nomenclature. A database of human genomic information, analogous to the model organism databases and including links to many types of phenotypic information, is needed.

b) *Develop better tools for data generation, capture, and annotation.* Large-scale, high-throughput genomics centers need readily available, transportable informatics tools for commonly performed tasks such as sample tracking, process management, map generation, sequence finishing, and primary annotation of data. Smaller users urgently need reliable tools to meet their sequencing and sequence analysis needs. Readily accessible information about the availability and utility of various tools should be provided, as well as training in the use of tools.

c) *Develop and improve tools and databases for comprehensive functional studies.* Massive amounts of data on gene expression and function will be generated in the near future. Databases that can organize and display this data in useful ways need to be developed. New statistical and mathematical methods are needed for analysis and comparison of expression and function data, in a variety of cells and tissues, at various times and under different conditions. Also needed are tools for modeling complex networks and interactions.

d) *Develop and improve tools for representing and analyzing sequence similarity and variation.* The study of sequence similarity and variation within and among species will become an increasingly important approach to biological problems. There will be many forms of sequence variation, of which SNPs will be only one type. Tools need to be created for capturing, displaying, and analyzing information about sequence variation.

e) Create mechanisms to support effective approaches for producing robust, exportable software that can be widely shared. Many useful software products are being developed in both academia and industry that could be of great benefit to the community. However, these tools generally are not robust enough to make them easily exportable to another laboratory. Mechanisms are needed for supporting the validation and development of such tools into products that can be readily shared and for providing training in the use of these products. Participation by the private sector is strongly encouraged.

Goal 8 — Training

The HGP has created the need for new kinds of scientific specialists who can be creative at the interface of biology and other disciplines such as computer science, engineering, mathematics, physics, chemistry, and the social sciences. As the popularity of genomic research increases, the demand for these specialists greatly exceeds the supply. In the past, the genome project has benefited immensely from the talents of nonbiological scientists, and their participation in the future is likely to be even more crucial. There is an urgent need to train more scientists in interdisciplinary areas that can contribute to genomics. Programs must be developed that will encourage training of both biological and nonbiological scientists for careers in genomics. Especially critical is the shortage of individuals trained in bioinformatics. Also needed are scientists trained in the management skills required to lead large data-production efforts. Another urgent need is for scholars who are trained to undertake studies on the societal impact of genetic discoveries. Such scholars should be knowledgeable in both genome-related sciences and in the social sciences. Ultimately, a stable academic environment for genomic science must be created so that innovative research can be nurtured and training of new individuals can be assured. The latter is the responsibility of the academic sector, but funding agencies can encourage it through their grants programs.

a) *Nurture the training of scientists skilled in genomics research. A number of approaches to training for genomics research should be explored. These include providing fellowship and career awards and encouraging the development of institutional training programs and curricula. Training that will facilitate collaboration among scientists from different disciplines, as well as courses that introduce scientists to new technologies or approaches, should also be included.*

b) *Encourage the establishment of academic career paths for genomic scientists. Ultimately, a strong academic presence for genomic science is needed to generate the training environment that will encourage individuals to enter the field. Currently, the high demand for genome scientists in industry threatens the retention of genome scientists in academia. Attractive incentives must be developed to maintain the critical mass essential for sponsoring the training of the next generation of genome scientists.*

c) *Increase the number of scholars who are knowledgeable in both genomic and genetic sciences and in ethics, law, or the social sciences. As the pace of genetic discoveries increases, the need for individuals who have the necessary training to study the social impact of these discoveries also increases. The ELSI program should expand its efforts to provide postdoctoral and senior fellowship opportunities for cross-training. Such opportunities should be provided both to scientists and health professionals who wish to obtain training in the social sciences and humanities and to scholars trained in law, the social sciences, or the humanities who wish to obtain training in genomic or genetic sciences.*

Considering the various programs that have been finished or are in progress, one notes that sequencing ventures are rapidly becoming more and more concentrated in nature. The network sequencing strategy developed by the Europeans for the *S. cerevisiae* genome proved its efficiency beyond all doubt. This original system was later used for other European sequencing ventures. However, it is not the strategy which today characterizes most of the worldwide programs. Instead of the dozens of European laboratories and 600 scientists and students deeply dedicated to particular aspects of yeast molecular biology to be found in the European yeast genome project, these programs involve large sequencing centers where the sequencing is as automated as possible and carried out by specialized scientists and technicians.

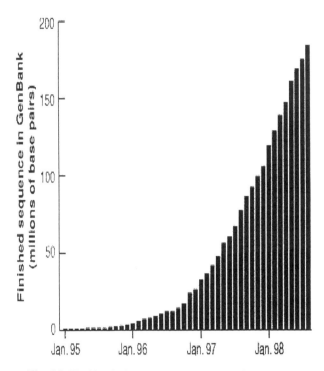

Fig. 35 Worldwide human genome sequencing progress (measured as base pairs of finished sequence deposited with GenBank)

In practice, all possible systems between these and the original network have been used, but it is clear that this industrial approach is supplanting the network approach even within Europe. All things considered, the field of sequencing is also affected by general trends and the key concepts of our time: financial viability, efficiency, automation and concentration. The 600 scientists and technicians and the many laboratories involved in the yeast project, which as mentioned above reinforced the cohesion of the yeast scientist community and generated a team spirit among them, particularly in

Table 38 Status of the human genome sequencing project
(*data generated 25 January 1999*)

	Size (Mb)	Redund %(estm)	Finished %(adj'd)	Finished (total bp)	Unfinished (total bp)	1999	1998
Chr. 1	263	1.0	3.4	9000416	11622031	18301	4752423
Chr. 2	255	3.0	0.6	1565841	92693	3926	538398
Chr. 3	214	1.9	0.9	1972220	384228	39848	650805
Chr. 4	203	1.9	2.9	6019449	10559000	0	3536032
Chr. 5	194	7.0	6.2	12980595	84383	163520	10354944
Chr. 6	183	1.9	10.7	19947700	15511894	55728	11006993
Chr. 7	171	0.4	23.5	40345216	2439496	341628	28625673
Chr. 8	155	0.2	2.6	4012513	116160	392926	3069761
Chr. 9	145	0.5	2.9	4156369	1122809	221311	1573552
Chr. 10	144	4.1	2.0	3061856	5145862	1369	2795893
Chr. 11	144	3.1	2.9	4280789	3493751	217258	613569
Chr. 12	143	1.4	3.2	4574780	1095724	33820	2008243
Chr. 13	114	4.1	1.5	1729741	497186	8220	62194
Chr. 14	109	3.6	1.6	1782489	0	18448	1403654
Chr. 15	106	0.3	0.6	678620	0	9203	40862
Chr. 16	98	0.2	16.1	15796798	2253860	17763	6865414
Chr. 17	92	1.2	27.2	25311426	5657022	411829	19842938
Chr. 18	85	0.0	1.3	1103410	391512	29436	1010225
Chr. 19	67	3.0	18.3	12674041	0	606643	9386020
Chr. 20	72	1.8	3.8	2817850	11445144	6670	2011824
Chr. 21	50	4.5	15.4	8068982	5567231	338842	6759678
Chr. 22	56	2.5	30.9	17778572	14885330	1183	3056472
Chr. X	164	1.0	25.3	41918176	11872521	526129	14632613
Chr. Y	59	0.8	2.5	1515962	320480	7162	753860
Subtotal:	3286	1.8	7.3	243093811	104558317	3471163	135352040
Chr. ??	n.a.	12.7	n.a.	37642391	61964091	514630	15017486
Total:	3286	3.2	8.3	280736202	166522408	3985793	150369526

Europe, used that cohesion and team spirit to share necessary data and ideas to rise to the challenge of deciphering the functions and interactions of newly discovered genes. Today's sequencing projects, however, involve fewer laboratories, far fewer scientists (and more technicians) and a large quantity of machines. Perhaps this evolution is inexorable. But in this mutation from genome projects to industrial enterprises, a certain spirit is lost for the researchers, and with it some of the pleasure and enthusiasm.

7.6 From Science to Economics

A New Era

The shock waves of the current biological revolution are washing over all the nations of the world. It is having as sizable and varied effects as the other revolutions that have changed the face of the world. Twenty years ago, the understanding of the genetic "text" of an organism was a dream. Today the list of genomes which have been sequenced grows day by day, to the point that it is increasingly difficult to follow current projects across the world as well as their results. It is risky to base perspectives on such an incomplete exercise, but a sketch of this multisectoral revolution and its future can still be roughed out.

Any genome research proceeds from the science of genetics, which is only a hundred years old, and its sister science of molecular biology. However, all genetic and molecular biology research is not automatically also genome work. It's often hard to see the borderline. It is clear that the genome is more than a simple string of genes and that DNA cannot operate without its environment. It's like a musical score; without instruments and musicians, it cannot be played. A lot of criticism can be made of the gradually building empire of genome research in biology, and it is valid. At the same time, one must recognize that if the molecular genetics aspects have been so privileged, especially in the 1980s, it is because they have proved the best way to clarify some of the fundamental mechanisms in all living things. The results prove beyond all doubt that this was the way to go.

Despite the dire predictions, the molecular approach seems to be successfully climbing the ladder of complexity. As research progresses, our lack of understanding of the complexity becomes more and more apparent. Genome research follows that rule too. It implies a particular degree of complexity, the need for long and systematic exploration which should integrate structure, function, relationships and interactions, ontogenesis and phylogenesis. Mutational changes are typical of genome research. This degree of complexity has many consequences. On analysis of the various programs underway, it is clear that this complexity has forced a greater concentration in the choice of targets, methods and techniques, and a concentration of research as well as the constitution of networks to harbor the collaboration, information exchange and training of researchers. At the international level, coordinating bodies have evolved and many formal and informal agreements have been achieved at various levels of the decision making process. A new biology is being born wherein the localized empirical approach at the laboratory level is being supplanted by national and international projects and plans with scientific, political, economic, ethical and religious effects.

No country can ignore the biotechnological revolution. Across the five continents, nations are launching their own genome conquest programs and pouring appreciable sums of money into them so as not to be left out in the coming world genome gold rush. For over a century, the advocates of biotechnology have been coming out with ever more sensational prophesies, and many of them have not come true. However, as our understanding improves, they become increasingly less improbable.

After stealing the fire of the nucleus, Promethean Man is appropriating the power to control and direct the evolution of Nature and his own species. With the techniques of transgenesis and his understanding of the structure of genomes, Man has acquired the power to modify the genomes of living things and orient biological evolution. By exploring and manipulating the heredity of acquired characteristics, Genetic Man is entering a time of Lamarkian biological evolution defined by three terms: molecularisation, informatisation and globalisation [176]. The harbingers of this time are those researchers who initiated and undertook the journeys through the genomes, patiently deciphering the text written at the core of life. In many ways, these journeys can be compared to those of 15th and 16th Century explorers. The genome journeys are also financed by government or international

---

[176] European Commission, 1982: *report of the FAST Program.*

organizations, their success depending on and stimulating the invention and development of many scientific, technological and social fields.

Like the early navigators, these researchers have discovered new territories to be mapped. Genome research will also have colossal effects; and although there are no countries to be conquered, there are new territories to annex, namely the new economic markets for biotechnological projects. The genome journeys are based on scientific, but especially political and commercial intentions. Their successes are celebrated not only for the new light brought to fundamental problems in biology and new discoveries made, but also for future research and applications. The conquest of new territories has often meant trade, prosperity and well-being for the conquerors, but there have also been unpleasant, counter-productive, perhaps poisonous secondary effects which, truth be told, are to be found with every invention. The most recent example of atomic research is a significant model from this point of view. What can we expect from this new biological era?

Beyond the Sequences

As Sydney Brenner says, "We've accumulated a large number of sequences from a variety of organisms. The problem is nobody knows what they do". The value of a gene is in the context of the physiology of an organism. To learn anything about physiology, development or evolution, scientists need to convert their laundry lists of genes into detailed maps of molecular pathways and interconnected networks of protein function.

The territories being explored are rich in discoveries, like those of the West Indies. As the number of completely sequenced genomes rises, it is becoming clear that the understanding gained through classical approaches to the contents of genomes was seriously lacking. Taking into account results obtained today in the various programs, between 30 and 40% of the genes discovered had no structural or functional similarities detectable in *in silico* analysis. This should keep the biologists occupied for several years. Current work is already providing an excellent estimation of the numbers of genes of different organisms in different evolutionary lineages (Table 39).

Table 39 Current prediction of the approximate number of genes and genome size in organisms of various evolutionary lineages

		GENES	GENOME IN SIZE (in Mb)
Procaryota	Mycoplasma genitalium	473	0.58
	Haemophilus influenzae	1,760	1.83
	Bacillus subtilis	3,700	4.2
	Escherichia coli	4,100	4.7
	Myxococcus xanthus	8,000	9.45
Fungi	Saccharomyces cerevisiae	6,300	13.5
Protoctista	Cyanidioschyzon merolae	5,400	11.7
	Oxytricha similis	12,000	600
Arthropoda	Drosophila melanogaster	12,000	165
Nematoda	Caernorhabditis elegans	14,000	100
Mollusca	Loligo pealii	>35,000	2,700
Chordata	Ciano intestinalis	N	165
	Fugu rubripes	70,000	400
	Danio rerio	N	1,900
	Mus musculus	70,000	3,300
	Homo sapiens	70,000	400
Plantae	Nicotina tabacum	43,000	4,500
	Arabidopsis thaliana	16,000–33,000	70–145

The number of bacterial genes varies between about 500 and 8,000 and overlaps that of unicellular eucaryotes. Eucaryotes with very different organizational complexity, such as the protozoans (*Caenorhabditis* and *Drosophila*), have a similar number of genes, between 12,000 and 14,000. A comparison with other organisms shows that a unicellular protozoan, a nematode and a fly develop and function with between 12,000 and 14,000 genes. These examples show that there can be enormous differences at the level of morphological complexity between organisms with a similar number of genes. The number of genes *per se* is, therefore, no measure of biological complexity. The increase of the average quantity of DNA inhabited by a genetic unit of 1 kb can be used as a measure of the necessary increase in regulatory elements in metazoan creatures. The number of biochemical routes and mechanisms is likely to be similar from metazoan to metazoan. The data on polysomal mRNA shows that *Loligo pealii* has at least 35,000 genes[177]. Later re-evaluations show that this plant might have as many as 43,000 genes[178]. So, if we exclude vertebrates, the variation in numbers of genes in multicellular eucaryotes is between 12,000 and 43,000 genes[179]. The human and mouse genomes have approximately 70,000 genes.

The question highlighted by these figures is why mammals have six times more genes than *C. elegans* and *D. melanogaster*. One possibility is that a significant fraction of this increase has occurred through polyploidisation, an evolutionary event common in many of the unicellular and metazoan lines[180]. It is thought that the evolution of the mammal genome included at least two of these events, duplicating an ancestral genome as well as the duplication of chromosomal sub-segments with a massive duplication of genes, giving rise to the various gene families. If the genome projects show this octuple nature of man and mouse genomes, then the basic number of genes in vertebrates could indeed be similar to that found in the worm and the fly, at 12,000 to 14,000 genes. It should be noted that *Ciona intestinalis* has a similar genome size and quantity of repetition to the fruit fly. If this is indicative of a fundamental genome for *Chordata*, then the number of biochemical routes and mechanisms should not be very different in flies, nematodes, the first *chordata* or humans. The duplicate routes and mechanisms in mammals are, however, likely to have adopted specialized biological expressions and functions.

---

[177] C.P. Capano *et al.*, "Complexity of nuclear and plysomal RNA from squid optic lobe and gill", *Journal of Neurochemistry*, vol. 46, 1986, pp. 1517–1521.

[178] J.C. Kamalay and R.B. Goldberg, "Regulation of structural gene expression in tobacco", *Cell*, vol. 19, 1980, pp. 935–946.

G. Maroni, "The organization of eukaryotic genes", *Evolutionary Biology*, vol. 29, 1986, pp. 1–19.

[179] F. Antegnera and A. Bird, "Number of CPG island and genes in human and mouse", *Proc. Natl. Acad. Sci*, USA, vol. 90, 1993, pp. 11995–11999.

F.S. Collins "Ahead of schedule and under budget: The genome project passes its fifth birthday", *Proc. Natl. Acad. Sci*, USA, vol. 92, 1995, pp. 1081–10823.

[180] P.W.H. Holland *et al*, "Gene duplications and the origins of vertebrate development", *Development*, 1994, supplement, pp. 125–133.

Table 40 Percentage of duplication

H. Influenzae	30% of the 1,760 genes
E. coli	46% of the 4,100 genes
Yeast	30% of the 5,800 genes

For yeast, the mouse and man, there is not enough data to provide a percentage estimate. Most of the mouse and man genes can be represented in multigene families, some of which have hundreds, if not thousands, of members. Estimate are of 2,000 kinase proteins and about 1,000 phosphatases, to be compared with the 117 kinase proteins and the 40 phosphatases of yeast. However, if mammal genes are indeed octuple, then a substantial proportion of the mouse and man genomes should have at least initially been created in duplication events. In multicellular organisms, functional duplication copies of a gene can be found in one genome, but if their expression patterns don't overlap, their products cannot be substituted for one another if one of the genes mutates. Information obtained with the help of databases are an essential guide for the analysis of the extent of potential compensation during the life cycle by providing detailed information on the expression sites of each gene. The consequences of certain genomic disruptions cannot be compensated for by normal epigenetic processes and result in the death of the organism. Table 41 shows the number of transcription units and genes *per* organism that, inactivated, are lethal.

Table 41 The estimated number of transcription units and lethal loci in different organisms (after Rubin)

Organism	Transcription units	Lethal loci
S. cerevisiae	6,300	1,200
C. elegans	14,000	2,700–3,500
D. melanogaster	12,000	3,600
A. thaliana	25,000	500
D. rerio	N	5,000
F. rubupes	70,000	N
M. musculus	70,000	5,000–26,000

The interpretation of inactivation and deletion data should, however, be taken with a grain of salt. In fact, it can be difficult to detect subtle phenotypic alterations in laboratory conditions. Furthermore, methods used to evaluate functions are often inadequate and little modifications in the creature's "fitness" are not usually measured. An example given by Steve Oliver clearly points out the difficulties:

"Although YCR32w (6,501 base pairs) is the longest ORF on the yeast chromosome III, its complete deletion had no obvious effect, and the mutant was exposed to many physiological challenges without results before it was discovered that it died when grown at low pH on glucose and challenged with acetic acid[181]. As YCR32w encodes a membrane protein, its product is probably an acetic acid exit pump. But can we reproduce this exercise for 3,000 more genes of unknown function once the yeast genome sequence is complete ? A more systematic approach must be found"[182].

In pluricellular organisms, it is not always possible to completely understand the phenotypic consequences of disrupting or disturbing a gene. In man, mono-genetic disease is rare. A complete understanding of phenotypic modifications caused by the mutation of one gene involves the understanding of the different cell types, the stages of development and the cell processes in which it functions, as well as the compensatory modifications that may appear to carry out the function in a different way. Two of the great challenges ahead will be to examine the inactivity of one or of many genes in different contexts and to isolate, map and characterize elements linked to the variations.

Almost all Gene Products are Expressed and Used in Various Places and Times during Development...

Classical genetics studies of the mouse and fruit fly have shown that some genes affect several aspects of the phenotype. These are called pleiotropic genes. In molecular terms, pleiotropy only happens when a protein or RNA is needed to function in different places or at different times, or both. A large-scale analysis of functional requirements was carried out on the fruit

---

[181] Y. Jia, *thesis*, Univ. Pierre et Marie Curie, Paris, 1993.

[182] S.G. Oliver, "From DNA sequence to biological function", *Nature*, vol. 379, 1996, pp. 597–600.

fly. Results showed that 75% of the 3,600 lethal loci of the fly genome are functionally required during ontogenesis. Analysis of the constitution and neural connections of the developing eye provides a similar result; 70% of the 3,600 lethal loci appear to be involved in the development of the eye. If the pleiotropy of lethal loci is not substantially different to that of non-lethal loci, then more than 70% of the genes in a genome are used in the construction of each of these organ systems.

Another indication of potential pleiotropy can be found while studying gene expression, which almost always reveals the expression of a gene in more than one place and time. Several examples of localized gene activity have been found, and it seems clear that all fruit fly genes are expressed in at least two different times or places during its development. Despite this, we cannot systematically assume that when a protein is expressed in a cell, it is there for functional reasons. Aspects of expression patterns might simply reflect the consequences of a dysfunction of the network or networks in which the gene participates.

The Databases on Structure and Expression Patterns will be very Important but Clearly not Enough to Allow the Deciphering of the Regulation Networks...

One of the approaches allowing functional evaluation of regulatory elements is the identification of regulatory regions maintained throughout the evolution of the organism, through interspecies comparisons and along with transgenic analysis. For example, comparison of DNA sequences of "promoter" regions of three different rhodopsin genes in *D. melanogaster* and *D. virilis* show a whole set of matching sequences, interchangeable with additional sequences conferring a type of cellular specificity. Detailed mutagenesis of 31 regulatory regions show that seven of the eight preserved sequences have their functions disturbed when they mutate, whereas none of the 23 non-preserved regions have disturbed functions when they are altered[183].

---

[183] M.E. Fortini and C.M. Rubin, "Analysis of cis-acting requirement of the Rh3 and Rh4 genes reveals a bipartite organization to rhodopsin promotors in *Drosophila melanogaster*", *Genes and Development*, vol. 4, 1990, pp. 444–463.

Computer analyses of different species show, without doubt, that a proportion of basic sequences are preserved and involved in regulation. Experimental analysis should show whether this remains valid within and between phyla.

To What Point Does the Understanding of Components of Regulations during Development Provide Information on the Forces of Molecular Interaction and the Thresholds for Normal Epigenetic or Physiological Response?

Most biological systems function in synergy, rather than on an on/off basis. Furthermore, there are threshold effects when the transcription factors are linked to other proteins as well as to high and low affinity DNA sites, or when the spacing between the DNA link sites is altered. Several protein/protein interactions also have significant effects on these affinities. In general, synergetic interactions can lead to large responses from small modifications in the concentration of transcriptional components, an effect also produced by the phosphorylation of transcription factors. The order in which the proteins are assembled within a transcription complex composed of several subunits is, of course, important, as are the protein/protein interactions and the regulatory function as a whole. However, neither the order nor the function of the network can be shown by *in silico* analysis based on the number and type of proteins active in a particular cell. The output of the many subunits are non-linear. A clarification of their nature cannot come directly from a combination of databases because it is not a property of the information itself, but rather a result of combinations of interactions that have to be analyzed at various levels. To get information on these interactions, thresholds and networks will require the development of transgenic organisms in which precise molecular alterations have been made.

Cellular Routes and Mechanisms are Largely the Same Across Model Organisms

The problem of understanding the processes of the development of organisms was put forward by the discovery that, on the one hand, genes and gene networks are preserved in organisms of families very different from each other, and on the other hand, by the revelation that certain genes and gene networks appear in one line and are absent in another. For example, the bacterial genome projects showed that *H. influenzae* has 68 genes for the biosynthesis of amino acids, but *Mycoplasma genitalium* only has one. Furthermore, most of the genes of the archaeobacteria *Methanococcus jannaschii* apparently have no equivalent in other organisms.[184] We still have no real clues as to the number of genes in the genomes of vertebrates, invertebrates, fungi, plants or protoctistans specific to one line. Certain major classes of genes are characteristic of particular lines. Genes linked to the immunoglobin of vertebrate immune systems cannot be found in the yeast, fly or worm genomes. Collagens are not to be found in unicellular eucaryotes, and tyrosine kinase receptors seem to be a metazoan invention.

On the other hand, and with varying degrees of sequence similarity, several million proteins can be found in several lines, and these proteins constitute most of the cell's machinery. Function also appears to be preserved at the highest level. In many cases, it is not just individual protein domains but entire complexes of multiple subunits and biochemical routes which are preserved. In some cases, the way in which these complexes and routes are used during development and in the organism's physiology are also maintained.

It is, however, a far more complicated job to evaluate the conservation of function. For structure, the various genome projects will provide the absolute base on which fundamental components such as protein domains, proteins and the subunit complexes can be compared in different evolutionary lines. However, to evaluate the preservation of functions, the determination of the function of a protein, or a route in different organisms, will have to be carried out. As mentioned above, understanding the network of genes and regulatory elements will require genetic and transgenic experimentation that will only be possible on a limited number of organisms. An important task

---

[184] C. Holden, "Genes confirm *Archae's* uniqueness", *Science*, vol. 271, 1996, p. 1061.

will be to determine to what point the innovative use of fundamental processes in a given line, compared to the invention of new molecular processes, has contributed to the development of morphological and biochemical novelties. The genome projects and transgenic data will not only determine to what point functional exchange at the gene level is possible between species, but will also help us choose what genes to use in inter-species transfer. The analysis of function loss due to mutation should be accompanied by a study of the effects of expression disruption. Most of our understanding of the processes involved in development in the fly, the worm and the mouse, and resulting from our studies of the cellular biology of yeast, have up to now been obtained by function loss analysis. The information obtained from attentive analysis of the resulting phenotypes seems to have been of great use in the clarification of genetic routes such as the cell cycle of yeast and the formation of initial patterns in the fruit fly embryo. Nevertheless, this approach is quickly stretched to its limits for several reasons, such as:

- Many genes have no phenotype for certain function losses that can be easily evaluated.
- Even when a particular phenotype can be observed, it only reflects the part of gene function that cannot be compensated for by other routes and genes. In most cases, this will only be a small part of the gene's function within the organism.
- The functional pleiotropy of genes complicates analysis. It is difficult to examine the role of a gene in cell processes if its mutations stop cell division and prevent the generation of a population of homozygotic mutant cells. If a mutation causes the death of the embryo, it is hard to study the role of the gene in the formation of an adult organ although sometimes it is possible to use temperature-sensitive mutations to overcome this problem. The use of recombination systems in specific sites in transgenic animals offers a general approach to obtain lines of specific mutants.

A spatial and temporal disturbance of the expression of individual genes provides an alternative means of disrupting networks. Another possible approach is to extract the region that separates a gene from its promoter.

In any case, these considerations show that future work will require powerful transgenic technologies which will allow increasingly fine control of the various networks linked to *in vivo* development. The yeast *Saccharomyces cerevisiae*, the fruit fly, the worm and the mouse are the only organisms for which techniques are now available to carry out such manipulations. Whilst yeast, the fly, worm, zebra fish, blowfish, *Xenopus*, frog and mouse will contribute to resolve common problems, they seem limited as a model for each other or for man.

These considerations provide reasonable indicators as to the diversity of information available in the not too distant future thanks to progress in techniques [185] and the structuring and coordination of international research.

---

[185] Several groups of researchers have developed techniques that will allow them to survey gene expression in cells by tagging and tracking RNA transcripts. New microtechnologies allow quantification of gene expression. P.O. Brown and his colleagues have developed a convenient tool a "DNA microarray", that uses nucleic acid hybridization to monitor thousands of genes at once. J.L. De Risi, V. Iyer and P.O. Brown, "Exploring the metabolic and genetic control of gene expression on a genomic scale", *Science*, 278, 1997 pp. 680–686. A new technique dubbed SAGE permits scientists to identify every gene that is transcribed in a cell. In the long run, scientists will want to look at proteins and determine where every protein acts inside the cell and which other proteins serve as its partners. Some researchers are working on tagging every protein in yeast. Others are modifying magnetic resonance imaging (MRI) techniques for visualising proteins in individual cells. And several groups of researchers are using variations on the yeast two-hybrid system to untangle protein-protein interactions. Like scientists at the NASA, molecular biologists aim to do things smaller, cheaper, faster and better. Throw in "more", and the list describes the DNA "chips" that some researchers are using to simultaneously monitor the expression of hundreds of genes. The chips, developed and manufactured by Affymetrix in Santa Clara, Calif. are actually large microarrays of cDNA oligonucleotides affixed to a small glass slide. Researchers extract mRNA from a cell, convert it into cDNA and label the sample with a fluorescent probe. Sequences complementary to the chip-bound probes hybridise to the slide and researchers then determine their relative amount by measuring the fluorescence of each spot. Now Davies is using chip technology to keep track of some 9,000 stains of yeast that sport a specific gene deletion. In each strain, the scientists have replaced the missing gene with a unique 20-base-pair sequence tag that serves as a molecular bar code to identify the deletion. Each chip should be able to carry 65,000 to 400,000 probes, says Gene Brown of Genetics Institute Inc. in Cambridge, Mass. But the chips also need to have a broad range of sensitivity in order to detect transcripts that may be expressed at levels anywhere from 10 to 10,000 copies per cell, as says Snyder. For a more detailed snapshot of global gene expression, other scientists are turning to SAGE (serial analysis of gene expression) a technique that allows them to simultaneously catalogue every transcript present in a cell at a given time. To capture the complement of genes expressed in *S. cerevisiae*, Victor Velculescu, Kenneth Kinzler, Bert Vogelstein and their colleagues at the Johns Hopkins University School of Medecine in Baltimore extract the mRNA transcripts from yeast and convert them to cDNA. The researchers then use an enzyme to excise a unique 10-base-pair tag from each cDNA that serves to

These functional research projects, undertaken in various organisms, are but a beginning, but they are very promising. Conscious of the importance of these questions, nations and industries are beginning to invest considerable sums to allow a real interpretation of genetic complexity, a sign that we are entering a post-genomic age. The age of sequencing will be succeeded by a time of transgenic biology involving modifications within one genome, an increasing number of genes transferred from one organism to another and greater use of natural differences within and between species to understand parts of the biological network.

Even integrated, the databases, although totally necessary in the preliminary analysis of sequences, are seriously limited. They have no information on non-linear responses or thresholds which support development or which can only be analyzed through *in vivo* experimentation. Furthermore, the fundamental questions of comparative morphogenesis and the comparison of brain function depend on a deeper understanding of spatial dependence, complexity and degeneration of biological systems.

What will the Short-term Contribution of Model Organisms to the Understanding of Human Biology be? And in what Way Can Information from the Genome Projects Help?

Of all the invertebrate model organisms currently having their genomes

---

identify the transcript. They hook the tags together, clone them into a vector and sequence them. To identify which genes in yeast are expressed, the researchers merely compare the tag sequences with the recently completed yeast genome sequence. Further, the number of time a tag sequence turns up indicates how many copies of its corresponding transcript were present in the cell. Now that the technique is up and running, Veculescu figures that it should take about a month to process 60,000 tags. In addition to tracking individual proteins, researchers need to chart protein interactions. To learn more about who's kissing whom, a number of researchers, including Roger Brent are using variations on the yeast two-hybrid systems to disrupt and identify specific protein interactions. Yeast two-hybrid systems rely on a set of complementary plasmids that give a signal when a target protein interacts with a bait protein inside a yeast cell. Roger Brent, Russel Finley of Wayne State University School of Medicine in Detroit and other researchers are all working on scaling up the two-hybrid system in order to construct a detailed global map of protein interactions. Once scientists have identified which proteins interact, then physically disrupting their coupling can reveal why Those interactions are biologically important. To selectively block specific protein interactions, Brent and his student Barak Cohen use peptide aptamers — peptide loops of 20 amino-acids attached to a stabilizing platform of bacterial thioredoxin.

sequenced, the fruit fly's genome has the highest number of structural similarities to the human genome[186]. It seems reasonable to assume that most of the elements of these biological processes and the manner in which they interact with each other would be carried across from the fly to man. Perhaps most surprising is the extent to which physiological functions in the development of the fundamental processes are preserved. As mentioned above, current experimental tools available for model organisms (but not for man) allow the genes to be assembled in networks. The genome projects on each of the model organisms will greatly facilitate the experimental work, and with the analysis of the human genome sequence, will allow this information to be applied to man. So the main contribution of model organisms to human biology, in the course of the next five years, will be the reduction of most of the 70,000 estimated genes of the human genome to a much smaller set of basic processes of known biochemical function involving several components. An understanding of the precise way in which these basic processes maintained during evolution are used in man, and the numerous ways in which their disruption can lead to pathological problems, can only come from studies in vertebrate models such as the mouse.

The post-sequencing age is enabling us to supersede the productions of evolutionary processes and to ask ourselves what we can create. The gene transfer approach could, eventually, be supplanted by a more radical approach to the puzzle of development by making new combinations of protein domains and regulatory motifs and by the construction of new gene networks and morphogenetic routes. We will then be capable not only of discerning the manner in which organisms are made up and have evolved, but more essentially, of estimating the potential for types of organisms that can be built.

---

[186] A. Sidow and W.K. Thomas, "A molecular evolutionary framework for eucaryotic model organisms", *Curr. Biol.*, vol. 4, 1994, pp. 596–603.

S. Artavanis-Tsakonas, K. Matsumo and M.E. Fortini, "Notch signalling", *Science*, vol. 268, 1995, pp. 225–232.

The Genome Saga

The genome saga is only beginning. It has widely varying consequences. Political and ethical debate on the terms "biotechnology", "genetic engineering" and "biodiversity" underline that these last few years have been typified by a rise in the power of biology and an unprecedented progression of understanding of the structure and function of living beings. But like all scientific advances, this sudden progression of knowledge and technique in biology has an ambivalent status — positive and negative. It is in, in any case, cumulative, irreversible, invasive and largely subversive.

This sudden increase in understanding has greatly overflowed into economics, in that it has been perceived as a new financial opportunity. It seems now that genetics is wasting its major trump cards. Of course, the sequencing of the genomes is only the beginning of a long process. The ways in which the genes are expressed and the proteins are involved in the appearance of illness, or specific properties in economically useful organisms, remain to be clarified. Certain researchers, such as Peter Goodfellow, Director of Pharmaceutical R&D at SmithKline Beecham, have no doubts as to the future of genetic engineering. Having resigned his post in genetics at Cambridge University in the summer of 1996, he declares that he has come to industry to make better medicines. Examples of researchers leaving university to exploit the opportunities offered by data available on the genomes are on the rise. Without a doubt, the progress made and the new function search programs have awakened the sleeping interest of industry; industrial platforms such as YIP show this clearly, as does the Merck and Co. intention, announced on 9 May 1997, of setting up a Merck Genome Research Institute, Inc. (MGRI) to support the development of technologies for linking human genetic traits and resolving the biological function of genes involved in illnesses. Although this institute is set up as a non-profit organization, it is another example of industry's interest in the results already obtained. To paraphrase C. Thomas Caskey of the MGRI, the mission of the institute meets a current scientific need to translate our understanding of the genome sequence into functions. Like the Merck Gene Index, the institute should ensure that the necessary gene technologies will be available to the entire biomedical community.

After years of expectation and important fluctuations, it seems that the time for a massive financial putsch has now come. The American biotechnological industries, which had taken a bad knock recently after a period of over-inflated hope and overrated stock, bounced back spectacularly in 1995. The period that has just ended was, according to Carl B. Felbaum, Director of the Biotechnology Industry Organization (BIO, Washington DC), one of the most important periods in this decade for the generation of profits, the availability of money and the formation of new partnerships. New England, which takes second place after California in biotech activity, did particularly well according to *Bioline* magazine. Although only 23 million dollars were available in public offer shares by one company in the first half of 1995, eight companies were together worth 279 million dollars by the third quarter of the same year. The editor of *Bioline*, Robert Gottlieb of Feinstein Partners Inc., pointed out that 1995 had seen the largest amount of funding available for biotechnology since 1991. Gottlieb attributes this renaissance to the fact that although venture capitalists have become more prudent, the pharmaceutical industry has now become the greatest source of funds for biotechnology, reaching 4.5 billions dollars in 1995; for example, two years ago, Pfizer invested 115 million dollars in four biotechnology companies. Certain firms don't hesitate to buy entire biotechnological firms (Table 42a and 42b). Some major scientific advances, of course, encourage this behavior, but there was also a fundamental step forward in 1995 in the regulatory context in the form of modifications adopted by the Food and Drug Administration to speed up the acceptance of medicines derived from biotechnology.

In all these *grandes manoeuvres*, the place of the genetics specialist is not a small one. The experts think that the genome programs, in particular the Human Genome Program, should overturn the way we understand medicine and the means of designing and producing medicines. This multiplication of alliances between large groups and specialists of genome research confirms the importance of the potential results (Table 43).

Table 42a *Joint ventures*

Companies 1	2	3	Millions of US$	Details
Genetics Institute (Cambridge, Massachusetts)	Chiron (Emeryville, California)	Genen-tech, (San Francisco)		Chiron and Genentech joined Genetics Institute's DiscoverEase protein development platform. DiscoverEase is based on the signal sequence trap which rapidly identifies and isolates genes coding for secreted proteins.
Monsanto (St. Louis)	Asgrow (Kalamazoo, MI)		240	Monsanto has signed a letter of intent to acquire Asgrow Agronomics from Seminis, a subsidiary of Empresas la Moderna.
Genzyme (Cambridge Massachusetts)	Diacrin (Charlestown, MA)		40	They have made a 50/50 venture to develop and commercialise products of neural cells for Parkinson's and Huntingdon's diseases.
Elan Corp, (Athlone, Ireland)	Cytogen Corp, (Princeton, NJ)		20	Will form a new firm, Targen Corp, to develop oncology products, separately from Elan and Cytogen (Elan has put US$20 million into Cytogen)
Antisoma Ltd (London)	Imperial Cancer Research Technology (London)			Have formed a firm called Cancer Therapeutics Ltd to develop anticancer products.
Visible Genetics (Toronto, Canada)	Univ. of Pittsburgh, USA	Univ. of Pittsburgh Medical Center		Will form a joint venture company called the Genetic Foundry to develop diagnostic tests based on DNA.
N.V. Organon, (Oss, NH)	Anergern, Redwood City, CA		2	N.V. Organon has taken a license on the Anergix Peptide used in rheumatoid arthritis therapies.
Tripos (St. Louis, MO)	Panlabs, (Bothell, WA)	Bristol-Myers Squibb, (New York)	1.5	Tripose and Panlabs received the funding from BMS in exchange for the development and synthesis of combinatorial libraries for the Bristol-Myers programme of screening new medicines.
Incyte Pharmaceuticals, Palo Alto (CAA)	Monsanto (St Louis, MO)			Incyte will determine the sequences and the information on gene expression for several Monsanto plant species in exchange for potential license rights and royalties on products developed.

Alliances multiplied fast. Table 42b compares the situation in different countries concerning the donor and recipients in genomic related alliances.

Table 42b Donor and recipient in genomics related alliances

USA 81%	European 79%
European 17%	USA 15%
Canada 2%	Japan 6%
Donors in genomics-related alliances by geographical distribution for 63 organizations. Donor organizations are recognized as those that undertake or donate the research for a recipient.	Recipients in genomics related alliances by geographical distribution (number of organizations 34).

(*Source: S.M. Thomas and Nicholas Simmonds, P-B Joly and Ph. Goujon, The industrial Use of Genome Resources in Europe, Report for the European Commission, Contract No. BIO4-CT96-0686.*)

Table 43 Current agreements between big pharmaceutical companies and genetics specialists

Laboratory	Firm	Investment	Subject
Biogen	Genovo	38M USD	Venture capital
Bristol Meyers Squibb	Somatic	10M USD	Capital participation
Ciba-Geigy	Chiron	2.1 billion USD	49.9 % capital participation
Glaxo Wellcome	Seguana, Magabios, Spectra	unknown	Research project
Merck	Vical	10M USD	Research project
Merck	HGS under way	50M USD	Access to data
Rhone Poulenc Rorer	14-firm partnership network	400M USD	Research project
Sandoz	Genetic Therapy	295M USD	Acquisition
Schering-Plough	HGS	55M USD	Access to data
Schering-Plough	HGS	unknown	Research project
SmithKline Beecham	HGS	125M USD	Access to data
Synthelabo	HGS	35M USD	Access to data

A real auction has been under way for access to the human genome. Synthelabo, the third largest French pharmaceutical firm, judged it appropriate to invest in access to the Human Genome Sciences (HGS) database on the human genome, a company devoted to sequencing the human genome ESTs for profit. At the same time, the American firm of Schering-Plough provided 55 million dollars. Merck was hoping to negotiate another agreement for 50 million dollars. These agreements allow industries to access the computer database of genetic coding sequenced by HGS, which since 1993, has had an exclusive deal with SmithKline Beecham (for which the latter paid 125 million dollars into the "Maryland Start-up"). In June 1995, the start-up had already signed a licensing agreement with the Japanese Takada Chemical Industries.

This is a general trend, and all the large pharmaceutical groups have signed similar contracts with gene race specialists. The interest in HGS was triggered by the promise of gene therapy from the early 1990s. At that time, sequencing was still the affair of scientific projects less interested in the question of intellectual property rights. The world scientific community, however, quickly changed its mind as the HGS-type start-ups were created with the intention of sequencing for profit, and the American government launched the patent race through the NIH.

Between 1995 and 1996, pharmaceutical companies made more than 40 agreements with firms specializing in genome work, resulting in a potential expenditure of 1.72 billion dollars. To this sum should be added the 489 million dollars raised in the stock market between 1993 and 1996 by these same genome firms. In total, investment in these new technologies over the last four years represents two thirds of public spending on the Human Genome Program in the last 15 years, that is to say, three billion dollars since 1990.

However, the 13% figure of the total expenditure on R&D of all American laboratories devoted to "genomics", masks a great reluctance amongst pharmaceutical firms. There has as yet been no return on the very heavy investment made in the early 1990s. Two years after the frenzy unleashed by SmithKline Beecham's 1993 purchase of exclusive rights to the HGS database, the large groups are facing a bitter closure of accounts. Sandoz, who spent 382 million dollars in 1992 on Californian Systemix, still has no return from this firm specialized in cell therapies. RP-Gencell, which registered the first encouraging results in gene therapy, has been forced to lay out 103 million dollars to restructure the *in vivo* part of its daughter company, Applied Immune Sciences (AIS). And SmithKline Beecham, which does not want to spend more on further exclusive rights to HGS's database, has begun, since 1995, to open the database to other groups, allowing the French firm of Synthelabo to conclude a contract with HGS, after having already allied itself with Genset to discover and sequence genes linked to prostate cancer.

Biotechnology remains a high-risk sector, and even though it appears to have reached a certain maturity, the promises of the genome seem to be hard to turn into fact. So, none of the great industrial groups truly believe in the "genetic miracle". For some, through these alliances with genomics firms, they intend to show their shareholders their vitality and agressivity. These agreements are also a good way to limit sizable financial risk. Despite the progress made, the 700 molecules produced by 170 biotechnology firms in phases I, II and III development[187], and the 28 gene therapy programs, the increasing amount of genetically modified organisms and the new technologies developed to attack the AIDS virus inside cells chosen by hybridoma, the second half of 1996 and the first three months of 1997 were financially very uncomfortable for the biotechnology companies. This was especially true in the United States, where the stock market can react so suddenly. Investors have retreated to more conventional shares, but the trend remains favorable to biotechnology in the long term. What is certain is that the United States will not loose its leading position (Tables 44 and 45). Currently, about one thousand firms are working in this field, and some 300 of these are quoted on NASDAQ. They employ 73,000 people, 22% more than in 1995. In 1996, more than 8 billion dollars of private funding were infused into the sector.

---

[187] 79 products have already been approved by the regulatory authorities and another 161 are in phase III, the last phase before commercialisation is permitted.

Table 44 Specialist Biotech Sector - EU/US Comparaison

Indicator	Europe	USA
Turnover (ECU million)	1,700	11,700
R&D Expenditure (Ecu million)	1,500	6,300
Number of compagnies	700	1,300
Number of Publicly Quoted Compagnies	50	300
Number of Employees	27,500	118,000

(source: Ernst and Young "European Biotech 97 – A New Economy", 1997)

Table 45 Location of biotech product development (% being developed in each region (1995))

	North America	Europe	Japan	Rest of the World
All Biotech Drugs	63%	25%	7%	5%
Gene Therapy Drugs	70%	22%	1%	7%

According to Pascal Brandys of Genset, Europe's biotech industry is one-tenth of that in the USA. European scientific expertise is as good as in the United States, but there is a problem with the translation of that expertise to the industrial sector. The root of the problem seems to be the economic environment, in particular, the health regimes. The price-fixing system is such that it is hard to recuperate costs. Another reason might be the European directives on patenting and genetically modified organisms (Table 46).

Table 46 Share of US patents by technological field[188]

	Biotechnology	Chemicals	Drugs
Europe	19.5	25.4	32.9
USA	59.4	59.1	51.0
Japan	16.6	12.1	12.0
Other	4.5	3.4	4.1
Total	100	100	100

---

[188] Source: Benchmarking the competitiveness of biotechnology in Europe, An independent report for Europabio by Business Decisions Limited and The Science Policy Research Unit, June 1997.

The fact that the United States, unlike Europe, has a certain dynamic venture capitalism and a financial market that is very receptive to innovation must also be taken into account to understand the European situation. Whereas before, the work was a stumble in the dark, now, with structure-function information, a more systematic form of research can be organized. Biotechnologies linked to genomics are undoubtedly going to overturn our lives more completely than any other technology this century. It remains to be seen how fast and how far this change will go.

Rules of the Game

As the first biotechnological products leave the laboratory, man acquires a mastery of life, sequences and information accumulate in the databases and new problems are emerging, becoming more defined. The use of biotechnology is not a mundane thing, even in our industrialized world. Preventive diagnostics, genetically modified organisms, gene therapies, transgenic crops, novel foods, novel drugs — all these advances result from the new techniques of genetic engineering and the new understanding of "the text of life". They are raising many points of debate on regulation, and at the same time, a fascination for and anxiety about the power of this new biology.

Trading Genes

Since the Asilomar meeting in January 1974, genetic engineering has come a long way. In less than 20 years we have progressed from a moratorium to avoid the potential risks for man and his environment to detailed legislation for genetic engineering tailored to the new understandings of life, which have become a major scientific and economic challenge.

Behind scientific and technical progress are darker battles than the *manoeuvres* of financial alliances, i.e., battles over patents for the commercial exploitation of discoveries. As Elke Jordan has pointed out, no firm will invest in a genetic invention if it isn't covered by a patent. However, despite the ethics committees and experts, and the laws and directives they inspire, great disorder reigns in this sector. Each protagonist plays with the words, challenges official decisions and threatens to take the matter to the highest courts — the Supreme Court in the United States, or the appeal court and its equivalents in Europe.

Some are even so worried as to throw in doubt the question of the patentability of living matter. "Nature cannot be invented", said a French judge in his critical comment on a verdict given by the US Supreme Court on 16 June 1980 in the matter of Diamond vs. Chakrabarty, concerning the patentability of a micro-organism as such (Amanda L. Chakrabarty, a researcher at General Electric, discovered the plasmids that, when inserted into a bacteria, could break down the components of petrol, and patented her invention as an agent for use in the fight against pollution). The number of cases linked to the patentability of the living continues to multiply in view of the challenges that this sort of patent presents.

One thing is certain; from single genes to whole animals, nothing is *not* patentable. It should be underlined that, in agreement with the group of advisors to the European Commission on the ethics of biotechnology[189], the practice of delivering patents on living material has been going on for some time. It certainly precedes genetic engineering. It was specifically approved by the 1961 UPOV Convention and then by the Strasbourg Convention in 1963. The first known patent on a living organism was granted in Finland in 1843. Louis Pasteur received a patent from the USA Patent Office for a yeast free of organic illness germs in 1873. After the micro-organisms came the animals; Myc Mice lab mice carry a human oncogene (Myc), which triggers mammary gland tumors; they were patented in 1988 by the USA Patent Office. The patent was later bought from Harvard by Du Pont De Nemours. In October 1991, the European Patent Office in Munich finally granted Myc Mice a license for all countries. Of course, in the early 1980s, when asked whether we should patent the living, most people would have said no. Lawyers considered plants and animals a universal heritage that

---

[189] Opinion of the group of ethical advisors for the Biotechnology programme of the European Commission on the ethical questions raised by the Commission's proposal for a Council Directive on the legal protection of biotechnological inventions, 30 September 1993, European Commission, p. 41.

Man should use to his benefit, not appropriate as his own invention. This conception prevailed in Europe particularly through the legacy of Catholicism.

But with the arrival of genetic engineering and molecular biology, a biological revolution caused a legal revolution which guaranteed that the industrial point of view would prevail, often despite the rules of ethics. The US Supreme Court decision in 1980 to grant invention status to bacteria that can digest certain components of petrol opened the door to the biotechnology lobbyists, who have rushed in.

It is true that man has been modifying the living since the dawn of livestock and crop farming by selecting within species. But it has never before been suggested that the results should be patented, although the law recognizes the work of the selective breeder[190]. Only living organisms resulting from biotechnological procedures can, in principle, be considered inventions, and therefore patentable. The introduction of one single gene in the genetic heritage of a creature allows a researcher to call it, and all its descendants, his own invention. National legislation is gradually falling in with this view, even through the interpretation of currently valid and ambiguous texts. A European Convention article from 1973 forbids the patenting of plant varieties or animal strains, except bacteria, yeasts and other products obtained by microbiological processes. This did not prevent the European Patent Office granting many derogations to this article. Patent requests involving several modified plant species were registered as early as 1989 (the Harvard Myc Mouse in 1991). The latter had been rejected in 1990 by the European Patent Office because it could be considered an animal strain. But the Du Pont de Nemours lawyers managed to change the judges' minds by showing that the transgenic rodent in question was not a strain but more of a microbiological product. The Venter affair (as mentioned earlier, in June 1991 Craig Venter registered a patent request for 347 and then 200 partial label sequences from complementary DNA banks) and similar patent requests made by the British MRC and private firms in the USA have shaken the research community and industry to the point where

---

[190] In this way, plant varieties obtained through crossing are protected by "certificates of vegetable origin" which do not prevent the farmer from replanting them himself.

many voices are being raised to denounce undue attempts to appropriate the human genome, which have been the trigger of renewed reflection on the matter.

In view of USA aggressiveness and the market waiting to be conquered, the European Commission in Brussels prepared, over several years, a directive for the legal protection of inventions in the field of biotechnology, a directive to clarify legislation for all living organisms except for man. The initial proposed directive, in 1988 [191], ruled out any objection to patenting life forms, and limited the exclusion to plant varieties proper (in the sense of the EPO's Ciba-Geigy decision of 1988)[192]. The adaptation of patenting law to living matter included the following provisions:

- Extension of product patenting: protection of patented genetic information once inserted into a plant variety extends to that plant variety if the information is essential for the applicability or utility of that variety (article 13).
- The rights of a procedure patent describing a genetic construction and covering the product resulting from that procedure were extended to identical products and variants from the same generation (article 12)
- The mechanism of dependent licensing does not seem to have created complete reciprocity of rights. If the patent holder of a new plant variety protected by COV wants to incorporate patented genetic materials, they must wait several years to be granted an average and

---

[191] Based on Article 100A of the Treaty of Rome, this Directive proposal (published in the Official Journal of the European Communities 10/3 of 13 January 1989) is submitted to the co-decision procedure set up by the Act of European Union.

[192] In 1973, when the CBE was being signed, vegetable varieties and essentially biological procedures of obtaining vegetables were excluded from patentability. This exclusion was justified by the existence of the UPOV Convention and the administrative and legal difficulties of delivering a patent on plants. The Chamber of Technical Appeal of the European Patent Office, in its decision for Ciba-Geigy of 26 July 1983, gave the first interpretation of Article 53b. It provided a definition of "vegetable variety" and stipulated that only plants whose characteristics are not modified through the successive cycles of reproduction can be excluded from patentability. In practice, the patent requested by Ciba was for a chemical treatment for seeds that could be applied to many crop plants. This reasoning could be extended to a plant coming from a genetic engineering technique, in that the purview of the patent, which originally covered a technique of genetic transformation, could extend to cover different plants growing from cells modified with the DNA sequence in question.

commercially viable license. Furthermore, they will have to prove that the new variety is indeed technical progress. If not, a plain paying license would be granted to the patent holder (article 14).
- The principle of reversing the burden of proof forces any suspected counterfeiter to prove that they have not used the patented product or process to create their variety (article 17).

From the very beginning, the legal commission of the European Parliament considered the ethical problems, the often thoughtless extension of patent rights to the detriment of the DOV, and farmers' privilege. Serious objections gave way to an attempted compromise with the legal commission in which the initial project made no mention of farmers' privilege. In January 1992, the legal commission of the European Parliament (CJPE) adopted the report of Rotheley, the German MEP, who had considerably modified the directive proposal without including farmers' privilege. The most notable changes were on the ethical questions, as well as:

- The removal of the article on the extension of patents to products containing essential patented information;
- The inception of a mandatory exploitation licensing system, licenses to be requested from a competent national authority, and if public interest required it, that the license be delivered to the patent holder himself or that of the COV;
- The rights conferred by the patent on a procedure to produce a biological material with intrinsic properties be applicable only to its first generation issue having identical properties (article 12).

Certain political groups in the European Parliament began a veritable campaign to mobilize the MEPs against the patenting of living things. This would force farmers to pay royalties to the patent holders for each generation of plants produced with biotechnology. The CJPE, therefore, inserted the clause on farmers' privilege into its proposal. This foresaw that the rights conferred by a patent would not apply to acts covered by that privilege. All the CJPEs amendments were retained during the 1992 April plenary session of the European Parliament. But the final vote was delayed to allow a compromise with the legal commission on the ethical problems and farmers' privilege. The latter, having been rejected by some of the industries, and under pressure from the agricultural lobby during the GATT and CAP (Common Agricultural Policy) negotiations, the Commission agreed to come to an agreement with the Parliament on the recognition of a limited privilege for the farmer, whilst protecting the patent holder's rights. Furthermore, the Commission wished to protect the rights granted under the directive. It finally accepted four compromise amendments in May 1992. The last of these, which involved farmers' privilege, permitted farmers to use, for propagation on their own fields, the seeds or other materials of multiplication they had gathered for use from seeds protected by patents. The compromise only involved the seeds, but the term "other materials of multiplication" was added after consultation with other European Parliament Commissioners.

The industrialists and the legal Commission expressed their disagreement with the terms of compromise. The proposal was submitted to a vote on a first reading. In October 1992, after more than three years of debate, the European Parliament finally adopted the third version of the Rotheley report by a weak majority. The modifications from the initial proposal of the directive were on the following points:

– The object of an invention would remain patentable as long as it was composed of biological material, used it, or was applied to it;
– Biological material, including plants and parts of plants, except for plant varieties proper, is patentable;
– The farmers' privilege was maintained as a high majority.

In December 1992, the European Commission presented a modified version of its proposal[193]. It maintained the patentability of transgenic plants on the fact that reproduction is not in itself an end to patent-holder's rights. It also pointed out that the protection of patented products extends to all biological material in which the genetic information is expressed and set up

---

[193] Modified Directive proposal published in the *Official Journal of the European Communities* no. 44/63 of 16 February 1993.

the regime of mandatory licensing (article 14) within the limits of public interest. The amended proposal maintained the limit of farmer's privilege to seeds, and without limiting the rights of the patent holders. Many industrial firms reaffirmed their opposition to farmers' privilege and mandatory licensing, preferring "legal rape" rather than the amended version.

From January 1993, the directive proposal forwarded to the Council of Ministers was studied in several experts' meetings in order to ensure the adoption of a common qualified majority position before the second reading in the European Parliament. Between July and December 1993, the Belgian Presidency of the European Union has tried to manage the political problem of farmers' privilege and the tensions about the ethical use of animals, which had increased with the European Parliament's resolution on the European patent delivered to the transgenic "oncomouse". A German proposal accepted the existence of farmers' privilege by dressing it legally as a specific case of patenting rights extinction, and this opened useful debate on its implementation. Some of the European industrial firms came out in favor of a tacit, case by case licensing process, allowing the patent holder to be remunerated and strictly limiting farmers' privilege.

After other related events, a directive proposal on the legal protection of biotechnological inventions was presented to the Parliament in early 1995. On 1 March 1995, the Parliament rejected it. This rejection was a move backwards on the part of the European Union, and the greatest beneficiary was the United States. It caused a strong reaction amongst European industrial firms who had been hoping for a more subtle text, especially on the patentability of certain genes associated with illnesses. Considering that this interdiction placed them in a position of inferiority vis-à-vis their competitors in the USA and Japan, some of them pointed out that the European Parliament, manipulated by the German Greens, had not understood the world challenge in the field of biotechnology. Some teams had even been thinking about leaving Europe to work in the United States, where legislation is much more favorable to the patent protection of results of biotechnology research. This rejection would also have consequences on investments; investors were hesitant to venture capital into the high-risk field of biotechnology research if the results were to be unpatentable.

In the light of all this, the European Commission prepared a legal protection project taking into account the objections of the industrial firms and the report of MEP Jean-Francois Mattei, who prepared new proposals showing that the status quo would be prejudicial to the EU's competitiveness with the USA and Japan. This project contained considerable improvements on the one rejected in March 1995. It considered that it was dangerous to leave biotechnological inventions without status under EU law while the EU patent resulting from the Strasbourg Convention in 1975 still hadn't come into force. This lack of legislation had a two-edged danger:

- There was a risk that the member states would decide what sort of protection they needed individually, decide what should be excluded from patentability and pass laws that might later be in contradiction to EU law.
- On the other hand, the EPO's attitude, at the time quite generous in the matter of patent delivery in the field of biotechnology, could change.

The need for a directive to legally protect biotechnological inventions is, therefore, quite clear, but which directive? The many projects and their amendments, show that the question is a tough one. The new report hinges on six chapters, I. Patentability, II. Extent of the protection, III. Mandatory licenses for dependents, IV. Deposition, access and new deposition of biological material, V. burden of proof and VI, final dispositions. Only Chapters I and II and Articles 2, 3, 4, 9, and 13 are of interest to us.

Article 2, which gives several definitions, is interesting, especially in its definition of a microbiological process.

Article 3 deserves particular attention since it concerns the exclusion patentability of the human body and its elements in their natural state. This exclusion is without reference to public order or moral standards, unlike the 1994 French bioethical law. Such an improvement recalls a fundamental principle in patenting law, which is that a discovery is not patentable but an invention is, as long as it upholds the criteria of novelty, inventive activity and industrial applicability. Of course, the expression "in their natural state" gives rise to ambiguity if one considers it to be other than a distinction between discovery and invention. For example, if a human gene is isolated from the body, characterized by its precise chemical structure, and used in an industrial application, it would be patentable[194]. But an unisolated gene in its natural state is unpatentable, as is a cloned and sequenced gene for which no industrial application has been found[195]. Furthermore, the question of the status of the embryo remains open.

Article 4 is just as important, even though it is in strict contradiction with Decision G3/95 of the EPO, which disallowed the patenting of plants or animals, since one of the major objectives of the directive is precisely to allow such protection.

Article 9 is much more problematic. The exclusion of the patentability of methods of genetic germ-line therapeutic treatment on man, aimed at also modifying the genome of his descendants, a method considered to be contrary

---

[194] Economic and industrial reality causes problems with the implementation of the idea of our common inheritance of the human genome, which was to be set before UNESCO in the autumn of 1997. This proposal, (which was adopted 11 November 1997) from the UNESCO International Bioethics Committee, presided over by Joelle Lenoir, solemnly proclaims that the human genome is a common inheritance of humanity. Article 6 defines the framework: "No one shall be subjected to discrimination based on genetic characteristics that is intended to infringe on or has the effect of infringing on human rights, fundamental freedoms and human dignity". About research on The Human Genome, Article 10 "No research or research its applications concerning the human genome, in particular in the fields of biology, genetics and medicine, should prevail over respect for the human rights, fundamental freedoms and human dignity of individuals or, where applicable, of groups of people." Article 11 "Practices which are contrary to human dignity, such as reproductive cloning of human beings, shall not be permitted. States and competent international organizations are invited to co-operate in identifying such practices and in taking, at national or international level, the measures necessary to ensure that the principles set out in this declaration are respected." Article 12 a) "Benefits from advances in biology, genetics and medicine, concerning the human genome, shall be made available to all, with due regard to the dignity and human rights of each individual. b) Freedom of research, which is necessary for the progress of knowledge, is part of freedom of thought. The applications of research, including applications in biology, genetics and medicine, concerning the human genome, shall seek to offer relief from suffering and improve the health of individuals and humankind as a whole. Of course a declaration is a formal and solemn instrument which has a more general and supple purview than a convention but is also more serious. This proposed declaration is important, but will it be enough? It isn't because there is a Declaration of the Rights of Man that it isn't flouted across the globe. The existence of industrial firms working on sequencing and building confidential databases or databases with confidential levels mortgages the application of the intentions of this declaration as it mortgages the dreams of the philosophers and ethicists of a status for the human genome which does not change as it crosses borders, as a genetic heritage and right of man".

[195] On this matter, see the patent requested by France's Bio Mérieux for the MSRV1 virus, which has been linked to multiple sclerosis (EPO 674004).

to public order or morals (whilst letting each state decide whether or not it is indeed contrary, and not the EPO), is acceptable. But to base an exclusion on these same motives for inventions covering procedures to modify the genetic identity of animals "in a way that would cause suffering or bodily handicap without substantial utility" for man or the animal in question seems very strange in a directive. The rejection of a patent request on the basis of such a motive during examination by a patenting office such as the EPO is inappropriate since it happens at the wrong time, in the wrong place and is decided by people untrained to make that decision. Furthermore, how can we quantify the suffering of an animal? At what point does this substantial utility become recognizable?

Article 13 examines the problem of farmers' privilege. This is an exception to the common law of patents which appears in Regulation 2100/94, but all the questions that accompany it have not been settled, especially with regard to the means of exploitation for small firms. Furthermore, the concept of farmers' privilege is not based on any text, and the animals concerned have not been defined. This provision, furthermore, raises the question of the applicability of the exclusion; what sort of animal variety? animal use? It is also surprising that the directive, a text with European union-wide effects, refers in this article to national provisions and practices.

Whilst we wait for the new amendments and for all the questions to be settled, substantial progress has been made. The new proposals made require that the member states protect the dignity of the human being in the application of the new biotechnologies. The United States seems to have no conscience in this matter. Even recently, the NIH attempted to patent 337 DNA fragments expressed specifically in the human brain, while the patenting of genetic information can only apply when it is specifically associated to an illness or an industrial application. To date, some 1,000 genes associated with illness have been patented worldwide[196].

---

[196] The growing interest pharmaceutical firms are showing in genetic studies of groups of human populations has caused an intense debate over the problem of the property rights over genetic information. The story of John Moore is an example. Moore, a Californian gentleman with leukemia, discovered that the biologists of the University of California, in association with Sandoz, had requested a patent on and

In the life sciences, the United States uses patents as strategic weapons. Such an attitude compounds the problem in that it tends to spread and threatens the principle of the free circulation of scientific knowledge and may slow down progress in genome research. There are the varied reactions of the ethical committees and the new directive proposals. But is it not too late already? Are we not already in a terrifying trap that no-one can stop from closing? Man has progressively altered all of nature, even the living. Is he not inexorably turning himself into a product and inventing a new form of slavery? Whatever may be, the challenge is very hard since, in the name of rational science, we are throwing back into debate the questions of the basic nature and the integrity of life of the human species. However, it is not progress in science and technology which is in itself dangerous but the use that is made of it. This use raises even more questions because the relationship between life and nature is a profoundly cultural problem. While Europeans have a unique and dominating cultural heritage, in the USA society has many cultural influences and is thus more open to innovation, more able to accept the uses and the risks linked to them, explains why, until very recently, few American states had taken measures to forbid discrimination in research regarding genetic testing. It is quite **understandable** and normal that societies limit experimentation on man, but it is equally abnormal that they consider products resulting from biotechnology to be more dangerous at the outset than those obtained with other technologies.

The fields opened up by genome research are fantastic; explorations of the genome are loaded with promise and discovery, but it's man's responsibility to exploit them well. What the sufferer and consumer require is that the products they use will not do them harm, whatever technology was used to produce them. The legislator must act on this basis and beware that the law does not unjustifiably impose constraints that could block useful technological innovation.

---

marketed cells taken from his spleen during an operation without telling anyone. The cells in question, which secrete anti-cancer substances, brought in hundreds of millions of dollars. Once he found out about it, Mr Moore took the biologists to court. After years of legal proceedings, the affair was settled out of court.

There is hope, but there is also legitimate concern. An explicit example of this is transgenic plants (Tables 47 and 48).

Table 47 Pending market applications under EEC/90/220

Originating CA	Applicant	Product	Current status
France	Ciba Geigy	Maize – insect resistant, herbicide tolerant, antibiotic resistant	Consent issued by the French CA, February 1997
France	Plant Genetic systems	Oilseed rape – male sterile, herbicide tolerant	Qualified majority vote in favor, December 1996.
France	Plant Genetic systems	Oilseed rape – male sterile, herbicide tolerant	Qualified majority vote in favor, December 1996.
UK	AgrEvo	Oilseed rape, herbicide tolerant	Objections received from some member states. Awaiting Commission decision.
France	AgrEvo France	Maize – herbicide tolerant	Objections received from some member states. Awaiting Commission decision.
France	Monsanto Europe	Maize – insect resistant and herbicide tolerant	Objections received from some member states. Awaiting Commission decision.
Finland	Valio Ltd	Streptococcus thermophilus antibiotic test kit	Objections received from some member states. Awaiting Commission decision.
UK	Northrup King Ltd	Maize – insect resistant and herbicide tolerant	Objections received from some member states. Awaiting Commission decision.
France	Pioneer Genetique	Maize – insect resistant and herbicide tolerant	Objections received from some member states. Awaiting Commission decision.
The Netherlands	Bejo Zaden	Chicory – male sterile, herbicide tolerant	Objections received from some member states. UK raised an objection due to insufficient information on toxicity. Awaiting Commission decision.
Germany	AgrEvo	Oilseed Rape – herbicide tolerant	Objections received from some member states. Awaiting Commission decision.
Belgium	Plant Genetic Systems	Oilseed Rape – herbicide tolerant and male sterile	Objections received from some member states. Awaiting Commission decision.

(source: June 1997 *Newsletter of the UK Advisory Committee on Releases to the Environment*)

Table 48 1996 Plantings of genetically-modified crops
(Plantings "000s acres)
Other countries are Argentina, China and Mexico – There is no significant planting in Europe

Crop	USA	Canada	Europe	Other	Total
Maize	470	0	0	0	470
Cotton	2,000	0	0	0	2,000
Rapeseed	0	350	0	0	350
Soyabean	1,000	0	0	375	1,375
Tomatoes	11	0	0	50	61
Potatoes	0	0	0	1	1
Tobacco	0	0	0	2,000	2,000
Total	3,481	350	0	2,426	6,257

After authorizing the importation of genetically modified soya, the European Commission backtracked. Under a barrage of criticism from ecological organizations and consumers, it recently delayed its green light for transgenic maize to await the opinions of scientific experts. This allowed MEPs to request compulsory labeling of these products, if commercialized, as well as the separation of genetically modified organisms, a request that came to be added to the current regulations (European Directives 90/219 and 90/220 respectively relative to the restricted use of genetically modified micro-organisms and the deliberate release of genetically modified organisms)[197]. On 15 May 1997, the new regulation on novel foods came into force[198] after many months of discussion. This text, which requires that all products containing GMOs and resulting from GMOs be labeled unless

---

[197] This legislation is based both on our inability to show that the manipulation and dissemination of foreign genes might be deleterious and the intention of certain parties to manage the risk on a preventative basis. While an analysis of the file shows that for the GMOs in question, no unacceptable risk is to be feared for this organism or that, and that the foreseeable risks can be managed in a satisfactory manner, the authorisation for release may be granted in accordance with the regulations. It is clear that these regulations will change as appropriate as we gain experience in these matters. Directive 90/200, besides its preventative nature, takes into account the "border-crossing" nature of GMOs and imposes a procedure of phased release with a progressive reduction of the confinement levels, possibly a public consultation, an exchange of information between member states and the establishment of a safety net procedure in case of risk.

[198] *Official Journal*, L43, 14 February 1997.

they are directly equivalent in substance to their traditionally made counterparts, brought strong reaction from researchers and industrial firms. Across the Atlantic, words are sharp. American policymakers did not appreciate European reserve, considering these administrative obstacles had been set up to gain the time to make good an alleged scientific gap...

One may be tempted to see in all of this a victory of reason by the consumer and ecological organizations, but will the legislation really be applicable? Dorothee Benoit Browaeys' article, "The labeling of novel foods is a red herring", shows that nothing is less certain. How can trustworthy means of control be put into place while European member states have no agreement on a protocol for investigation? Even if there were one, the most obvious means of control (PCR) is weak as it is only applicable to whole grains, fruit and vegetables. Processed products such as soya remnants after the oil extraction process or corn gluten feed, no longer have identifiable DNA molecules. The level is completely obscured in derived foodstuffs, in which the processing degrades proteins worse than DNA molecules. Transport conditions also damage the trustworthiness of checks (in a cargo or raw materials, batches often get mixed). Only the installation of independent production routes would allow labeling to be trustworthy. However, industry is very reticent towards this idea. As Gerard Pascal points out, "this requirement that we label GMOs looks very much like smoke in the eyes of the political opponents. In five years, all these requirements will be obsolete since there will be GMOs everywhere, even to the replacement of traditional plants with GMOs completely"[199].

There will always be risks associated with the use of the new biology and genomic discoveries and research, especially at this juncture, when applications are increasing every day and where, more than ever, biotechnology looks like our main solution to famine (Fig. 36) and preserving the environment throughout the next century.

---

[199] Quoted by Dorothy Benoit Browaeys in her article "L'étiquetage des nouveaux aliments, un leurre", *La Recherche,* no. 299, June 1997, p. 36.

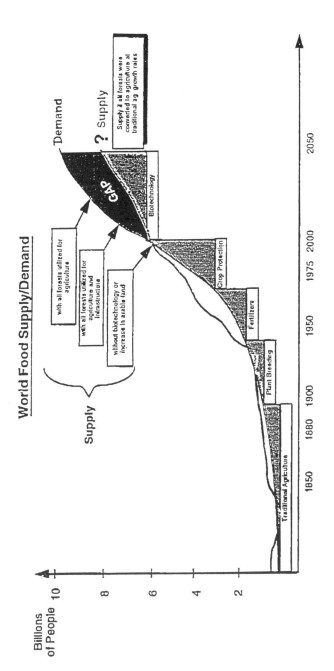

Fig. 36 Biotechnology may represent the main opportunity to prevent famine during the next century whilst preserving the environment.
(Diagram provided by the European Commission and A. Goffeau to the author)

Under pressure from industry and researchers, and against the economic, medical and scientific challenges[200], can reason prevail? Now that profit is king, it all looks very pessimistic. History teaches us that once a technique is available, it is used (the atom, artificial insemination and so on), and if genome research develops at an accelerated rate in industries and society, it is because people are interested in it. Aldous Huxley's *Brave New World*, written in answer to the prophecies of his brother Julian, will not come true tomorrow, but there is a biological revolution underway. The age of the biosociety is without a doubt upon us, and it depends on human wisdom as to whether it will be beneficial and whether genetic and genomic research will stave off the "Green Bomb". The future holds a lot more surprises for us, and society will have to learn to manage them.

---

[200] To counter the threatened constrictive regulations and hostility from Greens in the various parliaments, some firms have threatened or preferred to expatriate their laboratories to bluer skies, mostly to the USA and Japan. In addition, there is a problem in the southern countries where GMOs represent the solution to many problems and the geneticists are just in it for the adventure. During an international conference on biological safety of crop trials at Monterey, California, the contradictory interests of industrial countries and developing countries became all too evident.

8

Conclusion:
The Dreams of Reason or the New Biology's Dangerous Liaisons

8.1 Fascination but Anxiety Concerning Progress in the Life Sciences

John Burdon Sanderson Haldane wrote that physical and chemical inventions constitute blasphemy, and that each biological invention is a perversion. What would he have thought of the creations that will result from genome programs?

We have followed the origins and development of genome analysis activities as well as an increasing understanding manipulation of the processes at the heart of life. This progress is not only the concern of the philosopher and the biologist, but society as a whole. It invites us to deep reflection on the calculated risk of drawing humanity into an adventure with no guarantee as to how it will all end. If the economic prospectives of recent progress in genome studies and biotechnology are remarkable, the equally remarkable human and social pros and cons of biotechnology and genome research should not be forgotten.

In 1973, biologist Paul Berg's discovery unleashed the possibilities of genetic engineering, and in particular, recombinant DNA techniques. He warned the scientific community of the dangers of producing pathogenic material with this technique. As a result, between 1975 and 1976, scientists in the USA issued certain recommendations. However, in 1986, the OECD made a declaration that genetic engineering was not dangerous in itself.

These recommendations have been stretched to their limits. Scientists are no longer exploring routes for DNA recombination, but manipulating non-traditional viruses, mastering living matter through its genome and, currently, cloning living beings for therapeutic purposes. Some new thinking is going to have to be done, especially as, by definition, the researchers are working blind.

Now, in the new millennium, amazing progress in the life sciences simultaneously fascinates and frightens. As the genomes of more and more organisms are decoded, researchers are throwing themselves into the Herculean enterprise of decoding the human genome. Medical doctors are tracking hereditary diseases and hoping to cure them tomorrow. The first genetically modified vegetables are being consumed. The billions to be made from genetics are attracting giant industries, but the world's average citizens, scientists and governments are worried. Will the human species be modified? Is this a covert return to eugenics? Distinguishing fantasy from reality, nightmare from actual risk, is becoming a difficult task. Creative imagination must be defended and the irrational fought off. For this reason, science and medicine must continue to develop with a respect for individual life. This is where the problem lies. When it comes to biotechnology and genome research, it's not the science itself that should be controlled but the use which is made of it.

In this conclusion, it would be useless to try to sum up all the ethical problems linked to genome research and the new powers man has gained over the living. More modestly, I will simply ask, by critical, epistemological and philosophical analysis, what the new powers and pretensions of the new biology and its sponsors are, in particular regarding man, through the Human Genome Project. To conclude my history of the long journey "from Biotechnology to Genomes", I will cast into the arena some reflections that might seem a little subversive, but intended to demystify some biologists' dreams. These are opinions that have been reiterated throughout the history of biology. These reflections will also denounce and dissipate, when necessary, the dangers of what might be called the "genetic ideology", and expose the lacunae and justifications of this genocentric image of man and life itself.

Prometheus stole fire for man from Zeus by glorious trickery, but what happened next? Zeus, blinded by his anger, ordered his servants Force and Violence to seize Prometheus and carry him to the Caucasus where they chained him to the crest of a high cliff with unbreakable chains. We should make sure that DNA research does not become a chain that we can never break. From this point of view, the logo chosen for the publication sponsored by the US Department of Energy, the *Human Genome Newsletter*, which consists of a human form draped in a DNA double helix, is particularly symbolic. There are bars on the double helix.

One of the problems with evaluating genome projects and their spin-offs such as genetic engineering and genetic testing is that we are limited in our judgment because we cannot fully evaluate the technology's impact. Also, any evaluation should identify the institutions involved and the opinions of those who control the technology.

However, it has been said of the Human Genome Project[1] (HGP) that development in itself presents no new ethical concerns and perhaps only a few legal ones. Such problems as arise do so from the use and development of genetic tests and screening for therapeutic use, rather than the mapping and sequencing of genomes[2]. But John Maddox does not mince his words where the creation of new ethical dilemmas is concerned. "The availability of gene sequences, and ultimately of the sequence of the whole genome, will not create ethical problems that are intrinsically novel, but will simply make it easier, cheaper and more possible to pursue certain well-established objectives in the breeding of plants, animals and even people"[2].

One might object that it is in this recombination of plants, animals and people and the apparent assumption that (and here we find some of the prophecies of Haldane in *Daedalus*) we are handling nothing but linear progressions of scientific vectors, that the real challenges and problems lie. But the argument put forward by Maddox belies the clarity and transparency

---

[1] C. Clothier, *Report of the Clothier Committee on the Ethics of the New Gene Therapy*, 1992, Cm 1788.
[2] J. Maddox, "New genetics means no new ethics", *Nature,* vol. 364, 1993, p. 97.

that he had himself supported. He feels that, although the molecular causes of several diseases such as Huntington's and fragile X syndrome have been determined over the last century, this new knowledge has brought no new ethical problems, just ethical simplifications.

George Annas has no such formulations. He has suggested that the specific nature of genome projects, in particular the HGP, does not lie in the quest for knowledge. The history of science is filled with little else. What is particularly unique, he says, is that from the beginning, political and ethical problems have been raised by this research, and that preventive measures have to be taken in order to ensure "that the benefits of the project are maximized and the potential dark side is minimized"[3].

Annas has suggested that the HGP raises three levels of problems, the individual/family level, the social level and the species level. Most attention has been paid to the first level and questions of genetic testing, predictive medicine and consultation. Negligence in failing to offer or properly perform these tests has already resulted in lawsuits for wrongful birth[4] and wrongful life. Standards for genetic screening and counseling have indeed been discussed[5].

Problems at the second level involve society as a whole more directly. For Annas, the HGP raises three major social problems: population testing, the allocation of resources and eugenic commercialization. He asks more precisely, "To what uses should the fruits of the project be put in screening groups of people, such as applicants for the military, government workers, immigrants and others[6]?" How, if at all, should intellectual property laws such as patents be invoked ? What funding and resource allocation decisions are there to be made[7]?

---

[3] G. Annas, *Standard of Care*, Oxford University Press, New York, 1993, pp. 149–150.

[4] A.V. Bloomsbury, *Health authority*, 1 All ER 651, 1993.

R. Lee and D. Morgan, eds. *Birthrights: Law and ethics at the beginnings of life,* Routledge, London 1989.

[5] R. Chadwick, ed. Ethics, Reproduction and Genetic Control, London, Groom Helm, 1987.

[6] G. Annas, *Standard of Care*, Oxford University Press, New York, 1993.

[7] S. McLean and D. Giesen, "Legal and ethical consideration of the Human Genome Project", *Medical Law International*, vol. 1, 1994, pp. 159–176.

The genome projects' results could lead to improvements in prevention and treatment of illnesses through new methods and genetic exams. Access to genetic information can also improve the quality and efficiency of measures taken by the authorities, particularly in the fields of criminal justice, social security, public health and immigration. Each of these authorities has to deal with the difficulties of protecting personal privacy and dignity, and it is here that there will be difficulties for legislators[8].

Third-level issues are more speculative. They involve "how a genetic view of ourselves could change the way in which we think about ourselves[9]" (and, importantly, others). They may affect the way we view our relatedness, otherness and difference. In their research, which is closely linked to these issues, Marilyn Strathern and her colleagues have, in a cognate area, put forward that the deployment of reproductive technologies is affecting our views of family life and cultural practices[10]. The problems brought into play by the new technologies are undoubtedly immense, but I will stick to those that affect the three levels mentioned by Annas, in particular, the impacts of this new biology on how we conceive of life and man. It becomes particularly relevant, if we accept the post-modern reorientation of contemporary philosophy, the conclusions of our ethics are constructed, as opposed to emerging from contemplation as in the theoretical work of Plato and E. Kant. At a time when we are affirming that genetics will dissipate ignorance and elucidate matters, ethics is moving away from a subject that is rationally comprehensible and into a field that man is entirely responsible for. What

---

[8] L. Nielsen and S. Nespor, "Genetic tests, screening and the use of genetic data by public authorities in criminal justice, social security and alien and foreigner acts", *The Danish Center of Human Rights*, Copenhagen, 1993.

[9] G. Annas, *Standard of Care*, Oxford University Press, New York, 1993.

[10] M. Foucault, *L'ordre des choses*.

K. Thomas, *Religion and the Decline of Magic: Studies in popular beliefs in the 16th and 17th century;* Penguin, Harmandsworth, 1973.

M. Strathearn, "The meaning of assisted kinship" in ed. M. Stacey *Changing human reproduction*, Sage, London, 1992, pp. 148–160.

M. Strathearn *et al.*, *"Reproducing the Future: Anthropology, kinship and the new reproductive technologies,"* Manchester University Press, Manchester, 1993.

will emerge from these constructions? It all depends on how we manage our ethical freedom, our view of life and ourselves.

The genome projects already constitute a considerable contribution to science. The scientific and technical progress made is undeniable. The number of genes cloned is constantly increasing, as is the number of genes that cause hereditary illnesses have been located. The impact on bio-industry and medicine is beginning to show its effects more clearly. Gene therapy trials are underway. In no way would I want to deprive the genome programs, particularly the HGP, of any of their importance. The goal, in this conclusion, is to bring a critical and subversive point of view to bear against the enthusiastic, often excessive and not very publicly appetizing declarations that are made about the genome programs, in particular, the belief that the HGP will show us the nature of what we are and teach us everything about life. What will be criticized here will be the ideology of biological determinism, which, if it was not born with the genome programs and the HGP, has been considerably reinforced by them, with what could possibly be some very dangerous consequences. Should we accept that this knowledge can only lead to destruction, thus reinforcing a basic conflict between science and life that is too often allowed?

8.2 Reductionism *vis-a-vis* the Complexity of Life

Despite the self-evident successes of the genetic and molecular approaches to the study of life, successes that prove that the chosen way was and remains the right one, victory is manifestly incomplete, and contemporary biology remains scarred by conflicts of interpretation between what A. Rosenberg called the autonomist and the provincialist concepts of biology[11].

Some notable biologists who support extreme reductionism have not hesitated to declare that, "If they had a powerful enough computer and the total sequence of an organism, they could calculate the organism", meaning

---

[11] A. Rosenberg, "Anatomy and provincialism" in *The Structure of Biological Science*, University of Cambridge Press, 1985, pp. 11–35.

that it would be possible to completely describe its anatomy, its physiology and its behavior. Today, we have the complete DNA sequence of several organisms and it's not for all that that we are able to "calculate" them. There is no way to reduce a living organism to its DNA sequence, not even the organism calculates itself from its DNA...

A living organism, at any point in its lifespan, is the unique consequence of its history and development, resulting from the confrontation and determination of external and internal forces. External forces, which we usually refer to as the environment, are also partially the consequence of the activities of the organism, since it produces, assimilates and consumes the conditions of its own existence. Internal forces are not autonomous either, but respond to external forces. Part of the internal machinery of the cell only kicks in when external conditions require it. Furthermore, internal doesn't just mean genetic. Fruitflies have bristles under their wings, a different number under the left and right — a sort of fluctuating asymmetry. However, any individual fly has the same number of genes affecting its right and left sides. Moreover, the tiny size of a developing fruitfly and the environment it develops guarantee that both left and right sides have the same humidity and temperature. The differences between left and right sides are not caused by genetic factors or environmental differences but random variation in growth and cell division during development — a sort of developmental "white noise", in fact. The organism is much more than its genes, and its relationship to the environment is much more complex than is apparent. Organisms don't just enter the environment they develop in, they build it as they develop.

Epistemologically, when speaking of the living, the observer cannot change his point of view as he wishes. The object imposes absolute limits that make it impossible to move from a local to a global view or to process from bottom-up to top-down, trying to deduce the properties of all the component parts and preventing any other prediction than a global one. For very complex systems where networks, associations and connection make us give up a linear hierarchical structure without rejecting the inherent hierarchy of levels, and where the local is not the global and the whole is revealed only at the end because of this very complexity, the top-down and bottom-up approaches find difficulty in meeting. They leave a gap for the new and the unexpected (at least to the exterior observer) situations that remind one of the typical difficulties one meets when observing systems with several integrated levels, systems which force us to use different descriptive languages which do not overlap for each of the levels.

Even if this notion of levels, this problem of hierarchical organization, arises from our understanding of "something" which is connected to our theory of understanding and originates in our experiments, it remains that the notion is symptomatic of our inability to grasp all the levels with the same precision, i.e., to have direct access to the links between them. These links are exactly where the basic transformation of what is a distinction and separation at the lower level, a reuniting and unification in the upper level, and the emergence of new properties can be found. These new properties ensure that the organized system is at the same time whole and the sum of its parts. The links are also where, in information-theory terms, Henri Atlan[12] has located the birthplace of meaning in information transmitted from one level to the next. One might object that technical progress can, by providing new analytical and experimental tools, provide access to one of these link sites between two levels hitherto seen as negative; but even then, the problem has just moved on one jump because access to every interaction link will reveal other links with just as many problems lying between this level and the levels above and below it.

For example, molecular biology had a long history as it passed from chemistry to biology. During this transition, a separate description evolved with its own tools of theory and observation, distinct from those of the two levels it lurks between. A solution to this transition problem, it could be said, lies in discipline. However, this still pushes the transition problem back to a deeper level. Now there are two transitions where there was only one. As a condition of our observer status, we must relate to the system as to a hierarchical entity which can only be analyzed in a looping or interlacing of

---

[12] Henri Atlan, "L'émergence du nouveau et du sens" in Colloque de Cerisy, *L'Auto-organisation De la Physique au politique*, Sous la Direction de Paul Dumouchel et Jean-Pierre Dupuy, ed. du Seuil, 1983, pp. 115–138.

hierarchies that, at our level of understanding and approach, correspond to the use of different languages for each of the different levels.

I believe, like Pierre Livet[13], that it is impossible to graft our understanding of the object considered onto the understanding it would have of itself; in other words, to embed in the organization of the observed object the routes by which the observer has accessed the object. The reconstruction of the relationship between levels is not done through modeling the loop through the levels but by the production of a new access route, a new level requiring a new language, with its new methods of interpreting the relationships between the levels. This only pushes the problem of the transition into infinity since there are still two passages where before there was only one.

At an epistemological level, this limitation in our understanding, presented by the infinity of levels, means that we must freely choose the standpoint from which we observe an organism and the level on which we are going to examine it, even if certain experimental methods considerably limit the indefinite choice, even if you take into account criteria of efficiency, criteria which are always relative from our point of view. The understanding of the living cannot be separated from the logic of understanding the living.

Rejecting the possibility of understanding what is within oneself, i.e., rejecting the idea of complete understanding with the realization that there is only information in systems which determine their own semantic field; and that therefore, all living beings build their own reality, no reduction can be justifiable. Any reduction of reality to one level is but a random moment of abstraction from the complex whole. Any affirmation of a theoretical reducibility of the living being to its genome, to its "program", is therefore, and can only be, the consequence of a misunderstanding of the intrinsic complexity of life or the sign of an ideologically driven intention to mask this complexity by claiming that simplification is possible.

However, before dealing with the question of how justifiable a possible reducibility of a living being to its genome can be, and therefore the possible

---

[13] Pierre Livet, "La fascination de l'auto-organisation", in Colloque de Cerisy, *L'Auto-organisation De la Physique au politique*, Sous la Direction de Paul Dumouchel et Jean-Pierre Dupuy, ed. du Seuil, 1983, pp. 165–171.

definition of the latter by the base pair sequence of its DNA, a few more words need to be said on this question of reducibility, not about the general problem of reducing the living being to its genome, but the more specific one of the connection between biological and physico-chemical theories, specifically between classical Mendelian genetics and molecular genetics. Despite the great progress made in molecular genetics and the support of a considerable number of biologists, this reduction remains a problem.

D. Hull says, "Given our pre-analytic intuitions about reduction ...[the transition from Mendelian to molecular genetics] is a case of reduction, a paradigm case. [However on] ... the logical empiricist analysis of reduction [i.e., on the basis of the Nagel type of reduction model] ... Mendelian genetics cannot be reduced to molecular genetics. The long awaited reduction of a biological theory to physics and chemistry turns out not to be a case of reduction..., but an example of replacement"[14].

Hull also highlights that reduction is impossible: "Phenomena characterized by a single Mendelian predicate term can be produced by several different types of molecular mechanisms. Hence, any possible reduction will be complex. Conversely, the same types of molecular mechanism can produce phenomena that must be characterized by different Mendelian predicate terms. *Hence, reduction is impossible*"[15] (my emphasis).

Beyond Hull, Rosenberg also questions the connection between classical and molecular genetics. Rosenberg's argument is as follows:

> Molecular considerations explain the composition, expression, and transmission of the Mendelian gene. The behavior of the Mendelian gene explains the distribution and transmission of Mendelian phenotypic traits revealed in breeding experiments. But the identity of Mendelian genes is ultimately a function of these same breeding experiments. Mendelian genes are identified by identifying their effect: the phenotypes observationally identified in the breeding experiments. So, if

---

[14] D. Hull, *Philosophy of Biological Science*, Prentice-Hall, Englewood Cliffs, N.J., 1974, p. 44.
[15] *Ditto*, p. 39.

> molecular theory is to give the biochemical identity of Mendelian genes, and explain the assortment and segregation, it will have to be linked first to the Mendelian phenotypic traits; otherwise, it will not link up with and explain the character of Mendelian genes. But there is no manageable correlation or connection between molecular genes and Mendelian phenotypes, so there is no prospect of manageable correlation between Mendelian genes and molecular ones. The explanatory order is from molecules to genes to phenotypes. But the order of identification is from molecules to phenotypes and back to genes. So to meet the reductive requirement of connectability between terms, the concepts of the three levels must be linked in the order of molecules to phenotypes to genes, not in the explanatory order of molecules to genes to phenotypes. And this... is too complex an order of concepts to be of systematic, theoretical use[16].

Rosenberg is referring to problems similar to those brought up by Hull and points out that the extraordinary complexity of the biosynthetic pathways will resist any attempt to formulate a concise model of the relationships between classical genetics and molecular genetics. He concludes, like Hull, that "the relationship between the molecular gene and the Mendelian phenotype is not a one-to-one relationship, or a one-to-many relationship, or even a many-to-one relationship. It is a many-to-many relationship"[17].

Rosenberg affirms that because of this, "There cannot be actual derivation or deduction of some regularity about the transmission and distribution of phenotypes, say their segregation and independent assortment, from any complete statement of molecular genetics"[18]. He points out that "cannot be" is not a logical impossibility but one created by our limited power of expressing and manipulating symbols. The molecular premises required to carry out the deduction would be enormous and would contain frighteningly long and

---

[16] A. Rosenberg, *The Structure of Biological Science*, University of Cambridge Press, 1985, p. 97.
[17] *Ditto*, p. 105.
[18] *Ditto*, p. 106.

complicated clauses[19]. The illegitimate extension of a genetic paradigm from a relatively simple level of genetic coding to a complex level of cellular behavior represents, in fact, an epistemological error of the first order.

It could be said that Rosenberg is unduly pessimistic given the progress made in the molecular analysis of genomes and the increasingly important role of *in silico* work in sequence data analysis. Perhaps computer analysis really could lead to the sort of reduction that Hull and Rosenberg consider beyond our grasp. However, even if progress in the genome programs provides us with new perspectives, in practice, the connection between classical genetics and molecular genetics remains a problem. In no way does this justify an excessively optimistic interpretation of the logical practical simplicity of boiling down the mysteries of genetics to the molecular level. It would be particularly difficult to articulate the reduction relationships in a simple way since genetic traits result from the interactions of several genes. Absolute genetic determinism clearly cannot be supported as a theory.

In his work *Autonomy and Provincialism,* Rosenberg illustrates the problem of reduction, in particular, that of reducing function, using the example of hemoglobin. Hemoglobin (Hb) is a four-part macromolecule, a tetramere, with two α subunits and two β subunits. The shape of the entire α-hemoglobin molecule or β-hemoglobin molecule is exclusively determined by the order and properties of the amino acids that make them up.

There are four structural levels to the molecule:

- The primary structure — linear sequence of the amino acids;
- The secondary structure — the molecule's shape determined by the angles of the covalent bindings between amino acids;
- The tertiary structure — the shape imposed on the non-atomic, intermolecular bindings (hydrogen bridges) between amino acids unconnected in the primary, linear structure but brought into contact by the secondary structure;

---

[19] *Ditto.* For Rosenberg, another difficulty is linked to a reduction of the theory of the macro level to the theory of the microlevel. The language of the macrolevel indicates a subject which has no correlation in the microlevel. In fact, the concept of gene which sends us on to a phenotype function will never be used in a system of biochemical concepts since the gene function is usually a function of a register not present at the biochemical level.

– The quaternary structure — the articulation of the four α and β subunits into a general structure capable of trapping oxygen.

It can already be seen how the sequence of amino acids, through the reconstruction of various levels of structure, brings about the molecule's shape. And it is precisely that shape that allows hemoglobin to fix oxygen. The molecule of oxygen is, in fact, trapped inside the four subunits that form the hemoglobin. It is clear how the sequence of amino acids determines the shape and how the shape determines the function. This is the basic structure of reductionist explanation.

This analysis can be carried further by considering physiological properties. It is now known that the affinity of hemoglobin can be regulated by 2–3 diphosphoglycerate (DPG). DPG is absorbed in the tissues by hemoglobin instead of oxygen. This causes the hemoglobin to lose its affinity for oxygen to the point that the hemoglobin no longer traps the tissues' oxygen. In the lungs, the higher concentrations of oxygen reverse the process.

This property is important for the organism to function. It is also important for the development of the embryo. In fact, the primary sequence of fetal hemoglobin is slightly different to that found after birth. The difference is that fetal hemoglobin does not capture DPG. It has a higher affinity for oxygen than the mother's blood. This explains how fetal blood captures oxygen from the placental blood supply. Here again, is an example of reducing the physiological level to the biochemical level. This reduction goes even deeper. Genes have been identified whose activity stops at birth when the processes are taken over by a gene for adult hemoglobin. This reduces the genetic and physiological levels to the biochemical level.

Paradoxically, the antireductionists use the same example to show the gaps in this explanation. When the primary structure of hemoglobin determines the function of the tetramere, only nine of the 140 amino acids of each subunit are "conserved", or remain identical, across mammal species. The nine amino acids in question are to be found everywhere because they are the only ones that provide the same secondary and tertiary structures[20].

---

[20] Rosenberg, *The Structure of Biological Science*, University of Cambridge Press, 1985, p. 89.

This means that there is no particular amino acid sequence that provides hemoglobin. Structure can determine function, but a large number of different structures at the same level of the primary sequence can generate the same secondary structure and the same biological function. In a certain way, hemoglobin's function is independent of the particular chemical composition of the molecule. The anti-reductionist would continue the analysis and ask why the same nine amino acids are kept. The reductionist cannot answer that it is because of their contribution to the secondary structure and the way oxygen is transported because that's functional reasoning, which is precisely what the reductionists do not want to do.

In the end, both reductionist and anti-reductionist biologists call upon the mechanisms of natural selection, mechanisms that are very difficult, and perhaps impossible, to reduce to the laws of chemistry.

Is there not a certain mockery being made of knowledge and life? In his book, *La Connaissance de la Vie*, Georges Canguilhem says, "In man, the conflict is not between reflection and life, but between man and the world in the human conscience of life. Reflection is nothing other than the detachment of man from the world that allows him to step back, question and doubt the obstacle that has risen before him"[21]. This detachment is indeed necessary, as the analysis, but it highlights the impossibility of touching the object's reality since it is perceived from a determinist viewpoint. He points out the external nature of the subject and how it can sometimes lead to an erroneous perception of the object, which can be deformed by our grids of reference and our preconceptions.

The rhetoric that surrounds the genome programs celebrates the power of reduction i.e., that physico-chemical processes can be reduced to the play of atoms. Molecular biology has reduced heredity to the transmission of information encoded in the molecules of DNA. This rhetoric transforms the meaning of the explanation. It is no longer "explaining" in the sense of "bringing out the meaning" of what is being explained, but all the other consequences of the observer's determined viewpoint perceived along with

---

[21] Georges Canguilhem, *La connaissance de la vie*, ed. J. Vrin, 2nd edition, Paris, 1985.

the objects that reveal his very existence. Through this, even what is diverse can appear the same. Silence is being kept regarding the fact that the diversity explained doesn't usually exist before it has been explained, that it is less conquered than the product of a practical invention that comes to add itself to other practices. Reflection on the living must base its idea of what the living is from the living.

Goldstein held that for the biologist, whatever the importance of the analytical method in his research, it is his naive understanding, that accepts the facts in all simplicity, which is the basis of his knowledge. Today, we might well ask ourselves whether this is still the case. More than ever, as George Canguilhem has already written, "We suspect that in order to do mathematics it would suffice to be an angel, but in order to do biology, even with the help of intelligence, we would sometimes have to feel stupid." Is man in his pretentiousness still capable of that?

Scientists writing about the genome project are rejecting explicit determinism, but by their declarations they seem to recognize theoretical possibilities rather than speak from genuine conviction. If we take seriously the proposal that the internal and the external codetermine the life of the organism, as well as the inherent limits that affect any observation of the living and its complexity, we cannot really subscribe to the suggestion that the human genome sequence is the Holy Grail that will show us what it is to be human, change our philosophical point of view of ourselves and show us how life works. This is, though, the idea that the HGP have been putting about, and as far as the public is concerned, it is the HGP's justification. The fundamental importance of the project (and also more generally of all genome programs) lies less in what is revealed in biology and whatever successful therapy to whatever disease it may produce, but in the validation and reinforcement of biological determinism as an explanation of all social and individual variation. The medical model, which could, for example, begin with a genetic explanation for the extensive and irreversible degradation of the central nervous system characteristic of Huntington's chorea, could end by explaining human intelligence or why some people tend to drink or be depressed. A medical understanding of all human variation would allow the production of a model of human normality, including social normality, and permit the prevention or *a posteriori* corrective therapy to be applied to bring back to this normality any individual who might deviate from it.

This concept is becoming more and more widespread because biologists are continuously revealing to the public the discovery and cloning of genes that cause illnesses, predispositions and, in a vaguer manner, behavioral characteristics, to reinforce the promise of coming gene therapies as well as the importance of the HGP in this revelation of the secrets and causes of the anomalies of life. Having initially criticized the idea of absolute determinism, I must criticize the discussion that surrounds and reinforces the image of the HGP, often ideological[22], that could be called "geneticism", or the ideology of the "genetic all".

8.3 From Science to Ideology — The Dangers of "The Genetic All"

An important opening remark is that the hypothesis of genetic determinism (the possibility of a genetic understanding of the variables of human behavior) is subject to criticism. Several scientists have suggested, nevertheless that, it could be defended. D. Baltimore had no qualms in affirming, "That the HGP will allow us to examine human variability, for example, variations in mathematical ability or in what we call intelligence. Those variations are caused by the interaction of many genes. And certainly the best way that biologists have to unravel which genes are involved in complex traits is to find a set of markers that are linked to the disease and then find the genes associated with those markers. In other words, we need the linkage maps and the physical maps that will be generated by the Human Genome Project. Those maps will allow us to do new kinds of science"[23].

---

[22] In the remainder of this Conclusion, the notion of ideology is used in the sense of doctrine (or philosophy), resting on a scientific argument, endowed with excessive or unfounded credibility, caricatured so as to be easily understood and approved by the majority and, therefore, cautioned in the use which is made of it.

[23] D. Baltimore, declaration quoted in *The Human Genome Project: Deciphering the Blueprint of Heredity*, ed. by N. G. Cooper, University Science Books, 1994, p. 78.

Another example of the same tendency can be found in the well-publicized book written by Richard Herrnstein and Charles Murray, *The Bell Curve*. The authors affirmed that the intelligence quotient (IQ) is mostly determined by a person's genetics and that IQ differences between ethnic groups can be explained by genetic factors. They added that since IQ is genetically determined, programs to reduce learning inequalities were doomed to failure. They proposed to end all social programs, which for them were encouraging births in the lower brackets of society and leading, according to Herrnstein and Murray, to genetic decline. They underlined how urgent it was to develop social support programs that would encourage women of higher economic classes to have more children.

This thinking is a remarkable example of the dangers of the ideology of "the genetic all", which is developing at the moment and which is likely to be used for dangerous eugenic and political ends, unfounded, but nevertheless, based on this pseudoscientific concept of man. Media announcements of discoveries of a growing number of genetic factors that have a determinant effect on behavior (schizophrenia, alcoholism, agressivity, innovative tendencies, homosexuality), the success of the HGP in the identification of human illnesses, and the discovery of hereditary variations that seem to be linked to certain predispositions, help build this environment within which undue affirmation of the determination of other human traits will earn undeserving credit. Not only do the HGP's successes accentuate this tendency through the declarations scientists make of its perspectives, but it also leans on this widely shared credibility and tries to justify itself in social terms, as we can see. An anecdote that recently appeared in *La Recherche* underlines the problem:

"As yet not widely publicized, the discovery of a genetic mutation that predisposes people to racism is the most remarkable of events in scientific current affairs in recent weeks[24]. The team of Professor Hans Moltke of the molecular neurology department of the University of Heidelberg, can indulge in some self-congratulation. For as Moltke himself confirms, it is now possible to nip evil in the bud. The progress made in gene therapy makes it possible

---

[24] *Trends in Neurosciences*, September 1996.

to build a vector that can relatively easily go and replace the defective gene with a healthy one. In the rare cases where the mutation is hereditary, prevention can be implemented immediately by forcing the parent carrying the gene to procreate *in vitro* so that the healthy gene can be provided to the first cells of the embryo. But let us not cry victory too early! For gene therapy is only just learning to walk[25]. (...) To cure a person, one must first convince them to have their genome checked (...) To avoid the challenge of forcing them undemocratically, the best would be to look to the future and prepare a screening of all schoolchildren. Then the State would have to pay for it"[26]. The story ends with a question, "What proportion of our readers read this story thinking it was true? Admit that in any case it's certainly something to think about."

With declarations such as those of Baltimore, Herrnstein and Murray, and the perspectives of the HGP announced in the headlines, a badly informed person falls for this type of story and comes to believe that everything is indeed genetically determined, that whatever their behavior, identity or abilities, they are entirely without responsibility, and that the cause is "hereditary", "genetic", miracle words with powerful explanatory powers that, once spoken, dissipate mystery. When a symptom seems mysterious, it is inevitable that someone say that it is of genetic origin. Nothing seems to escape geneticism, and decoding the human genome seems to be the key to a new era for medicine and man, accompanied by a new understanding of man's identity and intelligence, and the possibility of being able to fight some 4,000 genetic diseases[27]. With this fiction, the review *La Recherche* stands as an example of the problematic effects of these announcements in reinforcing public belief that between progress in genetics and the traditional concepts we have of our identity such as freedom and will, there is a tightening knot, or a knot that seems to be tightening. But it only seems to

---

[25] *La Recherche*, May 1996.

[26] Quoted in *La Recherche*, vol. 292, November 1996, p. 5.

[27] Figure published in March 1994. In fact, over 7,746 genes, illnesses and hereditary weaknesses have been registered. (V.A. McKusick, *Mendelian Inheritance in Man, Catalogs of human genes or genetic disorders*). The figure of 7,746 genes from 26 January 1996.

be tightening. Leaving Baltimore aside, if we once more take up the assertions of Herrnstein and Murray on intelligence and its possible genetic determinants, we can see that they call on the authority of genetics to support that it is now beyond all technical doubt that cognitive capacity is substantially hereditary.

Research in this field is constantly evolving, and the studies quoted by Herrnstein and Murray have doubtful methodologies. The validity of their results is to be taken with a grain of salt[28]. Many geneticists have underlined the enormous scientific and methodological problems that affect any attempt to distinguish the influence of genetic factors from environmental ones, particularly given the interactions between genes and the environment in a human trait as complex as intelligence.

Furthermore, its relationship to value can be greatly questioned. The unique interaction between the organism and the environment can only be described in terms of the differences in levels of ability and even less in qualitative terms. It is true that if two genetically different organisms develop in the same environment they are different (developmental "white noise", genetic differences...), but these differences cannot be described as differences in ability since the genetic type that would be considered superior for one environment might turn out to be inferior in another.

Even if there were consensus on the hereditary transmissibility of cognitive capacity (which, as for all complex human traits, there certainly is not), the lessons of genetics have not been completely understood. The authors maintain that because cognitive capacity is largely hereditary, it is impossible to modify it. Therefore, any education program attempting to remedy deficiencies in capability is a waste of both time and money. This is not the correct message to be taken from genetics, nor a lesson for educational programming. Estimates of the transmission of heredity can

---

[28] For a rebuttal of this type of study, see the comments of R.C. Lewontin, S. Rose and L.K. Kamin, *Not In Our Genes*, Pantheon, New York 1984, pp. 101–106, the study carried out by Sir Cyril Burt, and more generally the work of L.J. Kamin, *The Science and Politics of I.Q.* Erlbaum, Potomac, MD, 1974. It points out that all the studies that try to emphasize the effects of genes on IQ are always confronted by experimental difficulties and are always scientifically doubtful at the methodological level. Results published in the article "IQ and race", *Nature*, 247, 1974, p. 316, contradict the affirmations of Herrnstein and Murray.

only be valid for the specific environment in which they are measured. If the environments change, the hereditability of traits can also change quite considerably. To say a trait is often inherited does not imply that it will not evolve. Height, for example, is a trait which is genetically determined, but it also depends on nutrition. Common conditions in which genetic factors undoubtedly play a role, such as diabetes or heart disease, can be attenuated by taking insulin or chemical substances that lessen the cholesterol levels or by appropriate diet. Pathologies associated with conditions linked to a single gene, such as phenylketonuria or Wilson's disease, can be prevented, or living conditions greatly improved for sufferers, by medical or nutritional therapies.

As the mysteries of human genetics unfold, its complexity is revealed to be greater and greater. It becomes increasingly obvious as more genes linked to genetic diseases are unveiled. We are only just beginning to explore the complex relationship between genes and the environment and between individual genes and the human genome. It is the lack of predictability on the basis of genetic information that is today the rule rather than the exception.

The simplistic, not scientifically founded, affirmations made by journalists who have run out of other things to write about, or by some scientists in interviews on the inheritance of complex human traits, are unjustifiable — just as unjustifiable as D. Baltimore's assertion that variations in cognitive capacity are caused by the interaction of numerous genes. These declarations are not based on science but on the ideology of the "genetic all" on which the HGP is widely based. The defendants of systematic sequencing and, again, the media, underline how important the HGP is for the discovery of the function of human genes through sequencing and locating the genes responsible for, or linked to, illnesses.

At the end of 1993, in *Gene*, J.D. Watson, the first Director of the National Center for Human Genome Research (NCHGR) described the Human Genome Project as follows:

"Its mission was not only to make much higher resolution genetic maps but also to assemble all the human DNA as overlapping cloned fragments running the entire length of all the human chromosomes. In their turn, these DNA pieces were to be sequenced and their respective genes revealed. Upon completion of the Human Genome Project we would know how many human genes exist and whether the then-estimated 100,000 number had been too low or too high"[29].

But building physical maps on several scales and sequencing the DNA will not provide us with the total number of genes in the human genome. The DNA linked to gene products (exons) is disseminated throughout an ocean of non-coding sections. Outside the regions that code for gene products, there are long pieces of polynucleotides that do not code for anything. Some of these fragments are switches that help regulate gene expression. The genes themselves are repeated twice or more throughout the genome on the same or different chromosomes. Sometimes, the same DNA fragment codes for two distinct gene products, depending on the way the message is read. Even if we know the coding sequences from end to end, we don't know how to partition the base pair sequence at the point where a structure gene begins or what successive base pair triplets constitute the reading framework for the product. In October 1993, F. Collins, who replaced Watson as NCHGR Director, and D. Galas, Associate Director of the Department of Energy's Office of Health and Environmental Research, wrote:

"Although there is still debate about the need to sequence the entire genome, it is now more widely recognized that the DNA sequence will reveal a wealth of biological information that could not be obtained in other ways. The sequence so far obtained from model organisms has demonstrated the existence of a large number of genes not previously suspected... Comparative sequence analysis has also confirmed the high degree of homology between genes across species. It is clear that sequence information represents a rich source for future investigations. Thus, the Human Genome Project must continue to pursue its original goal, namely, to obtain the complete human DNA sequence"[30].

---

[29] J.D. Watson, "Looking Forward", *Gene*, 135, 1993, p. 310.

[30] F. Collins and D. Galas, "A new five year plan for the US Human Genome Project", *Science*, 1993, p. 46.

Completing the sequencing work on the human genome will undoubtedly provide a huge quantity of data, but this doesn't mean that the data will be functional biological data. It will be purely physical information. Despite Collins' statement, sequences already obtained from simpler organisms do not show unexpected genes, because they cannot. Sequence data can only help locate genes that have already been indicated by using genetic techniques such as growth experiments. Finding sequence homologies between known model organism genes and parts of human DNA is one of the highest biological priorities. But for this to occur, the genes must first be found in the model system, located, sequenced and then hybridized with human genome DNA. The result of this process is the identification of the human gene, which can then be sequenced. The complete sequence of the human genome will play no role in this process and cannot help in dividing the human genome into genes. Except for the genes unique to our species (and there is no reason to believe that there will be any great quantity of them), it is not necessary to know the entire sequence of the human genome; and it is not enough to find the genes with. It certainly won't reveal how many genes there are.

In an article based on an interview with Director Collins, the *Science* journalist Leslie Roberts writes:

"Despite the slow progress, there is little sentiment for abandoning the goal of all-out sequencing... But some thought is begin given to a shortcut called one-pass sequencing. The original plan calls for sequencing the whole genome several times to ensure an error rate of 0.001%. "Suppose we try one pass coverage with 1% error rate but it only costs one tenth as much" asks Collins. The idea, then, would be to return to the really interesting regions and sequence them again"[31].

The questions this assertion brings to bear are the following: What are the regions worthy of interest? How do we identify them and what can we achieve from their identification? None of these questions will be resolved by the completion of sequencing work on the human genome. The interesting

---

[31] L. Roberts, "Taking Stock of the Genome Project", *Science*, 1993, p. 21.

regions are those where the products are involved in the production of proteins in the ribosomes, the DNA regions that produce mRNA for enzymes and other proteins, the regions that produce enzymes that control the expression of genes producing other enzymes and the parts of DNA that control the controllers... and so forth. How to identify these genes? Certainly not from the bottom-up approach because we don't know from the start which sequences have such functions. It would be better to start from the gene products back to the DNA sequences.

Once it is known where a functional gene's sequence is to be found on the chromosome, and it can be isolated, the sequence can tell us a lot about the product. However, the functional identification of a gene requires the identification of its product, the phenotype, which is shared in a family of organisms in accordance with the well-understood regularities of population genetics. Supposing that the phenotype includes an abnormality from a genetic defect, great therapeutic progress can be made without knowing the entire DNA sequence, where the defective gene is to be found, its length, how often it is repeated... all we need to do is isolate the message for a particular gene product, usually expressed in larger quantities in normal organs and tissues and for which we are trying to understand the defective function. Once large quantities of tissue-distinctive RNA have been identified, without returning to the gene, we can use reverse transcriptase to constitute cDNA clones and then use PCR to amplify them. Once inserted into the appropriate vectors, these synthetic genes provide enough proteins to put together products to treat, and hopefully cure, the sick.

Researchers are aware that to be able to produce cDNA from mRNA is a much easier route to the understanding of gene expression and regulation than sequencing the entire genome. In fact, this route is the base of a potentially lucrative sort of pharmacological research. After all, by hybridizing radioactively marked cDNA with the genome, molecular geneticists can focus on the five to 10% of DNA that really codes for products. So only a few researchers are asking for government funding to produce cDNA, most of them being out to patent them.

Of course, once a gene has been identified by its function, the phenotype, it can be located on a chromosome. What is necessary is the identification of abnormality detectable in a chromosome that co-varies systematically with a phenotype. Once this is obtained, it is possible to focus on the site on the chromosome and to sequence the DNA in that spot to discover the genetic cause. For this reason, physical maps of the genome already produced to a sufficiently high resolution are essential. However, such maps are far off the entire sequence. They are built from several thousand STSs. But to be useful, these physical maps and STSs must be preceded by genetic research that can reduce the field of sequencing for the entire human genome to several functional units.

Those who identify physical mapping as a main objective of the HGP implicitly relegate the entire sequence of the human genome to secondary status. Instead of being secondary in relation to mapping, it should not even be considered a goal of the project.

8.4 The Health Excuse — A New Utopia?

Aside from the argument that the HGP will allow us to site and allocate a function to all our genes, there is another argument often put forward to support this Big Science project, which is the search for genes linked to hereditary pathologies and the possibility of gene therapies.

If you follow the media and read popular science magazines, and even some of the scientific reviews, you might have the impression that the hereditary situation of man is declining. The number of genes linked to hereditary diseases is always increasing: the breast cancer gene (BRCA1), the gene predisposing to diabetes, the gene of muscular dystrophy, hemophilia, cystic fibrosis... the new genetics seems all-powerful, and with the perspectives linked to the HGP, seems to be opening up a new era of medicine. That is, at least, what we seem to be asked to believe, especially by scientists linked to the HGP. The figures put forward for several thousands of hereditary illnesses are not false, but they mean nothing in themselves and are only quoted to grab attention.

Illnesses for which we know the genetic cause are few; furthermore, they tend to affect only a small proportion of the population. The most frequent in Europe are cystic fibrosis and Duchenne's muscular dystrophy, these are those that the media are always telling us about. It is thought that one child in 2,000 either boy or girl, is born with cystic fibrosis, that is to say, each year, of the 700,000 children born in France, 350 will be affected. One boy in 3,500 is born with muscular dystrophy, 100 boys in France every year. Then there is hemophilia, which affects one male child in 50,000, 70 a year in France.

For other illnesses, genetic determinism is thought to occur but is less well understood. They are far greater in number but only affect a small proportion of the population (Huntingdon's chorea hits only one in 10,000). Their seriousness varies: some are benign, others quickly fatal. Lastly, there are illnesses which are only considered genetic predispositions. They only affect a tiny fraction of the population, unless you take the term "predisposition" to such a wide meaning that it no longer has any relevance.

Scientific theory is always used to explain more than it actually can; and for this reason, it often heads for elastic concepts for the vaguest science. Until now, the favorite elastic concept in molecular genetics has been the concept of information, a concept used sometimes in a physical sense, sometimes in a metaphorical one. Today, we must add predisposition to concepts used ideologically.

Predisposition is not a recent concept. A. Weismann was already using it. He separated the hereditary characteristics from the acquired ones and opened the door to the "genetic all" in his comment that an acquired characteristic is only acquired because the creature has a hereditary predisposition for acquiring it[32]. If this is true, it would be possible to find a predisposition for any characteristic, pathological or not. This notion of genetic predisposition has been widely used to increase the number of inherited pathologies[33].

It is greatly involved, as Dorothy Nelkin of the New York University points out, in popular culture. The belief in biological determinism, which

---

[32] A. Weismann, Essays on Heredity and Natural Selection.

[33] An example. One might speak of predisposition for cardio-vascular illness and cancer, and then respectively 200,000 and 150,000 more people in France per year would be dying of a "genetic" illness.

prevailed at the beginning of the century, has now perpetuated itself in the idea of genetic predisposition. A common theme is that people are not responsible for their behavior because it is predestined by their genes. Often the expression "programmed in the genes" is used. This theme also appears in discussions on alcoholism, gambling and other traits. It is a particularly widely used perspective in countries where the very basis of the democratic experiment is the perfectability of all human beings.

Contrary to what statistics say, the genetic situation of man has not worsened, and hereditary illness continues to touch a small proportion of the population. I am not suggesting that these people be denied their right to treatment. All must be done to improve their conditions of life. There are no priority illnesses, nor illnesses that affect too few people and should be ignored. New light should be brought to this frequently used and largely ideological discussion which surrounds genetic research, and more specifically systematic sequencing work and that of the HGP.

Gilbert and Watson have underlined how important the HGP is to the betterment of human health. In the perspective encouraged by the project, we will be able to locate all the genes responsible for hereditary illnesses on human chromosomes with maps and then, from their DNA sequences, the causal histories of the illnesses can be deduced and therapies created. In fact, a large number of incriminating genes have been mapped on the chromosomes. With the aid of molecular techniques, some of them have been closely located or even cloned, and for a far smaller group, the DNA sequenced. Generally, before they are even identified and their functions pinpointed, genes suspected of involvement in a hereditary or genetic illness are named after it, and they retain the name thereafter. People speak of the gene for muscular dystrophy, the gene for cystic fibrosis, terms which have no biological foundation and are completely incorrect. It could be said, at most, that there is a gene responsible for the synthesis of a protein, and that an alteration of that gene, altering that synthesis, causes an illness which shows up in the organs that need the protein to function on a normal basis. This identification between gene and illness reveals the aura that surrounds the word "gene", an aura continually reinforced by the geneticists' declarations. The illness is what brings the gene its functional definition, often through lack of physical delimitation and localization, but also through lack of a physiologically functional definition.

The simplifying, almost lying declarations made must not abuse the ears that hear them. When the discovery of a genetic predisposition or hereditary illness is announced, the mechanisms are simplified. The image the discoverers encourage and contribute to is of a sort of programming error, and that an easily determined physiological process can be followed through from the error to the illness.

In fact, this image is erroneous. Most ofter, it is from either statistical studies of families with the Mendelian characteristic of the transmission of the affliction, or the bringing to light of a correlation between the occurrence of an illness transmissible in this way and a peculiarity in the genome of sufferers, generally only detected and located more or less well with the new genome analysis methods[34]. The gene itself is rarely known as a determined functional or genetic unit. In their work, *Biologie Moleculaire et Medecine*, J.C. Kaplan and M. Delpech point out that the genetic alteration in itself has been located for only about 10% of the 5,000 known afflictions[35]. This does not mean that we understand the gene itself, and even less that we understand the mechanisms of the corresponding physiological disturbance[36]. Generally, we can just about say that such a more or less localized region of the genome probably has a particular but undetermined role in the extraordinarily complex physiological processes which, when modified in an undetermined fashion, can lead to such or such, a phenotypically characterized condition. Only for a few monofactorial genetic diseases do we have more precise information, and even in those cases, nothing is simple. Even if the HGP can contribute to the localization of defective genes, it is still not a miracle tool. To illustrate this, there are some hard facts.

---

[34] Which often themselves have a statistical component.

[35] J.C. Kaplan and M. Delpech, *Biologie moleculaire et medecine,* Flammarion, Paris, 1989, 2nd edition 1993, p. 266.

[36] The example of Duchenne's muscular dystrophy, the most common and well-understood of the muscle degeneration disorders, shows this well. The road from gene to illness is a long one. Since the identification in 1986 of the gene responsible for this affliction, a considerable amount of information has been gathered on the protein whose synthesis it governs, dystrophine, without allowing us to date to precisely determine the chain of mechanisms that lead to the appearance of muscular dystrophy.

At the protein level, there is an enormous amount of variation between "normal" individuals in the amino acid sequences of their proteins because a given protein can have a variety of amino acid components without any effect on its function. Each of us has two genes for each protein, one from our mothers and the other from our fathers. On average, the amino acid sequence specified by the gene inherited from the father and the amino acid sequence specified by the gene inherited from the mother vary every twelve genes.

Furthermore, all normal people have a large number of defective genes from one parent which are canceled out by the healthy copy from the other parent. So, any fragment of DNA sequenced will have a certain number of defective genes to add to the catalogue.

When the DNA of a person affected by a genetic or hereditary disease is compared to the DNA of a "standard" sequence, it is difficult to the point of impossibility to decide from simple cross checking which of the many differences between the two DNA strings is responsible for the illness. It is necessary to consider a large population of "healthy" and "ill" individuals and try to locate the differences they all have in common. But even this method can prove difficult to use. Some illnesses have many genetic causes, and sufferers can suffer for different reasons, even though those reasons are a consequence of genetic change. Today, we know of 120 mutations of the CFTR protein[37] that cause cystic fibrosis. For hemophilia, seven mutations that damage Factor VIII have been discovered. The list of illnesses for which several trigger mutations have been found is quite a long one.

There are also the types of illness that are paragons of simplicity; a single genotype (the same chance mutation) associated with a homogenous phenotype, but that causes a *clinically* different phenotype. Drepanocytosis

---

[37] The CFTR protein, according to recent results, seems to form a transmembrane channel that lets iron chlorates through. It is to be found in the membrane of epithelial cells.

is a classic example of just such an illness[38]. There are other variations of the relationship between the genotype and the phenotype:

One gene, several mutations, phenotypic variants

This is the case for both Duchenne's and Becker's muscular dystrophy, which despite their clinical differences, are linked to alterations in the same gene, the gene for dystrophine located on the short arm of chromosome X in the Xp21 position.

One gene, one mutation, several effects

The classical examples are some types of hemoglobinopathies, such as that of Knossos. It is caused by a mutation that causes a thalassemic phenotype (a reduction of the synthesis of beta-globin) and a hemoglobinopathy phenotype (an alteration in the biochemical properties of the mutant beta-globin which is formed).

One gene, several mutations, several diseases

An exceptional case of the androgen receptor gene — several mutations of the gene have been found as causes of insensibility to androgens, but it's now been shown that mutations in the 5' part of the gene can lead to a disease where the relationship with androgen sensitivity are but distant. This illness is a very rare affliction of the spinal bone marrow.

One disease, several genes

In terms of the correlation between phenotype and genotype, this field covers the classical notion of genetic heterogeneity: mutations of different

---

[38] Depranocytosis is a hemoglobin problem found all over the world. It manifests itself in a range of painful manners, and by the destruction of red blood cells in the small capillaries in bone, the hematopoeitic system, or the brain. This circulatory damage leads to a modification in their shape. At the genotype level, we know that the disease is always linked to a mutation of codon 6 of the gene of betaglobin. The biochemical phenotype is perfectly homogenous, with a presence of abnormal Hbs hemoglobin. This clear homogeneity in both the genotype and the biochemical phenotype cannot be found in the clinical phenotyep. On the one hand, the painful events due to local infarcts only occur in certain situations, when the patient's red blood cells together take on an abnormal form causing a slowdown of circulation in the capillaries. On the other hand, it is not certain that patients of the same sex and age in the same environment will have identical clinical symptoms. Does this variability in the phenotype imply that there are other genetic factors at work? Nobody knows.

genes can lead to the same phenotype. Marfan's disease is a good example. The syndrome causes a malformation typically involving a stretched aspect accompanied by scoliosis, anomalies in the crystalline and blood vessels, familial amyloidosis, and deposits of amyloidal substances in various organs or some collagen diseases. In most of the family forms of this disease, it seems it is the anomalies of the gene for fibrilline, to be found on chromosome XV, that are at fault. This location has been ruled out for other forms though, despite their great similarity of phenotype.

One gene, many ways to inherit it

The discovery that different means of heredity can exist for the same locus and the same gene was surprising, to say the least. It is apparently the case for the gene coding for the chloride channel. For some mutations, it's expressed as a dominant trait, producing a phenotype of congenital myotony. For others, it's expressed as a recessive which leads to a very different form of myotony, where the illness only occurs when the patient carries two mutant alleles. This worrying example does not leave us much hope of ever understanding such complex biochemical phenomena.

Mutations of unstable DNA

In their own way, these were also a surprise for the geneticists. Today, two frequent illnesses fit this description, the fragile X syndrome and Steiner's myotony, which causes lack of muscle tone. In both cases, repetitive mutations have affected the DNA, rendered unstable during meiosis. In this situation, the phenotype varies depending on how many repetitions there are.

Surprise phenotypes and genotypes

Syndromes of malformation are good examples of how difficult it is to predict the nature of a gene from a phenotype. Such is the case for piebaldism, a type of albinism linked to an anomaly in the migration of melanocytes. This dominant autosomic disease only affects the pigmentation system and none other. It is, however, due to an anomaly of a cellular proto-oncogene "c-kit", whose physiological role seems to be sited in the synthesis of blood tissues. However, no hematological sign has been found in patients homozygotic for the piebald mutation, who are normally severely mentally retarded.

Chromosomal anomaly phenotypes

This field is classically typified by chromosome 21 trisomy where mosaic 21 trisomies have been found expressed in only certain cells of affected children. It is difficult to establish correlations between genotype and phenotype. However, children carrying partial 21 trisomies have been described (here, the trisomy only involves a part of chromosome 21). These very rare cases have shown that a certain zone of chromosome 21, the q21-q22 band, is necessary and enough to cause the classical trisomic phenotype (mongolism) with its mental retardation and cardiopathy. Other elements of the phenotype would not be linked to the same chromosomal region but rather to adjacent regions.

So, whether a chromosomal disease or a monogenetic disease, a single gene or even a single mutation, the problem of correlation between phenotype and genotype remains one of the most complex problems to solve, despite the simplistic characterization of "one mutation, one gene, one illness".

In the case of polygenic afflictions determined by several genes and the environment, such as diabetes, arterial hypertension and cancer, the situation is even more complex. It makes no sense, except in certain cases such as retinoblastomas caused by the disappearance of part of chromosome 13, to speak of cancer genes. This is even more true for arterial hypertension.

It's clear how the representation of the power of genetics by scientists and the media is simplistic and caricatured. From the sequence to the genes, and in particular to genes linked to disease, there is no simple route. From this viewpoint, the media declarations in relaying some contributions such as those of Craig Venter on the so-called sequences and identifications of several tens of thousands of human genes, are causing confusion between the Expressed Sequence Tag labels and the genes themselves. Venter himself points out, "We have doubled the number of human genes identified with the help of DNA sequences"[39]. As Piotr Slonimski has said, "Any gene is a nucleic acid sequence. But not all nucleic acid sequences are genes. Knowing that only three to five percent of the human DNA sequence is

---

[39] Interview with C. Venter, *Biofutur*, April 1992, p. 18.

actually genes, it is hard to say how many genes are actually amongst the sequences labeled. It's very interesting as well as useful to have a label in the approach to the sequence and to stick along the sequence for mapping purposes. That is the fundamental contribution of Craig Venter and his team. But a label is not a sequenced gene. It is cheap to produce large quantities of labels, like producing ball-point pens when only one pen in a billion works. A small calculation shows that if there is five percent error per location, as Craig Venter thinks, then the probability of not making an error in a 400-nucleotide sequence is 0.95×400, that is to say $1,2 \times 10^9$. I can, therefore, bet a dollar against a billion dollars that there will not be more than one correct sequence in all those carried out in ten thousand years with this technique, at Craig's announced rate of 100,000 per year"[40].

Despite the optimism of certain journals, this work, and that being carried out under the DNA sequencing program initiated by Merck and run by R. Waterston[41], and the results of Genexpress (27,000 ESTs), does not allow us to judge the content of our genetic heritage. The analysis actually depends on 174,472 ESTs to which 118,406 sequences from the dbEST public database have been added. By comparing the 300,000 little sequences, containing 83 million nucleotides, with each other, Craig Venter has found some 300,000 groups of two or several ESTs that completely or partially overlap each other. Named THC, for Tentative Human Consensus Sequence, in principle, these units must be genes. This means of analysis, as pointed out by Piotr Slonimski, is fraught with uncertainties (5' or 3' sequences could begin at different transcription points, different origins of the banks, polymorphism, sequence errors) all of which increase the number of apparently different sequences. The real number of genes is, therefore, more likely to be less.

---

[40] P. Slonimski, "Une etiquette n'est pas un gene", *Biofutur*, April 1992, p. 5.

[41] The appearance in late September 1995 of a *Nature* supplement entitled Genome Directory marks an important step in the distribution of partial cDNA sequences. It was a political as well as a scientific event; after much controversy, Craig Venter finally published a large part of his results, revealing some 200,000 ESTs. Merck's initiative of sponsoring a large cDNA sequencing programme run by R. Waterston at St Louis, USA, was clearly no stranger to this decision. By the end of October 1995, this venture had submitted more than 200,000 human sequences in public databases on top of the 27,000 sequences provided by Genexpress and other genes of various sources, and this took away much of the value of Venter's treasured hoard.

Solitary sequences, of which there are more than 50,000, are even harder to estimate for gene content, for the same reasons. Lastly, some genes, seldom expressed (very seldom expressed or only expressed in a specific organ), at a precise moment in development, have escaped labeling because they are not to be found in the banks being used and have not been accounted for. This work, therefore, although it has its uses, is not telling us the number of genes in man's genetic heritage.

Despite these uncertainties, unforeseen problems and difficulties in knowing exactly which genes and sequence alterations lead to illness, and more generally, the difficulties in understanding the complexity of the mechanisms of heredity and the genome which is liquefying[42], becoming something changing, labile, involving the multiplication of regulation to the point where it is incredible. Molecular biology seems to be afflicted with the Ptolomean syndrome, and geneticists are not wasting their time telling the media this and of the fantastic progress in their discipline, the great successes of the HGP and the amazing perspectives that having the complete DNA sequence of the human genome will open up. According to the prophets of the HGP, such as Watson, Berg and Baltimore, we will be able to locate all defective genes on the chromosomes, and then from their DNA, deduce the causal links to the illnesses and produce therapies. In fact, the causal links are still missing and really efficient gene therapies do not exist yet, although they are precisely what is mentioned most often as the key to a better life, a life freed from all the illnesses that afflict us, and the HGP is hailed as a catalyst to the advent of that day.

Let us, however, look hard reality in the face. If our health depended on these gene therapies, we would all have been dead a long time ago. Without a doubt, our understanding of human physiology and anatomy has led to more efficient medical practices than these at the beginning of our own century. That progress, however, consists mainly in the development of methods of medical imagery, progress in surgical technique, obviously in the development of pragmatic means of correction of chemical imbalances,

---

[42] The image of the genetic message, of information, and the programs, is less and less appropriate, if it ever was.

and killing poisonous viruses and bacteria. None of these advances depend on deep understanding of the cell processes or the discoveries of molecular biology[43].

8.5 Behind Gene Therapy — The Dangerous Liaisons of the New Biology

Far from being reality, these therapies remain wishful thinking[44]. The truth is told in "Transfer of Genes to Humans: Early Lessons and Obstacles to Success" by R.G. Crystal[45], and the article entitled "La thérapie genique" by I. Verma[46]. The terms "potential", "we are beginning", "potential use", "the promise of" and "perspectives that can be opened in the near future" are often used.

I won't analyze the technical difficulties in bringing gene therapy to life, such as difficulties in mastering the precise insertion of the new DNA, inconsistent results, predictions based on experimental results taken from clinical studies on animals and not valid at the human level, problems with vector production, the impossibility of building a perfect universal vector (the ideal vector being different for each application) and difficulties in accurately targeting the DNA inserted into the cells[47]. I shall just point out that they are numerous and varied, and that the clinical trials under way, although they give cause for hope, have not until now, led to the resolution

---

[43] Cancer, diabetes, and cardio-vascular disease are always treated by aggressive chemical, physical or surgical intrusions, and by diet and medicinous therapies. Antibiotics were originally developed when we had no understanding of how they work. Diabetics continue to take insulin as they have been doing for 60 years, despite all the research on the cellular basis of pancreatic dysfunction.

[44] The first clinical study of a gene therapy began in September 1990. Michael Blase, French Anderson and their colleagues at the NIH introduced the gene for the enzyme adenosine desaminase to the genome of some children suffering from the rare severe immunodeficiency syndrome.

[45] R.G. Crystal, "Transfer of genes to humans: early lessons and obstacles to success", *Science*, vol. 270, 20 October 1995, pp. 404–401.

[46] I. Verma, "La therapie genique", *Pour la Science*, supplemental file, April 1974, pp. 90–97.

[47] The work being carried out by M. Capecchi at the University of Utah shows that these obstacles can be overcome.

of all these problems and remain of doubtful efficiency[48]. Furthermore, germ line gene therapy, although technically possible, has been left aside for the moment, because the new genes would be transmitted from generation to generation, causing serious ethical problems.

Until now, most of these treatments have not been "once and for all". They have side-effects, and are generally less efficient than some classical therapies. Far from being a panacea, as the media and scientists generally present them, gene therapies are in truth yet to come. Of course, progress has been made at the experimental level, and trials are under way, but if you sum them all up, it's clear that these therapies are advancing more slowly than expected and than is generally announced. Medicine will have really made a step forward when we know how to treat illnesses, once and for all, with vectors of proven innocuity (in particular, in view of the retroviruses). But even then, the treatments will not be as miraculous as planned, and in no way will they open the door to absolute health, since many illnesses are not only of genetic origin[49], but are also linked to environmental factors, such as cancer and high blood pressure, or even, for most of the illnesses affecting man, of completely environmental origin, such as microbial infections, viruses, insufficient hygiene and other non-genetic factors. In no way can these different gene therapies be used to justify the HGP.

We can, of course, continue to hope that these therapies will be improved, and they certainly will be. But this is where a second aspect kicks in. A gene therapy is hard to put together, and it is easier to make a pre-or postnatal diagnostic test. In all cases, these tests are, and will be, operational before therapies are available (since they require that the genes in question are located and identified, and these techniques of location and identification are just the ones used in genetic diagnostics). Russel Grieg, of SmithKline Beecham in the United States, predicts a change in the paradigm for pharmaceutical firms, leading from a focus on treatment to a focus on prediction and prevention. Dr Jim Havris (before at Glaxo but now at Sequana)

---

[48] R.G. Crystal, as above, and I. Verma, as above.

[49] And even when they are, we have seen that the link between phenotype and genotype is of astonishing complexity.

points out that despite the progress made in gene mapping, the great pharmaceutical industries are not ready to "get biological" for some time yet[50]. What does already happen is genetic diagnostics, and therefore, genetic testing[51]. And the media never speak of this, and rarely do scientists, in the same way that they never mention how many sufferers there are, the annual number of new cases, the mortality rate, the life expectancy or all the other things needed for a correct appreciation of the situation.

As for AIDS, scientists and journalists are prudent and avoid giving any illusions about possible therapies in exact opposition to the way the media has been vocal about hereditary diseases. The comparison between what is said in the media and the popular reviews, and what is read in specialist articles and scientific magazines, is very interesting in this respect. In the specialist articles, genetic diagnostics are mainly pushed to the fore, without any comments on their social application. It is, however, realized that for the most serious illnesses which still have no therapy, a pregnant woman may prefer to terminate pregnancy when the prenatal diagnostic reads positive. Only a few would refuse for religious or other reasons, and to that number we would have to add those who passed through the testing net.

When a prenatal test exists, and there is no therapy (which is the case for most if not all the genetic illnesses for which there are tests), there will, over time, be a rarefaction of new cases, a brake on the mortality rate and a drastic reduction, in the long term, on the number of sufferers. From this point of view, the therapies the media blithely announce are likely to be less and less needed, since the small number of sufferers that would need them are also being reduced by their lower life expectancy. The rarer and more serious the illness, the less likely the therapy will ever be developed, for the two following reasons:

First, the investment needed is enormous, and the large research firms are chary of pouring large amounts of money into a loss-making product;

---

[50] Quoted from the Second Annual Symposium in November 1993 on the Consequences of the Human Genome Project for Human Health Care.

[51] There are several tests already available, such as that for cystic fibrosis, Huntington's chorea, breast cancer, hemophilia and muscular dystrophy.

Second, the treatments are so complex that they cannot be produced except for a large number of sufferers and for long periods of time.

Furthermore, the diagnostics market is much larger than that for gene therapies. The latter only involves the ill, but diagnostics affects all adults who could be at risk, and with prenatal diagnostics, their children. Without taking into account the pressures of the pharmaceutical industry for a systematic application of these tests on all pregnancies, with loud announcements that a large number of genes for illnesses have been located, and in the ideology of the "genetic all", the fear of a deformed child and the dream of perfect health, these tests are already an important economic game and are promising radiant futures to the firms that are producing them.

Research on hereditary diseases, with the aid of information from the genome program, is more likely to provide a diagnostic rather than a gene therapy. Using therapy as the justification, what is being prepared is rather more eugenic in methodology[52], the birth of children who are biologically correct. As at the beginning of the century, the health argument is being put forward. I shall not take a side here, but point out that the scientists, media and industrialists should not dupe the general public. It is important that the ethical, social and political aspects of these tests are clearly discussed. Furthermore, given the vague nature of the notion of genetic predisposition, it is important that we don't underestimate or allow others to underestimate the risks linked to testing, not only of known hereditary diseases (since the market is not infinite) but also the various supposed predispositions[53].

One of the initiators of modern biology, J.D. Watson, said the following: "Strong opposition to programs aimed at preventing the birth of severely impaired children comes from individuals who believe that all human life is a reflection of God's existence and should be cherished and supported with all the resources at our disposal. They believe that genetically-impaired fetuses have as much right to exist as those destined for healthy, productive lives. Such arguments present no validity to those of us who see no evidence

---

[52] Eugenics is not only the intention of improving the genetic heritage of humanity, but by its etymology, also the science of good births.

[53] There are already tests for predisposition to several sorts of cancer.

for the sanctity (holiness) of life, believing instead that human as well as all other forms of life are the products not of God's hand but of an evolutionary process operating under the Darwinian principles of natural selection. This is not to say that humans do not have rights. They do but these have not come from God but instead from social contracts among humans who realize that human societies must operate under rules which allow for stability and predictability in day-to-day existence.

Foremost among these rules is the strict prohibition in virtually all societies against the killing of a fellow human being unless necessary self-defense is involved. Without this rule, our lives as functioning humans would be greatly diminished with no one able to count on the continued availability of those we love and depend upon. In contrast, the termination of a genetically-damaged fetus should not diminish the future lives of those individuals into whose world it would otherwise enter. In fact, the prevailing emotion must largely be one of relief at not being called to give love and support to an infant who can never have an existence whose eventual successes you can anticipate and share.

Thus I can see only unnecessary agony from laws that use the force of arbitrary religious revelations to impose the birth of genetically sick infants upon parents who would much prefer to terminate such pregnancies, hoping that their next conception leads to a healthy infant.

Using the name of God to let unnecessary personal tragedies occur is bound to upset not only those who follow less dogmatic guidelines for life but also many members of those religious groups whose leaders proclaim the absolute sanctity of all human life. In the long term, it is inevitable that those authorities who ask their followers to harm themselves in the name of God will increasingly find themselves isolated with their moral pronouncements regarded as hollow and to be ignored.

Nonetheless, we would not be surprised by increasing opposition to the Human Genome Project, seen as the most visible symbol of the evolutionary biology/genetics-based approach to human existence. But since the medical objectives of the genome project cannot easily be faulted, those who fear its implications will emphasize that its reductionist approach to human existence fails to acknowledge the overriding importance of the spiritual aspect of human existence, which they will argue is much more important than our genes in determining whether we are successes or failures in our lives. Under that argument, we would be making better use of our monies trying to improve the economic and moral environments of humans as opposed to the finding of genes that they believe will only marginally affect our health and social behavior.

With time, however, the truth must emerge that monies so spent have effectively no chance of rolling back the fundamental tragedies that come from genetic disease. So, I believe that over the next several decades, we shall witness an ever-growing consensus that humans have the right to terminate the lives of genetically unhealthy fetuses. But there remains the question as to who should make the decisions that lead to the termination of a pregnancy. Under no circumstances should these choices be assigned to the state, for even in our more homogeneous cultures, there exists wide divergence as to what form of future human life we should encourage. Instead, such decisions are best left solely in the hands of the prospective mother and father (if the father effectively plays this parental role).

Such unregulated freedom clearly opens up possibilities for irresponsible genetic choices that can harm all concerned. But we should not expect perfect results in handling genetic dilemmas any more than we can expect them from other aspects of human life. We have reason to hope that our genetic choices will improve as general knowledge of the consequences of bad throws of the genetic dice become better appreciated. Clearly, we must see that genetics assumes a much more prominent place in our educational curricula. Equally important, appropriate genetic screening procedures must become widely available to all our citizens regardless of their economic and social status.

At the same time, we must always be aware that the human society will only come to our genetic way of thinking in halting ways. Even many of our firmest supporters will worry at times that we are moving too fast in assuming roles that in the past we have assigned to the gods. Only they could predict the future as well as have the power to change our fate from bad to good or from good to bad. Today, we have some of those powers. Clearly, this is a situation which is bound to make many people apprehensive, fearing we will misuse our powers by helping create immobile, genetically-stratified societies that do not offer the prospect of hope and dignity for all their citizens. Thus, in so moving through genetics to what we hope will be better times for human life, we must proceed with caution and much humility"[54].

Watson's comments here, which are not scientific proposals but expressions of individual opinion, clearly illustrate the logical follow-on from the HGP and the genome programs in general. This logic could cause a depreciation of the way life is considered, and more and more, with the various systematic sequencing programs and the prowesses of genetic engineering, the genome will be considered machinery that can be manipulated at the whims, desires and needs of human beings.

Further from the concrete dangers, using genetic testing in health insurance, a subject already very hot in the United States, and in the near future in the field of employment, the accumulation of knowledge on the genome and the spread of predictive medicine are engendering a shift in ideological thinking in which we all, more or less unconsciously, participate. The shift is happening through an overly rapid slide from diagnostics based on the actual state of the sufferers to diagnostics based on the genotype. Researchers are partially guilty for this reduction of the phenotype to genetic determinism. It's an attitude that conforms with the scientific method; when trying to handle a reality which is too complex for us to understand, we try to work out the rules that govern it and understand more or less the way life works. To progress, we are chopping down the rules and changing the trends, smoothing out the curves and rounding up the numbers. It's necessary, but it is dangerous as soon as its limits are forgotten or ignored. It is also tempting to make genes the main explanation here too and to overestimate their importance. Recent history in experiments of gene inactivation in mice has brought many surprises, from the mdX mouse, which produces no dystrophine but does not get muscular dystrophy, to the mouse that produces no interleukin 2 but manages most immune reactions without it. Increasingly, it is being established that the equivalence between genes and phenotypes is a frightening simplification of matters, however convenient it may be.

---

[54] *La lettre du GREG*, no. 4, July 1995, Tribune libre de James D. Watson, p. 24.

8.6 Convenient Reductionism

It has to be acknowledged that this reductionism is following a trend in history. With the fall of ideologies, our societies are working more and more individualistically, competitively and commercially, which fits in beautifully with all that breaks up groups and isolates individuals. Genetic risk, and genetic predispositions are becoming tools by which people are classified, by which the most able are selected. The classification system used until now, of subjective appreciation, leaving room for will, motivation and experience already accumulated, the "objective" genetic data engraved in the genes revealed by a microsatellite or the detail of a given sequence takes away all responsibility and freedom from the individual. No more discussions on solidarity or equality, just the evaluation of genetic heritage with a battery of tests that will allow individuals to be hierarchized, handicaps affirmed, uncompetitive individuals pushed aside, and investment made in chosen persons. This isn't science fiction; the danger is real. The number of available tests is increasing every day. Society is beginning to believe in them in and use them. Most of all, the illusion of the absolute power of genetics, the illusion of a day when genetic illnesses have been conquered, is very prevalent amongst insurers, employers, politicians and doctors.

The notion of relative risk is very badly understood, and nothing is being done to help the public understand. The allele that makes the likelihood of having diabetes at some point in life rise from 0.5% to 3% is seen as the gene for diabetes, and its carriers "sick" people, although 97% of them will not have diabetes at all. In this process, researchers, who sometimes exaggerate the importance of their work or simplify it to caricature, are to be held responsible. Their work is on clearly defined organic afflictions with diagnostics on quite precise criteria. As for behavioral troubles, clinically or socially defined, of mental illness, the relationship between genetics and ideology is tenuous, to say the least, and it is easy to imagine the medical, social and political deformations of studies that try to find genetic causes for alcoholism, manic-depressive psychoses or homosexuality, behaviors for which the supposed revelation of a genetic link have all recently been the subject of comments in the best journals. This deformation of the "genetic all", besides its diagnostic dangers, could easily, in the current sociopolitical and cultural context, provide an alibi for and allow the public to ignore other decisive social parameters.

But of all this, little or nothing is said, the intention being to present consensus on ethical considerations, nothing being forgotten. Scientists and the media continue to sensationalize genes and illnesses, gene therapies (still to come) and to continue to widen the "genetic all". We should be asking ourselves why this argument, which after all is purely ideological, continues to prevail.

8.7 The Reasons Behind an Ideology

Why is the public being fed ever more sensational promises and ever more caricatural simplifications? Why are we still talking of genes for illness, building the HGP and the genome programs as the key to health, the Open Sesame to the understanding of human nature and the essence of man, and the future of societies, while reality is always frighteningly more complex than science will ever be able to reveal? Why should we continue to sequence the human genome?

The answer, in part, is that scientists and the public are so completely convinced of the ideology of single unitary causes that they believe in the complete efficiency of reductionism, so are not inquiring into the complex issues. But the first answer that comes to mind is that it's a way of providing funding to research through soliciting funds from charitable organizations and the ministries. The funds obtained are, of course, used to study hereditary disease, locate genes and do fundamental research (namely the megaproject of decoding the human genome) because they can all be useful in the fight against disease, for industry, and the economy, a fight entirely supported by the public.

From this point of view, it is, of course, preferable to bring forward the aspect of locating genes for illness and the possible production of therapies. The presentation of child sufferers ensures the emotional support of the general public and is completely representative of this. Public charity would be much less engaged by an appeal in favor of prenatal diagnostics. This financial aspect is not the only one to be taken into account. Participation in the HGP is considered to be a contribution to a "Big Science" project, like the Apollo or the Manhattan project; a project that will take 30 to 50 years and involve thousands of technicians and scientists. This is an attractive proposition for an ambitious biologist. High level careers will open up; Nobel prizes will be up for grabs. Important positions and fantastic equipment will be made available to the main leaders and protagonists of the project. More tempting for the leaders and scientists involved is in showing to politicians and public and funding committees, how large vistas are opened up by the understanding of the hereditary factors in disease — the gene therapies that will justify this project, which is so expensive at the private and public levels. The illnesses, and therefore the sequences, are of consensual interest at the public level, but also among pharmaceutical firms and with those in charge of health. It's always good to be able to show that research is good for something. This is a wonderful way to keep the machine running[55], especially since genetic engineering has not turned out to be as marvelous as expected. Furthermore, the HGP and other similar megaprograms, and gene therapies, are superb reasons to apply the new techniques that molecular biology has generated since the '70s and to breathe a new dynamism into certain industries. So, a not inconsiderable fraction of the millions or billions of dollars to be spent in the course of the genome programs and the HGO will go to the share dividends of companies providing chemical substances, computer software ... But the researchers will not be involved in this defense of the projects or the need to show the possible applications of research at a purely academic level. Among the molecular biologists who are university professors, there are some who are also involved to some degree in biotechnology companies. Technology is a major industry and a basic source of possible profits for venture capitalists. Globally, there is a whole marketplace gathered around the science of genetics. This is why it is

---

[55] In particular, at the beginning of the HGP, when it was considered an opportunity to reorient the activities of the large US national laboratories in which the activity had mainly been devoted to the development of strategic weapons.

important to reinforce research with appropriate publicity and emphatically underline the power of the new biology.

Studies on DNA are also, without a doubt, an industry with great public visibility, with an effect on public finance and the legitimacy of the science.

In fact, other than the commercial, academic, industrial and political aspects, the origin of the discussion and interest in gene therapies and genome decoding, especially the human genome, is also to be looked for within biology. The progress made by genetics and molecular biology has not only helped reveal the complexity of higher organism genes. The genes that, until now, were seen as clearly defined segments of the DNA string, each responsible for the synthesis of an equally clearly defined entity, a protein, have suddenly lost this clear definition. Their contours have become more vague, and their definition as entities refers more to the functional unit than the physical one.

The definition of the gene

Perhaps the first thing that ought to be said concerning the definition of the gene is that the concept has undergone alteration in some manner or another almost since its inception. This is to be expected, but it is important to realize that such change does not eliminate interlevel identification. Early in the development of genetics, in particular in Mendel's work, the gene is understood as a factor functionally (in the sense of causally) responsible for some phenotypic characteristic (in Mendel's words, "factors are acting together in the production of ... forms"). In the early 1900s, the gene was also constructed as a unit of mutation. The work of Sutton and Morgan on genetic maps further permitted the gene to be characterized as a unit of recombination. For many years, these three ways of characterizing the gene (as a unit of mutation, recombination and function) were thought to pick out the same basic unit, in spite of advances made in the area of biochemical genetics, such as the beadle-Tatum hypothesis of "one-gene-one enzyme". As organisms such as bacteria and bacteriophages began to become available for use in genetic research in the 1940s, however, some puzzling results concerning the size and properties of the unit of heredity began to surface. One difficulty that surfaced in the work of geneticists in the 1940s and early

1950s, for example, was that units of recombination and mutation appeared to be mapping within a unit responsible for a phenotypic characteristic. The interesting feature of the Watson-Crick discovery of DNA's tertiary structure was that it offered Seymour Benzer a means to unify these *prima facie* disparate gene concepts.

Benzer used a number of mutants of the bacteriophage T4 to construct a genetic map at a very high degree of resolution, and on the basis of this work, observed that "the classical 'gene', which served at once as the unit of genetic recombination, mutation and function is no longer adequate. These units require separate definition"[56].

Benzer introduced three concepts in terms of which to designate the gene, the *recon*, the *muton* and the *cistron*. He noted that there were, at that time, difficulties on both the genetic and molecular sides due to insufficient knowledge, but he felt that he could offer a "rough estimate" of the size of the three genetic units on the basis of his work with T4 and the Watson-Crick structure. He proposed that "the size of the *recon* [the unit of recombination] would be limited to no more than two nucleotides pairs", that the muton [the unit of mutation] was "no more than five" nucleotide pairs, and that the *cistron* or unit of function "turns out to be a very sophisticated structure" probably about one hundred times larger than the *muton*. Benzer's proposals have, by and large, been universally accepted. His term "*cistron*" is used widely, and though the terms "*recon*" and "*muton*" have not enjoy the same acceptance, the concepts, for which they stand, have become entrenched in the biomedical sciences. Further works on conceptual clarification of the gene notion continues.

Most geneticists accept a biochemical definition of the gene, which can be characterized as the "one-gene-one polypeptide principle". This involves regarding "the gene as equivalent to the cistron". At the same time, that there is a near universal acceptance of this definition of the gene, it is possible to observe that geneticists often find it inapplicable to work on higher plants or animals, essentially because not much is yet known about

---

[56] Seymour Benzer, "The elementary units of heredity", In a *symposium on the chemical basis of heredity*, ed. W.D. McElroy and B. Glass, Johns Hopkins University Press, John Hopkins, 1956, p. 116.

the DNA of these higher organisms. Thus, from a working point of view, geneticists still continue to use the Morgan recombination definition. This shows how difficult it is to define high level notions and bridge principles such as a "gene" unambiguously in molecular terms[57].

The genome, once seen as fundamentally invariant, has lost its image of an aperiodic crystal and has become something more changeable and motile. To compensate for this liquefaction, molecular biology must establish new rules, to the point where there are a fantastic number of them, especially from the evolutionary point of view.

Weismann's importance in the history of genetics comes from his concept of heredity as a structure, when everyone else was thinking in terms of function. By defining the gene as a more functional than physical unit, genetics is taking a step back to before his time. Molecular genetics still resists, but with increasing difficulty. In the evolution of any science, there is always a difficult time, when a theory becomes desuete, or at least insufficient, and the one that should replace it is not yet ready. This is perhaps where biology stands at the moment. The frenetic search for applications (the HGP and gene therapies and genetic engineering in general) looks a lot like the throes of a theory in crisis, or at least a theory stagnating.

According to Richard C. Strohman, "We have a complete theory of the gene that has exceeded its lawful boundary. As suggested by Jan Sapp, who has described the historical conflict between holistic (epigenetic) biology and monistic (genetic) biology, we need to go beyond the gene. There is a Kuhnian revolution going on; it's just taking a long time... and it suffers the

---

[57] Defining a gene as a segment of DNA will not do because there are some "genes" which, in point of fact, are never transcribed into mRNA. "Although our discussion of the biochemical gene has focused on the function of determining a polypeptide chain, it must not be overlooked that some functional units of DNA, instead of determining polypeptide chains, act as templates for transfer RNA or ribosomal RNA. The functional definition of the gene is usually taken to include such units, thus it would perhaps be more accurate to say that the gene is the length of DNA that codes for one functional product. In addition, some parts of the DNA molecule serve exclusively in the control of genetic transcription (such as the operator). These are usually called regions rather than genes. Besides the DNA in genes and control regions, there is much DNA of unknown function. Some of this nongenic DNA occurs between adjacent genes, but there are also other large regions of apparently nongenic DNA in chromosomes, especially near the centromeres. In some cases, as much as one fifth of all the chromosomal DNA may be nongenic." W.T. Keeton, *Biological Science*., W.W. Norton, 3d New York, 1980, p. 657.

worst of fates because, as he said, without a new paradigm waiting, we cannot give up the old one"[58]. As Feynman tells us, when your paradigm is running out of gas, and when you have not yet discovered the new one, you have to start guessing.

On the one hand, applications are exploited to help the theory live on. On the other hand, in decoding the genome, a long-term job is begun which will circumvent the need to address the holes in the theory, in the hope that afterwards, a new synthesis will emerge. It is no coincidence that now, as the dogma of molecular genetics is beginning to show cracks, the HGP, the other systematic genome sequencing programs, genes, hereditary pathologies and gene therapies are invading discussions, media and reviews. Faced with scary ignorance, the "genetic all" is a reassuring refuge. By associating a gene to an illness, the gene is assigned a unit, a thickness of reality. Genetics recovers from its uncertainties at the price of considerable reduction and simplification of the reality of living things. It recovers from its uncertainties, presenting itself to the credulous public as an all-powerful philanthropic entity (the HGP is man's route to knowledge for health) and in return, the public support and applaud the growth of technique[59], today affecting biology to the detriment of theory.

In molecular genetics, the several thousand hereditary diseases are presented philanthropically, like butterflies pinned in a case, to consolidate the weakening theory of programmed physiology. In return, the fact that they might reveal the mechanism of heredity is kept quiet, since these mechanisms (those of programming) are supposedly understood to the point where they can be used to create therapies. The important thing is to reinforce this image of a program for man and life in general, in its programmed physiology, because it will support molecular genetics in its current difficulties. With the

---

[58] Richard C. Strhman, "The coming Kuhnian revolution in biology", *Nature Biotechnology*, vol. 15, March 1997, p. 199.

[59] Wally Gilbert is among those who say that the Genome Project is not science because it's about improving the technology for doing things we already know how to do rather than about new ideas. But that's a rather naive view of what science is. As Sydney Brenner once said, 'In molecular biology there are technical advances, discoveries, and idea, and they usually occur in that order.' "Was von Leeuwenhoek doing science when he developed the microscope and realized how to use it for biology?" M. Olson in *The Human Genome Project, Deciphering the Blueprint of Heredity*, p. 75.

help of the notion of gene therapy, who can contradict our need to heal the sick and justify the systematic decoding of the human genome?

The genes for illnesses, gene therapy, the decoding of genomes, the dogma of genetics, all are brought together, and the effective approval of one of these terms implies that of the rest and brings reinforcement to the arguments and deterministic ideology held, even if the goals are less noble than they might appear (profit races by the large groups, long term transformations in agriculture with an industrial goal...). With a whole argument involving applications and perspectives, genetics reinforces its image and ensures a minimum stability through this difficult phase, brings together the funding for research and reinforces its orientation (technical, determinist and reductionist). This image is all the more welcome since it integrates perfectly into a world where the subject, the individual, the person, disappears slowly under the system that determines them, that takes away their responsibility. From this point of view, the genome program is not as politically neutral, as ideologically anodyne as we would like to believe. It is, in all cases, already widely used by politicians, businessmen and some scientists to bring a stop to questions that are above all social in nature and bring it all back to the simple question of genes. Barely has the gene found an entity with illness, than ideological deformations have grabbed it for the return of ideological determinism in behavior, the biology of the socially maladaptive and an increasingly ominous lack of consideration for life. Even without a new paradigm, the damage inflicted by the old one can no longer be scientifically supported.

EPILOGUE

Dreams or Nightmares? Man Reasoned out by His Genes

Far from seeing in the HGP, and more generally in the systematic sequencing programs, the worst of the new biology, I wanted, in this conclusion, to bring some subversive thought to bear on the ideology emerging from these programs, and more generally, genetics and molecular biology. That genetics should take support from real applications or just use it to compensate its theoretical weaknesses, gather funding or justify these technical ventures, is not too serious an issue, even if the HGP and similar programs encourage a negligence of theory and a restricted concept of life and the living. However, what is more serious is that this argument is exhibiting a preference for eugenic applications, even if this remains inexplicit, not by a taste for perversion but because these applications are very far from simple in their implementation.

The argument of the "genetic all" is dangerous, just as are the shortcuts that allow an excessively rapid passage from the gene to the illness, from the sequence to the essence of man or of the living, from the phenotype to the genotype. It is a dangerous argument since it reinforces the idea of a program for man, leaving him the playthings of his genes. More than the very real dangers of eugenics, or in any case a genetic classification of man and of the living, the argument which I condemn, the "programatist ideology" fashionable today, whose origins I have tried to illustrate, is more dangerous, since it traces an anthropological revolution, a genocentric and antihumanist revolution, which is reductionist, simplifying and mutilating.

The ideology and utopia of communication, the harbinger of a new era with a new man without interior thought, is today being superseded, not only in the ideology of the "genetic all", but also the utopia of geneticism. This ideology and utopia must be condemned, especially as the recession of religion and the great ideologies are leaving wide-open the territories of the imagination, leaving it a clear field as a possible belief. But when these territories are occupied with an unsound picture of man and life, not through their bodies but their genes and by the possibility that their entire being can be marked by their inheritance, there is the real danger, social and political, but also a deeper danger lurking in the image that man has of himself, that man accepts, since he believes the dreams that biologists and the media present to him. On this issue, the rays of light provided by the words of Pierre Legendre on the subject of the U-turn caused by the "Hitlerian implementation", which disarticulated the construction of the Western legal system "by an enactment of heredity as pure embodiment"[60] there, where

---

[60] P. Legendre, *Leçon VIII, le crime du Caporal Lortie*, Fayard, 1989, p. 20

the progress of humanity had passed "by the disembodiment of the means of entry in heredity".

If we don't want history to repeat itself, it is important to go beyond images and simplifications, beyond ideological dreams and discussions, not to get lost on the new roads opened by technical progress, and to always be aware that the image of life provided by science is only an image and will never coincide with the reality it represents. It is only in this manner, with a critical awareness of what there is behind the mirages, that dreams won't turn into nightmares. To abstain would be to let an official pseudo-genetics take its place, the results of which, although they won't be as bad as the Nazi atrocities, will still be frightening.

As Pierre Chambon reminds us[61], "Hung from a branch of the crown of the phylogenetic tree, most improbable fruit of the cosmic lottery, man is the only living being capable of representing himself as another, the only one capable of understanding his origins. It seems to me that this imposes particular duties upon him. Almost as a paradox, if one had to find in the history of the life sciences, as shown to us in biology and more particularly molecular genetics in the last 50 years, a source of ethical values, they would without doubt take into account respect for the biological universe ... An additional reason to incite "Genetic Man" to take the time to think". But can he still take that time to think? Does he wish to? The pious wishes of the ethical philosophers, the solemn declarations of international institutions, can they stave off this deformation of life being carried out by biotechnology in its service of economic interests of bioindustries and their lobbyists? Nothing is less certain. The use of information from the genome programs, the new opportunities being provided by genetic engineering and biotechnology should, in every case, be decided with the new generations which will have to handle these techniques. There is the sense of a reflection on ethics, but also, and perhaps most of all, education, which is today more than ever essential to ensure that biological inventions are not perversions of nature.

---

[61] "By invitation", P. Chambon, "Génétique moléculaire, évolution et société," *La Lettre du GREG*, vol. 3, December 1994, p. 20.

Bibliography

Chapter 1: The Creation of Biotechnology

Andreis E. "Il bureau per le ricerche biologiche e l'industria delle pelli", *La concesia*, Vol. 29, 1921, p. 164.

Berg Paul *et al.* "Potential biohazards of recombinant DNA molecules", *Science*, Vol. 185, 1974, p. 303.

Berg Paul *et al.* "Potential biohazards of recombinant DNA molecules", *Nature*, 1974.

Bernard Claude. *Leçons sur les phénomènes de la vie*, Vol. 1.

Biological Engineering Society, *The Lancet*, Vol. 2, 23 July 1960.

Bionics Symposium. *Living Prototypes — The Key to New Technology*, 13, 15, 15 September 1960, *Wadd Technical Report 60–600*, 30 rapports, 499 pages.

Boyd-Orr John. *As I recall*, McGibbon and Kee, London, 1966.

Branford V. V. and Geddes Patrick. *The Coming Polity*, Le Play House, London, 2d ed., 1919.

Brightman Rainald. "Biotechnology", *Nature*, Vol. 131, 29 April 1933, pp. 597–599.

Bud Robert. *The Uses of Life (A History of Biotechnology)*, Cambridge University Press, Cambridge, 1993, First Paperback edition, 1994.

Bush Vannevar. "The case for biological engineering", in *Scientists Face the World of 1942*, Rutgers University Press, New Brunswick, 1942.

Commission of the European Communities. "FAST subprogramme C: Bio-Society", FAST/ACPM/79/14-3E, 1979.

D'Arcy Thompson. *On Growth and Form*, Cambridge University Press, Cambridge, 1942.

Darwin Charles. *L'origine des espèces*, ed. Marabout Université, 1973.

Darwin Charles. *La descendance de l'homme et la selection sexuelle*, ed. Complexe, 1981.

Debru Claude. *L'esprit des protéines (Histoire et philosophie biochimique)*, éditions Hermann, Paris, 1983.

Delbrück Max. "Ueber hefe und gärung in der bierbrauerei", *Bayerische Bierbrauer*, Vol. 19, 1884, pp. 304–312.

Dronamaraju K.R. "Chronology of J.B.S. Haldane life", in Dronamraju K.R. *Haldane's Daedalus Revisited*, Oxford University Press, New York, 1995.

Dronamraju K.R. *Haldane's Daedalus Revisited*, Oxford University Press, New York, 1995.

Duclaux E. *Traité de microbiologie*, Vol. 1, Microbiologie générale, Paris, 1898.

Ereky Karl. "Die grobbetwebsmäbige entwicklung der schweïnemast in ungarn", *Mitteilungen der deutschen landwirtschafts-gesellschaft*, Vol. 34, 25 Août 1917, pp. 541–550.

Ereky Karl. *Nahrungsmittelproduktion und landwirtschaft*, Friedrich Kilians Nachfolger, Budapest, 1917.

Ereky Karl. *Biotechnologie der fleisch, Fett und milcherzeugung im landwirtschaftlichen groBbetriebe*, Paul Parey, Berlin, 1919.

Ewell Raymond. "The rising giant: the world food problem", in *Engineering of Unconventional Protein Production*, ed. Herman Bieber, Chemical Engineering Progress, Symposium series, n 93n, 65, 1969, pp. 1–4.

Finlay Mark R. "The German agricultural experiment stations and the beginnings of american agricultural research", *Agricultural History*, 62, 1988, pp. 41–50.

Fogel Lawrence J. *Biotechnology, Concepts and Applications*, Prentice Hall, Englewood Cliffs, 1963.

Forrest Derek W. *Francis Galton: The Life and Work of a Victorian Genius*, Taplinder, 1974.

Francé Raoul H. *Bios: Die Gesetze der Welt*, Hofstaengli, 1921.

Francé Raoul H. *Plants as Inventors*, Simpkin and Marschall, London, 1926.

Galton Francis. *Hereditary Genius, An Inquiry into its Laws and Consequences*, MacMillan and Co, London, 1869.

Galton. *Inquiries into Human Faculty and its Development*, ed. Macmillan, London, 1983.

Gayon Jean. *Darwin et l'après Darwin, (Une histoire de l'hypothèse de sélection naturelle)*, ed. Kimé, Paris, 1992.

Geddes Patrick. *Cities in Evolution: An Introduction to the Town Planning Movement and the Study of Civics*, Williams and Norgate, London, 1915.

Geddes Patrick and Thomson J.A. *Biology*, Home University Library, London, 1925.

Geison Gerald L. "Pasteur, Roux and Rabies: Scientific versus clinical mentalities", *Journal of the History of Medicine and Allied Sciences*, 45, 1990, pp. 341–365.

Giessler Alfred. *Biotechnik*, Zuelle and Meyer, Leipzig, 1939.

Haldane J.B.S. *The Causes of Evolution*, London, 1932.

Haldane J.B.S. *Hérédité et politique*, Vol. 1338, trad. P. Couderc, P.U.F., Paris, 1948.

Haldane J.B.S. *Daedalus or Science and Future*, in K.R. Dronamraju, *Haldane's Daedalus Revisited*, Oxford University Press, New York, 1995.

Hansen Emil Christian. *Practical Studies in Fermentation*, Trans. Alese K. Miller, Spon, London, 1896, p. 272.

Hartley Harold. "Agriculture as a source of raw materials for industry", *Journal of the Textile Institute*, Vol. 28, 1937, pp. 151–172.

Hartley Harold. "Chemical engineering at a cross-roads", *Transactions of the Institution of Chemical Engineers*, 30, 1952, pp. 13–19.

Hartley Harold. "Chemical engineering — The way ahead", *Transactions of the Institution of Chemical Engineers*, 33, 1955, pp. 20–26.

Hase Albrecht. "Ueber technische biologie: Ihre aufgaben und ziele, ihre prinzipielle und wirtschaftliche bedeutung", *Zeitschrift für technische biologie*, Vol. 8, 1920, pp. 23–45.

Hayduck F. "Max Gelbruck", *Berichte der Deutschen Chemischen Gesellschaft*, 53i, 1920, pp. 48A–62A.

Heden Carl-Goran. "Biological research directed towards the needs of underdevelopped areas", *TVF*, 7, 1961.

Hill A.V. "Biology and electronics", *Journal of the British Institution of Radio Engineers*, Vol. 19, 1959.

Hogben Lancelot. "The foundations of social biology", *Economica*, N. 31, Février 1931, pp. 4–24.

Hogben Lancelot. "Prolegomenon to political arithmetic", in *Political Arithmetic, A. Symposium of Population Studies*, ed. Lancelot Hogben, Allen and Unwin, London, 1938, pp. 13–46.

Huxley J.S. "Biology and Human Life", *2d Norman Lockyer Lecture*, British Science Guild, 23 November 1926.

Huxley J.S. "The applied science of the next hundred years, biological and social engineering", *Life and Letters*, Vol. 11, 1934, pp. 38–46.

Huxley J.S. and Haddon A.C. *We Europeans, a survey of racial problems*, Jonathan Cape, 1935.

Huxley J.S. *L'homme, cet être unique* (1941), trad. J. Castier, La presse française et étrangère, Paris, 1947.

Jungk R. *The Everyman Project: Resources for a Human Future*, Thames & Hudson, London, 1976.

Kennedy Max. "The evolution of the word biotechnology", *Trends in Biotechnology*, 9, 1991, pp. 218–220.

Kevles D.J. *Au nom de l'eugénisme*, ed. PUF, Paris, 1995.

Koraft J.A.R. "The 1961 picture of human factors research in business and industry in the United States of America", *Ergonomics*, Vol. 51, 1962, pp. 293–299.

Lamark *Philosophie zoologique*, Bibliothèque 10/18, 1968.

Latour Bruno. *Microbes: Guerre et paix; Suivi de irréduction*, A.M. Métilié, Paris, 1984.

Lederberd Joshua. "Experimental genetics and human evolution", *The American Naturalist*, 100, 1966, pp. 519–531.

Lederberg Joshua. "Biological future of man", in Wolstenholme ed., *Man and His Future*, J.A. Churchill, London, 1963, pp. 263–273.

Liebig Justus. *Die Organische chemie in ihrer anwendung auf physiologie und pathologie*, Braunschweig, 1842.

Lindner Paul. "Allgemeines aus dem bereich der biotechnologie", *Zeitschrift für Technische Biologie*, Vol. 8, 1920, pp. 54–56.

Mackaye Benton. *From Geography to Geotechnics*, University of Illinois Press, 1968, p. 22.

Masm F.A. "Microscopy and biology in industry", *Bulletin of the Bureau of Biotechnology*, Vol. 1, 1920, pp. 3–15.

McAuliffe Sharon and McAuliffe Kathleen. *Life for Sale*, ed. Coward, McLann and Georghegan, New York, 1981.

McMillen Wheeler. *New Riches from the Soil: The Progress of Chemurgy*, Van Nostrand, New York, 1946.

Multhauf Robert P. *The Origins of Chemistry*, Franklin Watts, New York, 1967.

Mumford Lewis. *Technics and Civilization*, ed. Harper, Brace and Workd, New York, 1934.

Novick A. and Szilard Leo. "Experiments with the chemostat on spontaneous mutations of bacteria", *Proceedings of the National Academy of Sciences*, Vol. 36, 1950, pp. 708–719.

Pauly A. *Darwinismus und Lamarckismus*, Reinhardt, Munich, 1905.

Pauly Philip. *Controling Life: Jacques Loeb and the Engineering Ideal in Biology*, Oxford University Press, 1987.

Pope William. "Address by the President", *Journal of the Society of Chemical Industry*, Vol. 40, 1921, pp. 179T–182T.

Rapp Friedrich. "Philosophy of technology: a review", *Interdisciplinary Science Reviews*, Vol. 10, 1985, pp. 126–139.

Russel John E. *A History of Agricultural Science in Great Britain*, Allen & Unwin, London, 1966.

Schling-Brodersen Ursula. *Entwicklung und Institutionalisierung der Agrikulturchemie im 19. Jahrhundert: Liebig und die landwirtschaftlichen Versuchsstationen, Braunschweiger veröffentlichungen zur Geschichte der Pharmazie, Braunschweiger Verschaften*, Vol. 31, University of Braunschweig, Braunschweig, 1989.

Schneider William H. *Quality and Quantity: The Quest for Biological Regeneration in Twentieth-Century France*, Cambridge University Press, 1990.

Shelley Mary. *Frankenstein*, ed. Garnier Flammarion, 1979.

Siebel E.A. *Western Brewer and Journal of Barley, Malt and Hop Trades,* January 1918, Vol. 25. "Some press comments", *Bulletin of the Bureau of Biotechnology*, Vol. 1, 1921, p. 83.

Spencer Herbert. *Les bases de la morale évolutionniste*, ed. G. Bullière, 1881.

Spencer Herbert. *L'individu contre l'Etat*, ed. Félix Alcan, 1901.

Strbanova Sona. "On the beginnings of biochimistry in Bohemia", *Acta Historiae rerum naturalium nec non technicarum*, Special Issue 9, 1977, pp. 149-221.

Szilard Léo. *The Voice of the Dolphins*, Simon and Schuster, 1961.

Tatum Adward L. "A case history in biological research", *Nobel Lecture*, 11 December 1958.

Tatum Edward L. "Perspectives from physiological genetics" in *The Control of Human Heredity and Evolution*, ed. T.M. Sonneborn, Macmillan, New York, 1965.

Taylor Craig L and Boelter L.M.K. "Biotechnology: A new fundamental in the training of engineers", *Life*, 24, 9 February 1948, pp. 85-87.

Teich Mikulas. "Science and the industrialisation of brewing", presented to the conference entitled "Biotechnology: long-term development", *The Science Museum*, London, February 1984.

Temkin Owsei. "Materialism in French and German physiology of the early nineteenth century", *Bulletin of the History of Medecine*, Vol. 20, 1946, pp. 322-327.

Thomson J. A. "Biology", *Encyclopedia Britannica, Supplément à la 11ème édition*, Vol. 1, 1926, pp. 383-385.

Tornier Gustav. "Ueberzählige bildungen und die bedeutung der pathologie für die biontotechnik (mit demonstrationen)" in *Verhandlungen des V. Internationalen Zoologen-Congresses zu Berlin*, 12-16 August 1901, ed. Paul Matschie, Gustav Fischer, Jena, 1902, pp. 467-500.

Treviranus Gottfried, Reinhold. *Biologie oder philosophie der lebenden nature für naturforscher und aertzte*, ed. Rower, Göttingen, 1802.

Tsuchiya H.M. and Keller K.H. "Bioengineering — is a new era beginning?", *Chemical Engineering Progress*, 61, May 1965, pp. 60-62.

Virey J. J. (1775-1846). "Pharmacist and philosopher of nature", *Bulletin of the History of Medicine*, Vol. 39, 1965, pp. 134-142.

Walgate Robert. *Miracle or Menace: Biotechnology and the Third World*, Panos Institute, London, 1990.

Waveguide, Dayton Section I.R.E. publication, 39 North Torrence Street, August/September 1960.

Weindling Paul. *Health, Politics and German Politics between National Unification and Nazism*, 1870-1945, Cambridge University Press, 1988.

Wells G.P. "Lancelot Thomas Hogben", *Biographical Memoirs of the Royal Society*, 24, 1978, pp. 183–221.

Wells H.G. *Anticipations of the Reaction of Mechanical and Scientific Progress upon Human Life and Thought*, Harper, London, 1902.

Wickenden W.E. "Technology and culture, commencement address", *Case School of Applied Science*, 29 May 1929 et 1933, pp. 4–9.

Wickenden W.E. "Final report of the director of investigations, June 1933", in *Report of the Investigations of Engineering Education, 1923–1929*, Vol. 1 and 2, Society for the Promotion of Engineering Education, Pittsburgh, 1934.

Wickenden W.E. *Technology and Culture*, Ohio College Association Bulletin.

Wiener Norbert. *Cybernetics or Control and Communication in the Animal and the Machine*, MIT Press, Cambridge, Massachusetts, 1948.

Wilkinson John. *From Farming to Biotechnology: A Theory of Agroindustrial Development*, Blackwell Publishers, Oxford, 1987.

Chapter 2: Political Interpretations of Biotechnology and the Birth of the First Research Programs

Advisory Council for Applied Research and Development, Advisory Board for the Research Councils and the Royal Society, *Biotechnology: Report of a Joint Working Party*, HMSO, Lonson, 1980.

Behrens Dieter, Buchholz K. and Rehm, H.J. *Biotechnology in Europe — A Community Strategy for European Biotechnology*, DECHEMA for European Federation of Biotechnology, Frankfurt, 1983.

Bernton Hall, Kovarik William and Sklar Scott. *The Forbidden Fuel: Power Alcohol in the Twentieth Century*, Boyd Griffin, New York, 1982.

BMFT. *Biotechnology, BMFT-Leistungplan, Plan Periode: 1979–1983*, BMFT, Bonn, 1978.

Bull Alan T., Ellwood Derek C. and Ratledge Colin. "The changing scene in microbial technology", in *Microbial Technology: Current State, Future, Prospects*, ed. A.T. Bull, D.C. Ellwood and C. Ratledge, Society of General Microbiology, Symposium No. 29, Cambridge University Press for the Society for General Microbiology, 1979.

Bull Alan T., Holt G. and Lilly M. *Biotechnology: International Trends and Perspectives*, OCDE, Paris, 1982.

Cantley Mark F. Plan by objective: *Biotechnology, XII-37/83/EN*, Commission of the European Communities, 1983.

Cantley Mark F. *The Regulations of Modern Biotechnology: A Historical and European Perspective.* A case study in how society copes with new knowledge in the last quarter of the twentieth century, chapter 18, Vol. 12: Legal economic and ethical dimensions, Traité en plusieurs volumes intitulé Biotechnology, ed. V.C.H., 1995.

Carrel A. *L'homme, cet Inconnu*, ed. Plon, Paris, 1941.

Carson Rachel. *Silent Spring*, Houghton Mifling, Boston, 1962, trad. fr. Le printemps silencieux, Plon, Paris, 1963; reed. Le livre de Poche, Paris, 1968.

Conseil Européen, Council decision of 25 July 1978 on a research programme of the European Economic Community on forecasting and assessment in the field of science and technology, *Official Journal of the European Communities*, No. L225/40 of 16 August 1978.

Cunningham P. "A bibiometric study of BEP publications (program of policy research in enginnering, science and technology)", University of Manchester, cité par D. de Nettancourt, "Biotechnologies communautaires: la dixième année", *Biofutur*, April 1991, p. 20.

de Duve Christian. *Cellular and Molecular Biology of the Pathological State*, EUR 6348, Office for Official Publication of the European Communities, 1979.

de Nettancourt Dreux, Goffeau André and Van Hoeck F. "Applied molecular and cellular biology — Background note on a possible action of the European communities for the optimal exploitation of the fundamentals of the new biology", *Commission of the European Communities*, DGXII, XII/207/77.E, 15 June 1977.

de Rosnay Joël. *Biotechnologies et Bio-Industries*, Jean-Paul Aubert, Microbiologie générale et appliquée.

DECHEMA, Biotechnology in Europe — *A Community Strategy for European Biotechnology: Report to the FAST Bio-society Project of the Commission of the European Communities*, DECHEMA, Frankfort, on behalf of the European Federation of Biotechnology, DECHEMA, 1983.

DECHEMA (Deutsche Gesellschaft für chemisches Apparatewesen e.v.) *Biotechnologie: Eine Studie über Forschung und Entwicklung-Möglichkeiten, Aufgaben und Schwerpunkten der Förderung*, DECHEMA, Frankfort, 1974.

Doc. Com. (77) 283 final, "The common policy in the field of science and technology", 30/6, 1977.

Ehrlich Paul. *The Population Bomb*, Ballantine, New York, 1968.

Ester ergebnisbericht des ad hoc ausschusses neue technologien, des beratenden ansschusses für forschungspolitik, Schriftenreihe forschungsplanung, Vol. 6, Bonn, BMBW, Décembre 1971.

European Commission. "Biotechnology in the community", *Com. (83) 672*, Commission of the European Communities, October 1983.

European Commission. "Biotechnology: the Community's role", *Communication from the Commission to the Council, Com (83) 328*, Vol. 8, June 1983.

European Commission. "Common policy for science and technology", *Bulletin of the European Communities*, supplement 3/77, 1977.

FAST. European Commission. *The FAST Programme: Vol. 1: Results and Recommendations*, 1982.

FAST. "A Community Strategy for Biotechnology in Europe", *FAST Occasional Papers*, no. 62, European Commission, 1983.

FAST. Commission des Communautés Européennes, Europe 1995: Mutations technologiques et enjeux sociaux, *Rapport FAST, Futurible*, Paris, 1983.

FAST. European Commission. Eurofutures: The challenges of innovation (the FAST report), Butterworths in association with the journal *Futures*, London, 1984.

FAST-Gruppe, *Kommission der europäischen, Gemeinschaften, Die Zukunft Europas: Gestaltung durch innovationen*, Springer-Verlag, Berlin, 1987.

"Genetic Engineering — Certainties and Doubts", *New Scientists*, 47, 24 September 1971, p. 614.

GMBF/GBF. Entwicklung eines Forschungsinstituts 1965–1975, Stifung Volkswgenwerke, Hannover, 1975.

Gros F., Jacob F. et Royer P. *Sciences de la vie et société, Rapport présenté à Monsieur le Président de la République V. G. d'Estaing*, La documentation française, Paris, 1979.

Japan Science and Technology Agency, *Outline of the White Paper on Science and Technology: Aimed at making technological innovations in social development*, February 1977. Trans. Forreign Press Center, pp. 176–178.

Kennet Wayland. *The Futures of Europe*, Cambridge University Press, 1976.

Lederberd Joshua. "Experimental genetics and human evolution", *The American Naturalist*, 100, 1966, pp. 519–531.

Lederberd Joshua. "A geneticist on safeguards", *New York Times*, 11 March 1975.

Magnien E. *Biomolecular Engineering in the European Community*, ed. Martinus Nijhoff, 1986.

Mateles R.I. and Tannenbaum S.R. *Single-Cell Protein*, MIT Press, Cambridge, Mass., 1968.

Meadows Donella H., Meadows Dennis L., Randers, Jorgen & Behrens, William W. *The Limits to Growth*, University Books, New-York, 1972, trad. fr.: Halte à la croissance, ed. Fayard, Paris, 1971.

Meeting du CGC-Biotechnology, 27 November 1986, point 3 de l'agenda, "BEP: evaluation of the final report by national delegations" (Doc. CGC).

Meeting du CGC-Biotechnology, Brussels, 7 March 1986, point 3: "Preliminary evaluation of the achievements of the biomolecular engineering programme-BEP (doc. CGC-IV-86/2).

Müller H.J. *Hors de la nuit, vues d'un biologiste sur l'avenir*, trad. J. Rostand, ed. Gallimard, Paris, 1938.

National Initiatives, Com (83) 328, European Commission, 1983.

National Science Board. *Science Indicators 1972: Report of the National Science Board*, US Government Printing Office, Washington DC, 1973.

OTA, Commercial Biotechnology: An International Analysis, OTA, Washington DC, 1984.

OTA, The Impacts of Applied Genetics: Applications to Microorganisms, Animals and Plants, OTA, Washington DC, 1981.

Raugel P.J. "Nothing ventured, nothing gained", *Biofutur*, June 1983.

Report on the Current State of Planning of Life Science Promotion in Japan in A. Rörsche "Genetic manipulation in applied biology: A study of the necessity, content and management principles of a possible community action", EUR 6078, Office for Official Publication of the EC, Luxembourg, 1979.

Rip A. and Nederhof A. "Between dirigism and Laissez-Faire: Effects of implementing the science policy priorities for biotechnology in the Netherlands", *Research Policy*, 15, 1986, pp. 253–268.

Rörsch A. Genetic manipulations in applied biology: a study of the necessity, content and management principles of a possible action, *EUR 6078,* Office for Official Publication of the European Communities, Luxembourg, 1979.

Rostand Jean. *L'homme, introduction à l'étude de la biologie humaine*, ed. Gallimard, Paris, 1926.

Sakaguchi Kin-ichiro and Ikeda Yonosuke. Applied microbiology in Japan: An outline of its historical development and characteristics", in *Profiles of Japanese Science and Scientists*, ed. Hideki Yukawa, Kodensha, Tokyo, 1970.

Sargeant K. and Evans C.G.T. Hazards involved in the industrial use of microorganisms: a study of the necessity, content and management principles of a possible community action, European Commission, *EUR 6349 EN,* 1979.

Schwartz Wilhelm. "Biotechnik und bioengineering", *Nachrichten aus Chemie und Technik*, Vol. 17, 1969, pp. 330–331.

Short draft report of the meeting of the CGC. Biotechnology, Brussels, 7 March 1986, OO2/IN2, le 12 March 1986, European Commission, Brussels.

Tatum Edward L. "A case history in biological research", *Nobel Lecture*, 11 December 1958.

Tatum Edward L. "Molecular biology, nucleic acids, and the future of medicine", *Perspectives in Biology and Medicine*, 10, 1966–1967.

Thomas D. Production of biological catalysts, stabilization and exploitation, *EUR 6079*, Office for Official Publication of the European Communities, Luxembourg, 1978.

Truchuya H.M. and Keller K.H. "Bioengineering — A new era beginning?", *Chemical Engineering Progress*, Vol. 61, May 1965, pp. 60–62.

Wada A. "One step from chemical automations", *Nature*, Vol. 257, 1975, pp. 633–634.

Wright Pearce. "Time for bug valley", *New Scientist*, 82, 5 July 1979, pp. 27–29.

Chapter 3: The Foundations of the Coming Revolution

Arber W. and Dussoix D. "Host specificity for DNA produced by *E. coli*. I. Host controlled modification of bacteriophage lambda", *J. Mol. Biol.*, Vol. 5, 1962, pp. 18–36.

Arber W. "Host controlled modification of bacteriophage", *Annu. Rev. Micro. Biol.*, Vol. 19, 1965, pp. 365–378.

Arber W. "DNA modification and restriction", *Prog. Nucl. Acid. Res. Mol. Biol.*, Vol. 14, 1974, pp. 1–37.

Avery O.T., McLeod C., McCarty M. "Studies on the chemical nature of the substance inducing transformation of pneumococial types. Induction of transformation by a desoxyribonucleic acid fraction isolated from Pneumococcus type III", *T. Exp. Med.*, Vol. 79, 1 February 1944, pp. 137–158.

Bateson W. *A Defence of Mendel's Principle of Heredity*, Cambridge University Press, Cambridge, 1902.

Bateson W. *Problems of Genetics*, University Press, Yale, New Haven, 1903.

Beadle G. "Différenciation de la couleur cinabar chez la drosophile (*Drosophilia melanogaster*)", *C.R. Acad. Sci.*, Paris, Vol. 201, 1935.

Beadle G. et Tatum E.L. "Genetic control of biochemical reactions in *Neurospora*", *Proc. Nat. Acad. Sci.*, USA, Vol. 27, 1941, p. 499.

Benzer S. "Fine structure of a genetic region in bacteriophage", *Proc. Nat. Acad. Sci.*, USA, Vol. 41, 1955, p. 344.

Berg Paul *et al.* "Potential biohazards of recombinant DNA molecules", *Science*, Vol. 185, 1974, p. 303.

Berg Paul *et al.* "Asilomar conference on recombinant DNA molecules", *Science*, Vol. 188, 1975, p. 991.

Bernon M. *Biofutur*, Vol. 50, 1986.

Beveri T. *Ergebnisse über die Konstitution des Chromatischen der Zelkerns*, Fischer, Iena, 1904.

Boivin A. *et al.* "Rôle de l'acide désoxy-ribonucléique hautement polymérisé dans le déterminisme des caractères héréditaires des bactéries. Signification pour la biochimie générale de l'hérédité", *Helvetica Chimica Acta*, Vol. 29, 1946, pp. 1338–1344.

Boivin A. *et al.* "L'acide thymonucléique polymérisé, principe paraissant susceptible de déterminer la spécificité sérologique et l'équipement enzymatique des bactéries: signification pour la biochimie de l'hérédité", *Experientia*, Vol. 1, 1945, pp. 334–335.

Boivin A. *et al. L.R. Acad. Sci.*, Vol. 226, 1948, p. 1061 et R. Vendrely, C. Vendrely, *Expérientia*, Vol. 4, 1948, p. 434.

Brown T.A. *Gene Cloning — An Introduction*, 2nd edition, Chapman and Hall, London, 1990.

Cantley Mark F. The regulation of modern biotechnology: A historical and european perspective. A case study in how societies cope with new knowledge in the last quarter of the twentieth century, chap. 18, Vol. 12, intitulé Legal, Economic and Ethical Dimensions, du traité Biotechnology, publié par V.C.H., 1995.

Chargaff E. "Chemical specificity of nucleic acids and mechanism of their enzymatic degradation", *Experientia*, Vol. 6, 1950, pp. 201–209.

Chargaff E. "Structure and function of nucleic acids as cell constituents", *Fed. Proc.*, Vol. 10, 1951, pp. 654–659.

Church G.M. and Kieffer-Higgins S. "Multiplex DNA sequencing", *Science*, Vol. 240, 1988, pp. 185–188.

Cohen S.N., Chang A.L.Y., Boyer H.W. and Helling. "Construction of biologically functional bacterial plasmids *in vitro*", *Proc. Nat. Acad. Sci.* USA, Vol. 70, 1973, pp. 3240–3244.

Cooper Necia Grant (ed.). *The Human Genome Project* (*Deciphering the Blueprint of Heredity*), University Science Book, Mill Valley, California, 1994.

Correns C. "G. Mendel's regel über das verhalten der nachkommenshaft der rassen bastarde", *Berichte der deutschen botanischen gesellschaft*, Vol. 18, 1900, pp. 158–168. Trad. anglaise sous le titre: "G. Mendel's law concerning the behaviour of varietal hybrids" in *The Origine of Genetics: A Mendel Source Book*, C. Stern and E.R. Sherwood eds., San Francisco, Freeman and Co, 1966, pp. 117–132.

Crichton Michael. *The Andromeda Strain*, Gentesis Corporation, 1969, trad. fr. La variété Andromède, ed. Pocket, 1995.

de Nettancourt Dreux, Goffeau André and Van Hoeck F. "Applied molecular and cellular biology: background note on a possible action of the European Communities for the optimal exploitation of the European Communities", *DGXII, DGXI/207/77-E*, 15 June 1977, table 1–2–3.

de Vries H. "Sur la loi de disjonction des hybrides", *CR Acad. Sc.* Vol. 130, Paris, 1900, pp. 845–847. de Vries H. *Espèces et variétés*, trad. Fr., ed. Alcan, Paris, 1909.

Debru Claude. *L'esprit des protéines*, ed. Hermann, 1983.

Durand Béatrice. "Séquençage de l'ADN", *Le Technoscope de Biofutur*, n. 23, Octobre 1988, pp. 3–13.

Edman Pehr. "A method for the determination of the amino acid sequence in peptides", *Acta Chemica Scandinavica*, Vol. 4, 1950, pp. 283–293.

Ephrussi B. *Quart. Rev. Biol.*, Vol. 17, 1942, p. 327.

Feulgen R. and Rossenbeck H. *J. Physiol.*, Vol. 135, 1924, p. 203.

Fincham J.R.S. and Ravetz J.R. *Genetically Engineered Organisms — Benefits and Risks*, Open University Press, Milton Keynes, 1991.

Franklin Rosalind E. and Gosling, R.G. "Molecular configuration in sodium thymonucleate", *Nature*, Vol. 171, 1953, pp. 740–741.

Freifelder D. *Molecular Biology.* 2nd edition, Jones and Bartlett, Boston, USA, 1987.

Fruton J.S. *Molécules and Life,* New York, 1972, p. 172.

Garrod A.E. *Inborn Errors of Metabolism*, Oxford University Press, 1909.

Gilbert Walter and Müller-Hill Benno. "Isolation of the lac repressor", *Proc. NAS*, Vol. 56, 1966, pp. 1891–1898.

Gilbert W. and Maxam A. "The nucleotide sequence of the *lac* operator", *Proceedings of the National Academy of Science* (USA), 70(12), 1973, pp. 3581–3584.

Gilbert W., Maxam A. and Mirzabekov A. "Contacts between the *lac* repressor and DNA revealed by methylation", in *Vinth Alfred Benzon Symposium*, Copenhagen: Control of ribosome synthesis, Nokjeldgaard and O. Maaloe ed., Academic Press, New York, 1976, pp. 139–143.

Gilbert W. "Why genes in pieces?", *Nature*, Vol. 271, 1978, p. 501.

Goujon Philippe. *Les voies de l'information la communication à la complexité*, Thèse de Doctorat, soutenue à Dijon, Juin 1993.

Griffith F. "The influence of immune serum on the biological properties of *Pneumococci*", Reports on Public Health and Medical Subjects, no. 18, His majesty's stationery office, London, 1923, pp. 1–13.

Griffith F. "The significance of pneumococcal types", *Journal of Hygiene*, Vol. 27, January 1928, pp. 141–144.

Haldane J.B.S. "The combination of linkage values, and the calculation of distances between loci of linked factors", *Journal of Genetics*, Vol. 8, 1919, pp. 299–309.

Hardy G.H. "Mendelian proportions in a mixed population", *Science*, Vol. 28, 1908, pp. 49–50.

Hershey A.D. and Chase M. "Independant functions of viral protein and nucleic acid in growth of bacteriophage", *J. Gen. Physiol.*, Vol. 36, 1952, pp. 39–56.

Hohan S. and Galibert F. "L'évolution des techniques de séquençage", *Biofutur, Spécial Génomes*, No. 94, Octobre 1990, pp. 33–37.

Holley R.W. *et al.* "Nucleotide sequence in the yeast alanine transfer ribonucleic acid", *Journal of Biological Chemistry*, Vol. 240, 1965, pp. 2122–2128.

Holley R.W. *et al.* "Structure of ribonucleic acid", *Science*, Vol. 147, 1965, pp. 1462–1465.

Hotchkiss Rollin D. "Portents for a genetic engineering", *Journal of Heredity*, Vol. 56, 1965, pp. 197–202.

Hunkapiller M.W. and Hood L.E. "A gas-liquid solid phase peptide and protein sequenator", *Journal of Biological Chemistry*, Vol. 256, 1981, pp. 7990–7997.

Hunkapiller M.W. and Hood L.E. "Protein sequence analysis: automated microsequencing", *Science*, Vol. 219, 1983, pp. 650–659.

Jacob François and Monod Jacques. "Genetic regulatory mechanism in the synthesis of proteins", *J. Mol. Biol.*, Vol. 3, 1961, pp. 318–356.

Jacob François. *La logique du vivant*, (une histoire de l'hérédité), ed. Gallimard, 1970.

Jacob François. "Le temps des modèles: la régulation binaire", in A. Lwoff et A. Ullmann, *Les origines de la biologie moléculaire*, ed. Etudes vivantes, Col. Academic Press, 1980.

Judson Horace, Freeland. *The Eighth Day of Creation*, Penguin Book, 1995, First published by J. Cape Ltd, 1979.

Kennedy Max. "The evolution of the word biotechnology", *Trends in Biotechnology*, Vol. 9, 1991, pp. 218–220.

Kevles J. Daniel and Hood Leroy. *The Code of Codes*, Harvard University Press, 1992.

Klug. "Rosalind Franklin and the discovering of the structure of DNA", *Nature*, Vol. 219, 24 August 1968, pp. 843–844.

Kolata G.B. "The 1980 Nobel Prize in Chemistry", *Science*, Vol. 210, 1980, pp. 887–889.

Kornberg A. "Biologie synthesis of deoxyribonucleic acide", *Science*, Vol. 131, 1960, pp. 1503–1508.

Kornberg A. *DNA Replication*, W.H. Freeman, New York, 1980.

Krimsky Sheldon. *Genetic Alchemy*, MIT Press, Cambridge, 1983.

Lederberg J. "Biological future of man", in Wolstenholme Gordon ed., *Man and His Future*, J.A. Churchill, London, 1963, pp. 263–273.

Luria S.E. "Host induced modification of viruses" in *Cold Spring Harbor Symp. Quant. Biol.*, Vol. 18, 1953, pp. 237–244.

Marx J.L. *A Revolution in Biotechnology*, ICSU / Cambridge University Press, Cambridge, 1989.

Maxam A.M. and Gilbert W. "A new method for sequencing DNA", *Proc. Nath. Acad. Sci.*, Vol. 74, USA, 1977, pp. 560–564.

McAuliffe Sharon and McAuliffe Kathleen. *Life for Sale*, Coward, McLann and Geoghegan, New York, 1981.

McCulloch Warren S. and Pitts Walter. "A logical calculus of the ideas permanent in nervous activity", *Bulletin of mathematical biophysics*, Vol. 5, December 1943, pp. 115–133.

Mendel G. "Versuche über Pflanzen-hybriden", *Verhandlungen des naturforschenden Vereines in Brünn*, Vol. 4, 1865, pp. 3–47, trad. fr. "Recherches sur des hybrides végétaux", *Bulletin Scientifique*, Vol. 41, 1907, pp. 371–419, republié dans *La découverte des lois de l'hérédité: une anthologie (1862–1900)*, Press Pocket, Paris, pp. 54–102.

Miescher F. "Uber die chemische zusammenzetzung der Eiterzellen", in *Hoppe Seyler's Medicenisch Gemische Unterschungen*, ed. August Hirschwald, Berlin, Vol. 4, 1871.

Miyahara M. "R&D on human gene analysis and mapping systems supported by the science and technology agency of Japan", *Report memorandum no. 135*, Tokyo Office of the US National Science Foundation, August 1931, 1987.

Monod Jacques and Jacob François. "Genetic regulatory mechanism in the synthesis of proteins", *J. Biol.*, No. 3, 1961, pp. 318–356.

Monod Jacques and Jacob François. "Teleonomic mechanism in cellular metabolism, growth and differentiation", *Cold Spring Harbor Symposia*, Vol. 26, 1961, pp. 389–401.

Morgan T.H. *Science*, Vol. 12, 1910, p. 120. T.H. Morgan, A.H. Sturtevant, H.J. Muller, C.B. Bridges, *The Mechanism of Mendelian Heredity*, Henry Holt, New York, 1915. Trad. Fr.: *Le mécanisme de l'hérédité mendélienne*, Malertin, Bruxelles, 1923. T.H. Morgan, "Sex limited inheritance in Drosophilia", *Science*, Vol. 32, 1910.

Muller Herman Joseph. *Studies in Genetics: The Selected Papers of H.J. Muller*, Indian University Press, Bloomington, 1962.

Mullis Kary. "L'amplification des gènes", *Pour la science*, dossier La génétique humaine, Avril 1994, pp. 18–25.

Nicholl Desmond S.T. *An Introduction to Genetic Engineering*, Cambridge University Press, Cambridge, 1994.

Olby. "Francis Crick, DNA and the Central Dogma", in *Daedalus*, Fall 1970, pp. 956–957.

Parenty M. "A l'ère de l'automatisation", *Biofutur, Spécial génomes*, n. 146, Juin 1995, pp. 34–38.

Pauling L. and Corey R.B. "A proposed structure for the nucleic acids", *Proc. NAS.*, Vol. 39, 1953, pp. 96–97.

Pontecorvo G. *Adv. in Enzymol*, Vol. 13, 1952, p. 221.

Primrose S.B. *Molecular Biotechnology*. 2nd edition, Blackwell, Oxford, 1991.

Punnet R.C. *Mimicry in Butterflies*, Cambridge University Press, Cambridge, 1915.

Rajnchapel-Messaï Jocelyne. "PCR, il faut payer pour amplifier", *Biofutur*, Mars 1994, pp. 53–56.

Richards B.M. and Carey N.H. "Insertion of benefical genetic information", *Searle Research Laboratories*, 16 January 1967.

Sanger F. "The free amino group of insulin", *Biochimical Journal,* Vol. 39, 1945, p. 514.

Sanger F. "Some chemical investigation on the structure of insulin", *Cold Spring Harbor Symposia*, Vol. 14, 1949, pp. 155–156.

Sanger F. "The arrangement of amino-acids", in *Protein Chemistry*, Vol. 7, 1952, pp. 55–57.

Sanger F. and Thompson E.O.P. "The amino-acid sequence in the glycyl chain of insulin 3, *Biochemical Journal*, Vol. 53, 1953, pp. 353–374.

Sanger F. The Croonian lecture, 1975: "Nucleotide sequences in DNA", *Proceedings of the Royal Society of London*, B191, 1975, pp. 317–333.

Sanger F. and Coulson A.R. "Rapid method for determining sequences in DNA by primed synthesis with DNA-polymerase", *Journal of Molecular Biology*, Vol. 94, 1975, pp. 441–478.

Sanger F., Nilken S. and Coulson A.R., "DNA sequencing with chain-terminating inhibitors", *Proceedings of the National Academy of Sciences* (USA), Vol. 74, 1977, pp. 5463–5468.

Sanger F. *et al.* "Nucleotide sequences of bacteriophage ΦX174", *Nature*, Vol. 265, 1977, pp. 685–687.

Sanger F. "Sequences, sequences and sequences", *Annual Reviews of Biochemistry*, Vol. 57, 1988, pp. 1–28.

Sasson Albert. "Les biotechnologies de la bio-industrie", *La recherche*, n. 188, Mai 1987.

Sayre Anne. *Rosalind Franklin and DNA*, ed. Norton, New York, 1975.

Schleif R. *Genetics and Molecular Biology*, Addison-Wesley, 1985.

Schrödinger Erwin. *What is Life?*, Cambridge University Press, Cambridge, Angleterre, Trad. fr. par Léon Keffler, *Qu'est-ce que la vie (L'aspect physique de la cellule vivante)*, ed. Christian Bourgois Editeur, 1986.

Smith Lloyd M. "DNA sequence analysis: past, present and future", *IBL*, October 1989, pp. 8–19.

Stadler J.L. *Science*, Vol. 120, 1954, p. 81.

Stokes A.B. and Wilson H.R. "Molecular structure of deoxypentose nucleic acids", *Nature*, Vol. 171, 1953, pp. 738–740.

Sun M. "Consensus elusive on Japan's genome plans", *Science*, Vol. 243, 31 March 1989, pp. 1656–1657.

Sutton W.J. "On the morphology of the chromosome group of brachystola Magna", *Biol. Bull.*, Vol. 4, 1902, pp. 24–39.

Swinbanks D. "Japan's human genome project takes shape", *Nature*, Vol. 351, 20 June 1991, p. 593.

Tatum E.L. "A case history in biological research", *Nobel Lecture*, 11 December 1958.

Tatum E.L. "Perspective from physiological genetics", in *The Control of Human Heredity and Evolution*, ed. T.M. Sonneborn, Mac Millan, New York, 1965.

Tatum E.L. "Molecular biology, nucleic acids and the future of medicine", *Perspectives in Biology and Medicine*, Vol. 10, 1966–67, p. 31.

Taylor G. Rattray. *The Biological Time Bomb*, New American Library, New York, 1968.

Todd Alexander and Brown D.M. "Nucleotides: part X, some observations on the structure and chemical behaviour of the nucleic acids", *Journal of the Chemical Society*, 1952, pp. 52–58.

US Congress, Office of Technology Assessment. *Impacts of Applied Genetics: Microorganisms, Plants and Animals*, OTA, HR-132, US Government Printing Office, Washington DC, USA, 1981.

US Congress, US House of Representatives, Committee on science and technology, Subcommittee on science, research and technology. *Science Policy Implications of DNA Recombinant Molecule Research*, 95th Cong., 1st Sess., 1977.

Von Tschermak E. "Ueber küntsliche kreuzung bei *Pisum sativum*", *Berichte Deuteschen botanischen gesellschaft*, Vol. 18, 1900, pp. 232–239. Trad. anglaise sous le titre: "Concerning artificial crossing in *Pisum sativum*, in: *Genetics*, Vol. 35, Suppl. au n. 5, "The birth of genetics", 2ème partie, 1950, pp. 42–47.

Wada A. and Soeda E. "Strategy for building an automatic and high speed DNA-sequencing system", in *Proceedings of the 4th Congress of the Federation of Asian and Oceanic Biochemists*, Cambridge University Press, London, 1986, pp. 1–16.

Wada A. and Soeda E. "Strategy for building an automated and high speed SNA-sequencing system", in *Integration and Control of Metabolic Processes*, Ol Kon ed., Cambridge University Press, New York, 1987, pp. 517–532.

Wada A. "The practicability of and neccessity for developping a large-scale DNA-base sequencing system: Toward the establishment of international super sequencer centers", *In Biotechnology and the Human Genonme: Innovations and Impact*, ed. A.D. Woodhead B.J. and Barnhart, Plenum, New York, 1988, pp. 119–130.

Watson J.D. and Crick F.H.C. "A structure for deoxyribose nucleic acid", *Nature*, Vol. 171, 1953, pp. 737–738.

Watson J.D. and Crick F.H.C. "Genetical implications of the structure of deoxyribonucleic acid", *Nature*, 30 May 1953, pp. 964–967.

Watson J.D. and Tooze J. *The DNA Story: A Documentary History of Gene Cloning*, W.H. Freeman and Company, San Francisco, 1981.

Watson J.D. *La double hélice*, Collection Pluriel, ed. Laffont, 1984.

Watson J.D., Hopkins N.H., Roberts J.W., Steitz J.A. and Weitner A.M. *Molecular Biology of the Gene*, 4th edition, Benjamin Cummings, Menlo Park, 1987.

Weaver W. and Shannon, C.E. *The Mathematical Theory of Communication*, the Board of Trustees of the University of Illinois, trad. fr. J. Cosnier, G. Dahan and S. Economides, *Théorie mathématique de la communication*, Retz-CEPL, Paris, 1975.

Weinberg N. "Ueber den nachweiss der vererbung beim menschen", *Jahreshriften des Vereins für vaterländische naturkunde in württemburg*, Vol. 64, 1908, pp. 368–382.

Weismann A. *Essais sur l'hérédité*, trad. française, Paris, 1892, p. 176.

Wiener Norbert, Rosenblueth Arturo and Bigelow Julian. "Behavior, purpose and teleology", *Philosophy of Science*, 1943, pp. 18–24.

Wiener Norbert. *Cybernetics or Control and Communication in the Animal and the Machine*, Librairie Hermann et Cie, Paris, 1948, (ed. suivantes: MIT Press, Cambridge, Mass).

Williams J.G. and Patient R.K. *Genetic Engineering*, IRL Press, Oxford, 1988.

Wolstenholme Gordon, ed., *Man and His Future*, J.A. Churchill, London, 1963.

Wu R. and Taylor E. "Nucleotide sequence analysis of DNA II. Complete nucleotide sequence of the cohesive ends of bacteriophage cambda DNA", *Molecular Biology*, Vol. 57, 1971, pp. 491–511.

Chapter 4: Taking the Genomes by Storm: The First Genetic and Physical Maps

Ackerman S. "Tackling the human genome", *American Scientist*, 76, January–February, 1988, pp. 17–18.

Angier Nathalie. *Les gènes du cancer*, ed. Plon, 1989.

Barinaga M. "Critics denounce first genome map as premature", *Nature*, 329, 15 October 1991, p. 571.

Botstein D., White R.L., Skolnick M. and Davis R.W. "Construction of a genetic linkage map in man using restriction fragment length polymorphysms", *American Journal of Human Genetics*, 32, 1980, pp. 314–331.

Botstein D. and Fink G.R. "Yeast: An experimental organism for modern biology", *Science*, 240, 10 June 1988, pp. 1439–1443.

Brenner S. "Genetics of behaviour", *British Medical Bulletin*, 29, 1973, pp. 269–271.

Brenner S. "Genetics of *Caenorhabditis elegans*", *Genetics*, 77 (1), 1974, pp. 71–94.

Burke D.J., Carle G.F. and Olson M.V. "Cloning of large segments of exogenous DNA into yeast by means of artificial chromosome vectors", *Science*, Vol. 236, 15 May 1987, pp. 806–812.

Caspersson T., Zech L. and Johansson C. "Differential banding of alkylating fluorochromes in human chromosomes", *Experimental Cell Research*, Vol. 60, 1970, pp. 315–319.

Caspersson T., Lomakka C.and Zech L. "Fluorescent banding", *Hereditas*, Vol. 67, 1970, pp. 89–102.

Caspersson T., Zech L., Johansson C. and Modest E.J. "Quinocrine mustard fluorescent banding", *Chromosoma*, Vol. 30, 1971, pp. 215–227.

Caspersson T. "The background for the development of the chromosome banding technique", *American Journal of Human Genetics*, Vol. 44, No. 4, April 1989, pp. 441–451.

Chumakov I., Rigault P., Guillou S., Foote S. *et al.* "Continuum of overlapping clones spanning the entire human chromosome 21q", *Nature*, 359, 1 October 1992, pp. 380–387.

Cook-Deegan Robert. *The Gene Wars (Sciences, Politics and the Human Genome)*, W.W. Norton & Company, New York, London, 1994.

Cooper Necia Grant (ed.). *The Human Genome Project (Deciphering The Blueprint of Heredity)*, University Science Book, Mill Valley, California, 1994.

Coulson A., Sulston J., Brenner S. and Karn J. "Toward a physical map of the genome of the nematode *Caenorhabditis elegans*", *Proceedings of the National Academy of Sciences* (USA), 83, October 1986, pp. 7821–7825.

Coulson A., Waterston R., Kiff J., Sulston J. and Kohara Y. "Genome linking with yeast artificial chromosomes", *Nature*, 335, 8 September 1988, pp. 184–186.

Davies K.E., Pearson P.L., Harper N.S. et al. "Linkage analyses of two cloned sequences flanking the Duchenne muscular dystrophy locus on the short arm of the human X chromosome", *Nucleic Acids Research*, 11, 1983, pp. 2302–2312.

Dib C. et al. "A comprehensive genetic map of the human genome based on 5,264 microsatellites", *Nature*, Vol. 380, 14 March 1996, pp. 152–154.

Donahue R.P., Bias W.B., Renwick J.H. and McKusick V.A. "Probable assignment of the duffy blood group locus to chromosome I in man", *Proceedings of the National Academy of Sciences*, Vol. 61, 1968, pp. 949–955.

Donis Keller H. et al. "A genetic linkage map of the human genome", *Cell*, 51, October 1987, pp. 319–337.

Feldmann H. and Steensman H.Y. chap III: "A detailed strategy for the construction of organized libraries for mapping and sequencing the genome of *Saccharomyces cerevisiae* and other industrial microbes", in *Sequencing the Yeast Genome, a Detailed Assessment, a posssible area for the Future Biotechnology Research Programme of the European Communities*, ed. A. Goffeau, Commission of the European Communities, Directorate General, Science, Research and Development, Directorate Biology, Division Biotechnology, BAP, 1988–1989, (XII/F/280).

Foote S., Vollrath D., Hilton A. and Page D.C. "The human Y chromosome: Overlapping DNA clones spanning the euchromatic region", *Science*, 258, 2 October 1992, pp. 60–66.

Goodman H.M., Olson M.V. and Hall D.B. "Nucleotide sequence of a mutant eukaryotic gene: The yeast tyrosine — Inserting ochre suppressor S.V.P. 4-0", *Proceedings of the National Academy of Sciences* (USA), 74, December 1977, pp. 5453–5457.

Grodzicker T., Williams J., Sharp P. and Sambrook J. "Physical mapping of temperature-sensitive mutants of adenovruises", *Cold Spring Harbor Sympresia on Quantitative Biology*, 39, 1974, pp. 439–446.

Gusella J.F., Wexler N.S., Conneally P.M. et al. "A polymorphic DNA marker genetically linked to Huntington's Disease", *Nature*, 306, 1983, pp. 234–248.

Ingram V.M. "Gene mutation in human haemoglobine: The chemical difference between normal and sickle cell haemoglobine", *Nature*, Vol. 180, 1957, pp. 326–328.

Jeffreys A.J. "DNA sequence variants in the G-γ, A-γ, δ and β globin genes of man", *Cell*, 18, 1979, pp. 1–10.

Jeffreys A.J. and Flavell R.A. "A physical map of the DNA regions flanking the rabbit beta globin gene", *Cell*, Vol. 12, October 1977, pp. 429–439.

Kan Y.W. and Dozy A.M. "Polymorphism of DNA sequence adjacent to human beta-globin structural gene relationship to sickle mutation", *Proceedings of the National Academy of Science* (USA) 75, 1978, pp. 5631–5635.

Kenyan L.J. "The nematode *Caenorhabditis elegans*", *Science*, 240, 10 June 1988, pp. 1448–1453.

Kohara Y. *et al.* "The physical map of the whole *E. Coli* chromosome", *Cell*, 50, 1987, pp. 495–508.

Lander E.S. and Waterman M.S. "Genomic mapping by fingerprinting random clones: A mathematical analyses", *Genomic*, 2, 1988, pp. 231–239.

Lindergren C.C., Lindergren G., Shult E.E. and Desborough S. "Chromosome maps of *Saccharomyces*", *Nature*, London, 183, 1959, pp. 800–802.

Little I. "Mapping the way ahead", *Nature*, 359, 1 October 1992, pp. 367–368.

Maniatis T., Jeffrey A. and Kleid D.G. "Nucleotide sequence of the rightward operator of phage lambda", *Proceedings of the National Academy of Science*, USA, Vol. 72, 1975, pp. 1184–1188.

Maniatis T., Hardison R.C., Lacy E. *et al.* "The isolation of structural genes from libraries of eucaryotic DNA", *Cell*, 15, October 1978, pp. 687–701.

Marx J.L. "Putting the human genome on a map", *Science*, 229, 12 July 1985, pp. 150–151.

Mohr J. "Search for linkage between Lutheran blood and other hereditary characters", *Acta Pathologica e Microbiologica Scandinavica*, Vol. 28, 1951, pp. 207–210.

Mortimer R.K. and Hawthorne D.C. "Genetic mapping in *Saccharomyces*", *Genetics*, 53, 1966, pp. 165–173.

Mortimer R.K. and Hawthorne D.C. "Genetic mapping in yeast", *Methods. Cell. Biol.*, 11, 1975, pp. 221–233.

Mortimer R.K. and Hawthorne D.C. "Genetic map of *Saccharomyces cerevisiae*", *Microbiol. Rev.*, 44, 1980, pp. 519–571.

Mortimer R.K. and Schild D. "Genetic map of *Saccharomyces cerevisiae*", pp. 224–233, in S.J. O'Brien (ed.), *Genetic Maps* 1984, Cold Spring Harbor Laboratory, Cold Spring Harbor, NY, 1984.

Murray Andrew and Szostak Jack. "Les chromosomes artificiels", *Pour la Science*, Janvier 1988, pp. 60–71.

Nathans D. and Smith H.O. "Restriction endonucleases in the analysis and restructuring of DNA molecules", in *Annual Reviews of Biochemistry*, 1975, pp. 273–293.

Neel J.W. "The inheritance of sickle cell anemia", *Science*, Vol. 110, 1949, pp. 64–66.

NIH/CEPH. Collaborative Mapping Group, "A comprehensive genetic linkage map of the human genome", *Science*, 258, 2 October 1992, pp. 67–68.

Olson M., Dutchik J.E., Graham M.Y. *et al.* "Random-clone strategy for genomic restriction mapping in yeast", *Proceedings of the National Academy of Sciences* (USA), 83, October 1986, pp. 7826–7830.

Olson M. *et al.* "A Common language for physical mapping of the human genome", *Science*, 245, 1989, p. 1434.

Pauling L., Itano H.A., Singer S.J. and Wells I.C. "Sickle cell anemia: a molecular disease", *Science*, Vol. 110, 1949, pp. 543–548.

Petes T.D., Botstein D. "Simple Mendelion inheritance of the reiterated ribosomal DNA of yeast", *Proceedings of the National Academy of Sciences* (USA), 74, November 1977, pp. 5091–5095.

Rigby P.W., Diechmann M., Rhodes C. and Berg, P. "Labelling deoxyribonucleic acid to high specific activity *in vitro* by nick translation with D.N.A. polymerase I", *Journal of Molecular Biology*, Vol. 113, 15 June 1977, pp. 237–251.

Riordan Jr., Ranmens J.M., Kerem B.S., Alan N. *et al.*, "Identification of the cystic fibrosis gene: Cloning and characterization of complementary DNA," *Science*, 245, 8 September 1989, pp. 1066–1072.

Roberts C. "The race for the cystic fibrosis gene" and "The race for the C.F. gene nears end", *Science,* 240, 8 April and 15 April 1988, pp. 141–144 and pp. 282–285.

Roberts L. "Flap arises over genetic map", *Science*, 238, 6 November 1987, pp. 750–752.

Royer B., Kunkel L., Monaco A. *et al.* "Cloning the gene for an inherited human disorder — Chronic granulomatous disease on the basis of its chromosomal location", *Nature*, 322, 1987, pp. 32–38.

Schwartz D.C. and Cantor, C.R. "Separation of yeast chromosome - Sized DNAs by pulsed field gel electrophoresis", *Cell*, 37, 1984, pp. 67–75.

Smith C.L., Econome J.G., Schutt A., Kico S. and Cantor C. "A physical map of the *Escherichia coli* K12 genome", *Science*, Vol. 236, 12 June 1987, pp. 1448–1453.

Smith H.O. "Nucleotide specificity of restriction endonucleases", *Science*, Vol. 205, 1979, pp. 455–462.

Solomon E. and Bodmer W.F. "Evolution of sickle cell variant gene", *Lancet*, April 1979, p. 923.

Southern E.M. "Detection of specific sequences among DNA fragments separated by gel electrophoresis", *Journal of Molecular Biology*, Vol. 38, 1975, pp. 503–517.

Stephens J.C., Cavanaugh M.L. et al. "Mapping the human genome: current status", *Science*, 250, 12 October 1990, pp. 237–244.

Sulston J.E. "Neuronal cell lineages in the nematode *C. elegans*", *Cold Spring Harbor Symposia on Quantitative Biology*, 48, 1983, pp. 443–452.

Sulston J.E., Schierenberg E., White J.G. and Tompson J.N. "The Embryonic cell lineage of the nematode *C. elegans*", *Developmental Biology*, 100, 1983, pp. 64–119.

Sulston J.E. and Horvitz H.R. "Post embryonic cell lineages of the nematode *C. elegans*", *Developmental Biology*, 56, 1977, pp. 110–156.

Sulston J., Du Z., Thomas K., Wilson R. et al. "The *C. elegans* genome sequency project: A beginning", *Nature*, 356, 5 March 1992, pp. 37–41.

Sulston J.E. and Brenner S. "The DNA of *C. elegans*", *Genetics*, 77 (1), 1974, pp. 95–104.

The Huntington's Disease Collaborative Research Group, "A novel gene containing a trinucleotide repeat that is expanded and unstable on Huntington's disease chromosomes", *Cell*, 72, 26 March 1993, pp. 971–983.

Vollrath D., Foote S., Hilton A., Brown L.G., Romero P., Bogan J.S. and Page D.C. "The human Y chromosome: A 43-interval map based on naturaly occuring deletions", *Science*, 258, 2 October 1992, pp. 52–59.

Weiss M. and Green H. "Human-mouse hybrid cell lines containing partial complements of human chromosomes and functioning human genes", *Proceedings of the National Academy of Sciences*, (USA), Vol. 58, 1967, pp. 1104–1111.

Weissenbach J., Gyapay G., Dip, C. et al. "A second generation linkage map of the human genome", *Nature*, 359, 29 October 1992, pp. 794–801.

White J.G., Southgate E., Thompson J.N. and Brenner S. "The structure of the nervous system of the nematode *C. elegans*", *Philosophical Transactions of the Royal Society of London*, Séries B, Vol. 314, 1986.

Wilson E.B. "The sex chromosomes", *Archiv für mikroskopie und anatomie entwicklungsmech*, Vol. 77, 1911, pp. 249–271.

Wyman A.R. and White R.L. "A highly polymorphic locus in human DNA", *Proceedings of the National Academy of Science* (USA), 77, 1980, pp. 6754–6758.

Chapter 5: The Human Genome Project and the International Sequencing Programmes

"Actualité des ESTs", *La lettre du Greg*, n. 6, Avril 1996, pp. 20–21.

Adams M.D. *et al.* "Complementary DNA sequencing: expressed sequence tags and human génome project", *Science*, 252, 1991, pp. 1651–1656.

Adams M.D. *et al.* "Sequence identification of 2,375 human brain genes", *Nature*, 355, 1992, pp. 632–634.

Allende J. "A view from the south", *FASEB Journal*, 5 January 1991, pp. 6–7.

Allende J. "Background on the human genome project", *Red Latinoamericana de Ciencias Biologicas*, 28 June, 1 July 1988.

"A long-range plan for the multinational coordinated *Arabidopsis thaliana* genome research project", *(NSF 90–80)*, 1990.

Anderson C. "U.S Genome Head faces charges of conflict", *Nature*, 356, 9 April 1992, p. 463.

Anderson C. "New French genome center aims to prove that bigger really is better", *Nature*, 357, 18 June 1992, pp. 526–527.

Anderson S. *et al.* "Sequence and organization of the human mitochondrial genome", *Nature*, 290, 1981, pp. 467–469.

Baer R. *et al.* "DNA sequence and expression of the B 95-8 Epstein-Barr virus genome", *Nature*, 310, 1984, pp. 207–211.

Baltimore D. "Genome sequencing: A small science approch", *Issues in Science and Technology*, Spring 1987, pp. 48–49.

Barnhart Benjamin J. *The Human Genome Project: a DOE Perspective*, OHER, DOE, Washington DC, 1988.

Beam A. and Hamilton J.O. "A grand plan to map the gene code", *Business Week*, 27 April 1987, pp. 116–117.

Beatty J. "Genetics in the atomic age: The atomic bomb casualty commission, 1947–1956", in *The Expansion of American Biology*, ed. K. Benson *et al.*, Rutgers University Press, 1 November 1991, pp. 284–324.

Begley S., Katz S.E. and Drew L. "The genome initiative", *Newsweek*, 31 August 1987, pp. 58–60.

Bitensky M., Burks C., Good W., Ficket J., Hildebrand E., Moyzis R., Deaven L., Cramm S. and Bell G. *Memo to Charles DeLisi*, DOE, LS-DO-85-1, 23 December 1986, Los Alamos National Laboratory.

Bodmer W.F. "Two cheers for genome sequencing", *The Scientist*, 20 October, 1986, pp. 11–12.

Cantor Charles R. "Orchestrating the human genome project", *Science*, Vol. 248, 6 April 1990, pp. 49–51.

Carey John. "The gene kings", *Business Week*, 8 May 1995, pp. 72–78.

Caskey T., Eisenberg R.S., Lander E.S., Strauss J. *HUGO Statement on Patenting of DNA Sequences*, ed. Dr. Belinda J.F. Rossitier, HUGO.

Catenhusen Wolf-Mickael. "Genome research from a politican's perspective", *B.F.E*, Vol. 7, No. 2, April 1990, pp. 136–139.

Chee M. and Burrell B. "Herpes virus: a study of parts", *Trends Genet.*, 6, 1990, pp. 86–91.

"China's rice genome research program", *Probe*, Vol. 3, No. 1/2, January–June 1993, pp. 7–8.

Claverie J.M. *Cartographie, Séquençage génomique, et bases de données biotechnologiques aux USA situations et enjeux stratégiques*, Janvier 1990 (rapport rédigé suite à une mission aux Etats Unis en Décembre 1989).

Cold Spring Harbor Symposia on Quantitative Biology, *The Molecular Biology of Homo sapiens*, 51, 1986.

Collins F. and Galas D. "A new five-year plan for the US human genome project", *Science*, Vol. 262, 1 October 1993.

Concor D. "French find short cut to map of human genome", *New Scientist*, 23 May 1992, p. 5.

Consiglio Nazionale della Recherche, "Progetto Strategico CNR: Mappaggio e Sequenzimento del Genoma Humano", 1987.

Cooper Necia Grant (ed.). *The Human Genome Project (Deciphering the Blueprint of Heredity)*, University Science Book, Mill Valley, California, 1994.

Damerval T. and Therre H. un entretien avec J.C. Venter "La conquête du génome", *Biofutur*, 110, Avril 1992, pp. 18–20.

Davies K.E. "The worm turns and delivers", *Nature*, Vol. 356, 5 March 1992, pp. 14–15.

Davis Bob. "US scientist escheus tact in pushing genetic project", *The Wall Street Journal Europe*, 18 June 1990.

del Guercio G. "Designer genes", *Boston Magazine*, August 1987, pp. 79–87.

DeLisi C. "The human genome project", *American Scientist*, 76, 1988, pp. 488–493.

"Des mutations héréditaires affectent les victimes de Tchernobyl", *Le Monde* des 28–29 Avril 1996.

Devine K.M. and Wolfe Ken. "Bacterial genomes: a TIGR in the tank", *TIG*, Vol. 11, No. 11, November 1995, pp. 429–431.

Dib C., Faure S., Fizanies C. *et al.* "A comprehensive genetic map of the human genome based on 5,264 microsatellites", *Nature*, Vol. 280, 14 March 1996, pp. 152–154.

Dickson David. "Britain launches genome programme", *Science,* 245, 31 March 1989, p. 1657.

Dickson David. "Wellcome sets sequencing project in motion", *Nature*, Vol. 378, 9 November 1995, p. 120.

Dietrich W.F., Muller J., Steen R. *et al.* "A comprehensive genetic map of the mouse genome", *Nature*, Vol. 380, 14 March 1996, pp. 149–152.

Dulbecco Renato. "A turning point in cancer research: Sequencing the human genome" in *Viruses and Human Cancer*, ed. Alan R. Liss, 1987, pp. 1–14.

Dulbecco Renato. "A turning point in cancer research: Sequencing the human genome", *Science*, Vol. 231, 7 March 1986, pp. 1055–1056.

"Elements of molecular medicine: Human genome research", Federal Ministry of Education, Science, Research and Technology, *BMBF*, p. 20.

European Science Foundation, Report on genome research, *ESF*, 1991.

Fiers W. *et al.* "Complete nucleotide sequence of S.V40 DNA ", *Nature*, 273, 1978, pp. 113–120.

First Report on the State of Science and Technology in Europe, Commission des Communautés Européennes, 1989.

Fleischmann R.D. *et al. Science*, 269, 1995, pp. 496–512.

Fraser C. *et al.* "The minimal gene complement of *Mycoplasma genitalium*", *Science*, October 1995, pp. 397–403.

Galibert F. *et al.* "Nucleotide sequence of the hepatitis B virus genome (subtype ayw) cloned in *E. coli*", *Nature*, 281, 1979, pp. 646–650.

Galloway John. "Britain and the human genome", *New Scientist*, July 1990, pp. 41–46.

"Genome monitoring (The organization to monitor the human genome project should be welcomed)", *Nature*, Vol. 335, 22 September 1988, p. 284.

Genome Projects: How Big, How Fast?, OTA, OTA -B.A-373, Government Printing Office, Washington, DC, April 1988.

Gilbert W. "Two cheers for genome sequencing", *The Scientist*, 20 October 1986, p. 11.

Gilbert W. "Genome sequencing: Creating a new biology for the twenty-first century", *Issues in Science and Technology*, 3, 1987, pp. 26–35.

Gilbert W. "Towards a paradigm shift in biology", *Nature*, Vol. 349, 10 January 1991, p. 99.

Goffeau André. Rapport de mission Goffeau les 3 et 5 Septembre 1990. "Biologie moléculaire de Bacillus" à l'Institut Pasteur", *AG/AA - nm* - Paris, Bruxelles, 19 Novembre 1990.

Goffeau André. Rapport de mission, Symposium on "Human genome research strategies and priorities", 29–31 January 1990, *Unesco Headquarters*, Paris, C.E.C., 15 February 1990.

Goffeau André. "Life with 482 genes", *Science*, Vol. 270, 20 October 1995, pp. 445–446.

Gruskin K.D. and Smith T.F. "Molecular genetics and computer analyses", *CABIOS*, 3, no. 3, 1987, pp. 167–170.

Hall S.S. "Genesis: The sequel", *California*, July 1988, pp. 62–69.

Harrois-Monin Françoise. "Génétique: la mine d'or, *Le vif/L'express*, 13 Octobre 1995, pp. 79–80.

Hilts P.J. "Head of gene map threatens to quit", *New York Times*, 9 April 1992, A. 26.

Hodgkin J. *et al.* "The nematode *C. Elegans* and its genome", *Science*, Vol. 270, 20 October 1995, pp. 410–414.

Holzmann D. "Mapping the genes, inside and out", *Insight*, 11 May 1987, pp. 52–54.

Hood L. and Smith L.S. "Genome sequencing, how to proceed", *Issues in Science and Technology*, 3, 1987, pp. 36–46.

Human Genome News, Vol. 7, n. 3 et 4, September–December 1995.

Human Genome Workshop, "Notes and conclusions from the human genome workshop", May 24–26, 1985, reprinted in Sinsheimer, "The Santa Cruz workshop", *Genomics*, 5, May 1985, pp. 954–956.

Issues in Sciences and Technology, 3, 1987, p. 37.

Johnston M. *et al.* "Complete Nucleotide sequence of *Saccharomyces cerevisiae* chromosome VIII", *Science*, 265, 1994, pp. 2077–2082.

Jordan B. "Le Programme Français Genome Humain", *Medecine, Science*, 9, 1990.

Jordan B. "Chercheur et Brevet, La fin d'un blocage", *Biofutur*, Juin 1995, pp. 86–88.

Jordan B. "Chronique d'une mort annoncée?", *La lettre du GREG*, No. 7, July 1996.

Kanigel R. "The genome project", *New York Times Magazine*, 13 December 1987.

La lettre du GREG, Séquençage systématique des génomes, N. 4, Juillet 1995, pp. 18–20.

"La recherche sur le génome en Allemagne", *Biofutur*, Juillet–Août 1994, pp. 35–37.

Le GIP-GREG, Origine, définition et missions", *La lettre du GREG*, N. 1, Avril 1994, pp. 1–6.

Left D.N. "Massive D.N.A automation sequences mycoplasma smallest total genome", *Bioworld Today*, Vol. 6, No. 202, 20 October 1995, p. 1, 4.

Lewin R. "Proposal to sequence the human genome stirs debate", *Science*, 232, 27 June 1986.

Lewin R. *Science*, 232, 1986.

McAuliffe K. "Reading the human blueprint", *US News and World Report*, 28 December 1987 and 14 January 1988, pp. 92–93.

McCarthy S. *USDA's Plant Genome Research, Agricultural Libraries Information Notes*, Vol. 17, No. 10, October 1991, pp. 1–10.

McKusick Victor A. "HUGO news, the human genome organisation history, purposes and membership", *Genomics*, 5, 1989, pp. 385–387.

McKusick Victor A. "The morbid anatomy of the human genome", *Howard Hughes Medical Institute*, Science, 14 October 1988, p. 229, *Gen. Eng. News*, July/August 1989, p. 30.

McLaren D.J. Human genome research: A review of European and international contributions, *M.R.C*, Royaume-Uni, Janvier 1991.

Mervis Jeffrey. "Renowed bioengineer picked to head Lawrence Berkeley human genome center", section: "NIH dominance", *The Scientist*, 3, 11 July 1988.

National Research Council. *Mapping and Sequencing the Human Genome*, National Academy Press, Washington D.C, 1988.

Oddo G. "More business involement in C.N.R's finalized programs", *Il Sole 24 Ore*, Milan, 14 July 1987, p. 6.

Olson M., Hood L., Cantor C. and Botstein D. "A common language for physical mapping of the human genome", *Science*, Vol. 245, 29 September 1989, pp. 1434–1435.

Palca J. "The genome project: Life after Watson", *Science,* 256, 15 May 1992, pp. 956–958.

Pines M. "Mapping the human genome (occasional paper number one) ", *HHMI*, December 1987.

Rajnchapel - Messaï J. "le BCRD version 1995", *Biofutur*, 139, 1994, p. 75.

Pines M. "Shall we grasp the opportunity to map and sequence all human genes and create a "human gene dictionary?" *Bethesda, M.D: HHMI, Prepared for a Meeting of the Trustees of HHMI*, 1986.

Reddy V.B. *et al.* "The genome of simian virus 40", *Science*, 200, pp. 494–502.

Redonder R., à Watson J.D. lettre du 15 Septembre 1989.

Research News, "Getting the bugs worked out", *Science*, Vol. 267, 13 January 1995, pp. 172–174.

Roberts C. "Large scale sequencing trials begin", *Science,* 250, 9 November 1990, pp. 756–757.

Roberts Leslie. "Who owns the human genome?", *Science*, Vol. 237, 24 July 1987, pp. 358–370.

Roberts Leslie. "Agencies vie over Human Genome Project", *Science*, Vol. 237, 31 July 1987, pp. 486–488.

Roberts Leslie. "Carving up the human genome", *Science*, 242, 2 December 1988, pp. 18–19.

Roberts Leslie. "Large scale sequencing trials begin", *Science*, 250, 7 December 1990, pp. 1336–1338.

Roberts Leslie. "Friends say Jim Watson will resign soon", *Science*, 256, 10 April 1992, p. 171.

Roberts Leslie. "Why Watson quit as project head", *Science* 256, 17 April 1992, pp. 301–302.

Sanger F. *et al.* "Nucleotide sequences of bacteriophage X174", *Nature*, 265, 1977, pp. 687–695.

Sanger F. *et al.* "Nucleotide sequence of bacteriophage lambda DNA", *JMB*, 162, 1982, pp. 729–773.

Sanger F. "Sequences, Sequences and Sequences", *Annual Reviews of Biochemistry*, 57, 1988, pp. 1–28.

Schoen A. "Les infortunes de HUGO ou le poids des nations", *Biofutur*, *Spécial Génomes*, No. 146, Juin 1995, pp. 92–94.

Schwartz D.E. *et al.* "Nucleotide sequence of Rous Sarcoma virus", *Cell*, 32, 1983, pp. 9–17.

Science, 2 January 1981, p. 33. "Human protein catalogue — Idea of human genome project." Sehmeck, H.H. "DNA pioneer to tackle biggest genome project ever", *New York Times*, 4 October 1988, C 1, C 16.

Sinsheimer R. *Letter, R. Sinsheimer to David Gardner*, University of California at Santa Cruz, 19 November 1984.

Sinsheimer R. "The Santa Cruz workshop", *Genomics*, 5, 1989, pp. 954–956.

Smith T.F. "The history of the genetic sequences databases", *Genomics,* (6), April 1990, pp. 701–707.

Sturtevant A. "Social implication of the genetics of man", *Science*, 120, 1947, p. 407.

Subcommittee on the Human Genome, Health and Environnemental Research Advisory Committee, *Report on the Human Genome Initiative*, Prepared for the Office of Health and Environmental Research, Office of Energy Research, Department of Energy, Germantown, MD, DOE, April 1987.

Sulston J. et al. "The *C. elegans* genome sequencing project: a beginning", *Nature*, Vol. 356, 5 March 1992, pp. 37–41.

The Human Genome Project - Deciphering the Blueprint of Heredity, ed. by Necia Grant Cooper, section: What is the Genome Project?, University Science Books, Mill Valley, 1994, pp. 71–84.

"The human genome project in Japan: The second five years", *Genome Digest*, Vol. 2, No. 4, October 1995, p. 6

US Congress. *Commercial Biotechnology, an International Analysis*, OTA, BA 218, Government Printing office, OTA, January 1984.

US Congress. *Mapping our Genes - Genome Projects: How Big? How Fast?*, OTA-OTA B.A 373. Washington DC Government Printing Office, 1988.

US Congress. *Technologies for Detecting Heritable Mutations in Human Beings* (OTA - H - 298), Washington DC Government Printing Office, OTA, Washington, 1986.

US House of Representatives, Departments of Labor, Health and Human Services, Education and related Agencies Appropriations for 1993. *Commitee and Appropriations*, 25 March 1992, Part 4, pp. 607–608.

Understanding Our Genetic Inheritance, US. Department of Health and Human Services, and *Department of Energy, NIH Publication No. 90-1590*, Washington, 1990, p. 9.

USDA's High-Priority Commitment to the Plant Genome Research Program. *Probe (Newsletter for the USDA Plant Genome Research Program)*, Vol. 3, No. 1/2, January–June 1993, pp. 1–3.

Wada N. "The complete index to man", *Science*, 211, 2 January 1981, pp. 33–35.

Wain S., Holson et al., "Nucleotide sequence of the AIDS Virus LAV", *Cell*, 40, 1985, pp. 9–17.

Watson J.D. *Letter to Dedonder* R. lettre datée du 9 Novembre 1989.

Watson J.D. "The human genome project: Past, present and future", *Science*, Vol. 248, 6 April 1990.

Watson J.D. and Cook Deegan R.M. "Perspectives on the human genome project", *Biofutur*, October 1990, pp. 65–67.

Weinberg R.A. "DNA sequencing: an alternative strategy for the human genome project", *Biofutur*, October 1990, pp. 20–21.

Wilkie Tom. *Perilous Knowledge, The Human Genome Project and its Implications*, Faber and Faber, London, 1994.

Chapter 6: The European Biotechnological Strategy and the Sequencing of the Yeast Genome

Barrel B.G. *Lettre à A. Goffeau*, datée du 12 Janvier 1989.

Biofutur, n. 78, 1989.

Brenner S. *Map of Man, (First-draft communiqué à B. Loder)*. Commission of the European Communities, 10 February 1986.

Clark-Walber D. *Lettre à A. Goffeau*, datée du 5 Avril 1988.

Com. (86) 272 final.

Comm. EC. *FAST Programme, Results and Recommendations*, December 1982. *FAST Occasional Paper*, no. 62, A Community Strategy for Biotechnology in Europe, March 1983.

Comm. EC. *Biotechnology: The Community's Role, Communication from the Commission to the Council*, COM (83) 328, 8 June 1983.

Comm. EC. *Biotechnology in the Community, Communication from the Commission to the Council, Com. (83) 672*, October 1983.

Comm. EC. *Towards a Market-Driven Agriculture: A Consultation Document from CUBE*, XII/951/December 1985.

Comm. EC. *Towards a Market-Driven Agriculture.* (a contribution from the viewpoint of biotechnology). A discussion paper from CUBE, XII, February 1986.

Comm. EC. *Biotechnology in the Community: Stimulating Agro-Industrial Development,* discussion paper of the Commission, *Com.* (86) 221/2, April 1986.

Comm. EC. Proposal for a Council decision to adopt a future multi-annual program (1988–93) for biotechnology based agro-industrial research and technological development: Eclair, *Com.* (87) 667, December 1987.

Comm. EC. Proposal for a Council decision to adopt a multi-annual R&D program in food science and technology, 1989 to mid 1993, *Com.* (88) 351, June 1988.

Compte Rendu de la Table Ronde sur le Séquençage des Génomes Humain et Microbiens, December 1987.

Compte Rendu de l'IRDAC Round-Table on Bio and Agro-Industrial Technology, February 1989.

Davies R. *Lettre au Dr. Magnien*, lettre datée du 6 Mai 1987.

Davies R. *Lettre à André Goffeau*, datée du 22 Juillet 1987.

de Nettancourt Dreux, Goffeau André and Van Hoeck F. *Applied Molecular and Cellular Biology*, background note on a possible action of the European Communities for the optimal exploitation of the fundamentals of the new biology, *Report D.G. XII*, Commission of the European Communities, XII/207/77-E, 15 June 1977.

de Nettancourt Dreux. *Mission Report,* discussion with the Office Federal Suisse de l'Education et de la Science "on the possible participation of Swiss laboratories to Community R.&.D. biotechnology programs", 29 June 1987 *XII/87*, D.N/ Km, 28/87 D.N 40, 1 July 1987.

Dennet Wayland. *The Future of Europe*, Cambridge University Press, 1976.

Detailed draft agenda — Meeting of the CGC Biotechnology, 9, September 1987, European Commission, Brussels.

Doc. C.G.C - IV/86/3. *Revision of the Biotechnology Action Programme - BAP.*

Dulbecco Renato. "A turning point in cancer research, sequencing the human genome", *Science*, 231, 1986, pp. 1055–1056.

European Commission. Common Policy for Science and Technology, *Bulletin of the European Communities, Supplement 3/77*, 1977; OJ N C7, 29 January 1974, p. 2.

European Commission. Proposal for a Council decision adopting a multinational research programme of the European Economic Community in the field of biotechnology, *Com.* (84) 230, 1984.

European Commission. Proposal for a Council directive on the protection of biotechnological inventions, communication from the commission to the council, *Com.* (88) 496, October 1988.

European Parliament (April 1989). *Report on Consumers and the Internal Market*, 1992 (the Abber Report).

European Parliament (February 1987). Resolution on biotechnology in Europe and the need for an integrated policy (the Viehoff Report); Doc A2-134/86, *Official Journal of the European Communities*, C 76/25-29, 23 March 1987.

Flandrog L. "Un bilan largement positif", *Biofutur*, Novembre 1989, pp. 30–34.

Franklin J. *The Role of Information Technology and Services in the Future Competitiviness of Europe's Bio-Industries*, report for Comm. EC, 1988.

Frontali Clara. *Lettre adressée à M. Cantley*, datée du 8 Juillet 1987.

Green Paper, Com 85 333 final.

Goffeau André. *Sequencing the Yeast Genome,* (Preliminary Draft for Assessment and Proposal of a New Research and Development Programme), January 1987.

Goffeau André. *Mission Report, Preparation of A Yeast Sequencing Program*, 14 to 19 June 1987, XII/87 - AG/hl, Brussels.

Goffeau André. *Lettre adressée à Olson*, datée du 11 Août 1987.

Goffeau André. *Mission Reports, Preparation "Sequencing the Yeast Genome"*, discussion NOVO with Mrs H. Nielsen and Niels-Fill à Copenhague, 20 Août 1987, Brussels, 30 Septembre 1987, XII/87 - A.G/H.L, 1987.

Goffeau André. *Preparation Human Frontier*, Oviedo, 21-23 August 1987, réf. XII/87, A.G/Kb, Brussels, 30 September 1987.

Goffeau André. *Mission Report,* visit to Dr S. Oliver (UMIST) for preparation of "Sequencing Yeast Genome", Manchester, 13 August 1987, XII/87, A.G/kl, Brussels, 30 September 1987.

Goffeau André. *Mission Report, "Réunion de concertation sur le séquençage du Génome Humain avec B. Hess, F. Gros et G. Valentini"*, Paris, 18 Novembre 1987, Brussels, XII/87 - A.G/Kb, 1987.

Goffeau André (ed.). *Sequencing the Yeast Genome, a Detailed Assessment*, CEC, DG XII, Directorate Biology, Division biotechnology, BAP: 1988–1989. Sequencing the Yeast Genome, a Possible Area for the Future Biotechnology Research Programme of the European Communities, 1988.

Goffeau André. *Lettre à B. Dujon,* datée du 23 Août 1992.

Hacquin F. (CRIM, France) *Séquençage du Génome de la levure et communication*, DGXII, 27 Juillet 1987.

Hartwell L. *The Cetus, UCLA Symposium on Yeast Cell Biology*, Keystone, Colorado, 1985.

IRDAC opinion on future R&D programmes in the field of biotechnology. Final version of the document approved in the last meeting of the IRDAC-WP5 Biotechnology, 24 November 1987.

Kingsman A.J. *Lettre à A. Goffeau en réponse au questionnaire envoyé par A. Goffeau*, datée du 19 Août 1987.

Kingsman A.J. *Lettre au Dr. M. Probert* (MRC Grande-Bretagne), datée du 19 Août 1987.

Linder P. *Lettre à A. Goffeau*, datée du 20 Mai 1988.

Magnien E., Aguilar A., Wragg P., de Nettancourt Dreux. "Les laboratoires Européens sans murs", *Biofutur*, Novembre 1989, pp. 17–29.

Martin W.J. and Davies, R. *Biotechnology*, 4, 1986, pp. 890–895.

Mewes H.W. *Lettre à A. Goffeau*, datée du 7 Juin 1987.

Mewes H.W. and Pfeiffer F. *Lettre à A. Goffeau en réponse au questionnaire envoyé par A. Goffeau*, datée du 7 Juin 1987.

Mortimer R.K. and Schild David. "Genetic Map of *Saccharomyces cerevisiae*", Edition 9", *Microbiological Reviews*, September 1985, pp. 181–212.

Munck L. and Rexen F. *Cereal Crops for Industrial Use in Europe*, Comm. E.C. Eur. 9617 EN, 1984.

Narjes K.H. (November 1985), *The European Commission's Strategy for Biotechnology in Industrial Biotechnology in Europe: Issues for Public Policy*, ed. D. Davies CEPS, Brussels, 1987.

Nieuwenhuis B. *Note for the File, BAP net on Eurokom*, 14 July 1987.

OJ. No. C 208, 4 August 1983.

OJ. No. C12, 14 January 1985.

OJ. No. C25, 28 January 1985.

OJ. No. C7, 29 January 1974.

OJ. No. L203, 78/636/E.E.C, 27 July 1978.

OJ. No. L305, 8 November 1983.

OJ. No. L375, 20 December 1981.

OJ. No. L83, 25 March 1985.

Oliver S. *Lettre à A. Goffeau en réponse au questionnaire envoyé par A. Goffeau.*

Olson M., Dutchnik J., Graham M. *et al. Proc. Natl. Acad. Sci.*, USA, 1986.

Olson M. *Lettre à A. Goffeau*, datée du 7 Octobre 1987.

Philipson L. *Lettre adressée à Goffeau*, datée du 28 Juillet 1987.

Pierad A. and Glansdorff N. *Lettre au Dr. J. de Brabander*, 16 December 1987.

Pohl F.M. and Beck S. *Embo J.*, 3, 1984, pp. 2905–2929.

Pohl. *Lettre à A. Goffeau en réponse au questionnaire adressé par A. Goffeau sur le projet de séquençage*, datée du 22 Mai 1987.

Rapport du meeting of the CGC biotechnology, Brussels, 7 March 1986.

Rapport du meeting du CGC biotechnology, 31 March 1987.

Research evaluation, *Report No. 32*, Eur 11833 EN, 29/88 DN46, 29 September 1988 rev. 2 February 1989.

Rose M.S. *Lettre à A. Goffeau en réponse au qestionnaire envoyé par A. Goffeau*, datée du 3 Juin 1987.

Simichen G. *Lettre à A. Goffeau*, datée du 13 Avril 1989.

Thomas D.Y. *Lettre adressée à A. Goffeau*, datée du 13 Juillet 1987.

Tubb Roy. *Lettre au Dr. M. Cantley*, datée du 2 Septembre 1987.

UK Royal Society, ACARD, ABRC, *Biotechnology: Report of a Joint Working Party*, the "Spinks Report", March 1980.

Van Der Meer R. *Revision of BAP I, Preparation of BAP 2*, 28 February 1986.

Van der Meer R., Magnien E., de Nettancourt Dreux. "Laboratoires Européens sans murs: Une recherche précompétitive ciblée", *Biofutur*, Juillet-Août, 1988, pp. 53–56.

van Hoeck F. *Note à l'attention de Mr P. Fasella concernant "l'engagement Pour l'exécution des 7 études en vue de la préparation du programme BRIDGE"*, 30 Octobre 1987.

van Hoeck F. *Note à l'attention de Mr P. Fasella, "Engagement pour l'exécution d'une étude en vue de la préparation du programme BRIDGE"*, Bruxelles, 1 Decembre 1987.

Vassarotti A. and Magnien E. *Biotechnology R&D in the EC (BAP: 1985–1989)*, Vol. 1: Catalogue of Achievements, ed. Elsevier for the Commission of The European Communities, Paris, 1990.

Vassarotti A. and Magnien E. *Biotechnology R&D in the EC (BAP: 1985–1989)* Vol. II: detailed final report, ed. Elsevier for the Commission of The European Communities, Paris, 1990.

Wada A. *Nature*, 1987.

Washington University Record, Vol. 12, No. 15, pl. December 1987.

Chapter 7: Life Deciphered

A *Consortium of European Laboratories Sequences an Entire Chromosome of Yeast*, Brussels, 15 May 1992.

Aguilar A. *Mission Report — Participation in the ELWW Meeting on* Bacillus subtilis, Jouy en Josas (F), 13–14 March 1989, Brussels, 6 April 1989, WXII/89 - AA/vc.

Ajioka J.W. *et al.* "*Drosophila* genome project: One hit coverage in yeast artificial chromosome", *Chromosoma*, Vol. 100, 1991, pp. 495–509.

Amjad M. *et al.* "An *Sfi*1 restriction map of the *Bacillus* 168 genome", *Gene*, 101, 1990, pp. 15–21.

Anagnostopoulos C., Piggot P.J. and Hoch J.A. "The genetic map of *Bacillus subtilis*", in *Bacillus and Other Gram-Positive Bacteria: Biochemistry, Physiology and Molecular Genetics*, ed. A.L. Sonenshein, J.A. Hoch and R. Losick, *American Society for Microbiology*, Washington DC, 1993, pp. 425–461.

Antequera F. and Bird A. "Number of CPG islands and genes in human and mouse", *Proc. Natl. Acad. Sci*, USA, Vol. 90, 1993, pp. 11995–11999.

Arabidopsis thaliana genome research project, *Progress Report*, Year Four, National Science Foundation, 1994.

Artavanis-Tsakonas S., Matsumo K. and Fortini M.E. "Notch signaling", *Science*, Vol. 268, 1995, pp. 225–232.

Azevzdo V., Alvares E., Zumstein E., Damiani G., Sgaramella V., Ehrlich and Seror P. "An ordered collection of *Bacillus subtilis* DNA segments cloned in yeast artificial chromosome", *Proc. Natl. Acad. Sci.*, USA, 90, 1993, pp. 6047–6051.

Bains William. "Sequence — so what?", *Biotechnology*, Vol. 10, July 1992.

Barrel B.G. *et al.* "The nucleotide sequence of *Saccharomyces cerevisiae* chromosome XIII", *Yeast Genome Directory, Nature*, Vol. 387, 29 May 1997, pp. 90–93.

Barrell B.G. *et al.* "The nucleotide sequence of *Saccharomyces cerevisiae* chromosome IX", *Yeast Genome Directory, Nature*, 1997, pp. 84–87.

Bassett D.E., Boguski M.S. and Hieter P. "Yeast genes and human disease", *Nature*, Vol. 379, 15 February 1996, pp. 589–590.

Bietry Michel. "Le génome humain, patrimoine commun de l'humanité", *Le Figaro*, Lundi, 7 Octobre 1996.

"Biotech on a roll", *Science*, Vol. 271, 12 January 1996, p. 151.

Biotechnology Newswatch, 10, 6, March 19, 1990.

Blomberg A. *et al.* "Interlaboratory reproducibility of yeast protein patterns analysed by immobilised pH gradient two dimensional gel electrophoresis", *Electrophoresis*, 1995.

Bock P. *et al.* "Comprehensive sequence analysis of the 182 predicted Open Reading Frames of yeast chromosome III", *Protein Sci.*, Vol. 1, 1992, pp. 1677–1690.

Bolotin-Fukuhara M. *Functional Studies of Yeast Genes, Proposal for Mass Construction and Screening of Descriptant Strains*, 1991 (doc. communiqué par A. Goffeau).

Borde Valérie. "La mosaïque canadienne", *Biofutur*, 153, Février 1996, pp. 14–19.

Bork Peer *et al.* "What's in a genome?", *Nature*, Vol. 358, 23 July 1992.

Boucherie H. *et al.* "Two dimensional protein map of *S. cerevisiae*: Construction of a gene-protein index", *Yeast*, Vol. 11, 1995, pp. 601–613.

Boutry M. "Arabidopsis: une mauvaise graine sort de l'ombre", *Biofutur*, Vol. 94, 1990, pp. 55–57.

Brenner S. *et al.* "Characterization of the pufferfish (fugu) as a compact model vertebrate genome", *Nature*, 366, 1993, pp. 265–268.

Brenner S., Hubbard T., Murzin A. and Chothia C. "Gene duplication in *H. influenzae*", *Nature*, 378, 1995, p. 140.

Bridges C. "Correspondence between linkage maps and salivary chromosome structure, as illustrated in the tip of chromosome 2R of *Drosophila melanogaster*", *Cytologia Fugii Jubilee Volume*, 1937, pp. 745–755.

Broach J.R., Pringle E.W. Jone (Eds.), *The Molecular and Cellular Biology of the Yeast* Saccharomyces, Vol. 1, *Genome Dynamics, Protein Synthesis and Energetics*, Cold Spring Harbor Laboratory Press, Cold Spring Harbor, NY, 1991.

Browaeys Dorothée Benoit. "L'étiquetage des nouveaux aliments, un leurre", *La Recherche*, N. 299, Juin 1997, p. 36.

Bruschi C.V., McMillan J.N., Coglievina M. and Esposito M.S. "The genomic instability of yeast cdc6-1/cdc6-1 mutants involves chromosome structure and recombinaison", *Mol. Genet.*, 249, 1995, pp. 8–18.

Bult Carol J. *et al.* "Complete genome sequence of the methanogenic archaeon, *Methanococcus jannaschii*", *Science*, Vol. 273, No. 5278, 23 August 1996, pp. 1058–1074.

Burns N.S., Grimwade B., Ross-Macdonald P.B., Choi E.Y., Finberg K., Roeder G.S. and Snyder M. "Large scale analysis of gene expression, protein localization, and gene disruption in *S. cerevisiae*", *Genes and Development*, 8, 1994, pp. 1087–1105.

Bussey Howard *et al.* "The nucleotide sequence of chromosome I from *Saccharomyces cerevisiae*", *Proc. Acad. Sci., USA*, Vol. 92, 1995, pp. 3809–3813.

Bussey Howard. "Chain of being", *The Sciences*, March/April 1996, pp. 28–33.

Caboche M., Delseny M. and Lescure B. "*Arabidopsis* tient ses promesses", *Biofutur*, September 1994, pp. 25–30.

Cai H., Kiefel J., Yee J. and Duncan I. "A yeast artificial chromosome clone map of the *Drosophila* genome", *Genetics*, Vol. 136, 1994, pp. 1385–1401.

Capano C.P. *et al.* "Complexity of nuclear and polysomal RNA from squid optic lobe and grill", *Journal of Neurochemistry*, Vol. 46, 1986, pp. 1517–1521.

Carey J. *et al.* "The gene kinds", *Business Week*, May 1995, pp. 72–77.

Ceccaldi Paula. "Un arsenal législatif pour les OGM", *Biofutur*, 154, Mars 1996, pp. 21–25.

Chausson L., Nouaille. "Etats Unis: Biotechnologies et réglementations", *Biofutur*, Mars 1985, pp. 60–61.

Chu W.S., Magee B.B. and Magee P.T. *J. Bacteriol.*, Vol. 175, No. 6637, 1993.

Chotia Cyrus. "One thousand families for the molecular biologist", *Nature*, 357, 18 June 1992, pp. 543–544.

Choulika A., Perrin A., Dujon B. and Nicolas J.F. "Induction of homologous recombination in mammalian chromosomes by using the I-*Sce*I system of *Saccharomyces cerevisiae*", *Molecular and Cellular Biology*, April 1995, pp. 1968–1973.

Colleaux L. *et al. Cell.*, Vol. 44, 1986, pp. 521–533.

Colleaux L. et al. *Proc. Natl. Sci.*, Vol. 85, 1988, pp. 6022–6026.

Colleaux L., Rougeulle C., Avner P. and Dujon B. "Rapid physical mapping of YAC inserts by random integration of I-*Sce*I sites", *Nucleic Acids Research*, Vol. 20, No. 21, 1993, pp. 265–271.

Collins F.S. "Ahead of schedule and under budget: the genome project passes its fifth birthday", *Proc. Natl. Acad. Sci.*, USA, Vol. 92, 1995, pp. 10821–10823.

Commission of the European Communities. *Report of the FAST Programme (Forecasting and Assessment in Science and Technology)*, 1982.

Communiqué de presse. *A Consortium of European Laboratories Sequences an Entire Chromosome of Yeast*, Commission of the European Communities, Directorate-General XII, Brussels, 15 May 1992.

Communiqué de presse. *Une première mondiale gâce à la recherche communautaire: le séquençage complet du génome d'un organisme complexe*, Commission of the European Communities, Directorate-General XII, Brussels, 24 Avril 1996 (IP/96/344).

Cooper Necia Grant (ed.). *The Human Genome Project (Deciphering the Blueprint of Heredity)*, University Science Book, Mill Valley, California, 1994.

Coppée et al. "Construction of a library of 73 individual genes deletions in view of the functional analysis of the new genes discovered by systematic sequencing of S. cerevisiae chromosome III", *Yeast*, 1995.

Crépin K. and Lints F. "De l'évaluation du risque à sa tolérance", *Biofutur*, 153, Février 1996, pp. 20–24.

Danchin A. "Complete genome sequencing: future and prospects", in A. Goffeau, *Sequencing the Yeast Genome: A Detailed Assessment*, CEC, 1989.

Danchin A. *"Séquence totale du génome de* Bacillus subtilis. *Participation française"*, document archive A. Goffeau.

de Nettancourt Dreux. "The T projects of BRIDGE, a new tool for technology transfer in the community", *Agro-Industry High Tech.*, Vol. 3, 1991, pp. 3–9.

de Nettancourt Dreux. Commission of the European Communities Directorate - General XII, Science, Research and Development, *BRIDGE Report: Sequencing the Yeast Genome*, 1992.

De Risi J.L., Iyer V. and Brown P.O. "Exploring the metabolic and genetic control of gene expression on a genemic scale", *Science*, 278, 1997 pp. 680–686.

Deroin Philippe. "Surenchère pour l'accès au génome humain", *L'Usine Nouvelle*, N. 2557, 11 Juillet 1996, p. 24.

Devine K.M. "The *Bacillus subtilis* genome project: Aims and progress", *TibTech*, Vol. 13, June 1993, pp. 210–216.

Dickson D. and Abbot Alison. "EU research ministers lean on Brussels", *Nature*, Vol. 386, 20 March 1997, p. 205.

Diehl B.E. and Pringle J.R. "Molecular analysis of *Saccharomyces cerevisiae* chromosome I: Identification of additional transcribed regions and demonstration that some encode essential functions", *Genetics*, Vol. 127, 1991, pp. 287–298.

Dietrich F. *et al.* "Status of the sequencing of chromosomes V et VI", in *Report of Manchester Conference of the Yeast Genome Sequencing Network*, Commission of the European Communities, DG XII, Biotechnology, February 1994, p. 23.

Dietrich F.S. *et al.* "The nucleotide sequence of *Saccharomyces cerevisiae* chromosome V", *Yeast Genome Directory*, *Nature*, Vol. 387, 29 May 1997, pp. 78–81.

Dujon B. "Altogether now, sequencing the yeast genome", *Current Biology*, 2, 1992, pp. 279–281.

Dujon B. "Le séquençage systématique du génome de la levure *Saccharomyces cerevisiae*: Résultats du programme Européen", *Bull. Soc. Fr. Microbiol.*, 8, 3, 1993, pp. 151–155.

Dujon B. "Mapping and sequencing the nuclear genome of the yeast *Saccharomyces cerevisiae*: strategies and results of the European enterprise", *Cold Spring Harbor Symposium on Quantitative Biology*, Vol. LVIII, Cold Spring Harbor Laboratory Press, 1993, pp. 357–366.

Dujon B. "Mapping and sequencing the nuclear genome of the yeast *Saccharomyces cerevisiae*: Strategies and Results of the European Enterprise", *Cold Spring Harbor Symposia on Quantitative Biology*, Volume LVIII, Cold Spring Harbor Laboratory Press, 1993, p. 357.

Dujon B. "Mapping and sequencing the nuclear genome of the yeast *Saccharomyces cerevisiae*: strategies and results of the European enterprise", *Cold Spring Harbor Symposium on Quantitative Biology*, Vol. LVIII, Cold Spring Harbor Laboratory Press, 1993, pp. 357–366.

Dujon B. *et al.* "Complete DNA sequence of yeast chromosome XI", *Nature*, Vol. 369, 2 June 1994, pp. 371–378.

Dujon B. *et al.* "Complete DNA sequence of yeast chromosome XI", *Nature*, Vol. 389, 1994, pp. 371–378.

Dujon B. "The yeast genome project: What did we learn?", *Trends in Genetics*, Vol. 12, No. 7, July 1996, pp. 263–270.

Dujon B. "Le génome de la levure", *m/s, N spécial*, Vol. 12, Octobre 1996, pp. 30–31.

Enayati E. "Intellectual property under GATT", *Biotechnology*, Vol. 13, May 1995, pp. 460–462.

Fairhead C., Dujon B. "Consequences of unique double-stranded breaks in yeast chromosomes: death or homozygosis", *Mol. Gen. Genet.*, 240, 1993, pp. 170–180.

Fairhead C., Heard E., Arnaud D., Avner P. and Dujon B. "Insertion of unique sites into YAC arms for rapid physical analysis following YAC transfer into mammalian cells", *Nucleic Acids Research*, Vol. 23, No. 10, 1995, pp. 4011–4012.

Fairhead C., Llorente B., Denis F., Soler M. and Dujon B. "New vectors for combinatorial deletions in yeast chromosomes and for gap-repair cloning using "split–marker" recombinaison", *Yeast*, 1996.

Feldmann H. *et al.* "Complete DNA sequence of yeast chromosome II", *EMBO J.*, Vol. 13, 1994, pp. 5795–5809.

Feldmann H. *et al.* "Complete sequence of chromosome II of *Saccharomyces cerevisiae*", *EMBO J.*, Vol. 13, 1994, pp. 5795–5809.

Fey Stephen J., Nawrocki Arkadiusz and Larsen Peter Mose. "Function analysis by 2D gel electrophoresis: From gene to function, yeast genome sequencing network", 8–10 June 1995, DGXII, Commission of the European Communities, *BIOTECH Programme Books of Abstracts*, p. 214.

Final European Conference of the Yeast Genome Sequencing Network, Trieste, Italy, Congress Centre Stazione Marittima, 25, 28 September 1996, Commission of the European Communities, DGXII, Brussels.

Fink G.R. "Pseudogenes in yeast?", *Cell*, Vol. 49, 10 April 1987, pp. 5–6.

First Report on the State of Science and Technology in Europe, CE, 1989.

Flandroy L. and Vincent Ch. "Patrimoine héréditaire: Droits, pouvoirs et Devoirs de l'Humanité", *Biofutur*, Janvier 1990, pp. 18–21.

Fleischmann R.D. *et al.* "Whole-genome random sequencing and assembly of *Haemophilus influenzae* Rd", *Science*, Vol. 269, No. 5223, 1995, pp. 496–512.

Fleishmann R.D. et al. *Science*, Vol. 269, 1995, pp. 496–512.

Fortini M.E. and Rubin C.M. "Analysis of cis-acting requirement of the Rh3 and Rh4 genes reveals a bipartit organization to rhodopsin promoters in *Drosophila melanogaster*", *Genes and Development*, Vol. 4, 1990, pp. 444–463.

Fraser C.M. et al. "The minimal gene complement of *Mycoplasma genitalium*", *Science*, Vol. 270, 20 October 1995, pp. 397–403.

Galibert F. et al. "Complete nucleotide sequence of *Saccharomyces cerevisiae* chromosome X", *EMBO J.*, Vol. 15, 1996, pp. 2031–2049.

Gallochat A. "Protection des inventions biotechnologiques: Un sursaut européen?", *Les Echos*, jeudi 18 Avril 1996.

Garza D., Link A.J., Duncan I.W. and Hartl D.L. "*Drosophila* genome project: one-hit coverage in yeast artificial chromosomes", *Chromosoma*, 100, 1991, pp. 495–509.

Gatti M. and Pimpinelli S. "Functional elements in *Drosophila melanogaster* heterochromatin", *Ann. Rev. Genet.*, 26, 1992, pp. 239–275.

Giraudat J. "*Arabidopsis*: état des lieux", *Biofutur*, Vol. 95, 1990, p. 54.

Goeble M.G. and Petes T.D. "Most of the yeast genomic sequences are not essential for cell growth and division", *Cell*, Vol. 46, 1986, pp. 983–992.

Goffeau André and Vassarotti A. *Rapport de mission, Participation in the BAP Meeting "Yeast sequencing strategies: From BAP to BRIDGE"*, Louvain-la-Neuve, Belgique, Brussels, 3 April 1989, DGXII, AV/1H/03.

Goffeau André and Vassarotti A. *Rapport de Mission, Participation in the BAP Meeting "Sequencing of the yeast chromosome III", Tutzing (D)*, 31 October – 2 November 1989, Brussels, 23 November 1989, DGXII, AV/1H/09.

Goffeau André. *Rapport de Mission, Symposium on "Human Genome Research Strategies and Priorities,* 29–31 January 1990, UNESCO Headquarters, Brussels, 15 February 1990.

Goffeau André and Vassarotti A. *Rapport de mission, Report on the BAP Meeting on "The Sequencing of the Yeast Chromosome III"*, Brussels, 8 October 1990 (AG-nm-).

Goffeau André. *Rapport de mission, Représenter M. Fasella à la conférence Internationale sur le Génome de* Bacillus subtilis, Paris, les 2–3 Septembre 1990, et présenter les activités communautaires dans le domaine du séquençage (Paris, 5 le Septembre 1990), Brussels, 19 November 1990, AG/AA-nm-Paris.

Goffeau André. *Rapport de mission*, 9–10 May 1991, *Plant Genetic Engineering: Europe and Italy*, Brussels, 7 June 1991.

Goffeau André. *Rapport de mission,* San Francisco, 23–28 May 1991, *Yeast Genetic and Molecular Biology*, Brussels, 7 June 1991, Sanfr. rap.

Goffeau André. *Rapport de mission, Présentation du réseau de séquençage levure,* BRIDGE, Rome, 9–10 May 1991, Brussels, 14 June 1991, AG-Is-Rome.

Goffeau André. *Rapport de mission, Meeting on Genome Analysis in the EC, Elounda, Crete,* 14–17 May 1991, Brussels, 19 June 1991, AG-Is-Crete.

Goffeau André. *Rapport de mission, NATO Workshop on "Genome Organization Function and Evolution", Spetsai, Grèce,* 16–22 Septembre 1992, Brussels, 8 Octobre 1992, ML/RAP.COM. grecerap.

Goffeau André and Vassaroti A. *Rapport de mission, 1) Participation in the 2nd BRIDGE meeting "Sequencing the Yeast Genome"* (Munich, 18–20 October 1992) and 2*) Participation in the YIP Meeting* (Munich, / Penzberg 20–21 October 1992), Brussels, 4 December 1992, AV/at/miss08.

Goffeau André, Slonimski P., Nakai K. and Risler J.L. "How many yeast genes code for membrane-spanning proteins?", *Yeast*, Vol. 9, 1993, pp. 691–702.

Goffeau André, Nakai K., Slonimski P. and Risler J.L. "The membrane proteins encoded by yeast chromosome III genes", *FEBS*, Vol. 325, No. 1, 2, 1993, pp. 112–117.

Goffeau André. "Genes in search of functions", *Nature*, Vol. 369, 12 May 1994, pp. 101–102.

Goffeau André, Mordan Philippe and Vassarotti Ati. "Le génome de la levure bientôt déchiffré", Europe, Supplément au N. 26 de *La Recherche*, 276, Mai 1995, pp. 23–25.

Goffeau André *et al.* "Life with 6,000 genes", *Science*, Vol. 274, October 1996, pp. 546–567.

Goodman C.S., Eckers J.R. and Dean C. "The genome of *A. thaliana*", *Proc. Natl. Acad. Sci. USA*, 92, 1995, pp. 10831–10835.

Grivell L.A. and Planta R.J. "Yeast: The model eucaryote?", *TIBTECH*, Vol. 8, September 1990, pp. 241–243.

Grunstein M. et Hogness D.S. "Colony hybridization: A method for the isolation cloned DNAs that contain a specific gene", *Proc. Nat. Acad. Sci., USA*, Vol. 72, 1975, pp. 3961–3965.

Harris S.D. and Pringle J.R. "Genetic analysis of *Saccharomyces cerevisiae* chromosome I: On the role of mutagen specificity in delimiting the set of genes identifiable using temperature-sensitive-lethal mutation", *Genetics*, Vol. 127, 1991, pp. 279-285.

Harrois-Monin F. "Génétique: La mine d'or", *Le Vif/l'Express*, 13 Octobre 1995, pp. 79-80.

Hartl D.L. *et al.* "Genome structure and evolution in *Drosophila*: Applications of the framework P1 map", *Proc. Natl. Acad. Sci., USA*, Vol. 91, 1994, pp. 6824-6829.

Himmebreich *et al. Nuc. Acid. Res.*, Vol. 24, 1996, pp. 4420-4449.

Hodgkin Jonathan, Plasterk, Ronald H.A., Waterston, Robert H. "The nematode *Caenorhabditis elegans* and its genome", *Science*, Vol. 270, 20 October 1995, p. 414.

Hodson John. "Sequencing and mapping efforts in model organisms", *Biotechnology*, Vol. 10, July 1992, pp. 760-762.

Hodson John. "Form for Euroinformatics?", *Biotechnology*, Vol. 10, July 1992, pp. 755-756.

Hohiesel J.D. *et al.* "High-resolution cosmid and P1 maps spanning the 14-Mbp of the fission yeast *Schizosaccharomyces pombe*", *Cell*, Vol. 73, 1993, pp. 109-120.

Holden C. "Genes confirm Archae's uniqueness", *Science*, Vol. 271, 1996, p. 1061.

Holland P.W.H. *et al.* "Gene duplications and the origins of vertebrate development", Development, 1994, *Supplement*, pp. 125-133.

Huang Meng-ER, Chuat J.C., Thierry A., Dujon B. and Galibert F. "Construction of a cosmid contig and of an *AcoR1* restriction map of yeast chromosome X, DNA Sequence", *The Journal of Sequencing and Mapping*, Vol. 4, 1994, pp. 293-300.

Intellectual Property in Genome Mapping Programmes. A report of an EC workshop, 20-22 November 1992.

Isono Katsumi, Yoshikawa Akikazu and Tanaka Seiji. "An approach to genome analysis of *Saccharomyces cerevisiae*", in 2nd meeting of the European Network in charge of *Sequencing the Yeast Chromosome III*, Den Haag (NL), 26-27 July 1990, *Book of Abstracts*, CEC, DGXII, pp. 5-7.

Jacq C. et al. "The nucleotide sequence of chromosome IV from *Saccharomyces cerevisiae*", *Yeast Genome Directory, Nature*, Vol. 387, 29 May 1997, pp. 75–78.

JOEC. 10/3, 13 Janvier 1989.

JOEC. L43, 14 Février 1997.

John B. and Milkos G.L.G. *The Eukaryote Genome and Evolution*, Allen and Unwin, London, 1988.

Johnston Mark et al., "Complete nucleotide sequence of *Saccharomices cerevisiae* chromosome VIII", *Science*, Vol. 265, 1994, pp. 2077–2082.

Jonhston Mark. "Towards a complete understanding of how a simple eukaryotic cell works", *TIG*, Vol. 12, No. 7, 1996, pp. 242–243.

Johnston Mark et al. "The complete sequence of *Saccharomyces cerevisiae* chromosome XII", *Yeast Genome Directory, Nature*, Vol. 387, 29 May 1997, pp. 87–90.

Johnston Mark. "DNA chip stuff", transmis à André Goffeau pour l'auteur.

Joly P.B. and Mangematin V. "Is the yeast model exportable?", manuscrit soumis à Nature pour le numéto spécial sur la levure et transmis à l'auteur.

Kaback D.B., Oeller P.W., Steensma H.Y., Hirshman J., Ruezinsky D., Colemen K.G. and Pringle J.R. "Temperature-sensitive lethal mutations on yeast chromosome I appear to define only a small number of genes", *Genetics,* Vol. 108, 1984, pp. 67–90.

Kamalay J.C., and Goldberg, R.B. "Regulation of structural gene expression in tobacco", *Cell*, Vol. 19, 1980, pp. 935–946.

Kahn Axel. "Encadrer sans entraver", *Biofutur*, Avril 1991, p. 54.

Kalogeropoulos A. "Linguistic analysis of chromosome III DNA sequence of *Saccharomyces cerevisiae*", *Yeast*, 9, 1993, pp. 889–905.

Kaneko et al., *DNA Res.*, Vol. 3, 1996, pp. 109–136.

Karlin S., Blaisdell B.E., Sapolsky R.J., Cadon L. and Burge C. "Assessment of DNA inhomogeneies in yeast chromosome III", *Nuc. Ac. Res.*, 21, 1993, pp. 703–711.

Karpen G.H. and Spradling A.C. "Reduced DNA polytenization of a minichromosome region undergoing position-effect variegation in *Drosophila*", *Cell*, Vol. 63, 1990, pp. 97–107.

Koonin E. et al. "Yeast chromosome III, new gene functions", *EMBO J.*, Vol. 13, 1994, pp. 493–503.

Krawetz S.A. "Sequence errors described in GenBank: a means to determine the accuracy of DNA sequence interpretation", *Nuc. Ac. Res.*, 17, 1989, pp. 3957–3957.

Kunst F., Vassarotti Ati and Danchin Antoine. "Organization of the European *Bacillus subtilis* genome sequencing project", *Microbiology*, 141, 1995, pp. 249–255.

La lettre du Greg, N. 8, Decembre 1996, p. 12.

Lalo Stettler S., Mariotte S., Slonimsky P., Thuriaux P. "Two yeast chromosomes are related by a fossil duplication of their centromeric regions", *C.R. Acad. Sc.*, Paris, 316, 1993, pp. 367–373.

"La longue marche des propositions communautaires", *Biofutur*, Janvier 1994, pp. 19–21.

Larsen Peter Mose and Fey Stephen J. "Two dimensional gel electrophoresis and yeast genetics", comments to the meeting held at the *CEC* on 14 June, Vol. 21, June 1991.

Le M.H., Diricka D. and Karpen G.H. "Islands of complex DNA are widespread in *Drosophila melanogaster* centric heterochromatin", *Genetics*, 141, 1995, pp. 283–303.

Levy J. "Sequencing the yeast genome: an international achievement", *Yeast*, Vol. 10, 1994, pp. 1689–1706.

Little Peter. "The worm turns out and delivers", *Nature*, Vol. 356, 5 March 1992, pp. 14–15.

Louis E.J. "A nearly complete set of marked telomeres in S288C for mapping and cloning", *in Reports of Manchester Conference of the Yeast Genome Sequencing Network*, Commission of the European Communities, DGXII, Biotechnology, February 1994.

Louis E.J. "Corrected sequence for the right telomere of *Saccharomyces cerevisiae* chromosome III", *Yeast*, Vol. 10, 1994, pp. 271–274.

Madueno E.G. et al. "A physical map of the chromosome X of *Drosophila melanogaster*: Cosmid contigs and sequence tagged sites", *Genetics*, Vol. 139, 1995, pp. 1631–1647.

Magnien E., Bevan M. and Planqué K. "A European "BRIDGE project to tackle a model plant genome", *Trends Biotech*, Vol. 10, 1992, pp. 12–15.

Maier E., Hoheisel J.D., McCarthy L., Mott R., Grigoriev A.V., Monaco A.P., Larin Z. and Lehrach H. "Yeast artificial chromosome clones completely spanning the genome of *Schizosaccharomyces pombe*", *Nature Genet.*, 1, 1992, pp. 273–277.

Marini A.M., Vissers S., Urrestarazu A. and André B. "Cloning and expression of the MEP1 gene encoding an ammonium transporter in *Saccharomyces cerevisiae*", *The EMBO Journal*, 13, 1994, pp. 3456–3463.

Maroni G. "The organization of eukaryotic genes", *Evolutionary Biology*, Vol. 29, 1996, pp. 1–19.

Mewes H.W. *Lettre à Goffeau*, datée du 15 October 1990.

Mewes H.W. "European science", *Science*, Vol. 256, 1992, p. 1378.

Mewes H.W., Doelz R. and George D.G. "Sequence databases — An indispensable source for biotechnological research", *J. Biotec.*, 35, 1994, pp. 239–256.

Miklos G.L.G. "Emergence of organizational complexities during metazoan evolution: Perspectives from molecular biology, palaenology and neo-darwinism", *Memoirs Australasian Assn. Palaeontologists*, 15, 1993, pp. 7–41.

Miklos G.L.G. and Rubin G.M. "The role of the genome projectin determining gene function: Insights from model organisms", *Cell*, 86, 1996, pp. 521–529.

Millet Annette. "La ruée vers les gènes", *Biofutur*, Juin 1995, pp. 78–83.

MIPS News, Number two, July 1992.

Mizukami T. *et al. Cell*, Vol. 73, No. 109, 1993, p. 121.

Monteilhel C. *et al. Nuc. Acids Res.*, Vol. 16, 1990, pp. 1407–1413.

Mortimer R.K., Contopoulou C.R. and King J.S. "Genetic and physical maps of *Saccharomyces cerevisiae*, Edition XI", *Yeast*, Vol. 8, 1992, pp. 817–902.

Moszer I., Glaser P. and Danchin A. "Subtilist: a relational database for the *Bacillus subtilis* genome", *Microbiology*, Vol. 141, 1995, pp. 261–268.

Murakami Y. *et al.* "Analysis of the nuclotide sequence of chromosome VI from *Saccharomyces cerevisiae*", *Nature Genet.*, Vol. 10, 1995, pp. 261–268.

Murphy T. and Karpen G.H. "Localization of centromere function in a *Drosophila* minichromosome", *Cell*, Vol. 82, 1995, pp. 599–609.

Nature Biotechnology, Vol. 14, No. 11, November 1996.

Navarre C., Ghislain M., Leterme S., Ferroud C., Dufour J.P. and Goffeau André. *J. Biol. Chem.*, Vol. 267, 1992, pp. 6425–6428.

Newlon C.S. *et al.* "Structure and organization of yeast chromosome III", *UCLA Symp. Mol. Cell. Biol. New Series*, Vol. 33, 1986, pp. 211–223.

Newlon C.S. *et al.* "Analysis of a circular derivative of *S. cerevisiae* chromosome III: A physical map and identification and location of ARS elements", *Genetics*, Vol. 129, 1991, pp. 343–357.

Ogasawara N. *et al.* "Systematic sequencing of the *Bacillus subtilis* genome: progress report of the Japanese group", *Microbiology*, Vol. 141, 1995, pp. 257–259.

Oliver S. *et al.* "The complete sequence of yeast chromosome III", *Nature*, Vol. 357, 1992, pp. 38–46.

Oliver S. *Biotechnology Action Programme, Progress Report*, Period: 1 January, 1989 – 30 June, 1989.

Oliver S. "From DNA sequence to biological function", *Nature*, Vol. 379, 1996, pp. 597–600.

Oliver S. "A network approach to the systematic analysis of yeast gene function", *TIG*, Vol. 12, No. 7, July 1996, pp. 241–242

Olson M. *et al.* "Random-clone: Strategy for genomic restriction mapping in yeast", *Proc. Natl. Acad. Sci.*, USA, Vol. 83, 1986, pp. 7826–7830.

Olson M. *et al.* "Genome structure and organization in *S. cerevisiae*", in ed. Broach, J.R. Pringle, E.W. Jone, *The Molecular and Cellular Biology of the Yeast* Saccharomyces, Vol. 1, Genome Dynamics, Protein Synthesis and Energetics, Cold Spring Harbor Laboratory Press, Cold Spring Harbor, N.Y., 1991, pp. 1–40.

Palazzolo M.L. *et al.* "Optimized strategies for STS selection in genome mapping", *Proc. Natl. Acad. Sci.*, Vol. 88, 1991, pp. 8034–8038.

Pardue M.L. *et al.* "Cytological localisation of DNA complimentary to ribosomal RNA in polytène chromosome of diptera", *Chromosoma*, Vol. 29, 1970, pp. 268–290.

Perez Alain. "Les liaisons dangereuses de la médecine génétique", *Les Echos*, Mercredi, 4 Septembre 1996.

Perrin Anne, Buckle Malclm and Dujon B. "Asymetrical recognition and activity of the I-*Sce*I endonuclease on its site and on intron-exon junctions", *The EMBO Journal*, Vol. 12, No. 7, 1993, pp. 2939–2947.

Perrin A., Thierry A. et Dujon B. "La méganucléase I-*Sce*I, premier exemple d'une nouvelle classe d'endonucléases utiles pour les grands génomes", 2 Decembre 1991 (texte communiqué par B. Dujon).

Petes T.D., Malone E.R. and Symington L.S. in *The Molecular and Cellular Biology of the Yeast Saccharomyces: Genome Dynamics, Protein Synthesis and Energetics*, eds. Broach J.R., Pringle J.R. and Jones E.W. Cold Spring Harbor Laboratory, New York, 1991, pp. 407–521.

Pfeffer S.R. "Clues to brain function from baker's yeast", *Proc. Natl. Acad. Sci. USA*, Vol. 91, March 1994, pp. 1987–1988.

Philippsen P. *et al.* "The nucleotide sequence of *Saccharomyces cerevisiae* chromosome XIV and its evolutionary implications", Yeast Genome Directory, *Nature*, Vol. 387, 29 May 1997, pp. 93–98.

Philipson, L. *Nature*, Vol. 351, 9 May 1991.

Picket F.B. and Meeks-Wagner D.R. "Seeing double: appreciating genetic redundancy", *The Plant Cell*, 7, 1995, pp. 1347–1356.

Plessis Anne, Perrin Arnaud, Haber James E. and Dujon Bernard. "Site-specific recombination determined by I-*Sce*I, a mitochondrial group I intron-encoded endonuclease expressed in the yeast nucleus", *Genetics*, 130, March 1992, 451–460.

Plessis Anne and Dujon Bernard. "Multiple tandem integrations of transforming DNA sequences in yeast chromosomes suggest a mechanism for integrative transformation by homologous recombination", *Gene*, 134, 1993, pp. 41–50.

Press book. Conférence de presse, 24 April 1996. *Le Génome de la Levure*, CGI Europe, 1996.

Proposition modifiée de directive publiée, *JOCE* 44/36, 16 February 1993.

Report of Manchester Conference of the Yeast Genome Sequencing Network, Commission of the European Communities, DGXII, Biotechnology, February 1994.

Report of the meeting held at Martinsried, 2/3 November 1989, to consider the feasibility of including the sequencing of the *Bacillus subtilis* genome as a T project in the BRIDGE program, CEC, Brussels.

Richard G.F., Fairhead C. and Dujon B. "Cross-hybridization and transcriptional maps of yeast chromosome XI show a high degree of genomic redundancy and trends to expression clustering" (Submitted to *J. Mol. Biol.*: Running title: "Transcript map and duplications of yeast chromosome XI").

Richard G.F. and Dujon B. "Distribution and variability of trinucleotide repeats in the genome of the yeast *Saccharomyces cerevisiae*" (submitted *Gene*: Running title: "Trinucleotide repeats in yeast").

Rieger Klaüs-Jörg. "Basic physiological analysis key-step between genomic sequencing and protein function", in *Final European Conference of the Yeast Genome Sequencing Network*, Trieste, Italy, Congress Centre Stazione Maritima, 25, 28 September 1996. (Commission of the European Communities, DGXII, Brussels).

Riles *et al.*, *Genetics*, Vol. 134, 1993, pp. 81–150.

Rinati T., Bolotin-Fukuhara M. and Frontali, L. "A *Saccharomyces cerevisiae* gene essential for viability has been conserved by evolution", *Gene*, 160, 1995, pp. 135–136.

Roberts L. *Science*, Vol. 245, 1989, p. 1439.

Ross-Macdonald P. Aarwal, Erdman S., Burns N., Sheehan A., Malczynski M., Roeder G.S. and Snyder M. "Large functional analysis of the *S. cerevisiae* genome", *EUROFAN meeting*, 1996.

Rossion P. "La vie en pièces détachées", *Science et Vie*, Août, 1992, pp. 49–51 et p. 163.

Rothstein R.J. *J. Meth. Enzym.*, Vol. 101, 1983, pp. 202–211.

Rubin G.M. *et al.* "The chromosomal arrangement of coding sequences in a family of repeated genes", *Prog. Nucleic Acid. Res. Mol. Biol.*, Vol. 19, 1976, pp. 221–226.

Rubin G.M. "Around the genomes: The *Drosophila* genome project", *Genome Research*, 6, 1996, pp. 71–79.

Sanders J.P.M., Gist-Brocades, N.V. "Will industry benefit from the yeast DNA sequence analysis?", *1st BRIDGE meeting on "Sequencing the Yeast Chromosome II and XI*, BRIDGE, Brugge (B), 22–24 September 1991, p. 106.

"Sequencing the yeast genome and the functional analysis of new genes", *CEBEC* (China - EC Biotechnology Center), pp. 43–46.

"Science in Europe", *Science*, 24 April 1992, p. 462.

Sharp P. and Lloyd A. "Regional base composition variation along yeast chromosome III. Evolution of chromosome primary structure", *Nucleic Acides Res.*, Vol. 21, 1993, pp. 179-183.

Siden I., Saunders R.D.C. *et al.* "Towards a physical map of the *Drosophila melangaster* genomic divisions", *Nucleic Acids Res.*, Vol. 18, 1990, pp. 6261-6270.

Sidow A. and Thomas W.K. "A molecular evolutionary framework for eukaryotic model organisms", *Curr. Biol.*, Vol. 4, 1994, pp. 596-603.

Sidow A. and Thomas W.K. *Curr. Biol.*, Vol. 4, No. 596, 1994.

Sikorski S.R. and Hieter P. "A system of shuttle vectors and yeast host strains designed for efficient manipulation of DNA, in *Saccharomyces cerevisiae*", *Genetics*, Vol. 122, 1989, pp. 19-27.

Slonimski P. "Les génomes, des clefs pour la structure, la fonction et l'évolution", éditorial, *Biofutur*, Novembre 1993, p. 5.

Slonimski P. and Brouillet S. "A data base of chromosome III of *Saccharomyses cerevisiae*", *Yeast*, Vol. 9, 1993, pp. 941-1029.

Smith Victoria, Chou Karen, Lashkari Deval, Botstein David and Brown Patrick. "Large-scale investigation of gene function in yeast using genetic footprinting", *Eurofan*, LLN, 28-31 March 1996, *Books of Abstracts*, CEC, DGXII, Biotechnology, EC4.

Smoller D.A. *et al.* "Characterization of bacteriophage P1 ubiary containing inserts of *Drosophila* DNA of 75-100 kilobase pairs", *Chromosoma*, Vol. 100, 1991, pp. 487-494.

Struhl K. "High-frequency transformation of yeast: Autonomous replication of hybrid DNA molecules", *Proc. Nat. Acad. Sci.*, USA 76, 1979, pp. 1035-1039.

Sturtevant A.H. "The linear arrangement of six sex-linked factors in *Drosophila*, as shown by their mode of association", *J. Exp. Zool.*, Vol. 14, 1913, pp. 43-59.

Sulston J. *et al.* "The *C. elegans* genome sequencing project: A beginning", *Nature*, Vol. 356, 5 March 1992, pp. 37-41.

Tettelin H., Thierry A., Fairhead C., Perrin A. and Dujon B. "*In vitro* fragmentation of yeast chromosomes and yeast artificial chromosomes at artificialy inserted sites and applications to genome mapping", *Methods in Molecular Genetics*, Vol. 6, 1995, pp. 81-107.

Tettelin H. et al. "The nucleotide sequence of *Saccharomyces cerevisiae* chromosome VII", *Yeast Genome Directory, Nature*, Vol. 387, 29 May 1997, pp. 81–84.

The multinational coordinated *Arabidopsis thaliana* genome research project. *Progress Report*. Year Five, National Science Foundation, 1995.

The Yeast Genome Directory, Nature, supplement Vol. 387, 29 May 1997.

Thierry A. "Cleavage of yeast and bacteriophage T7 genomes at a single site using the rare cutter endonuclease I-*Sce*I", *Nucleic Acids Research*, Vol. 19, No. 1, 1991, pp. 189–190.

Thierry A. and Dujon B. "Nested chromosomal fragmentation in yeast using the meganuclease I-*Sce*I: a new method for physical mapping of eukaryotic genomes", *Nucleic Acids Research*, Vol. 20, No. 21, 1992, pp. 5625–5631.

Thierry A., Gaillon L., Galibert F. and Dujon B. "Construction of a complete genomic library of *Saccharomyces cerevisiae* and physical mapping of chromosome XI at 3.7 KB resolution", *Yeast*, Vol. 11, 1995, pp. 121–135.

Tubb Roy. "The long-term benefits of genome sequence data to the yeast-dependent industries", in *Sequencing the Yeast Genome: A detailed assessment*, ed. A. Goffeau, CGC XIII-88/5J.

Vassarotti A. *Mission Reports: 1) Meeting of Key ESSA Representatives* to Discuss the BIOTECH Work Programme (Martinsried, 21 January 1993) - *2) Meeting of US Scientists* to discuss the bioinformatics needs of *Arabidopsis* Community (Dallas, 5–6 June 1993), Brussels, 5 August 1993, XII/93 - AV/ch Miss 11.

Vassarotti A., Goffeau A., Magnien E., Loder B. and Fasella, P. "Genome research activities in the EC", *Biofutur*, Vol. 85, 1990, pp. 1–4.

Vassarotti A. *Rapport de mission, YIP meeting,* Helsinki, Finlande, 1–2 June 1992, Brussels, 5 August 1992, AV/at/miss06.

Vassarotti A. and Goffeau A. "Sequencing the yeast genome: the European effort", *Trends Biotechnol.*, Vol. 10, 1992, pp. 15–18.

Vassaroti A. and Goffeau André. "A second dimension for the European yeast genome sequencing project", *Biopractice*, Vol. 1, 1992.

Vassarotti A. *et al.* "Structure and organization of the European yeast genome sequencing network", *Journal of Biotechnology*, Vol. 41, 1995.

Wach A., Brachat A. and Philippsen P. "Plasmids for simultaneous testing of yeast essential alleles and strength of transcription (pSt-Yeast) a strategic proposal" in *Report of Manchester conference of the Yeast Genome Sequencing Network,* Commission of the European Communities, DGXII, Biotechnology, February 1994, p. 127.

Wassarman D.A., Therrien M. and Rubin G.M. "The Ras signaling pathway in Drosophila", *Current opinion in Genetics and Development,* 5, 1995, pp. 44–50.

Watson J.D. *Science,* Vol. 237, 31 July 1987.

Wensink P. et al. "A system for mapping DNA sequences in the chromosomes of *Drosophila melanogaster*", *Cell,* Vol. 3, 1974, pp. 315–325.

Wicksteed B.L., Collins I., Dershowitz A., Stateva L.I., Green R.P., Oliver S., Brown A.J.P. and Newlon C.S. "A physical comparaison of chromosome III dans 6 souches de *Saccharomyces cerevisiae*", *Yeast,* Vol. 10, 1994, pp. 39–57.

Williams Niegel. "Yeast genome sequence ferments new research", *Science,* Vol. 272, No. 5261, 26 Avril 1996, p. 481.

"Yeast ferments genome-wide functional analyses", Editorial, *Nature Genetics,* Vol. 14, No. 1, September 1996, pp. 1–2.

Yoshikawa Akikazu and Isono Katsumi. "Chromosome III of *S. cerevisiae*: An ordered clone bank, a detailed restriction map and analysis of transcripts suggest the presence of 160 genes", *Yeast,* Vol. 6, 1990, pp. 383–401.

Yoshikawa Akikazu, Isono Katsumi. "Chromosome III of *Saccharomyces cerevisiae*: A ordered clone bank, a detailed restriction map and analysis of transcripts suggests the presence of 160 genes", *Yeast,* 1991.

Yoshikawa H., Ogasawara N., Kunst F. and Danchin A. "The systematic sequencing of *Bacillus subtilis* genome", in *Final European Conference of the Yeast Genome Sequencing Network,* Trieste, 25, 28 September 1996, CEC, DGXII, *Book of Abstracts.*

Conclusion: The Dreams of Reason

Annas George. *Standard of Care,* Oxford University Press, New York, 1993.

Bloomsbury Allen V. *Health Authority,* 1 All ER 651, 1993.

Canguilhem Georges. *La connaissance de la vie,* ed. J. Vrin, deuxième édition, Paris, 1985.

Chadwick R. (ed.) *Ethics, Reproduction and Genetic Control*, London, Groom Helm, 1987.

Chambon Pierre. "Génétique moléculaire, évolution et société", *La lettre du GREG*, No. 3, Décembre 1994, p. 20

Clothier C. *Report of the Clothier Committee on the Ethics of Gene Therapy*, 1992, Cm 1788.

Collins F. and Galas D. "A new five-year plan for the US human genome project", *Science*, Vol. 262, 1 October 1993.

Cooper Necia, Grant (ed.). *The Human Genome Project: Deciphering the Blueprint of Heredity*, University Science Books, 1994.

Crystal R.G. "Transfer of genes to humans: Early lessons and obstacles to success", *Science*, Vol. 270, 20 October 1995, pp. 404–410.

Entretien avec C. Venter, *Biofutur*, Avril 1992, p. 18.

Foulcaut M. *L'ordre des choses.*

"I.Q. and race", *Nature*, 247, 1974.

Hull D. *Philosophy of Biological Science*, Prentice-Hall, Englewood Cliffs, N.J., 1974.

Kamin L.J. *The Science and Politics of I.Q.*, Erlbaum, Potomac, M.D., 1974.

Kaplan J.C. et Delpech, M. *Biologie moléculaire et médecine*, Flammarion, Paris, 1989, 2nd ed., 1993.

La recherche du No. 297, Avril 1997, rubrique intitulée "Science et société", p. 14.

La Recherche, Mai 1996.

La Recherche, No. 292, November 1996, p. 5.

Lee R. and Morgan D. (eds.) *Birthrights: Law and Ethics at the Beginnings of Life*, Routledge, London, 1989.

Legendre Pierre. *Leçon VIII, le crime du Caporal Lortie*, Fayard, 1989, p. 20.

Lewontin R.C. Rose S. and Kamin L.K. *Not in our Genes*, Pantheon, New York, 1984.

Maddox J. "New genetics means no new ethics", *Nature*, Vol. 364, 1993, p. 97.

McKusick V.A. *Mendelian Inheritance in Man. Catalogs of Human Genes on Genetic Disorder.*

McLean S. and Giesen D. "Legal and ethical consideration of the Human Genome Project", *Medical Law International*, Vol. 1, 1994, pp. 159–176.

Nielsen L. and Nespor S. "Genetic tests, screening and use of genetic data by public authorities in criminal justice, social security and alien and foreigner acts", *The Danish Center of Human Rights*, Copenhagen, 1993.

Rifkin Jeremy. *The Biotech Century: Harnessing the Gene and Remaking the World.* Jeremy P. Tarcher / G.P. Putman's Sons, New York, 1998.

Roberts L. "Taking stock of the Genome Project", *Science*, 1993.

Rosenberg A. "Autonomy and provincialism", in *The Structure of Biological Science*, Cambridge University Press, Cambridge, 1985, pp. 11–35.

Rosenberg A. *The Structure of Biological Science*, University of Cambridge Press, Cambridge, 1985.

Second annual symposium in November 1993 on the consequence of the H.G.P. for human health-care.

Slonimski P. "Une étiquette n'est pas un gène", *Biofutur*, Avril 1992, p. 5.

Strathern M. "The meaning of assisted kinship", in ed. M. Stacey, *Changing Human Reproduction,* Sage Publications, London, 1992, pp. 148–169.

Strathern M. et al. *Reproducing the Future: Anthropology, Kinship and the New Reproductive Technologies*, Manchester University Press, Manchester, 1993.

Thomas K. *Religion and the Decline of Magic: Studies in Popular Reliefs in Sixteenth and Seventeenth Century England*, Penguin, Harmondsworth, 1973.

Trends in Neurosciences, September 1996.

Vrema I. "La thérapie génique", *Pour la Science*, dossier hors-série, Avril 1994, pp. 90–97.

Watson J.D. "Ethical implications of the Human Genome Project", *La lettre du GREG*, No. 4, Juillet 1995.

Watson J.D. "Looking forward", *Gene*, 135, 1993.

Weismann A. *Essais sur l'hérédité et la sélection naturelle*, trad. H. de Varigny, Reinwald, Paris, 1982.

Acronyms

ABCC (Atomic Bomb Casualty Commission)
ABRC (Advisory Board for the Research Council)
ACARD (Advisory Council for Applied Research and Development)
ACC (Actions Concertées Coordonnées)
ACE (Atomic Energy Commission)
AFM (Association Française contre les Myopathies)
AIDA (Agro-Industrial Demonstration Action)
AMFEP (the Association of Microbial Food Enzyme Producers)
ASTI (Institut d'Etudes du Travail)
BAP (Biotechnology Action Program)
BCRD (Budget Civil de la Recherche et Développement)
BEP (Biomolecular Engineering Program)
BICEPS (Bio-Informatics Collaboration European Program and Strategy)
BMFT (Bundesministerium für Forschung und Technologie)
BMFT (Bundesministeruim für Forschung und Technik),
BRIC (The Biotechnology Regulation Inter-Service Committee)
BRIDGE (Biotechnology Research for Innovation, Development and Growth)
BSC (Biotechnology Steering Committee)
cDNA (complementary DNA)
CEC (Commission of the European Communities)
CEFIC (The Council of the European Chemical Industry)
CEPH (Centre d'étude du Polymorphisme Humain)
CIAA (the Confederation of Agro-Food Industries)
CNR (Consiglio Nazionale delle Richerche)
CNRS (Centre National de la Recherche Scientifique)
CODEST (Committee for the European Development of Science and Technology)

CORDI (Advisory Committee on Industrial Research and Development)
CST (Comités Scientifiques et Techniques)
CUBE (Concertation Unit for Biotechnology in Europe)
DDT (dichlorodiphényltrichloréthane)
DECHEMA (Deutsche Gesellschaft für chemisches Apparatewesen)
DFG (Deutsche Forschung Gemeinschaft)
DG (Divisions Générales)
DNA (deoxyribonucleic acid)
DNFB (DiNitroFluoroBenzène)
DOE (Department of Energy)
ECDP (Engineering Council for Professional Development)
ECLAIR (European Collaborative Linkage of Agriculture and Industry through Research)
ECPE (the European Center for Public Enterprises)
ECRAB (the European Committee on Regulatory Aspects of Biotechnology)
EFB (European Federation of Biotechnology)
EFPIA (the European Federation of Pharmaceutical Industry Associations)
ELSI (Ethical, Legal, and Social Implications)
ELWWs (European Laboratories Without Walls)
EMBL (European Molecular Biology Laboratory)
EMBO (European Molecular Biology Organization)
EPO (European Patient Office)
ERDA (Energy Research and Development Administration)
ETUC (the European Trade Union Confederation)
FAST (Forecasting and Assessment in Science and Technology)
FDA (Food and Drug Administration)
FEICRO (the Federation of European Industrial Cooperative Research
Flair (Food-Linked Agro-Industrial Research)
GBF (Gesellschaft für Biotechnologische Forschung)
GIBIP (the Green Industry Biotechnology Industrial Platform)
GIFAP (the International Association of Agrochemical Industries)
GIP (Groupement d'Intérêt Public)
GMBF (Gesellschaft für Molekularbiologische Forschung)
GREG (pour Goroupement de Recherchs et d'Etudes sur les Génomes)
GTC (Genome Therapeutics Corp)
HERAC (Health and Environmental Research Advisory Committee)
HGMC (Human Genome Mapping Committee)
HGP (Human Genome Project)
HHMI (Howard Hughes Medical Institute)
HLA (Human Leucocyte Antigen)
HUGO (Human Genome Organisation)

Acronyms

ICI (Industries Chimiques Impériales Britanniques)
ICRF (Imperial Cancer Research Fund)
IGEN (The Integrated Gene Expression Network)
INA (National Agronomical Institute)
INRA (Institut National de la Recherche Agronomique)
IRDAC (Industrial Research and Development Advisory Committee)
IVA (The Royal Swedish Academy of Engineering Sciences, ou Ingenjövetenskapsakademien)
LCL (lignées cellulaires lymphoblastoides)
LNC (Ligue National contre le Cancer)
MENRT (Ministère de l'Education Nationale, de la recherche et de la Technologie)
MGI (Microbial Genome Initiative)
MIT (Massachusetts Institute of Technology)
MITI (Ministry of International Trade and Industry)
MRC (Medical Research Council - GB)
MRD (Microbiological Research Department)
MRE (Microbiological Research Establishment)
NAFC (National Alcohol Fuels Commission)
NAS (National Academy of Science)
NCI (National Cancer Institute)
NCR (National Research Council)
NIAR (National Institute for Agrobiological Resources)
NIGMS (National Institute of General Medical Sciences)
NIH (National Institute of Health)
NIMR (National Institute of Medical Research)
NRC (National Research Council)
NSERC (National Sciences and Engineering Research Council)
NSF (National Science Foundation)
OECD (Organization for Economic Co-operation and Development)
OHER (Office of Health and Environmental Research)
OHGR (Office of Human Genome Research)
OMB (Office of Management and Budget)
OMS (Organisation Mondiale de la Santé)
ORFs (Open Reading Frames)
OTA (Office of Technology Assessment)
pb (paires de bases (pb))
PCR (Polymerase Chain Reaction)
PTO (Patent and Trademark Office)
RANN (Research Applied to National Need)
rDNA (recombinant DNA)
RERF (Radiation Effects Research Foundation)
RFLP (Restriction Fragment Length Polymorphism)
RGP (Rice Genome Project)
RNA (ribonucleic acid)
SCP (Single-Cell Protein)

SERC (Science and Engineering Research Council)
SKB (Schweiz-Koordinationsstelle für Biotechnologie)
SRC (Science Research Council- GB)
SSC (Superconducting Super Collider)
STA (Science and Technology Agency)
STAFF (Society for Technology-Innovation of Agriculture Forestry and Fisheries)
STOA (Science and Technology Options Assessment)
STS (Sequence-Tagged Sites)
TCR (T. Cell Receptor)
TIGR (The Institute for Genomic Research)
UCLA (University of California at Los Angeles)
UCSC (University of California at Santa Cruz)
UMIST (the University of Manchester Institute of Science and Technology)
UNESCO (United Nations Educational, Scientific and Cultural organization)
UNICE (The Union of Industries of the European Community)
UPOV (International Union for the Protection of New Varieties of Plants)
USDA (US Department of Agriculture)
YAC (Yeast Artificial Chromosomes)

Author Index

Ackerman, S. 196
Adams, M.D. 257
Adler, Reid 257
Aguilar, A. 351–353
Ahmed, A. 574
Aigle, Michel 391, 409, 412
Ajioka, J.W. 576
Alon, N. 199
Alberghina, L. 412
Albert, A. 338
Albertini, A. 338
Alexandraki, D. 496
Allen 39
Allen, G. 340
Allen, Woody 166
Allende, J. 323
Allende, Jorge 322
Altman-Jöhl, R. 522
Altmann, Thomas 567
Anderson, C. 259, 292
Anderson, Norman 221
Anderson, S. 267
Andre, Bruno 497
Andreis, E. 17
Angier, N. 186
Annas, G. 646, 647
Annas, George 646

Ansorge, Wilhelm 265
Antegnera, F. 610
Appleyard, Ray 93, 114
Aram, R.H. 338
Arber, W. 151
Arendt, F. 338
Artavanis-Tsakonas, S. 619
Ashburner, M. 574
Atlan, Henri 650
Avery, Oswald Theodore 123
Avner, P. 453
Azarwal, S. 510

Bacon, Francis 42
Baer, R. 266
Baker, K. 340
Balbiani 120
Ballabio, Andrea 256
Baltimore, David 195, 240, 246
Banting, Frederick 156
Barataud, Bernard 199
Bardeen, John 144
Barinaga, M. 196
Barnhart, Benjamin J. 236–238
Barrel, Bart 224, 416, 519
Bateman, J.E. 362
Bateson, William 119

Bayev, Alexander Drovich 281
Bayh, Birch 67
Beach, David 439, 519
Beadle, George Wells 117, 121
Beam, A. 226
Beatty, J. 232
Begley, S. 226
Behrens, Dieter 331
Beide, B.A. 338
Bell, J. 574
Bell, George 235
Benoit, Dorothy 640
Benzer, Seymour 122, 208, 687
Berg, Paul 2, 154, 167, 243, 643
Bergson, Henri 24
Berman, Alex 24
Bernard, Claude 9
Bernhauer, Karl 80
Bernon, M. 172
Bernstein, F. 117
Bernton, Hal 66
Bertinchampts 93
Berzélius 12
Best, Charles 156
Betz, A. 338
Bevan, Mike 563, 564, 571
Beveridge, William 38
Bias, W.B. 182
Bienfet, R. 338
Bigelow, Julian 133
Bigot, B. 291
Binder, N. 338
Bird, A. 610
Bitensky, Mark 235
Blackburn, Elisabeth 211
Blattner, Frederick 261, 267
Bloomsbury, A.V. 646

Bock, P. 443
Bodmer, Walter 189, 243, 252, 277
Boelter, L.M.K. 46
Bogan, J.S. 215
Boivin, A. 124
Bolotin-Fukuhara, Monique 489, 493, 496
Borris, R. 548
Bostock, Judy 238
Botstein, David 190, 224, 240
Boucherie, Helian 437
Boutry, Marc 554, 556
Boyd-Orr, John 41
Boyer, Herbert 151, 154, 156
Brabandere, J. de 338, 402
Brachat, A. 488
Brachet, Jean 371
Bragg, Sir Lawrence 129
Brandford, Victor 32
Brandt, Willy 79
Branscombe, Elbert 237
Braun, Fernand 330
Breitenbach, M. 497
Brenner, Sydney 207, 208, 276, 327, 608, 689
Brent, Roger 618
Brewer, Herbert 37
Bridges, C. 575
Brightman, Rainald 39
Brillouin, Léon 132
Broda, N. 400
Bron, Sierd 548
Brosch, R. 528
Brouillet, S. 443
Browaeys, Dorothee Benoit 640
Brown, Alistair 412, 427
Brown, D.M. 125

Brown, L.G. 215
Brown, R. 340
Bruschi, Carlo 497
Bulher, J.M. 412
Bull, Alan T. 88
Bult, C.J. 512
Burke, David 210
Burks, C. 235
Burns, N. 489, 510
Burrell, B. 266
Bush, Vannevar 45, 51
Bussey, Howard 439

Cadoré, Bruno xx
Cai, H. 576
Canguilhem, Georges 656
Cantley, Mark xix, 84, 107, 108, 115, 170, 329, 370, 392, 396
Cantor, Charles R. 212, 247
Capano, C.P. 610
Capek, Karel 22
Carbon, John 210
Carey, John 258
Carey, Norman 167
Carignani, Giovanna 412
Carle, Georges 210
Caro, Heinrich 9
Carpentier, Michel 415
Carrano, Anthony 237, 256, 261
Carrel, Alexis 62
Carson, Rachel 64, 93
Carter, James 66
Carvalho Guerra, F-J. A. 338
Caskey, C. Thomas 256, 620
Caspersson, Torbjorn O. 184
Castell, W.M. 340
Catenhusen, Wolf-Mickael 296

Cavanaugh, M.L. 199
Cellarius, Richard A. 58
Chain, Ernst 42
Chakrabarty, Amanda L. 628
Chakravarti, Aravinda 248, 583
Chambon, Pierre 692
Champagnat, Alfred 59
Chargaff, Erwin 125
Charles, Enid 40
Chase, Marta 125
Chee, M. 266
Chen, D. 292
Chen, Ellson 571
Chen, Ren-Biao 318
Chen, Zhu 316, 318
Cherry, Mike 565, 570
Chirac, Jacques 288
Chizuka, Tadami 308
Chou, K. 510
Chu, Jiayou 317
Chumakov, I. 215
Church, George 224, 234
Ciba-Geigy 373, 390, 391, 432, 630, 623
Clarke, Louise 210
Clark-Walber 416
Claude, Albert 371
Claverie, J.M. 326
Clement, Philippe 433
Clutter, Mary 553
Cohen, Barak 618
Cohen, Daniel 199, 200, 215
Cohen, Stanley 151, 152
Cole, Stewart T. 528
Coleman, K.G. 443
Colleaux, L. 452, 453
Collins, Francis S. 248, 583, 610

Collins, I. 447
Compton, Arthur Holly 230
Compton, Karl 45
Conde, J. 412
Conneally, P.M. 192, 198
Connel, Mc 412
Connerton, I. 548
Connstein, W. 19
Contopoulou, C.R. 456
Cook-Deegan, Robert 264
Corey, Robert B. 130
Correns, C. 119
Cotton, Richard G.H. 256
Coulson, Alan 205, 269, 522
Cowan, Sir James 276
Cox, David R. 256
Craig Venter, J. 589
Cramm, S. 235
Cranston, Alan 221
Crichton, Michael 166
Crick, Francis Harry Compton 126
Crouzet, Marc 412
Crystal, R.G. 676
Cuenot, Lucien Claude 121
Cullum, J. 400
Cunningham, P. 105, 338
Curie, Marie 144, 612
Curien, Hubert 289

d'Arcy Thompson 31
Dahrendorf, Ralf 106
Dalsager (Commissaire Européen) 112
Damerval, T. 258
Danchin, A. 545, 549, 551
Dangl, J. 557
Dani, Ginger 211

Darwin, Charles 30
Daum, G. 497
Dausset, Jean 117, 197, 200, 289
Davies, Julian 392
Davies, Kay 193
Davignon, Etienne 108, 330
Davis, Bob 248
Davis, Ron 518, 530
Davis, Ronald W. 190, 194, 224, 261
de Bretagne, Yves xx
de Duve, Christian 102
de Kerckhove d'Exaerde, Alban xx
de Nettancourt, Dreux vii, xix, 76, 93, 388
de Vries, Hugo 119
Dean, C. 557
Deaven, L. 235
Debru, Claude xx, 8, 9, 127
Dedonder, Raymond 265
Delbruck, Max 298
Delgado, Marco 433
Delhey, G. 340
DeLisi, Charles 231, 234, 235, 246, 273
Delpech, M. 669
Delweg, Hanswerner 80
Denis, F. 489
Denizot, F. 548
Dershowitz, A. 447
Desborough, S. 205
Devine, K.M. 270
Dib, C. 270, 292
Dickson, D. 277
Diechmann, M. 187
Diehl, B.E. 467
Dietrich, F. 269, 467
Dietrich, W.F. 269

Djerassi, Carl 166
Dobzhansky, Theodosius 165
Donahue, Roger 182
Donis-Keller, Helen 194, 224
Doolittle, Ford 304, 305
Dornauer, H. 340
Douzou, P. 338
Dozy, Andrees M. 188
Drew, L. 226
Dronamraju, Krishna R. 31, 34, 35
du Pont de Nemours 21, 59, 71, 628, 629
Du, Chuan-Shu 320
Du, Ruo-Fu 317
Du, Z. 212
Dubois, Evelyne 409
Ducan, Ian 269
Duclaux, Emile 9
Dufour, J.P. 443
Dujon, Bernard xix, 412, 424, 463, 501
Dulbecco, Renatto 219
Durand, Béatrice 139
Dussoix, D. 151
Düsterhoft, A. 520
Dutchik, J.E. 205
Dutchnik, J. 377
Dutertre, Bruno 433

Ecker, Joe 563, 564, 571
Econome, J.G. 213
Economidis, I. 351, 409
Edgar, Robert 220, 223, 225
Edman, Pehr 139
Ehrlich, Paul 64
Ehrlich, S.D. 548, 550
Eisenberg, R.S. 255
Elgar, Greg 279

Ellwood, Derek C. 88
Endo, Isao 306
Engelhardt, Vladimir Alexandrovitch 281
Enström, Axel 51
Entian, Karl-Dieter 496, 497, 503, 548
Ephrussi, Boris 121
Erdman, S. 510
Ereky, Karl 17, 18
Errington, J. 548
Evans, Charles 102
Eyben, D. 373

Fairhead, Cécile 489
Falise, Michel xx
Fan, Jun 530
Farben, I.G. 21
Fasella, Paolo 331, 335, 491, 546
Faure, S. 292
Feckman, F. 574
Federspiel, Nancy 571
Feillet, Pierre 354
Felbaum, Carl B. 621
Feldmann, Horst 401, 404, 451, 453, 456, 477, 497
Feltz, Bernard xix
Ferroud, C. 443
Feulgen, R. 123
Fey, Stephen J. 490, 492–494
Ficket, J. 235
Field, S. 511
Fields, Bernard 195
Fiers, W. 266, 412
Fillon, François 290, 291
Fink, G.R. 206
Finley, Russel 618
Finn, Robert 83

Finucane, B. 338
Fischer, Emil 10
Fizanies, C. 292
Flavell, R.A. 187
Fleer, Reinhard 433
Fleischmann, R.D. 270, 514
Flemming 119
Fogel, Lawrence J. 43
Foote, S. 215
Forrest, Derek W. 30
Forrester, Jay W. 65
Forssman, Sven 53
Fortini, M.E. 613, 619
Foucault Michel 647
Foury, Françoise 412, 498
Francé, Raoul 26
Frankenstein 7, 8
Franklin, J. 361, 415
Franklin, Rosalind 125, 130
Fraser 130, 271, 272, 512, 513
Fraser, C. 271
Freud, Sigmund 26
Friesen, James 304
Fromageot, Claude 136
Frontali, Clara 392
Frontali, Laura 412
Fruton, J.S. 135
Frycklund, Linda 433
Fujiki, Norio 308
Fukuhara, Hiroshi 391, 412

Gabor, Dennis 65
Gaden, Elmer 54, 56, 57
Gaillardin, Claude 520
Galas, D. 267, 663
Galibert, Francis 149, 266, 477, 514, 463

Galizzi, A. 548
Gallibert, Francis 520
Galloway, John 277
Galton, Francis 30, 62
Garcia-Ballesta, J.P. 412
Gardner, David 223
Gareis, Hans-Georg 331
Garrels, James I. 511
Garrod, Archibald Edward 117, 121
Gascuel, Olivier 544
Gasson, N. 409
Gaston, Thorn 329
Gayon, Jean xx, 128
Geddes, Patrick 31, 32, 115
Gelbart, W. 574
Gendre, Frédéric 433
Gent 412, 427
Geoff, Walker 354
George Radda 523
Gesteland, Raymond 248, 261, 583
Geynet, P. 340
Ghislain, M. 443
Giesen, D. 646
Giessler, Alfred 29
Gilbert, Walter 145, 147, 159, 195, 219, 224, 225, 241, 269, 282, 283
Gilson, Keith 278
Giraudat, J. 554
Giscard d'Estaing, Valérie 90
Glansdorff, N. 402, 412
Glaser, P. 549
Gloecker, R. 433
Glover, D. 574
Goeble, M.G. 489
Goeddel, Kurt 161

Goffeau, André vii, xi, xiv, xix, xx, 76, 175, 207, 264, 268, 271, 327, 352, 369–371, 373, 374, 376, 377, 380, 381, 386, 392, 398, 399, 403, 409, 413, 414, 416, 441, 458, 461, 463, 483, 487, 492, 509, 511, 545, 546, 556
Goldberg, R.B. 610
Goldscheid, Rudolf 26, 27, 62, 115
Goldstein 657
Goldway, M. 435
Good, W. 235
Goodfellow, Peter 620
Goodman, H.M. 188
Gorbachov, Mikhail 283, 284
Gosling, Raymond 130
Gottlieb, Robert 621
Gough, Mike 234
Goujon, Philippe vi, vii, 133
Graham, M. 377
Gralla, Jay 146
Grande, H.J.J. 338
Grandi, G. 548
Grant, Madison 36
Green, H. 184
Green, R.P. 447
Greenhalf 427
Gregory, R.A. 40
Grenson, M. 412
Grieg, Russel 677
Griffith, Frederich 123
Griffith, John 125
Grignon, Thivernahal 520
Grisolia, Santiago 320
Grivell, Leslie 412
Grodzicker, T. 187
Gros, François 90, 91, 138, 399

Groot, G. 374
Grunstein, M. 575
Gruskin, K.D. 226
Guercio, G. del 226
Guignard, M. 338
Guillou, S. 215
Gusella, James 192, 193
Gyapay, G. 199

Hadden, A.C. 36
Haiech, J. 548
Haldane, John Burdon Sanderson 34, 115, 121, 643
Hale, William J. 20
Hall, B.D. 188
Hall, S.S. 221
Hamilton, J.O. 226
Hanoune, Jacques 289
Hansen, Emil Christian 5
Hardison, R.C. 203
Hardy, Godfrey Harold 120
Haritos, A. 493
Harper, P.S. 198
Harris, S.D. 467
Hartl, Dan 269
Hartley, Harold 42, 59, 87, 115
Hartwell, L. 375
Hartzell, G. 417
Harvey-Jones, Sir John 331
Harwood, C. 548
Harzell, G. 268
Hase, Albrecht 19
Havris, Jim 677
Hawthorne, D.C. 205
Hayduck, F. 19
Hayes, William 122
Heden, Carl-Göran 52, 57

Hegemann, J.H. 503
Heidegger, Martin xiv
Helling 154
Hennessy, K. 268, 417
Hériard, Bertrand xx
Hermington, John 237
Herrero, A. 340
Herrnstein, Richard 659
Hershey, Alfred Day 125
Heslot, Henri 371, 391
Hieter, P. 487, 488
Hildebrand, E. 235
Hilger, François 412
Hill, A.V. 49
Hilton, A. 215
Hilts, P.J. 259
Hinchcliffe, Edward 432
Hinnen, A. 373, 496
Hirshman, J. 443
Hirsinger, M. 340
Hitler, Adolf 123
Hoch, Jim 545
Hodgetts, R.B. 305
Hodgkin, Jonathan 524
Hoeck, Fernand van vii, 76
Hoffman, Max 222
Hogben, Lancelot 34, 38–40, 62
Hogness, David 575
Hohan, S. 149
Hoheisel, Joerd 462, 463, 519
Holden, C. 615
Holland, P.W.H. 610
Hollenberg, C.P. 412
Holley, Robert W. 140
Holt, Geoffrey T. 90
Holzmann, D. 226
Hood, Leroy 139, 169, 219

Hoover, Herbert 44
Hooykaas, P.J.J. 558
Hori, Masaaki 306, 307
Horvitz, H.R. 208
Hotchkiss, Rollin 124, 166
Hull, D. 652
Hunkapiller, M.W. 139
Hurford, Nigel 433
Hutchinson, Fred 211
Huxley, Aldous 36, 37, 642
Huxley, Julian 31, 36, 40, 62, 115
Huxley, Thomas 30

Ikawa, Koji 180
Ikeda, Joh-E. 306
Ingram, Vernon 186
Iserentant, Dirk 432
Isono, Katsumi 268
Itakura, Keiichi 157
Itano, H.A. 186

Jackman, P. 412
Jackson, John 331
Jacob, François 90, 91, 118, 122, 145
Jacq, Claude 412, 461
Jacquet, Michel 412
Jaklevic, J. 574
Jarry, B. 340
Jeambourquin, R. 340
Jeantet, Louis 276
Jeffreys, Alec 189
Jensen, Orla 15, 16
Jia, Y. 612
Jimenez, Antonio 412
Jimenez, Martinez, J. 520
Johanssen, Wilhelm 120
Johansson, C. 184

Johnston, Mark 473
Joly, Pierre B. 441, 466, 623
Jonas, R.A. 338
Jordan, B. 258, 292
Jordan, Elke 248, 583, 627
Jorgensen, Alfred 15, 16
Joris, B. 548
Judson, Horace Freeland 127, 130, 132, 138

Kaback, David B. 467
Kafatos, Fotis 354
Kalman, Sue 530
Kamalay, J.C. 610
Kamehisha, Minoru 311
Kamin, L.J. 661
Kamin, L.K. 661
Kammener, Paul 37
Kan, Yuet Wai 188
Kanehisa, Minoru 312
Kanigel, R. 226
Kant, Emmanuel 647
Kaplan, J.C. 669
Kapp, Ernest 23
Karl, Pearson 30
Karn, J. 205, 210
Karpen, G.H. 579
Karumata, D. 548
Katz, S.E. 226
Kaufman, T. 574
Kavanagh, T. 557
Keiden 373
Keller, K.H. 115
Kempers, Jan 433
Kendrew, John 129
Kennedy, John F. 64
Kennedy, Max 163

Kennet (Lord) 107
Kennet, Wayland 107, 332
Kenyon, C.J. 209
Kepler 416
Kerem, B.S. 199
Keto, John E. 47
Kevles, Daniel J. 28, 37, 169
Kico, S. 213
Kiefel, J. 576
Kieffer-Higgins, S. 149
Kielland-Brandt, M. 373
Kiff, J. 212
Kimmel, Bruce 581
King, Alexander 65
King, J.S. 446, 456, 467
Kingman, A.J. 389
Kinzler, Kenneth 617
Kirschstein, Ruth 247
Kiss, I. 574
Kisselev, Lev 256, 285, 286
Kleid, D.G. 187
Klein, H. 338
Klis, Frans M. 504
Klug, Aaron 282
Knapp, F.C. 117
Koeman, Jan 354
Kohara, Ynji 213, 311
Kolata, G.B. 147
Koonin, E. 443
Koraft, J.A.R. 46
Korhola, Matti 432
Kornberg, Arthur 141
Koshland, Daniel 241
Kossen, N. 340
Kourilsky, Philippe 289
Kovarik, William 66
Kravitz, Kerry 190

Krimsky, Sheldon 169
Krupp 33
Kuczynski, René 39
Kuhn, Thomas 229
Kunkel, L. 198
Kunst, Frank 512, 546, 549, 551

Lacy, E. 203
Lamarck, Jean-Baptiste de Monet, chevalier de) 24
Lander, E.S. 212, 255
Landsteiner, Karl 117
Lang, B.F. 305
Lashkari, D. 510, 511
Lasko, P. 574
Latour, Bruno 6
Lazar, Philippe 289
Le Play, P.G.F. 32
Lederberg, Joshua 122, 152, 165
Leeuw, de 374
Left, D.N. 271
Legendre, Pierre 691
Lehrach, Hans 256, 283, 299
Lelong, M. 338
Lemieux, B. 305
Lener, M. 338
Lerman, Leonard 224
Lesaffre, S.I. 389, 431
Leterne, S. 443
Leuchtenberger, W. 340
Levine, P. 117
Lewin, R. 243
Lewis, S. 574
Lewontin, R.C. 661
Linder, P. 416
Liebermann, C.T. 9
Liebig, Justus 8

Lilly, Malcom M. 90
Lindergren, C.C. 205
Lindergren, G. 205
Lindner, Paul 18, 19
Lindsay, D.G. 338
Linneus 417
Little, I. 215
Liu, Guoyang 319
Livet, Pierre 651
Llorente, B. 489
Lloyd, A. 446
Locke, J. 574
Loder, B. 327, 450
Loeb, Jacques 24
Lomakka, C. 184
Lopez, A. 526
Lott, A.F. 338
Louis, Edward J. 464, 504
Louis, K. 574
Lucchini, Giovanna 412, 496
Lüdecke, K. 19
Ludwig, Robert 220, 223
Luke, David L. 562
Luria, Salvador Edward 122, 128, 151
Lwoff, André 122
Lyssenko, Trofim Denisovich 117, 282, 284

Mackaye, Benton 33
Macq, P. 418
Maddox, John 645
Madueno, E.G. 577
Magee, B.B. 521
Magee, P.T. 521
Magni, G. 338, 409
Magnien, Etienne vii, 94, 409
Malczynski, M. 510

Malmborg, Charlotte Af 354
Malone, E.R. 446
Mangematin, V. 441, 466
Mangold, P. 340
Maniatis, Tom 203
Marathe, Rekha 530
Marcker, K.A. 338
Margolskee, J. 401
Marks, John 240
Marr, Tom 519
Martienssen, Rob 553, 561–563, 571
Martin, C. 574
Martin, Robert 504
Martin, W.J. 362, 383
Marx, J.L. 201
Mason, F.A. 17
Mason, Frederik 17
Massen, R. 362
Masson 39
Mateles, Richard I. 85
Mathews, K. 574
Matsubara, Ken-Ichi 306, 311
Matsumo, K. 619
Mavel, C. 548
Maxam, Allan 145, 282
Mayor, Frederico 321
McAuliffe, Kathleen 171
McAuliffe, Sharon 2, 171
McCarty, Maclyn 123, 124
McCombie, W. Richard 554
McConnel, David 373, 391, 409
McCullock, Warren S. 133
McDonnel, James 227
McDermid, H. 574
McKusick, Victor 243, 252, 278, 321
McKusker, F. 511
McLaren, D.J. 297

McLean, S. 646
McLeod, Colin 124
McMillen, Wheeler 20
Mcsweeney, B. 340
Meadows, Dennis L. 65
Meadows, Donella H. 65
Meinke, David 564, 571
Meischer, Johann Friedrich 119
Mellado, R.P. 548
Mendel, Johann (en relig. Gregor) 118
Mendeleyev 416
Mendelsohn, Mortimer 233
Mervis, Jeffrey 247
Mewes, Werner 380, 392, 396, 424
Meyer, P. 558
Michael Cherry, J. 571
Michaelis, George 493
Mignoni, G. 340
Miklos, George F. 573
Miller, J.H. 270
Miller, M. 340
Minier, M. 400
Mirzabekov, Andrei 146, 282, 285, 286
Miyahara, M. 179
Mizobuchi, Kiyoshi 268
Mizukami, T. 519
Modest, E.J. 184
Mohr, Jan 183
Molodell, Dundee, J. 574
Moltke, Hans 659
Monaco, A. 198
Monod, Jacques 122, 145
Moore, John 636
Mordant, Philippe 498, 502
Moreno, S. 520
Moretti, M. 338
Morgan, Thomas Hunt 120, 182

Mori, Wataru 310
Morin, Edgar xviii
Mortimer, Bob 424
Mortimer, Robert K. 205, 446, 456, 467
Mose Larsen, Peter 490, 492–494
Moses, Julian 28
Moszer, I. 549
Moyzis, R. 235
Muller, Herman 132
Muller, Hugo 24
Müller-Hill, Benno 145
Mulligan, B. 557
Mulligan, J.T. 268, 417
Mullis, Kary 162
Mumford, Lewis 32, 43, 115
Munck, L. 343
Murakami, Y. 306, 471, 477
Murphy 16, 17
Murphy, T. 579
Murray, Andrew 211
Murray, Charles 659

Nakai, K. 445, 447
Nakamura, Yusuke 256
Narjes, K.H. 343
Narla, Mohandas 261
Nash, D. 574
Nathans, D. 187
Navarre, C. 443
Nawrocki, A. 494
Neal, James V. 233, 234
Needham, Joseph 37
Nelkin, Dorothy 667
Nelson, Jerry 221
Nespor, S. 647
Neumann, Caspar 7
Newlon, Caroll 401, 448

Nicolas, Alain 497
Niederberger, Peter 432
Nielsen, Hilmer 331, 340
Nielsen, L. 647
Niels-Fill 389
Nietzsche, Friedrich 31
Nilken, S. 141
Nilsson-Tilgreen 373
Nixon, Richard 69
Noller, Harry 220, 223
Nombela, Cesar 497
Nomine, G. 340
Novick, A. 57
Nuesch, P. 390
Nurse, Paul 522

O'Grady, J. 338
Oddo, G. 273
Oeller, P.W. 443
Oesterling, Tom 195
Ogasawara, Naotake 547, 548, 550, 551
Ohki, Misao 311, 312
Oishi, Michio 309
Okada, Kiyotaka 571
Okayama, Hiroto 306
Oliver, Steve 392, 395, 400, 403, 405,
  412, 424, 426, 427, 612
Olson, Maynard 205, 210, 234,
  383, 386
Osinga, Klaus 433
Ostwald, Wilhelm 26
Oudega, B. 548
Ouslow, M. 121
Ozbekhan, Hasan 65

Page, David 215
Palazzolo, M.L. 577

Palca, J. 259, 260
Palmieri, Thomas 238
Pandolfi 335
Panetta, Leon 225
Pardo, Daniel 433
Pardue, M.L. 575
Parenty, M. 149
Parkill, J. 528
Pasiphae 35
Pasteur, Louis 6, 628
Pataryas, T. 493
Patrinos, Ari 248, 583
Pauling, Linus 129, 144, 186
Pauly, August 26
Pauly, Philip 24
Pearson, P.L. 198
Peccei, Aurelio 65
Pelsy, G. 338
Peltonen, Leena 256
Perkin, William 9
Perrin, A. 453
Perutz, Max 129, 282
Petersen, I. 338
Petes, T.D. 187, 446, 489
Petrella, Ricardo 329
Pfeiffer, F. 380, 396, 406
Philippsen, Peter 412, 463
Philipson, Lennart 439
Pickles, H. 338
Pierard, A. 402
Pilgrim, D. 574
Pincus, Gregory 24, 165
Pines, Maya 245
Pirie, N.W. 42
Pitts, Walter 133
Planqué, K. 547
Planta, Rudi J. 401, 412, 497

Plasterk, Ronald 524
Plato 647
Platt, John R. 58
Pohl, Fritz M. 362, 380, 383, 391, 396, 407, 412
Pohl, Peter 149
Pohl, Thomas 383
Pontecorvo, Guido 121
Pope, Sir William 20
Poustka, Annemarie 256
Povey, Susan 256
Prieels, Anne Marie xix, 338, 434, 496, 502
Pringle, J.R. 443, 446, 467
Printz, P. 338
Probert, M. 392
Prometheus 34, 35, 645
Pryklung 432
Punnet, R.C. 120

Quertier, Francis 564, 566, 571

Radda, George 523
Rajnchapel-Messaï, Jocelyne 162, 290
Randers, Behrens 65
Randers, Jorgen 65
Ranmens, J.M. 199
Rapoport, G. 548
Rapp, Friederich 23
Ratledge, Colin 88
Rattray Taylor, G. 166
Raugel, P.J. 70
Reagan, Ronald 235
Reddy, V.B. 266
Rehm, H.J. 80
Reinhart, R. 520
Reiss, B. 558

Renwick, J.H. 182
Revilla Redreira, R. 338
Rexen, F. 343
Reynaud, J.P. 340
Rhodes, C. 187
Richards, Brian 167
Rieger, Klaus-Jörg 494
Rieger, Michael 520
Riezmann, Howard 504
Rigault, P. 215
Rigby, P.W. 187
Riggs, Arthur 157
Riles, Linda 470, 472, 514
Riordan, Jr. 199
Risler, J.L. 445, 447
Roberts, Leslie 239, 244, 664
Roberts, Rich 561
Rockefeller 38, 49, 123
Rodriguez-Pousada, Claudina 483
Roeder, G.S. 510
Romero, P. 215
Rörsch, A. 76, 102
Rose, A.M. 305
Rose, M.S. 389
Rose, S. 661
Rosenberg, A. 648, 653
Rosenblueth, Arturo 133
Rosnay, Joël de 91
Rossenbeck, H. 123
Ross-Macdonald, P. 510
Rostand, Jean 62
Rothstein, R. 487, 488, 490
Rouguelle, C. 453
Rounsley, Steve 564, 571
Roux, Wilhelm 23
Roy, K. 574
Royer, B. 198

Royer, Pierre 90, 91
Rubin, C.M. 613
Rubin, Gerald M. 261, 581
Rubio, V. 340
Ruddle, Frank 235, 236
Ruezinsky, D. 443
Ruml, Beardsley 38
Russel, E.S. 31
Russell, P. 522
Rüttgers, Jürgen 299
Ryan, J. 338

Sainsbury, Lord 523
Sakaki, Yoshiyuki 256, 310, 311, 313
Sakharov, Andrei 282
Salanoubat, Marcel 571
Sambrook, J. 187
Sanders, J.P.M. 481
Sanger, Frederick 135
Sapp, Jan 688
Sargeant, Ken 102, 107, 329, 342
Sasson, Albert 172
Saunders, R.D.C. 576
Savakis, B. 574
Saviotti, Pier Paolo 354
Sayre, Anne 130
Scheller, William 66
Scheibel, Elmar 497
Schierenberg, E. 208
Schild, D. 205
Schlessinger, David 256, 261
Schmeck, H.M. 262
Schmidt-Kastner, Gunther 354
Schoen, A. 252
Schopenhauer, Arthur 31
Schrödinger, Erwin 126, 129
Schutt, A. 213

Schwartz, David 212, 224, 234
Schwartz, Wilhelm 81
Scrimshaw, Nevin S. 59
Scriver, Charles 303, 304
Seikosha, Daini 175
Sekeris, C.E. 338
Sekiya, Takeo 307
Senez, Jacques 59
Sentenac, André 391
Seror, S. 548
Sgouros, John 427, 428
Shannon, Claude E. 132, 133
Sharp, P. 187, 409, 446
Shaw, Duncan 521
Sheehan, A. 510
Sheets T.A. 172
Shejbal, E. 340
Shelley, Mary 8
Shen, Yan 318, 320
Shimizu, Nobuyoshi 306, 311
Shult, E.E. 205
Siden-Kiamos, Inga 580
Sidow, A. 524, 619
Siebel, Emil A. 16
Siebel, John Ewald 16
Sikorski, S.R. 487, 488
Simchen, G. 416
Simpson, K. 340
Singer, Maxime 167
Singer, S.J. 186
Sinsheimer, Robert 220
Sklar, Scott 66
Skolnick, Mark 190
Slonimski, Piotr 372, 392, 409, 412, 424, 442, 673, 674
Smith, C.L. 213
Smith, David 235

Smith, Hamilton O. 187
Smith, Lloyd M. 219
Smith, Michael 304
Smith, Temple F. 221, 226
Smith, Victoria 510
Smoller, D.A. 577
Snyder, Mike 510
Soeda, Eiichi 471
Soler, M. 489
Soll, Dieter 167
Solomon, Ellen 189
Solvay, Jacques 331
Somerville, Chris 571
Southern, E.M. 187
Southgate, Eileen 209
Spencer, Herbert 30, 32
Spradling, A.C. 579
Springer, Patricia 554
Stadler, Lewis John 121
Stahl, Georg Ernst 7
Stahl, U. 374
Stateva, L.I. 447
Stavropoulos, Alexander 331
Steele, Jack E. 47
Steen, R. 269
Steensma, H. Yde 412
Stephens, J.C. 199
Stephens, Richard 530
Stokes, Alexander R. 130
Strathearn, M. 647
Strauss, J. 255
Strohman, Richard C. 688
Struhl, K. 425
Stucka, R. 456
Sturtevant, Alfred 232
Sulston, John 205, 207, 208, 224, 258, 259, 269, 275, 278, 522

Sun, M. 180
Sundaresan, Venkatesan 554
Suter, B. 574
Sutherland, Grant 256
Sutton, Walter Stanborough 120
Sverdlov, Evgeny 287
Swanson, Robert 154, 156
Swinbanks, D. 180
Symington, L.S. 446
Szilard, Leo 57, 58
Szostak, Jack 211

Tabata, Satoshi 564, 571
Takanami, Mitsuru 309
Tanaka, S. 438
Tannenbaum, Steven R. 85
Tatum, Edward Lawrie 117, 121, 164, 165, 166
Taylor, Craig G. 43, 46
Taylor, E. 140
Temkin, Owsei 24
Terrat, R. 396, 407
Tettelin, Hervé 463
Thatcher, Margaret 277
Theologis, Sakis 571
Therre, H. 258
Thiemann, Hugo 65
Thierry, Agnes 463
Thiers, G. 338
Thireos, George 412
Thomas, D.Y. 390
Thomas, Daniel 102, 436
Thomas, Karen 212, 647
Thomas, W.K. 524, 619
Thompson, E.O.P. 137
Thompson, J. Arthur 32
Thompson, J.N. 208, 209

Thompson, Nichol 209
Thorn, Gaston 108, 329
Todd, Alexander 125
Tonegawa, Susuma 309
Tooze, John 169
Tornier, Gustave 23
Treviranus, Gottfried Rheinhold 10
Trivelpiece, Alvin 237
Truchya, H.M. 115
Tsaftaris, A.S. 338
Tschermark, Erich Von 119
Tsiftsoglou, A. 409
Tsui, Lap-Chee 256
Tubb, Roy S. 430

Ullman, Agnès 145

van Beneden 120
van den Berg, Johan A. 374
van den Bosch, M.C.F. 338
van der Meer, R. 338, 340, 352, 359, 360, 372
van der Platt, J. 373, 374
van Elzen, P. 557
van Hoeck, Fernand vii, 76
van Ommen, Gert-Jan 256
van Veldhuysen, J. 340
Vandecandelaere, Gaston xx
Vassarotti, Alessio 556
Veenhuis, M. 496
Vela, C. 340
Velander, Edy 51
Velculescu, Victor 617
Veldkamp, E. 338
Vendrely, Colette 124
Vendrely, Roger 124

Venter, J. Craig 589
Verbakel, John M.A. 433
Verma, I. 676
Vezzoni, Paolo 274
Vickers, Tony 278
Vidal, M. 511
Virey, Jean-Jacques 24
Viadescu, Barbu 432
Vogelstein, Bert 617
Volkaert, Guido 412
Vollrath, D. 215
Von Baeyer, Adolf 9
Von Borstel, J. 436
Von Wettstein, Dieter 373, 412
Vos, P. 557

Wada, Akiyoshi 177, 217, 305
Wain, S. 267
Wain-Hobson, Simon 545
Wald, Salomon 89
Waldeyer, Wilhelm 120
Walkers, T.K. 86
Walsh, James 240
Walters, LeRoy 248, 583
Wandel, R. 338
Warmuth, E. 338
Watanabe, Itaru 309
Waterman, Michael 224
Waterston, Robert 205, 224, 256, 259, 261, 269, 522, 524
Watson, James Dewey 126, 128–131, 169, 239, 244, 246–248, 264, 439, 662, 663, 679, 681
Weaver, Warren 133
Weinberg, Robert 240
Weinberg, Wilhem 120
Weindling, Paul 27

Weinstock, G. 271
Weismann, August 120, 667
Weiss, M. 184
Weissenbach, Jean 199, 256, 295
Weizmann, Chaim 19, 86
Weldon, W.F.R. 30
Wells, Herbert Georges 34
Wells, I.C. 186
Wensink, P. 575
Westerhoff, H.V 496
Wexler, Nancy 192, 262
White, John 209
White, Owen 571
White, Ray 233
White, Raymond L. 192
Whittier, Bob 567
Wickenden, William 43, 44
Wicksteed, B.L. 447
Wiener, Norbert 47, 48, 133
Wiesner, Julius 11
Wilkie, Tom 222, 248
Wilkins, Maurice Hugh F. Frederick 125, 130
William, W. 65
Williams, J. 187
Williams, N. 509
Williams, Shirley 88
Williamson, Robert 193
Willis, Thomas 7
Willmitzer, L. 558
Wilson, Alan 219
Wilson, Edmund B. 182
Wilson, H.R. 130
Wilson, Rick 571
Wimbush, James 195
Winsor, Barbara 504
Witunski, Michael 227

Wohler, Freiderich 8
Wolf, Werner 433
Wolfe, Ken 270
Wollmann, Eugène 122
Woodford, F.P. 338
Worton, Ronald 303, 304
Woteki, Catherine 553
Wragg, P. 352, 353
Wu, R. 140
Wyman, Arlene 197
Wyngaarden, James 223, 246, 321

Xavier, A. 338

Yanagida, M. 519
Yankovsky, N. 287
Yee, J. 576

Yisuke, Nakamura 311
Yoshikawa, Akihiro 426, 438, 445
Yoshikawa, Hiroshi 449, 472, 547, 551
Young, Lord Kennet Wayland 107
Yu, Long 319
Yuan, Jiang-Gang 318

Zabarovski, Eugene 286
Zabeau, M. 557
Zakin, Virginia 211
Zanten, D. von W. 340
Zech, L. 184
Zeng, Yi-Tao 319
Zimmermann, K.F. 391, 412
Zinder, Norman 247
Zworykin, Wladimir 49

Subject Index

a philosophy of technology   23
acetic acid   20, 612
acetone   19, 20
agricultural produce   17, 20
Agrobacterium   105, 558
alcaptonuria   121
Alzheimer's disease   193, 248
antihumanist revolution   691
antituberculosis drugs   527
aperiodic crystal   144, 688
applied chemistry   10, 14, 84
applied immune sciences   625
applied microbiology   53, 75, 81, 86, 87, 91
Aquifex aeolicus   512
Arabidopsis thaliana   272, 292, 306, 323, 328, 375, 376, 413, 414, 419, 539, 552–557, 572, 584
Archaebacteria   v
Archaeoglobus fulgidus   512
asilomar (meeting)   169, 627
Association of Biotechnology Companies   341
autosomes   183, 576

B. burdorferi   272

Bacillus subtilis (genome sequencing European program)   265, 404, 414, 419, 420, 449, 512, 543, 545, 546, 549
bacteriological biotechnics   53
bacteriophage lambda   140, 151, 266
bacteriophage phi X174   142
bacteriophages   122, 128, 168, 185, 203, 546, 686
BAP proposal and objectives   349
Beecham   84, 258, 433, 620, 624, 625, 677
Beowulf Genomics   280, 518, 521
BEP program and project   350
Big Science   ix, x, 218, 241, 441, 666, 684
bio-society   xv, 107, 108, 642
bioengineering   47, 48, 53, 54, 55, 81, 115
bioethics   171, 635
biofuels   66, 89, 91, 163
biogas   58, 172
biogen   72, 114, 159, 161, 174, 195, 225, 295, 296, 476, 496
bioindustrial revolution   90

biological determinism 648, 657, 667
biological engineering 4, 37, 40, 42, 45, 46, 48–55, 81, 82, 86, 91, 114, 115, 117, 164–167
Biomass Policy Office 175
biomolecular engineering 92, 102, 105, 106, 345
biontotechnik 23
bioprocess technology 55
bioreactors 4, 75, 84, 89, 91, 95, 100, 104, 175, 346, 351
bioresources 63, 64
BIOTECH I and II programs 417, 449
biotechnical chemistry 15, 16
biotechnical era 44
biotechnie 24
biotechnique 15, 25, 42, 44
biotechnological era 116
biotechnological inventions 335, 628, 633, 634
biotechnology vii, x, xi–xv, xvii, xix, 1–7, 13, 14, 16–25, 27, 31, 33, 40–43, 46–48, 50, 53–58, 60–64, 68, 71–73, 77–86, 88–92, 94, 95, 99, 101–116, 128, 132, 144, 154, 156, 159, 161, 163, 164, 169, 170, 172–177, 181, 213, 218, 225, 227, 259, 268, 273, 275, 283, 290, 296, 302, 303, 326, 328–335, 338–341, 343–350, 352, 353, 355–359, 361, 363–366, 368–370, 373, 375, 404, 423, 430–432, 450, 491, 530, 531, 536, 548, 549, 553, 556, 557, 584, 585, 607, 626, 642–644, 685, 692
biotechnology program vii, x, xv, xix, 61, 82, 92, 94, 102, 172, 175, 268, 318, 328, 335, 344, 350, 353, 355, 356, 359, 363, 369, 370, 390, 423, 461, 530, 531, 628
bioteknik 50, 52, 53
Borrelia burgdorferi 512
BRIDGE program 359, 360, 362, 365, 366, 368, 369, 382, 384, 395, 396, 397, 413, 415, 419, 420, 424, 430, 436, 438, 449, 450, 546, 556, 558
butanol 19

Caenorhabditis elegans (genome sequencing) 182, 376, 449, 524–526, 584
Canada, genome research in 62, 256, 262, 303, 304, 305, 326, 385, 390, 397, 435, 437, 439, 466, 467, 507, 509, 623, 639
cancer genetic approach to 242
Candida albicans (genome sequencing) 520, 521
centromere 210, 401, 435, 446, 456, 566, 578, 579, 688
chemistry 6–16, 20, 21, 39, 40, 45, 73, 75, 76, 80–84, 86, 90, 117, 128, 129, 135, 137–140, 144, 147, 162, 286, 287, 290, 312, 400, 591, 602, 650, 652, 656
chemurgy 20, 21
China, genome research in 320
Chlamydia trachomatis 271, 529, 545
chromomeres 120
chromosome committee 253
chromosomes x, 2, 118, 120, 123, 124, 152, 162, 171, 182–185, 188, 192, 193, 205, 207, 209–212, 215,

218, 231, 237, 244, 247, 248, 250, 251, 253, 262, 269, 295, 299, 306, 308, 315, 317, 372, 374–376, 378, 385, 390, 392, 393, 394, 396, 401, 404, 405, 415, 420, 425, 426, 436, 439, 451–453, 457, 461–463, 466, 467, 473, 476–480, 489, 493, 515, 516, 519, 523, 536, 554, 555, 557, 559, 575, 576, 578, 583, 662, 663, 668, 675, 688
classical genetics   116, 117, 191, 201, 206, 551, 612, 653, 654
cloning xi, 105, 147, 149, 154, 166, 169, 180, 198, 199, 203, 205, 209, 211, 229, 263, 266, 277, 293, 306, 317, 319, 323, 373, 375, 381, 401, 404, 426, 442, 447, 464, 485, 489, 525, 539, 540, 551, 558, 575, 595, 598, 635, 644, 658
Club of Rome   65, 66, 93, 107
Cold Spring Harbor Laboratory   205, 240, 259, 424, 426, 446, 456, 553, 561, 562
coli bacillus   544
colonic polyposis   193
computer networks   229, 386, 407
concertation (activity, politics, units) 111–113, 292, 330, 341, 346–349, 356, 357, 361, 363, 366, 367, 399
concertation activity   348
Congress, US   171
Council of Ministers (European)   633
cybernetics   47, 48, 49, 133
cystic fibrosis   193, 198, 199, 262, 298, 526, 666–668, 670, 677
Czechoslovakia   81

Daedalus   31, 34–37, 125, 645
DDT   64
DECHEMA   81, 83, 84, 88, 108
Denmark, genome research in   xx, 15, 17, 62, 301, 302, 326, 338, 373, 383, 394, 493, 549
DG XII (and the biotechnology policy)   xiv, xix, 107, 112, 114, 329, 334, 335, 354, 464, 467, 488, 502, 510
Diamond vs. Chakrabarty   628
DNA sequencer   179, 472
DNA sequencing   115, 140, 141, 149, 154, 162, 178–180, 201, 220, 221, 224, 228, 234, 235, 238, 244, 254, 257, 264, 282, 305, 309, 310, 362, 385, 402, 403, 529, 535, 561, 583, 584, 590, 591, 598, 674
DOE, genome project   xiv, xx, 235–239, 245, 246, 248, 384, 386, 518, 547, 556, 561, 562, 572, 583, 585, 586, 589, 590, 599
double helix   2, 132, 137, 138, 185, 209, 246, 587, 645
Drosophila melanogaster   121, 182, 325, 328, 375, 376, 413, 414, 419, 552, 573, 575–577, 580, 583, 584, 613
Drosophila melanogaster program   573
Duchenne's muscular dystrophy   193, 198, 199, 298, 303, 526, 667, 669

ectogenesis   35–37
ELSI research program   262
Engelhardt Institute   282, 284–287

enzyme engineering 4, 67, 74, 77, 89, 91, 94, 97, 99, 102, 104
Epstein-Barr virus 266
ergonomics 46, 48, 53, 54
Escherichia coli 182, 213, 375, 399, 512, 584
ethical problems 305, 375, 631, 632, 644, 645, 646, 676
Eubacteria v
eucaryotic v, 205, 375, 388, 397, 453, 508, 524, 619
eugenic 2, 21, 28, 30, 34, 36–38, 40, 41, 62, 122, 164, 165, 166, 296, 644, 646, 659, 679, 691
eugenics 2, 28, 30, 34, 36–38, 40, 62, 122, 164–166, 296, 644, 679, 691
eugenics law 62
Euratom 92, 93, 95, 371, 387
EUROFAN projects (yeast and the) and the industry 279, 495, 497–510
Europe+30 106, 107
European biotechnology industry 364
European Commission on the ethics of biotechnology 628
European Community 92, 94, 99, 105, 109, 217, 229, 279, 329, 331–340, 345, 352, 383, 389, 394, 399, 411, 416, 419, 434–436, 449, 511, 546, 576
European Economic Community (see also CUBE) x, 95, 107, 346, 417
European genome research 369, 550
European Parliament 95, 108, 328, 332, 344, 370, 419, 421, 631–633
European Parliament Legal Commission of the 631
European Patent Office 628–630

European Stock Center 558
European yeast chromosome III sequencing program 416
eutechnics 33
eutopia 33
extremophile genomes 274

FAST group 109, 114, 329, 330
FAST program 106, 108, 112, 329, 330, 332, 607
fermentation 5–7, 9–15, 19, 41, 46–48, 50, 52, 55–58, 67, 74, 76, 77, 79–81, 84, 85, 88, 175, 177, 205, 374, 385–387, 400, 431, 508, 537
Finland, genome research in 303
France, genome research in 291
Frankenstein 7, 8
functional analysis program 445, 501

GenCorp 227, 229
gene mapping 162, 182, 183, 185, 196, 289, 295, 301, 302, 419, 423, 677
Genentech 71, 114, 156, 157, 158, 159, 161, 171, 173, 174
genes (the problem of the definition) 3, 4, 24, 26, 35–37, 100, 101, 105, 117, 120, 121, 126–129, 137, 138, 140, 149, 152, 154, 158, 159, 162, 165, 166, 170, 171, 176, 178, 179, 183–191, 193–199, 201–204, 206, 210, 211, 213, 217, 218, 221, 221, 224, 226, 228, 231, 233, 236, 239–242, 245, 254–258, 260, 262, 263, 266, 268, 270, 271, 286–288, 291, 293, 295, 297, 301, 302, 305, 307, 308, 312, 315, 317–319, 323, 326, 359,

373, 375, 376, 385, 390, 393, 397–399, 411, 413, 419, 420, 424–426, 430, 437, 438, 442–447, 454–457, 467, 472, 475, 476, 478–490, 492–495, 497, 500, 501, 503–505, 507–511, 514, 520–523, 525–529, 536–543, 546, 549, 550, 552–558, 561, 562, 569, 570, 572, 573, 575, 577–580, 586, 587, 592–597, 605–613, 615–620, 625, 627, 628, 633, 636, 639, 648, 649, 652, 653, 654, 655, 658, 660–666, 668–671, 673–678, 680, 682–684, 686, 688–691

Généthon 199, 200, 292, 295

genetic engineering 1–5, 25, 35, 64, 84, 89, 91, 94, 95, 97, 99–102, 104, 109, 114–116, 156, 158–161, 163–166, 169–175, 211, 212, 217, 296, 345–347, 350, 361, 508, 620, 627–630, 643, 645, 682, 685, 688, 692

genetic engineering techniques 1, 3, 350

genetic ideology (and its reasons) 644

genetic linkage map 190, 196, 199, 201, 204, 214, 220, 224, 288

genetic markers 191–193, 269, 318

genetic revolution 116

genexpress 674

genome therapeutics 261, 270, 512

genoscope (in France) 293–295, 382

genset 625, 626

geotechnical age 33

geotechnics 33

Germany, genome reasearch in 5, 6, 9, 12, 14, 15, 17, 19, 24, 29, 30, 39, 62, 73, 77, 78, 80–82, 84, 85, 89, 92, 102, 122, 256, 262, 287, 296, 297–299, 302, 326, 357, 360, 362, 380, 383, 388, 394, 395, 398, 409, 421, 435, 461, 463, 473, 493, 496, 497, 501, 502, 509

gist brocades 391, 431, 432, 547

Glaxo 88, 677

glycerol 19

GMOs' labeling 639

Guinness, Arthur, Son and Co 431

Hakkô 74, 75

Harvard Myc Mouse 629

Heamophilus influenzae 270

Helicobacter pylori 270, 512

hemochromatosis 190

hemoglobinopathies 189, 671

hepatitis B 173, 266, 267

hibakusha 231–234

Hoffman Foundation 222–224

Hoffman-La Roche 80, 159, 162, 549

holy grail (Gilbert's metaphor) 219, 225, 226, 657

Homo faber 25

human cytomegalovirus 266, 417

human engineering 48, 53

human frontier science program (Japanese) 174, 176, 180, 360, 400, 413

human genome 4, 143, 178–181, 187, 189–191, 194, 196–201, 204, 214–220, 222–226, 228, 230, 231, 234, 236–267, 269–280, 285–292, 294, 296–300, 302–307, 310–313, 315–320, 321, 323–328, 335, 370, 372, 374, 385, 387, 388, 397–400, 403, 413, 414, 419, 420, 423, 424, 439, 452, 471–473, 483, 507, 509, 537,

543, 547, 549, 561, 563, 583–587, 589, 592, 596–599, 604, 605, 619, 621, 624, 630, 635, 644–646, 657, 658, 660, 662–664, 666, 675, 677, 680, 684, 686, 689, 690
human genome (mapping the)   199, 215, 245
human genome newsletter   645
human genome project   4, 179, 180, 187, 198, 199, 201, 204, 216–220, 222, 225, 228, 234, 236–239, 242, 244, 246–252, 254, 257, 260–262, 264, 266, 267, 270, 272, 273, 280, 285, 286, 294, 299, 300, 310, 316, 321, 323–325, 327, 372, 374, 387, 420, 471, 507, 509, 549, 561, 563, 583, 584, 589, 644–646, 658, 662, 663, 677, 680, 689
human leukocyte antigen   190
Hungary   17
Huntington's disease   192, 193, 197, 198, 199, 324

in silico analysis and research   257, 259, 438, 442, 480, 503, 507, 514, 608, 614
information (concept and theory)   25, 63, 74, 95, 102, 106–108, 111, 112, 114, 117, 123, 126, 132, 133, 135, 137, 138, 141, 147, 149, 167, 173, 174, 176, 177, 178, 180, 181, 196, 206, 208, 217, 219, 221, 226, 229, 239, 250–252, 254, 255, 258, 261, 263, 272, 278–280, 284, 287, 301, 308, 311, 312, 318, 321, 329, 336–339, 348, 349, 353, 356, 361, 366, 368, 376, 377, 379, 381, 384, 385, 388, 390, 393, 396, 398, 406–408, 410, 411, 415, 422, 423, 427, 432, 434–436, 443, 448, 450, 459, 474, 478, 481, 487, 491, 492, 495, 498, 501, 502, 508, 511, 522, 527–529, 537, 542, 556, 560, 561, 566, 568–570, 572, 573, 580–582, 586, 589, 592, 596, 597, 599, 600–602, 607, 611, 614, 616–619, 627, 630, 631, 632, 636, 639, 647, 650, 651, 656, 662–664, 667, 669, 675, 679, 692
Institut Pasteur   15, 264, 289, 293, 449, 452, 527, 528, 545, 547, 549
interferon   72, 159, 161, 173
International Organizations (and genome program) (and international cooperation)   59, 320, 336, 635
IRDAC Working Party 5, the opinion of   397
Italy, genome research in   59, 92, 229, 242, 256, 273, 274, 281, 288, 302, 304, 326, 338, 383, 387, 388, 395, 409, 494, 496, 497, 549, 551

James McDonnel Foundation   227
Japan, genome research in   305

Karolinska Institute (Stokolm)   52, 286
Keck Telescope   221, 222

Labimap project   288
Laboratory of Molecular Biology (Cambridge)   269, 522
lactic acid   12, 369
Leishmania (the parasite)   301, 324, 552

life for sale 2, 171
limits to growth 65, 93
LKB-Pharmacia 265
Los Alamos and Lawrence Livermore Laboratories 231
Los Alamos National Laboratory 235, 261

M. avium 274
Manhattan project 232, 241, 685
markers 185, 188–194, 196–200, 204, 215, 231, 250, 251, 269, 286, 300, 308, 317, 318, 576–578, 593, 598, 658
mauveine 9
mechatronics 74
medical electronics 49, 50
Methanobacterium thermoautotrophicum 512
Methanococcus jannaschi 271, 512, 615
microbiology 1, 6, 11–14, 46, 52–55, 57, 74–77, 80–84, 86–88, 91, 373, 374, 435, 493, 512, 538, 548, 549
Minamata's disease 74
mitochondrial *DNA* 267, 317
molecular biology 25, 57, 58, 69, 83, 86, 95, 97, 98, 100, 102, 116, 126, 132, 137, 138, 140–142, 145, 149, 154, 164–167, 170, 171, 178, 180, 185–187, 190, 191, 195, 202, 207, 217, 221, 226, 229, 231, 233, 235, 240, 241, 243, 244, 264, 269, 271, 274–276, 281–286, 295, 298, 299, 303, 305, 306, 323, 325, 327, 362, 372, 400, 413, 416, 417, 424, 440, 522, 556, 558, 559, 580, 604, 606, 629, 650, 656, 675, 685, 686, 688–690
molecular genetics 117, 130, 154, 163, 164, 170, 182, 185, 186, 208, 217–219, 226, 276, 286, 295, 296, 323, 327, 372, 375, 411, 413, 425, 435, 441, 452, 453, 493, 508, 518, 544, 545, 596, 606, 652–654, 667, 688, 689, 692
monbu-sho 312, 313
Monsanto 72, 156, 331
muscular dystrophy 193, 198–200, 298, 303, 323, 526, 666–669, 671, 677, 682
Mycobacterium leprae 274, 518
Mycobacterium tuberculosis 271, 281, 293, 512, 525, 526, 528
Mycobacterium tuberculosis (genome sequencing) 528
Mycoplasma capricolum 228, 269
Mycoplasma genitalium 271, 512, 513, 615

nematode genome 209, 278
neurofibromatosis 193, 526
Neurospora crassa 117
NIH guidelines 169
Novo Industries 340
nuclear power programs 230
nuclear weapons 230, 231, 235, 556

Ohio State University 512
Ohio Stock Center 558, 567
oncogenes 186, 242, 266, 482
organic chemistry 9–11, 135, 286, 287

orphan genes 445, 480, 487
orthogenesis 26

patents 99, 105, 162, 255, 331, 508, 626, 627, 628, 631, 632, 636, 637, 646
PCR technique 162
penicillin 42, 55, 80, 87
pharmaceutical firm 624, 636, 677, 685
phenylketonuria 298, 662
phi-X-174 (virus) 142, 222, 266
physical mapping 162, 187, 201, 204, 205, 207, 209, 212–214, 220, 223, 234, 237, 238, 244, 257, 276, 278, 288, 302, 401, 419, 425, 453, 454, 521, 536, 542, 557, 566, 666
physiological genetics 126–128, 163, 166
physiology 6, 8–12, 15, 24, 46, 49, 52, 163, 167, 206, 219, 346, 364, 371, 373, 408, 481, 525, 539, 541, 555, 557, 558, 608, 615, 649, 675, 689
plasmid 150, 152, 154, 156, 168, 192, 203, 210, 375, 402, 426, 488, 502, 503, 518, 529, 581, 618, 628
Plasmodium falciparum (genome sequencing) 271, 525, 528, 529
polymorphic markers 185, 188, 598
polyomic virus 266
Porton Down 55, 85, 87, 88, 107, 277
prenatal test 287, 678
Prometheus 34, 35, 645
protein sequencing 139, 141, 146
psychotechnics 45
Ptolomean syndrome 675
Pyrococcus furiosus 271

R&D policy in biotechnology 99
recombinant DNA 2, 3, 60, 61, 83, 84, 89, 115, 139, 149, 152, 154, 156, 159, 162, 167–169, 172, 175, 176, 202, 262, 333, 347, 364, 575, 643
reductionism (and the complexity of life) 648, 682, 684
restriction endonucleases (and enzymes) 153, 187
restriction fragment length polymorphism (RFLP) 189
RFLP map 192, 194, 198, 201, 234
RFLP mapping 201, 234
ribonucleic acid 138, 140
rice genome project 308, 313, 316
RIKEN Institute 179, 306, 435, 439
Rous' Sarcoma 267
Russia, genome research in 29, 58, 262, 284–288, 326, 527
Russian HGP 285–287

Saccharomyces cerevisiae 205, 234, 268, 328, 362, 369, 375, 377, 408, 423, 424, 434, 438, 443, 445, 456, 464, 467, 477, 487, 488, 493, 582, 584, 617
Saccharomyces cerevisiae (genome project) 234
Salk Institute 157, 219, 242
Sanger Center 278–280, 293, 472, 473, 509, 513, 518, 519, 521, 522, 527, 529, 580, 586
Schizosaccharomyces pombe (genome sequencing) 473, 519, 520
sequencing DNA (methods) 139, 147, 251, 397, 413
sequencing proteins (insulin) 140

Subject Index 781

shotgun sequencing   472, 514, 567, 568, 581
sickle cell anemia   188, 189
Silent Spring — by Rachel Carson   64, 93
single cell proteins (SCPs)   56, 85
SmithKline Beecham   258, 433, 620, 624, 625, 677
social biology   38, 39
somatostatine   157
Soviet Union, molecular biology in   28, 57, 59, 66, 73, 147, 229, 235, 236, 248, 322
soyabean   639
Stanford University   154, 190, 261, 272, 439, 529, 530, 562, 564
Streptococcus pneumoniae   123
Streptomyces   388, 400
Synechocystis sp.   512
synkaryon   183
syntex   166
synthesize   3, 8–10, 40, 100, 141–143, 166, 402, 481

T-project   424, 450, 546, 547, 557, 558
technology assessment   110, 114, 171, 217, 332
telescope project   222
telomeres   210, 211, 462, 464, 566, 578
the bioindustry office   175
the BIOTECH programs   461
the health excuse   666
the perfect gentleman sequencer (the deontological rules)   457, 458
the PIGMAP project, the EUROFAN project   279
thymosin alpha-1   159
TIGR (and genome sequencing programs)   258, 270–272, 420, 512–516, 518, 562, 564–566, 568, 571, 589
topological genetics   127
transgenic maize   639
Treponema pallidum   271, 513

United Kingdom, genome research in   xx, 62, 84, 274, 275, 277–279, 281, 288, 296, 297, 302, 309, 325, 326, 338, 421
University of California   45, 151, 158, 188, 219, 220–225, 236, 240, 261, 270, 448, 530, 636
urea   8
US Human Genome Project (new goal for 1998–2003)   248–250, 267, 563, 583, 663
US patent law   255
utopia   21, 33, 40, 666, 691

venture capital   69–71, 154, 157, 173, 195, 227, 621, 627, 633, 685
vitamins   28, 41

Washington University   261, 269, 325, 394, 472, 511, 522, 562, 564
Wellcome Trust   278–280, 293, 309, 325, 473, 518, 519, 525–529, 586
Whitehead Institute   205, 215, 240, 261, 272
whole earth catalog   64

Wilson's disease 319, 526, 662

X-ray crystallography 129

yeast v, vii, x–xv, xix, 13, 14, 19, 59, 80, 121, 140, 172, 178, 182, 187, 188, 190, 202, 204–207, 210–213, 215, 220, 234, 244, 246, 251, 264, 268, 269, 277, 305, 306, 312, 327, 328, 351, 357, 362, 364, 369–378, 380–390, 392, 393, 396–401, 403–409, 412–417, 419–421, 423–432, 434–454, 456, 458, 462, 464, 466, 467, 470–474, 476–481, 483, 485–495, 500, 501, 507–511, 514, 519, 521, 523, 527–530, 535, 536, 545, 546, 551, 552, 554, 556, 557, 561, 573, 576, 598, 603, 604, 611, 612, 615–618, 628, 629

yeast cosmid banks (construction and use of) 452
yeast database 428
yeast industry platform (YIP), activities and companies xix, 430, 431, 509
yeast network (the European yeast genome sequencing) xi, 421, 430, 492, 494
yeast sequencing (organization of, publication of data, links with industry) 382, 383, 387, 415, 435, 440, 451, 458

zootechnics 28
Zymotechnical Institute 16
zymotechnology 5–7, 10, 12–17, 20, 21, 41, 42, 114
zymotecnia 7